Crystallography and Crystal Defects

Third Edition

Crystallography and Crystal Defects

Third Edition

ANTHONY KELLY and KEVIN M. KNOWLES

Department of Materials Science & Metallurgy
University of Cambridge
UK

WILEY

This edition first published 2020
© 2020 John Wiley & Sons Ltd

Edition History
John Wiley & Sons (1e, 2000)
John Wiley & Sons (2e, 2012)

The right of Anthony Kelly and Kevin M. Knowles be identified as the authors of this work has been asserted in accordance with law.

Registered Offices
John Wiley & Sons, Inc., 111 River Street, Hoboken, NJ 07030, USA
John Wiley & Sons Ltd, The Atrium, Southern Gate, Chichester, West Sussex, PO19 8SQ, UK

Editorial Office
The Atrium, Southern Gate, Chichester, West Sussex, PO19 8SQ, UK

For details of our global editorial offices, customer services, and more information about Wiley products visit us at www.wiley.com.

Wiley also publishes its books in a variety of electronic formats and by print-on-demand. Some content that appears in standard print versions of this book may not be available in other formats.

Library of Congress Cataloging-in-Publication Data

Names: Kelly, A. (Anthony), author. | Knowles, Kevin M., author.
Title: Crystallography and crystal defects / Anthony Kelly
  (University of Cambridge), Kevin M. Knowles (University of
  Cambridge).
Description: Third edition. | Hoboken, NJ : John Wiley & Sons, 2020. |
  Includes bibliographical references and index.
Identifiers: LCCN 2019035382 (print) | LCCN 2019035383 (ebook) | ISBN
  9781119420170 (hardback) | ISBN 9781119420156 (adobe pdf) | ISBN
  9781119420163 (epub)
Subjects: LCSH: Crystallography. | Crystals–Defects.
Classification: LCC QD931 .K4 2020 (print) | LCC QD931 (ebook) | DDC
  548/.8–dc23
LC record available at https://lccn.loc.gov/2019035382
LC ebook record available at https://lccn.loc.gov/2019035383

Cover Design: Wiley
Cover Image: Courtesy of Kevin M. Knowles, © oxygen/Getty Images

Set in 10/12pt and Times LT Std by SPi Global, Chennai, India

Printed and bound by CPI Group (UK) Ltd, Croydon, CR0 4YY

10 9 8 7 6 5 4 3 2 1

Contents

Preface to the Third Edition   **xiii**

Companion Website   **xv**

Part I Perfect Crystals   **1**

1 Lattice Geometry   **3**
  1.1   The Unit Cell   3
  1.2   Lattice Planes and Directions   8
  1.3   The Weiss Zone Law   11
  1.4   Symmetry Operators   14
        1.4.1   Translational Symmetry   15
        1.4.2   Rotational Symmetry   15
        1.4.3   Centre of Symmetry   16
        1.4.4   Reflection Symmetry   16
  1.5   Restrictions on Symmetry Operations   17
  1.6   Possible Combinations of Rotational Symmetries   21
  1.7   Crystal Systems   25
  1.8   Space Lattices (Bravais Lattices)   26
        Problems   37
        Suggestions for Further Reading   40
        References   40

2 Point Groups and Space Groups   **43**
  2.1   Macroscopic Symmetry Elements   43
  2.2   Orthorhombic System   49
  2.3   Tetragonal System   52
  2.4   Cubic System   53
  2.5   Hexagonal System   55
  2.6   Trigonal System   60
  2.7   Monoclinic System   63
  2.8   Triclinic System   65
  2.9   Special Forms in the Crystal Classes   68
  2.10   Enantiomorphous Crystal Classes   69

2.11	Laue Groups	69
2.12	Space Groups	70
	2.12.1 Screw Axes	70
	2.12.2 Glide Planes	71
	2.12.3 Combinations of Symmetry Operations to Form Space Groups	72
	2.12.4 The Relationship Between the Space Group Symbol and the Point Group Symmetry for a Crystal	74
2.13	The 17 Two-Dimensional Space Groups	74
2.14	Nomenclature for Point Groups and Space Groups	79
2.15	Groups, Subgroups, and Supergroups	80
2.16	An Example of a Three-Dimensional Space Group	81
2.17	Frequency of Space Groups in Inorganic Crystals and Minerals	82
2.18	Magnetic Groups	83
	Problems	85
	Suggestions for Further Reading	87
	References	87
3	**Crystal Structures**	**89**
3.1	Introduction	89
3.2	Common Metallic Structures	90
	3.2.1 Cubic Close-Packed ($Fm\bar{3}m$)	90
	3.2.2 Hexagonal Close-Packed ($P6_3/mmc$)	93
	3.2.3 Double Hexagonal Close-Packed ($P6_3/mmc$)	96
	3.2.4 Body-Centred Cubic ($Im\bar{3}m$)	96
3.3	Related Metallic Structures	97
	3.3.1 Indium ($I4/mmm$)	97
	3.3.2 Mercury ($R\bar{3}m$)	98
	3.3.3 β-Sn ($I4_1/amd$)	98
3.4	Other Elements and Related Compounds	99
	3.4.1 Diamond ($Fd\bar{3}m$)	99
	3.4.2 Graphite ($P6_3/mmc$)	99
	3.4.3 Hexagonal Boron Nitride ($P6_3/mmc$)	101
	3.4.4 Arsenic, Antimony, and Bismuth ($R\bar{3}m$)	101
3.5	Simple MX and MX_2 Compounds	102
	3.5.1 Sodium Chloride, NaCl ($Fm\bar{3}m$)	102
	3.5.2 Caesium Chloride, CsCl ($Pm\bar{3}m$)	103
	3.5.3 Sphalerite, α-ZnS ($F\bar{4}3m$)	104
	3.5.4 Wurtzite, β-ZnS ($P6_3mc$)	104
	3.5.5 Nickel Arsenide, NiAs ($P6_3/mmc$)	105
	3.5.6 Calcium Fluoride, CaF_2 ($Fm\bar{3}m$)	106
	3.5.7 Rutile, TiO_2 ($P4_2/mnm$)	107
3.6	Other Inorganic Compounds	108
	3.6.1 Perovskite ($Pm\bar{3}m$)	108
	3.6.2 α-Al_2O_3 ($R\bar{3}c$), $FeTiO_3$ ($R\bar{3}$) and $LiNbO_3$ ($R3c$)	109
	3.6.3 Spinel ($Fd\bar{3}m$), Inverse Spinel and Related Structures	110

	3.6.4	Garnet ($Ia\overline{3}d$)	111
	3.6.5	Calcite, $CaCO_3$ ($R\overline{3}c$)	113
3.7	Interatomic Distances		114
3.8	Solid Solutions		114
3.9	Polymers		118
3.10	Additional Crystal Structures and their Designation		120
	Problems		123
	Suggestions for Further Reading		125
	References		126

4 Amorphous Materials and Special Types of Crystal–Solid Aggregates **129**
4.1	Introduction		129
4.2	Amorphous Materials		130
4.3	Liquid Crystals		131
	4.3.1	Nematic Phases	133
	4.3.2	Cholesteric Phases	135
	4.3.3	Smectic Phases	135
4.4	Geometry of Polyhedra		135
4.5	Icosahedral Packing		140
4.6	Quasicrystals		142
	4.6.1	A Little Recent History and a New Definition	143
4.7	Incommensurate Structures		143
4.8	Foams, Porous Materials, and Cellular Materials		144
	Problems		145
	Suggestions for Further Reading		146
	References		146

5 Tensors **149**
5.1	Nature of a Tensor	149
5.2	Transformation of Components of a Vector	150
5.3	Dummy Suffix Notation	153
5.4	Transformation of Components of a Second-Rank Tensor	154
5.5	Definition of a Tensor of the Second Rank	156
5.6	Tensor of the Second Rank Referred to Principal Axes	157
5.7	Limitations Imposed by Crystal Symmetry for Second-Rank Tensors	160
5.8	Representation Quadric	163
5.9	Radius–Normal Property of the Representation Quadric	166
5.10	Third- and Fourth-Rank Tensors	168
	Problems	168
	Suggestions for Further Reading	170
	References	170

6 Strain, Stress, Piezoelectricity and Elasticity **171**
6.1	Strain: Introduction	171
6.2	Infinitesimal Strain	172
6.3	Stress	176

6.4		Piezoelectricity	182
	6.4.1	Class 2	184
	6.4.2	Class 222	185
	6.4.3	Class 23	186
	6.4.4	Class 432	186
	6.4.5	The Converse Effect	187
6.5		Elasticity of Crystals	187
	6.5.1	Class $\bar{1}$	191
	6.5.2	Class 2	191
	6.5.3	Class 222	192
	6.5.4	Class 23	192
	6.5.5	Reduction in the Number of Independent c_{ij} and s_{ij} for Other Crystal Classes	193
	6.5.6	Isotropic Media	194
6.6		Elasticity of Cubic Crystals	200
	6.6.1	Transformation of the Stiffness and Compliance Tensors for Cubic Materials	200
	6.6.2	Shear Stiffnesses and Compliances on the (111) Plane of Cubic Crystals	201
		Problems	201
		Suggestions for Further Reading	204
		References	204

Part II Imperfect Crystals — **207**

7   Glide — **209**
7.1	Translation Glide	209
7.2	Glide Elements	213
7.3	Independent Slip Systems	218
7.4	Large Strains of Single Crystals: The Choice of Glide System	227
7.5	Large Strains: The Change in the Orientation of the Lattice During Glide	231
	Problems	237
	Suggestions for Further Reading	239
	References	239

8   Dislocations — **243**
8.1	Introduction	243
8.2	Dislocation Motion	249
8.3	The Force on a Dislocation	251
8.4	The Distortion in a Dislocated Crystal	255
8.5	Atom Positions Close to a Dislocation	260
8.6	The Interaction of Dislocations with One Another	263
	Problems	267
	Suggestions for Further Reading	269
	References	269

9   Dislocations in Crystals   **271**
9.1   The Strain Energy of a Dislocation   271
9.2   Stacking Faults and Partial Dislocations   279
9.3   Dislocations in C.C.P. Metals   281
9.4   Dislocations in the Rock Salt Structure   288
9.5   Dislocations in Hexagonal Metals   291
9.6   Dislocations in B.C.C. Crystals   296
9.7   Dislocations in Some Covalent Solids   297
9.8   Dislocations in Low Symmetry Crystal Structures   302
9.9   Dislocations in Other Crystal Structures   303
         Problems   303
         Suggestions for Further Reading   304
         References   304

10   Point Defects   **309**
10.1   Introduction   309
10.2   Point Defects in Ionic Crystals   313
10.3   Point Defect Aggregates   314
10.4   Point Defect Configurations   316
10.5   Experiments on Point Defects in Equilibrium   321
10.6   Experiments on Quenched Metals   325
10.7   Radiation Damage   328
10.8   Anelasticity and Point Defect Symmetry   329
         Problems   333
         Suggestions for Further Reading   335
         References   335

11   Twinning   **339**
11.1   Introduction   339
11.2   Description of Deformation Twinning   341
11.3   Examples of Twin Structures   346
         11.3.1   C.C.P. Metals   346
         11.3.2   B.C.C. Metals   346
         11.3.3   Sphalerite (Zinc Blende)   347
         11.3.4   Calcite   349
         11.3.5   Hexagonal Metals   349
         11.3.6   Graphite   351
11.4   Twinning Elements   352
11.5   The Morphology of Deformation Twinning   358
11.6   Friedel's Classification of (Growth) Twinning   361
         11.6.1   Twinning by Merohedry   362
         11.6.2   Twinning by Reticular Merohedry   364
         11.6.3   Twinning by Pseudomerohedry   365
         11.6.4   Twinning By Reticular Pseudomerohedry   365
11.7   Atomistic Modelling of Twin Boundaries   365
         Problems   367
         Suggestions for Further Reading   369
         References   370

12 **Martensitic Transformations** **373**
12.1 Introduction 373
12.2 General Crystallographic Features 374
12.3 Transformation in Cobalt 376
12.4 Transformation in Zirconium 379
12.5 Transformation in Indium–Thallium Alloys 385
12.6 Transformations in Steels 388
12.7 Transformations in Copper Alloys 392
12.8 Transformations in Ni–Ti-Based Alloys 392
12.9 Magnetic Shape Memory Alloys 393
12.10 Transformations in Non-metals 393
12.11 Crystallographic Aspects of Nucleation and Growth 395
12.12 The Shape Memory Effect and Superelasticity 397
12.12.1 The One-Way Shape Memory Effect 397
12.12.2 The Two-Way Shape Memory Effect 398
12.12.3 Superelasticity 399
12.13 Modern Theories of Martensitic Transformations 400
12.13.1 The Topological Model 400
12.13.2 Non-linear Elasticity Model 401
Problems 401
Suggestions for Further Reading 403
References 403

13 **Grain Boundaries** **407**
13.1 The Structure of Surfaces and Surface Free Energy 407
13.2 Structure and Energy of Grain Boundaries 413
13.3 Equivalent Geometrical Descriptions of High-Angle Grain
Boundaries 425
13.4 Interface Junctions 429
13.5 The Shapes of Crystals and Grains 434
Problems 439
Suggestions for Further Reading 442
References 443

14 **Interphase Boundaries** **445**
14.1 Boundaries Between Different Phases 445
14.2 Interphase Boundaries Between C.C.P. and B.C.C. Phases 450
14.3 Strained Layer Epitaxy of Semiconductors 456
Problems 460
Suggestions for Further Reading 461
References 462

15 **Texture** **463**
15.1 Texture 463
15.2 Euler Angles 469

15.3	Microtexture	471
	15.3.1 Rodrigues Vectors and Rodrigues Space	472
	Problems	472
	Suggestions for Further Reading	474
	References	474
Appendix 1	**Crystallographic Calculations**	**477**
	A1.1 Vector Algebra	477
	A1.1.1 The Scalar Product	478
	A1.1.2 The Vector Product	480
	A1.2 The Reciprocal Lattice	482
	A1.3 Matrices	484
	A1.4 Rotation Matrices and Unit Quaternions	489
	References	491
Appendix 2	**The Stereographic Projection**	**493**
	A2.1 Principles	493
	A2.2 Constructions	497
	A2.2.1 To Construct a Small Circle	497
	A2.2.2 To Find the Opposite of a Pole	499
	A2.2.3 To Draw a Great Circle through Two Poles	500
	A2.2.4 To Find the Pole of a Great Circle	501
	A2.2.5 To Measure the Angle Between Two Poles on an Inclined Great Circle	502
	A2.3 Constructions with the Wulff Net	502
	A2.3.1 Two-Surface Analysis	505
	A2.4 Proof of the Properties of the Stereographic Projection	507
	References	510
Appendix 3	**Interplanar Spacings and Interplanar Angles**	**511**
	A3.1 Interplanar Spacings	511
	A3.1.1 Triclinic	512
	A3.1.2 Monoclinic	512
	A3.1.3 Orthorhombic	512
	A3.1.4 Trigonal	513
	A3.1.5 Tetragonal	513
	A3.1.6 Hexagonal	513
	A3.1.7 Cubic	513
	A3.2 Interplanar Angles	514
	A3.2.1 Orthorhombic	514
	A3.2.2 Hexagonal	515
	A3.2.3 Cubic	515

Appendix 4   Transformation of Indices Following a Change of Unit Cell   517
    A4.1   Change of Indices of Directions   517
    A4.2   Change of Indices of Planes   519
    A4.3   Example 1: Interchange of Hexagonal and Orthorhombic
          Indices for Hexagonal Crystals   520
    A4.4   Example 2: Interchange of Rhombohedral and Hexagonal
          Indices   522

Appendix 5   Slip Systems in C.C.P. and B.C.C. Crystals   525
    A5.1   Independent Glide Systems in C.C.P. Metals   525
        A5.1.1   Example: Slip Along $[1\bar{1}0]$ on the (111) Slip Plane   525
        A5.1.2   Number of Independent Glide Systems   526
    A5.2   Diehl's Rule and the OILS Rule   527
        A5.2.1   Use of Diehl's Rule for $\{111\}\langle 1\bar{1}0\rangle$ Slip (such as
              C.C.P. Metals)   528
        A5.2.2   Use of Diehl's Rule for $\{1\bar{1}0\}\langle 111\rangle$ Slip (such as
              B.C.C. Metals)   528
        A5.2.3   The OILS Rule   529
    A5.3   Proof of Diehl's Rule and the OILS Rule   529
        References   530

Appendix 6   Homogeneous Strain   531
    A6.1   Simple Extension   532
    A6.2   Simple Shear   532
    A6.3   Pure Shear   532
    A6.4   The Relationship Between Pure Shear and Simple Shear   533

Appendix 7   Crystal Structure Data   535
    A7.1   Crystal Structures of the Elements, Interatomic Distances and
          Six-Fold Coordination-Number Ionic Radii   535
    A7.2   Crystals with the Sodium Chloride Structure   540
    A7.3   Crystals with the Caesium Chloride Structure   542
    A7.4   Crystals with the Sphalerite Structure   542
    A7.5   Crystals with the Wurtzite Structure   543
    A7.6   Crystals with the Nickel Arsenide Structure   543
    A7.7   Crystals with the Fluorite Structure   544
    A7.8   Crystals with the Rutile Structure   544

Appendix 8   Further Resources   545
    A8.1   Useful Web Sites   545
    A8.2   Educational and Information Resources   545
    A8.3   Computer Software Packages   546

Brief Solutions to Selected Problems   **547**

Index   **555**

Preface to the Third Edition

While there are a plethora of books devoted to crystallography and, in particular, X-ray crystallography and electron crystallography of varying levels of sophistication, ranging from undergraduate primers to advanced research level texts, there is still no other textbook which cover the topics of crystallography and crystal defects in as integrated and complete a manner as previous editions of this book. As a consequence, unless there is one of these previous editions to hand, or, better still, this new edition, students and researchers interested in crystallography and crystal defects will have to have one or more sources for their crystallography, and one or more sources for their understanding of defects in crystalline materials and how the microscopic behaviour of the movement of dislocations or formation of twins leads to macroscopic plasticity in crystalline materials.

In this third edition, two new chapters have been created to extend coverage on texture and interphase boundaries, so that the book can continue to be both a useful learning resource and a source reference for senior undergraduates, graduate students and researchers on core aspects of crystallography and crystal defects of particular relevance to materials science. A number of sections of the book have also been either added or extended, such as the sections on magnetic groups, the elasticity of cubic crystals, dislocations in low symmetry crystal structures, atomistic modelling of twin boundaries, the description and classification of growth twins, modern theories of martensitic transformations and the mechanism of the shape memory effect in alloys which exhibit martensitic transformations. Original source references to key crystallographic terms familiar to materials scientists such as Miller indices, Bravais lattices, the Weiss zone law, and Miller–Bravais indices and Weber indices for hexagonal and trigonal materials have been included for completeness. New questions have been added to the end of each chapter where appropriate. As with the Second Edition, brief solutions to all the questions are given after the appendices, while full worked solutions are available at the password protected Wiley Web page accompanying this book at www.wiley.com/go/kelly/crystallography3e. In addition, a number of optical and electron micrographs illustrating various aspects of the microstructure of materials which follow from the theory outlined in this book have been uploaded to this password protected Wiley Web page.

Sadly, the senior author, Anthony Kelly, died on 3 June 2014, and so the responsibility for preparing this extended third edition has fallen entirely on my shoulders as the junior author. Therefore, any typographical errors either introduced during the preparation of this third edition or missed from the first two editions of this book are ultimately my responsibility. It will be evident from reading this book, and prior editions of this book, that the book is unashamedly not an introductory textbook to crystallography and crystal defects, even though it addresses fundamental topics in materials science. However, the aim in writing this third edition has been to continue to communicate both general background knowledge and specialist research knowledge in as accessible a way as possible to an audience educated to degree level, highlighting specialist textbooks, review papers and original research papers for those interested in a deeper understanding of the various concepts introduced. The contents of this third edition have been successfully used by me for two recent week-long courses in the GIAN program run by the Ministry of Human Resource Development of the Government of India, one on crystallography for materials scientists aimed at advanced undergraduates and new postgraduates, and a more focused one on interfaces in materials for Ph.D. students and post-doctoral research workers. Discussions with Professors Dipankar Banerjee at the Indian Institute of Science Bangalore, Rajesh Prasad at the Indian Institute of Technology, Delhi and Anjan Sil at the Indian Institute of Technology, Roorkee arising from these GIAN courses have all been invaluable in deciding what to include in this third edition, and at what level of sophistication. Rajesh Prasad deserves a special mention for highlighting a number of typographical errors in the second edition that I have now corrected in this third edition.

There are two colour images on the front cover of this book, both related to the crystallography of devitrite, $Na_2Ca_3Si_6O_{16}$. The larger of these is a low magnification photograph of devitrite needles nucleated on the surface of a block of soda–lime–silica float glass after a heat treatment of 17 hours at 850 °C observed in transmitted polarized light with a sensitive tint at 45° to the polarizer and analyzer, which are aligned vertically and horizontally, respectively. The needles are in a thin section cut perpendicular to the surface of the glass block. The edge of the sample is below the bottom of this photograph. I am grateful to Dr Robert P. Thompson from the University of Cambridge, Department of Materials Science and Metallurgy for permission to use this photograph. The second smaller colour image of atomistic modelling of a type II twin boundary in devitrite was produced by Prof. Bin Li, now at the School of Materials Science and Engineering, Sun Yat-sen University, Guangzhou, People's Republic of China. I am also grateful to Prof. Li for permission to use this image for the front cover.

Finally, as with the second edition, it is hoped that the reader whose understanding of crystallography and crystal defects goes well beyond what is described here will nevertheless find parts where his or her knowledge has been enriched.

Kevin M. Knowles
March 2020

Companion Website

This book is accompanied by a companion website:

www.wiley.com/go/kelly/crystallography3e

The Website includes:

- Solutions
- Computer programs for crystallographic calculations
- PPT slides of all figures from the book
- Optical and electron micrographs illustrating various aspects of the microstructure of materials

Scan this QR code to visit the companion website.

Part I
Perfect Crystals

1

Lattice Geometry

1.1 The Unit Cell

Crystals are solid materials in which the atoms are regularly arranged with respect to one another. This regularity of arrangement can be described in terms of symmetry operations; these operations determine the symmetry of the physical properties of a crystal. For example, the symmetry operations show in which directions the electrical resistance of a crystal will be the same. Many naturally occurring crystals, such as halite (sodium chloride), quartz (silica), and calcite (calcium carbonate), have very well-developed external faces. These faces show regular arrangements at a macroscopic level, which indicate the regular arrangements of the atoms at an atomic level. Historically, such crystals are of great importance because the laws of crystal symmetry were deduced from measurements of the interfacial angles in them; measurements were first carried out in the seventeenth century. Even today, the study of such crystals still possesses some heuristic advantages in learning about symmetry.

Nowadays the atomic pattern within a crystal can be studied directly by techniques such as high-resolution transmission electron microscopy. This atomic pattern is the fundamental pattern described by the symmetry operations and we shall begin with it.

In a crystal of graphite the carbon atoms are joined together in sheets. These sheets are only loosely bound to one another by van der Waals forces. A single sheet of such atoms provides an example of a two-dimensional crystal; indeed, recent research has shown that such sheets can actually be isolated and their properties examined. These single sheets are now termed 'graphene'. The arrangement of the atoms within a sheet of graphene is shown in Figure 1.1a. In this representation of the atomic pattern, the centre of each atom is represented by a small dot, and lines joining adjacent dots represent bonds between atoms. All of the atoms in this sheet are identical. Each atom possesses three nearest neighbours.

Crystallography and Crystal Defects, Third Edition. Anthony Kelly and Kevin M. Knowles.
© 2020 John Wiley & Sons Ltd. Published 2020 by John Wiley & Sons Ltd.
Companion website: www.wiley.com/go/kelly/crystallography3e

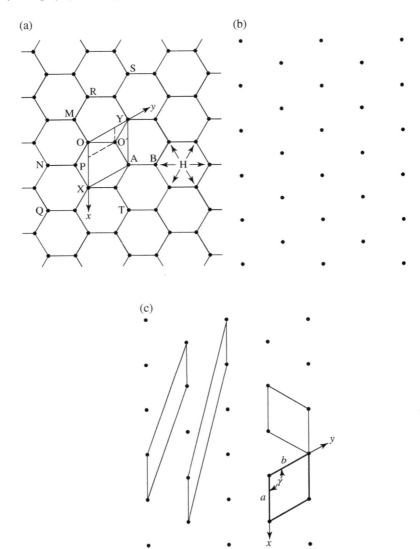

Figure 1.1 (a) The arrangement of the atoms in graphene, a single sheet of graphite in which the centre of each atom is represented by a dot, some of which are labelled using capital letters, (b) the lattice of graphene, (c) various choices of primitive unit cells for graphene. In (c), the conventional unit cell is outlined in heavy lines and the corresponding x- and y-axes are marked. In (a), H represents the position of an axis of six-fold rotational symmetry (see Section 1.4.2)

We describe this by saying that the coordination number is 3. In this case the coordination number is the same for all the atoms. It is the same for the two atoms marked A and B. However, atoms A and B have different environments: the orientation of the neighbours is different at A and B. Atoms in a similar situation to those at A are found at N and Q; there is a similar situation to B at M and at P.

It is obvious that we can describe the whole arrangement of atoms and interatomic bonds shown in Figure 1.1a by choosing a small unit such as OXAY, describing the arrangement of the atoms and bonds within it, then moving the unit so that it occupies the position NQXO and repeating the description and then moving it to ROYS and so on, until we have filled all space with identical units and described the whole pattern. If the repetition of the unit is understood to occur automatically, then to describe the crystal we need only describe the arrangement of the atoms and interatomic bonds within one unit. The unit chosen we would call the 'unit parallelogram' in two dimensions (in three dimensions, the 'unit cell'). In choosing the unit we always choose a parallelogram in two dimensions or a parallelepiped in three dimensions. The reason for this will become clear later.

Having chosen the unit, we describe the positions of the atoms inside it by choosing an origin O and taking axes Ox and Oy parallel to the sides, so that the angle between Ox and Oy is $\geq 90°$. We state the lengths of the sides a and b, taking a equal to the distance OX and b equal to the distance OY (Figure 1.1a), and we give the angle γ between Ox and Oy. In this case $a = b = 2.45\,\text{Å}$[1] (at 25 °C) and $\gamma = 120°$. To describe the positions of the atoms within the unit parallelogram, we note that there is one at each corner and one wholly inside the cell. The atoms at O, X, A and Y all have identical surroundings.[2]

In describing the positions of the atoms we take the sides of the parallelogram, a and b, as units of length. Then the coordinates of the atom at O are $(0, 0)$; those at X $(1, 0)$; those at Y $(0, 1)$; and those at A $(1, 1)$. The coordinates of the atom at O$'$ are obtained by drawing lines through O$'$ parallel to the axes Ox and Oy. The coordinates of O$'$ are therefore $\left(\frac{1}{3}, \frac{2}{3}\right)$. To describe the contents fully inside the unit parallelogram – that is, to describe the positions of the atoms – we need only give the coordinates of the atom at the origin, $(0, 0)$, and those of the atom at O$'$. The reason is that the atoms at X, A and Y have identical surroundings to those at O and an atom such as O, X, A or Y is shared between the four cells meeting at these points. The number of atoms contained within the area OXAY is two. O$'$ is within the area, giving one atom. O, X, A and Y provide four atoms each shared between four unit cells, giving an additional $4 \times \frac{1}{4} = 1$. A second way of arriving at the same result is to move the origin of the unit cell slightly away from the centre of the atom at O so that the coordinate of the O atom is $(\varepsilon_1, \varepsilon_2)$ and the coordinate of the O$'$ atom is $\left(\frac{1}{3} + \varepsilon_1, \frac{2}{3} + \varepsilon_2\right)$, where $0 < \varepsilon_1, \varepsilon_2 \ll 1$. Under these circumstances the centres of the atoms at X, A and Y lie outside the unit cell, so that the atom count within the unit cell is simply two. We note that the minimum number of atoms which a unit parallelogram could contain in a sheet of graphene is two, since the atoms at O and O$'$ have different environments.

To describe the atomic positions in Figure 1.1a we chose OXAY as one unit parallelogram. We could equally well have chosen OXTA. The choice of a particular unit parallelogram or unit cell is arbitrary, subject to the constraint that the unit cell tessellates, that is, it repeats periodically, in this case in two dimensions. Therefore, NQPM is not a permissible choice – although NQPM is a parallelogram, it cannot be repeated to produce the graphene structure because P and M have environments which differ from those at N and Q.

---

[1] $1\,\text{Å} = 10^{-10}$ m.

[2] In choosing a unit parallelogram or a unit cell, the crystal is always considered to be infinitely large. The pattern in Figure 1.1a must then be thought of as extending to infinity. The fact that O, X, A and Y in a finite crystal are at slightly different positions with respect to the boundary of the pattern can then be neglected.

The corners of the unit cell OXAY in Figure 1.1a all possess identical surroundings. We could choose and mark on the diagram all points with surroundings identical to those at O, X, A and so on. Such points are N, Q, R, S and so on. The array of all such points with surroundings identical to those of a given point we call the mesh or net in two dimensions (a lattice in three dimensions). Each of the points is called a lattice point.

A formal definition of the lattice is as follows: *A lattice is a set of points in space such that the surroundings of one point are identical with those of all the others. The type of symmetry described by the lattice is referred to as translational symmetry.* The lattice of the graphene crystal structure drawn in Figure 1.1a is shown in Figure 1.1b. It consists of a set of points with identical surroundings. Just a set of points: no atoms are involved. Various primitive unit cells are marked in Figure 1.1c. A primitive unit cell is defined as a unit parallelogram which contains just one lattice point. The conventional unit cell for graphene corresponding to OXAY in Figure 1.1a is outlined in Figure 1.1c in heavy lines and the corresponding x- and y-axes are marked.

In general, the conventional primitive unit cell of a two-dimensional net which shows no obvious symmetry is taken with its sides as short and as nearly equal as possible, with γ, the angle between the x- and y-axes, taken to be obtuse, if it is not equal to 90°. However, the symmetry of the pattern must always be taken into account. The net shown in Figure 1.1b is very symmetric and in this case we can take the sides to be equal, so that $a = b$ and $\gamma = 120°$.

Comparisons of Figures 1.1a and b emphasize that the choice of the origin for the lattice is arbitrary. If we had chosen O' in Figure 1.1a as the origin instead of O and then marked all corresponding points in Figure 1.1a, we would have obtained the identical lattice with only a change of origin. The lattice then represents an essential aspect of the translational symmetry of the crystal however we choose the origin.

In three dimensions the definition of the lattice is the same as in two dimensions. The unit cell is now the parallelepiped containing just one lattice point. The origin is taken at a corner of the unit cell. The sides of the unit parallelepiped are taken as the axes of the crystal, x, y, z, using a right-handed notation. The angles α, β, γ between the axes are called the axial angles (see Figure 1.2). The smallest separations of the lattice points along the x-, y- and z-axes are denoted by a, b, c, respectively and called the lattice parameters.

Inspection of the drawing of the arrangement of the ions in a crystal of caesium chloride, CsCl, in Figure 3.17 shows that the lattice is an array of points such that $a = b = c$, $\alpha = \beta = \gamma = 90°$, so that the unit cell is a cube. There is one caesium ion and one chlorine ion associated with each lattice point. If we take the origin at the centre of a caesium ion, then there is one caesium ion in the unit cell with coordinates $(0, 0, 0)$ and one chlorine ion with coordinates $\left(\frac{1}{2}, \frac{1}{2}, \frac{1}{2}\right)$. A projection along the z-axis is shown in Figure 1.3. It must be remembered that the caesium ions at the corners of the unit cell project on top of one another; thus, ions at elevations 0 and 1 are superimposed, as indicated in Figure 1.3. Since translational symmetry dictates that if there is an ion at $(x, y, 0)$, there must be an ion with exactly the same environment at $(x, y, 1)$, the ions at elevation 1 along the z-axis can be omitted by convention. It is also standard practice when drawing projections of crystal structures not to indicate the elevations of atoms or ions if they are 0.

It is apparent from this drawing of the crystal structure of caesium chloride that the coordination number for each caesium ion and each chlorine ion is eight, each ion having eight

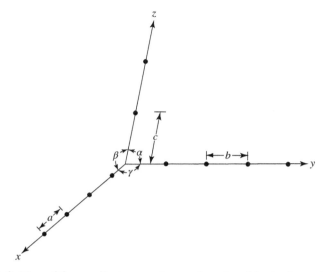

Figure 1.2 Definition of the smallest separations *a*, *b* and *c* of the lattice points along the *x*-, *y*- and *z*-axes respectively, together with the angles α, β and γ between the axes for a lattice in three dimensions

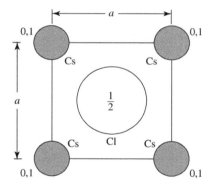

Figure 1.3 The numbers give the elevations of the centres of the atoms, along the *z*-axis, taking the lattice parameter *c* as the unit of length

of the other kind of ion as neighbours. The separation of these nearest neighbours, *d*, is easily seen to be given by:

$$d = \left[\left(\frac{a}{2} \right)^2 + \left(\frac{b}{2} \right)^2 + \left(\frac{c}{2} \right)^2 \right]^{1/2} = \frac{\sqrt{3}a}{2} \tag{1.1}$$

since $a = b = c$.

The number of units of the formula CsCl per unit cell is clearly 1.

1.2 Lattice Planes and Directions

A rectangular mesh of a hypothetical two-dimensional crystal with mesh parameters *a* and *b* of very different magnitude is shown in Figure 1.4. Note that the parallel mesh lines OB, O'B', O"B", and so on all form part of a set and that the spacing of all lines in the set is quite regular; this is similar for the set of lines parallel to AB: A'B', A"B" and so on. The spacing of each of these sets is determined only by **a** and **b** (and the angle between **a** and **b,** which in this example is 90°). Also, the angle between these two sets depends only on the *ratio* of *a* to *b*, where *a* and *b* are the magnitudes of **a** and **b**, respectively. If external faces of the crystal formed parallel to O"B and to AB, the angle between these faces would be uniquely related to the ratio *a* : *b*. Furthermore, this angle would be independent of how large these faces were (see Figure 1.5). This was recognized by early crystallographers, who deduced the existence of the lattice structure of crystals from the observation of the constancy of angles between corresponding faces. This law of constancy of angle states: *In all crystals of the same substance the angle between corresponding faces has a constant value.*

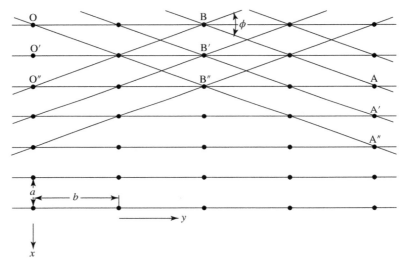

Figure 1.4 A rectangular mesh of a hypothetical two-dimensional crystal with mesh parameters *a* and *b* of very different magnitude

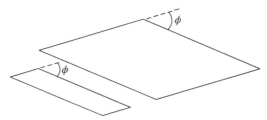

Figure 1.5 Demonstration of the law of constancy of angles between faces of crystals: the angle ϕ between the faces is independent of the size of the faces

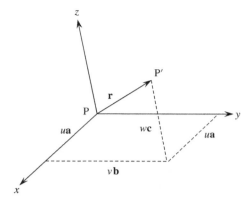

Figure 1.6   A vector **r** written as the sum of translations along the *x*-, *y*- and *z*-axes

The analogy between lines in a mesh and planes in a crystal lattice is very close. Crystal faces form parallel to lattice planes and important lattice planes contain a high density of lattice points. Lattice planes form an infinite regularly spaced set which collectively passes through all points of the lattice. The spacing of the members of the set is determined only by the lattice parameters and axial angles, and the angles between various lattice planes are determined only by the axial angles and the ratios of the lattice parameters to one another.

Prior to establishing a methodology for designating a set of lattice planes, it is expedient to consider how directions in a crystal are specified. A direction is simply a line in the crystal. Select any two points on the line, say P and P'. Choose one as the origin, say P (Figure 1.6). Write the vector **r** between the two points in terms of translations along the *x*-, *y*- and *z*-axes so that:

$$\mathbf{r} = u\mathbf{a} + v\mathbf{b} + w\mathbf{c} \qquad (1.2)$$

where **a**, **b** and **c** are vectors along the *x*-, *y*- and *z*-axes, respectively, and have magnitudes equal to the lattice parameters (Figure 1.6). The direction is then denoted as [*uvw*] – always cleared of fractions and reduced to its lowest terms. The triplet of numbers indicating a direction is always enclosed in square brackets. Some examples are given in Figure 1.7. Negative values of *u*, *v* and *w* are indicated in this figure by a bar over the appropriate index.

If *u*, *v* and *w* are integers and the origin P is chosen at a lattice point, then P' is also a lattice point and the line PP' produced is a row of lattice points. Such a line is called a *rational* line, just as a plane of lattice points is called a rational plane.

We designate a set of lattice planes as follows (see Figure 1.8). Let one member of the set meet the chosen axes, *x*, *y*, *z* at distances from the origin of A, B and C, respectively. We can choose the origin to be a lattice point. Vectors **a**, **b** and **c** then define the distances between adjacent lattice points along the *x*-, *y*- and *z*-axes, respectively. The Miller[3] indices

---

[3] Indices for planes are termed Miller indices after William Hallowes Miller (1801–1880) who popularized the use of these indices through his book *A Treatise on Crystallography* published in 1839 [1]. Interestingly, Miller was not the first to advocate the use of the (*hkl*) notation for planes – as Miller himself acknowledged in a preface to his book, the Rev. William Whewell had previously noted that planes can be represented by indices *h*, *k* and *l* in this manner in 1825 [2].

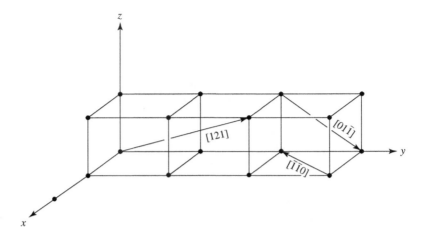

Figure 1.7   Examples of various lattice vectors in a crystal

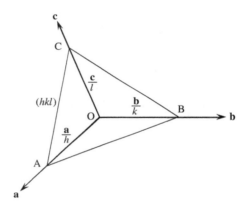

Figure 1.8   The plane (*hkl*) in a crystal making intercepts of **a**/*h*, **b**/*k* and **c**/*l* along the *x*-, *y*- and *z*-axes, respectively

(*hkl*) of the set of lattice planes are then defined in terms of the intercepts A, B and C so that the length OA = *a*/*h*, where *a* is the magnitude of **a**; likewise, OB = *b*/*k* and OC = *c*/*l*.

Thus, for example, the plane marked Y in Figure 1.9 makes intercepts on the axes of infinity, 2*b* and infinity, respectively. Taking the reciprocals of these intercepts gives $h = 0$, $k = \frac{1}{2}$ and $l = 0$. Clearing the fractions gives $h = 0$, $k = 1$ and $l = 0$. Hence, the set of lattice planes parallel to Y is designated (010). The triplet of numbers describing the Miller index is always enclosed in round brackets. Similarly, the plane marked P in Figure 1.9 has intercepts 1*a*, 2*b* and $\frac{1}{3}c$. Therefore, taking the reciprocals of these intercepts, $h = 1$, $k = \frac{1}{2}$ and $l = 3$. Clearing the fractions, we have (216) as the Miller indices. The indices of a number of other planes are shown in Figure 1.9. Negative values of the intercepts are indicated in the Miller index notation by a bar over the appropriate index (see the examples in Figure 1.9).

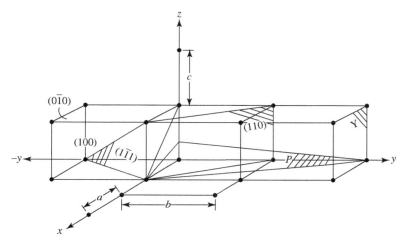

Figure 1.9 Examples of various lattice planes in a crystal. The indices of the planes P and Y are discussed in the accompanying part of the text in Section 1.2

The reason for using Miller indices to index crystal planes is that they greatly simplify certain crystal calculations. Furthermore, with a reasonable choice of unit cell, small values of the indices (*hkl*) belong to widely spaced planes containing a large areal density of lattice points. Well-developed crystals are usually bounded by such planes, so that it is found experimentally that prominent crystal faces have intercepts on the axes which when expressed as multiples of *a*, *b* and *c* have ratios to one another that are small rational numbers.[4]

1.3   The Weiss Zone Law

This law expresses the mathematical condition for a vector [*uvw*] to lie in a plane (*hkl*). This condition can be determined through elementary vector considerations. Consider the plane (*hkl*) in Figure 1.10 with the normal to the plane $\overrightarrow{OP}$.

A general vector, **r**, lying in (*hkl*) can be expressed as a linear combination of any two vectors lying in this plane, such as $\overrightarrow{AB}$ and $\overrightarrow{AC}$. That is:

$$\mathbf{r} = \lambda\overrightarrow{AB} + \mu\overrightarrow{AC} \tag{1.3}$$

for suitable λ and μ. Hence, expressing $\overrightarrow{AB}$ and $\overrightarrow{AC}$ in terms of **a**, **b** and **c**, it follows that:

$$\mathbf{r} = \lambda\left(\frac{\mathbf{b}}{k} - \frac{\mathbf{a}}{h}\right) + \mu\left(\frac{\mathbf{c}}{l} - \frac{\mathbf{a}}{h}\right) = -\frac{(\lambda+\mu)}{h}\mathbf{a} + \frac{\lambda}{k}\mathbf{b} + \frac{\mu}{l}\mathbf{c} \tag{1.4}$$

---

[4] Formally, the *law of rational indices* states that all planes which can occur as faces of crystals have intercepts on the axes which, when expressed as multiples of certain unit lengths along the axes (themselves proportional to *a*, *b*, *c*), have ratios that are small rational numbers. A rational number can always be written as *p/q*, where *p* and *q* are integers.

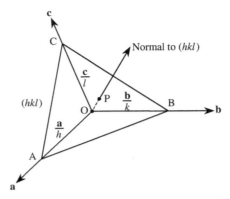

Figure 1.10   The plane (*hkl*) in a crystal making intercepts of **a**/*h*, **b**/*k* and **c**/*l* along the *x*-, *y*- and *z*-axes, respectively, together with the vector normal to (*hkl*)

If we re-express this as $\mathbf{r} = u\mathbf{a} + v\mathbf{b} + w\mathbf{c}$ – a general vector [*uvw*] lying in (*hkl*) – it follows that:

$$u = -\frac{(\lambda + \mu)}{h}, \quad v = \frac{\lambda}{k}, \quad w = \frac{\mu}{l} \tag{1.5}$$

and so:

$$hu + kv + lw = 0 \tag{1.6}$$

which is the condition for a vector [*uvw*] to lie in the plane (*hkl*): **Weiss zone law**.[5] It is evident from this derivation that it is valid for arbitrary orientations of the *x*-, *y*- and *z*-axes with respect to one another.

Frequently, a number of important crystal lattice planes all lie in the same zone; that is, they intersect one another in parallel lines. For instance, in Figure 1.9 the planes (100), ($0\bar{1}0$) and (110) are all parallel to the direction [001]. They would be said to lie in the zone [001], since [001] is a common direction lying in all of them. The normals to all of these planes are perpendicular to [001]. This is not an accident – the normals are constrained to be perpendicular to [001] by the Weiss zone law.

To see why, we can make use of elementary vector algebra relationships discussed in Appendix 1, Section A1.1. Consider the plane (*hkl*) shown in Figure 1.10. The vector normal to this plane, **n**, must be parallel to the cross product $\overrightarrow{AB} \times \overrightarrow{AC}$. Hence:

$$\mathbf{n} \parallel \left(\frac{\mathbf{b}}{k} - \frac{\mathbf{a}}{h} \right) \times \left(\frac{\mathbf{c}}{l} - \frac{\mathbf{a}}{h} \right), \quad \text{i.e.,} \quad \mathbf{n} \parallel \left(\frac{\mathbf{b} \times \mathbf{c}}{kl} - \frac{\mathbf{b} \times \mathbf{a}}{hk} - \frac{\mathbf{a} \times \mathbf{c}}{hl} \right) \tag{1.7}$$

and so, after some straightforward mathematical manipulation, making use of the identities

$$\mathbf{a} \times \mathbf{b} = -\mathbf{b} \times \mathbf{a} \quad \text{and} \quad \mathbf{c} \times \mathbf{a} = -\mathbf{a} \times \mathbf{c},$$

it is apparent that **n** is parallel to the vector $h\mathbf{a}^* + k\mathbf{b}^* + l\mathbf{c}^*$. That is:

$$\mathbf{n} = \xi(h\mathbf{a}^* + k\mathbf{b}^* + l\mathbf{c}^*) \tag{1.8}$$

---

[5] The name Weiss zone law for this equation was first used by William James Lewis (1847–1926) in the form *Weiss's zone-law* on pages 39 and 40 of his 1899 book on crystallography [3]. Lewis's reasoning was that Christian Samuel Weiss (1780–1856) had been the first to call attention to this relationship through his 1804–1806 translation into German of René Just Haüy's 1801 *Traité de Minéralogie*. Miller knew about this equation as well – it is derived geometrically on page 10 of [1].

for a constant of proportionality, ξ. The vectors $\mathbf{a}^*$, $\mathbf{b}^*$ and $\mathbf{c}^*$ in Eq. (1.8) are termed *reciprocal lattice vectors*, defined through the equations:

$$\mathbf{a}^* = \frac{\mathbf{b} \times \mathbf{c}}{\mathbf{a} \cdot [\mathbf{b} \times \mathbf{c}]}, \quad \mathbf{b}^* = \frac{\mathbf{c} \times \mathbf{a}}{\mathbf{a} \cdot [\mathbf{b} \times \mathbf{c}]}, \quad \mathbf{c}^* = \frac{\mathbf{a} \times \mathbf{b}}{\mathbf{a} \cdot [\mathbf{b} \times \mathbf{c}]} \tag{1.9}$$

If the normal to the (*hkl*) set of planes is simply taken to be the vector

$$\mathbf{n} = h\mathbf{a}^* + k\mathbf{b}^* + l\mathbf{c}^* \tag{1.10}$$

the magnitude of $\mathbf{n}$ is inversely proportional to the spacing of the *hkl* planes; that is, it is inversely proportional to the distance OP in Figure 1.10 (Section A1.2), irrespective of the orientations of the *x*-, *y*- and *z*-axes with respect to one another.

Furthermore, it is evident that the scalar product of a normal to a set of planes, $\mathbf{n}$, with a vector $\mathbf{r} = [uvw]$ lying on one of these planes must be zero. That is, $\mathbf{r} \cdot \mathbf{n} = 0$. Writing out this dot product explicitly, we obtain the result:

$$hu + kv + lw = 0$$

which is the Weiss zone law. This demonstrates that the Weiss zone law is a scalar product between two vectors, one of which lies in one of a set of planes and the other of which is normal to the set of planes.

Given the indices of any two planes, say $(h_1k_1l_1)$ and $(h_2k_2l_2)$, the indices of the zone [*uvw*] in which they lie are found by solving the simultaneous equations:

$$h_1 u + k_1 v + l_1 w = 0 \tag{1.11}$$

$$h_2 u + k_2 v + l_2 w = 0 \tag{1.12}$$

Since it is only the ratios $u : v : w$ which are of interest, these equations can be solved to give:

$$\begin{aligned} u &= k_1 l_2 - k_2 l_1 \\ v &= l_1 h_2 - l_2 h_1 \\ w &= h_1 k_2 - h_2 k_1 \end{aligned} \tag{1.13}$$

There are other methods of producing the same result. For example, we could write down the planes in the form:

$$\begin{matrix} h_1 & k_1 & l_1 & h_1 & k_1 & l_1 \\ h_2 & k_2 & l_2 & h_2 & k_2 & l_2 \end{matrix}$$

We then cross out the first and the last columns and evaluate the 2×2 determinants from (i) the second and third columns, (ii) the third and fourth columns, and (iii) the fourth and fifth columns:

$$\begin{matrix} \cancel{h_1} & k_1 & l_1 & h_1 & k_1 & \cancel{l_1} \\ \cancel{h_2} & k_2 & l_2 & h_2 & k_2 & \cancel{l_2} \end{matrix} \tag{1.14}$$

Therefore, we find $[uvw] = [k_1l_2 - k_2l_1, l_1h_2 - l_2h_1, h_1k_2 - h_2k_1]$. A third method is to evaluate the determinant

$$\begin{vmatrix} \mathbf{a} & \mathbf{b} & \mathbf{c} \\ h_1 & k_1 & l_1 \\ h_2 & k_2 & l_2 \end{vmatrix} \tag{1.15}$$

to determine $[uvw]$. The result is $[uvw] = (k_1l_2 - k_2l_1)\,\mathbf{a} + (l_1h_2 - l_2h_1)\,\mathbf{b} + (h_1k_2 - h_2k_1)\,\mathbf{c}$.

Thus, for example, supposing $(h_1k_1l_1) = (112)$ and $(h_2k_2l_2) = (\bar{1}43)$, we would have:

$$\frac{\cancel{1}\quad 1\quad 2\quad 1\quad 1\quad \cancel{2}}{\cancel{\bar{1}}\quad 4\quad 3\quad \bar{1}\quad 4\quad \cancel{3}}$$

and so $[uvw] = [-5, -5, 5] \equiv [11\bar{1}]$. Likewise, given two directions $[u_1v_1w_1]$ and $[u_2v_2w_2]$, we can obtain the plane (hkl) containing these two directions by solving the simultaneous equations:

$$hu_1 + kv_1 + lw_1 = 0 \tag{1.16}$$

$$hu_2 + kv_2 + lw_2 = 0 \tag{1.17}$$

Using a similar method to the one used to produce Eq. (1.14), we draw up the three 2×2 determinants as follows:

$$\frac{\cancel{u_1}\quad v_1\quad w_1\quad u_1\quad v_1\quad \cancel{w_1}}{\cancel{u_2}\quad v_2\quad w_2\quad u_2\quad v_2\quad \cancel{w_2}} \tag{1.18}$$

to find that $(hkl) = (v_1w_2 - v_2w_1, w_1u_2 - w_2u_1, u_1v_2 - u_2v_1)$. The method equivalent to Eq. (1.15) is to evaluate the determinant:

$$\begin{vmatrix} \mathbf{a}^* & \mathbf{b}^* & \mathbf{c}^* \\ u_1 & v_1 & w_1 \\ u_2 & v_2 & w_2 \end{vmatrix} \tag{1.19}$$

It is also evident from Eqs. (1.11) and (1.12), the conditions for two planes $(h_1k_1l_1)$ and $(h_2k_2l_2)$ to lie in the same zone $[uvw]$, that by multiplying Eq. (1.11) by a number m and Eq. (1.12) by a number n and adding them, we have:

$$(mh_1 + nh_2)u + (mk_1 + nk_2)v + (ml_1 + nl_2)w = 0 \tag{1.20}$$

Therefore the plane $(mh_1 + nh_2, mk_1 + nk_2, ml_1 + nl_2)$ also lies in $[uvw]$. In other words, the indices formed by taking linear combinations of the indices of two planes in a given zone provide the indices of a further plane in that same zone. In general m and n can be positive or negative. If, however, m and n are both positive, then the normal to the plane under consideration must lie between the normals of $(h_1k_1l_1)$ and $(h_2k_2l_2)$: we will revisit this result in Section 2.2.

1.4   Symmetry Operators[6]

The symmetrical arrangement of atoms in crystals is described formally in terms of operations of symmetry. The symmetry arises because an atom or group of atoms is

---

[6] A discussion of whether these should be termed *symmetry operators* or *symmetry elements*, the term used in previous editions of this book, is given in Section 1.2.3 of [4]. Current International Union of Crystallography nomenclature is to use the term *symmetry operators*, reserving the word 'element' to define a member of a set, such as a group (Section 2.14). Readers should be aware that both terms are used in practice in the crystallographic literature.

repeated in a regular way to form a pattern. Any operation of repetition can be described in terms of one of the following four different types of pure symmetry element or symmetry operator.

1.4.1 Translational Symmetry

This describes the fact that similar atoms in identical surroundings are repeated at different points within the crystal. Any one of these points can be brought into coincidence with another by an operation of translational symmetry. For instance, in Figure 1.1a the carbon atoms at O, Y, N and Q occupy completely similar positions. We use the idea of the lattice to describe this symmetry. The lattice is a set of points each with an identical environment which can be found by inspection of the crystal structure, as in the example in Figure 1.1b. We can define the arrangement of lattice points in a three-dimensional crystal by observing that the vector **r** joining any two lattice points (or the operation of translational symmetry bringing one lattice point into coincidence with another) can always be written as:

$$\mathbf{r} = u\mathbf{a} + v\mathbf{b} + w\mathbf{c} \qquad (1.21)$$

where u, v, w are positive or negative integers, or equal to zero. Inspection of Figure 1.11 shows that to use this description we must be careful to choose **a**, **b** and **c** so as to include all lattice points. We do this by, say, making **a** the shortest vector between lattice points in the lattice, or one of several shortest ones. We then choose **b** as the shortest not parallel to **a** and **c** as the shortest not coplanar with **a** and **b**. Thus **a**, **b** and **c** define a primitive unit cell of the lattice in the same sense as in Section 1.1. Only one lattice point is included within the volume $\mathbf{a} \cdot [\mathbf{b} \times \mathbf{c}]$,[7] which is the volume of the primitive unit cell; under these circumstances **a**, **b** and **c** are called the lattice translation vectors.

1.4.2 Rotational Symmetry

If one stood at the point marked H in Figure 1.1a and regarded the surroundings in a particular direction, say that indicated by one of the arrows, then on turning through an angle of $60° = 360°/6$ the outlook would be identical. We say an axis of sixfold rotational symmetry passes normal to the paper through the point H. Similarly, at O′ an axis of threefold symmetry passes normal to the paper, since an identical outlook is found after a rotation of $360°/3 = 120°$. A crystal possesses an n-fold axis of rotational symmetry if it coincides with

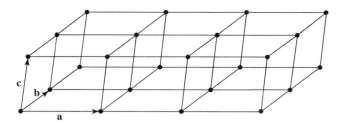

Figure 1.11   Translation symmetry in a crystal

---

[7] The volume $\mathbf{a} \cdot [\mathbf{b} \times \mathbf{c}]$ is equal to $abc \sin \alpha \cos \varphi$, where φ is the angle between **a** and the normal to the plane containing **b** and **c**. See also Equation (1.42). If $a \le b \le c$ and α, β and γ are all $<90°$, the primitive unit cell described in this manner is a positive reduced form or Type I cell in the nomenclature of Santoro and Mighell [5]; if $a \le b \le c$ and α, β and γ are all $\ge 90°$, the primitive unit cell is a negative reduced form or Type II cell in the same nomenclature. Both these forms are examples of unit cells termed *reduced cells* or *Niggli reduced cells* after Paul Niggli's original work on this topic [6].

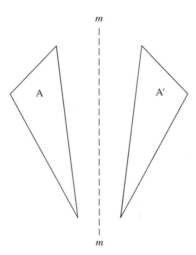

Figure 1.12   Reflection symmetry

itself upon rotation about the axis of $360°/n = 2\pi/n$ radians. In crystals, axes of rotational symmetry with values of n equal to one, two, three, four and six are the only ones found which are compatible with translational symmetry. These correspond to repetition every 360°, 180°, 120°, 90° and 60° and are called monad, diad, triad, tetrad and hexad axes, respectively. The reasons for these limitations on the value of n are explained in Section 1.5.

1.4.3   Centre of Symmetry

A *centre of symmetry* is a point where inversion through that point produces an identical arrangement. The operation of inversion moves the point with coordinates (x, y, z) to the position $(-x, -y, -z)$. For instance, in Figure 1.3 (see also Figure 3.17), if we stood at the centre of the unit cell (coordinates $\frac{1}{2}, \frac{1}{2}, \frac{1}{2}$) and looked in any direction $[uvw]$, we would find an identical outlook if we looked in the direction $[\overline{uvw}]$. Clearly, in a lattice, all lattice points are centres of symmetry of the lattice. This follows from Eq. (1.21), since u, v, w can be positive or negative. Of course, since the origin of a set of lattice points can be arbitrarily chosen in a given crystal structure, we must not assume that, in a given crystal with a centre of symmetry, the centres of symmetry necessarily lie where we have chosen to place the lattice points.

1.4.4   Reflection Symmetry

The fourth symmetry is reflection symmetry. The operation is that of reflection in a mirror. Mirror symmetry relates, for example, our left and right hands. The dotted plane running normal to the paper, marked m in Figure 1.12, reflects object A to A′, and vice versa. A′ cannot be moved about in the plane of the paper and made to superpose on A. The dotted plane is a mirror plane. It should be noted that the operation of inversion through a centre of symmetry also produces a right-handed object from a left-handed one.

1.5 Restrictions on Symmetry Operations

All crystals show translational symmetry.[8] A given crystal may, or may not, possess other symmetry operations. Axes of rotational symmetry must be consistent with the translational symmetry of the lattice. A onefold rotation axis is obviously consistent. To prove that in addition only diads, triads, tetrads and hexads can occur in a crystal, we consider just a two-dimensional lattice or net.

Let A, A′, A″, … in Figure 1.13 be lattice points of the mesh and let us choose the direction AA′A″ so that the lattice translation vector **t** of the mesh in this direction is the shortest lattice translation vector of the net. Suppose an axis of n-fold rotational symmetry runs normal to the net at A. Then the point A′ must be repeated at B by an anticlockwise rotation through an angle $\alpha = A'AB = 2\pi/n$. Also, since A′ is a lattice point exactly similar to A, there must be an n-fold axis of rotational symmetry passing normal to the paper through A′. This repeats A at B′ through a clockwise rotation, as shown in Figure 1.13. That these two rotations are in opposite senses does not matter – they are both a consequence of the n-fold axis of rotational symmetry under consideration.

B and B′ define a lattice row parallel to AA′. Therefore, the separation of B and B′ by Eq. (1.12) must be an integral number times **t**. Call this integer N. From Figure 1.13 the separation of B and B′ is $(t - 2t \cos \alpha)$. Therefore, the possible values of α are restricted to those satisfying the equation:

$$t - 2t \cos \alpha = Nt$$

or:

$$\cos \alpha = \frac{1 - N}{2} \tag{1.22}$$

where N is an integer. Since $-1 \leq \cos \alpha \leq 1$ the only possible solutions are shown in Table 1.1. These correspond to onefold, sixfold, fourfold, threefold, and twofold axes of rotational symmetry. No other axis of rotational symmetry is consistent with the translational symmetry of a lattice and hence other axes do not occur in crystals.[9]

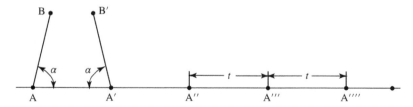

Figure 1.13 Diagram to help determine which rotation axes are consistent with translational symmetry

---

[8] The attention of the reader is, however, drawn to Section 4.6.1 for a discussion on the contemporary definition of crystals in the light of the discovery of quasicrystals and other materials in which there is ordering in an aperiodic manner.

[9] Of course, this does not preclude local rotation axes having unusual rotation symmetries, for example the occurrence of local icosahedral coordination found in compounds such as $MoAl_{12}$ and in individual C_{60} molecules in buckminsterfullerene (Section 4.5). However, when these units pack together in crystals, the fivefold rotational axes of the individual units are lost when describing the symmetry of the crystal.

Table 1.1   Solutions of Eq. (1.22)

N	-1	0	1	2	3
$\cos \alpha$	1	$\frac{1}{2}$	0	$-\frac{1}{2}$	-1
α	$0°$	$60°$	$90°$	$120°$	$180°$

Corresponding to the various allowed values of α derived from Eq. (1.22), three two-dimensional lattices, also known as nets or meshes, are defined. These are shown as the first three diagrams on the left-hand side of Figure 1.14. It should be noted that the hexad axis and the triad axis both require the same triequiangular mesh, the unit cell of which is a 120° rhombus (see Figure 1.14c).

In the same way that the possession of rotational symmetry axes perpendicular to the net places restriction on the net, restrictions are placed upon the net by the possession of a mirror plane: consideration of this identifies the two additional nets shown in Figures 1.14d and e. To see this, let A and A′ be two lattice points of a net and let the vector **t** joining them be a lattice translation vector defining one edge of the unit cell. A mirror plane can be placed normal to the lattice row AA′, as in Figure 1.15a, or as in Figure 1.15b. It cannot be placed arbitrarily anywhere in between A and A′. It must either lie midway between A and A′, as in Figure 1.15a, or pass through a lattice point, as in Figure 1.15b. Since AA′ determines a row of lattice points, a net can be built up consistent with mirror symmetry by placing a row identical to AA′ parallel with AA′, but displaced from it. There are just two possible arrangements, which are both shown in Figure 1.16, with the original lattice vector **t** indicated and *all of the mirror planes* consistent with the arrangement of the lattice points marked on the two diagrams. Hence, the spatial arrangements shown in Figure 1.16 give rise to the nets shown in Figures 1.14d and e.

Returning to Figure 1.14a, the left-hand diagram shows the net of points consistent with a twofold axis of symmetry normal to the net and with no axis of higher symmetry. This net corresponds with the solutions $N = -1$ or 3, $\alpha = 0$ or 180° in Table 1.1 and is based on a *parallelogram*. The lengths of two adjacent sides of the parallelogram (a, b in Figure 1.14a) and the value of the included angle γ can be chosen arbitrarily without removing the consistency with twofold symmetry. If a motif showing twofold symmetry normal to the net were associated with each lattice point, then twofold symmetry axes would be present at all the points shown in the right-hand diagram of Figure 1.14a.

All regular nets of points are consistent with twofold symmetry axes normal to the net because such a net of *points* is necessarily centrosymmetric and *in two dimensions* there is no difference between a centre of symmetry and a twofold or diad axis. The net corresponding to $N = 1$, $\alpha = 90°$ in Table 1.1 is based upon a square, shown in the left-hand diagram of Figure 1.14b. If a two-dimensional crystal possesses fourfold symmetry, it must necessarily possess this net. In addition, the atomic motif associated with each of the lattice or net points must also possess fourfold rotational symmetry. Provided the motif fulfils this condition, there must be a fourfold axis at the centre of each of the basic squares of the net and twofold axes at the midpoints of the sides, as shown in the right-hand diagram of Figure 1.14b. A two-dimensional crystal possessing fourfold rotational symmetry

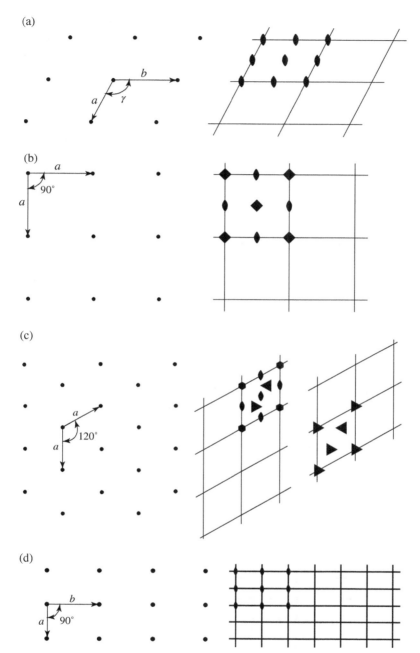

Figure 1.14   The five symmetrical plane lattices or nets. Rotational symmetry axes normal to the paper are indicated by the following symbols: ⧫ = diad; ▲ = triad; ■ = tetrad; ⬢ = hexad. Nets in (d) and (e) are both consistent with mirror symmetry, with the mirrors indicated by thick lines

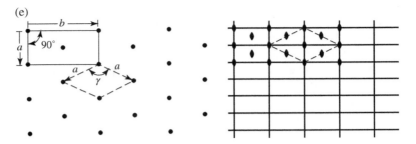

(e)

Figure 1.14 (*Continued*)

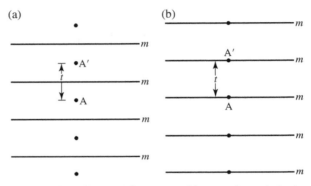

Figure 1.15 Restrictions placed on two-dimensional lattices through the imposition of mirror planes. These can be placed normal to a lattice row AA′ only as in (a) or (b)

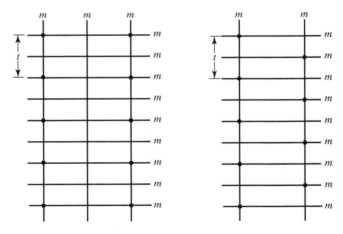

Figure 1.16 The two possible arrangements of nets consistent with mirror symmetry

cannot possess fewer symmetry operations than those shown in the right-hand diagram of Figure 1.14b.[10]

The net consistent with both $\alpha = 60°$ and $\alpha = 120°$, corresponding to the possession of hexad symmetry and triad symmetry, respectively, is the triequiangular net shown in Figure 1.14c. The primitive unit cell of this net has both sides equal and the included angle is necessarily 120°. It must be noted clearly that such a mesh of *points* is always consistent with sixfold symmetry. If the atomic motif associated with each lattice point is consistent with sixfold symmetry then diad and triad axes are automatically present, as shown in the central diagram of Figure 1.14c. A two-dimensional crystal will possess threefold symmetry provided the atomic motif placed at each lattice point of the lattice shown in Figure 1.14c possesses threefold symmetry. The only symmetry operations *necessarily* present are then just the threefold axes arranged as indicated in the far right-hand diagram of Figure 1.14c.

The simple rectangular net shown in Figure 1.14d has a primitive cell with a and b not necessarily equal. The angle between a and b is 90°. This *net* of points is consistent with the presence of diad axes at the intersection of the mirror planes, as is the mesh shown in Figure 1.14e. The simplest unit cell for the net in Figure 1.14e is a rhombus, indicated with the dotted lines. This has the two sides of the cell equal, and the angle between them, γ, can take any value. When dealing with a net based on a rhombus, it is, however, often *convenient* to choose as the unit cell a rectangle which contains an additional lattice point at its centre. This cell, outlined with full lines in the left-hand diagram of Figure 1.14e, has the angle between a and b necessarily equal to 90°. Hence it contains an additional lattice point inside it, which is called a *non-primitive* unit cell. The primitive unit cell is the dotted rhombus. The non-primitive cell clearly has twice the area of the primitive one and contains twice as many lattice points. It is chosen because it is naturally related to the symmetry, and is called the *centred rectangular cell*. This feature of choosing a non-primitive cell, because it is more naturally related to the symmetry operations, is one we shall meet often when dealing with the three-dimensional space lattices. The arrangements of diad axes and mirror planes consistent with the rectangular net and with the centred rectangular net are shown in the right-hand diagrams of Figures 1.14d and 1.14e, respectively.

1.6 Possible Combinations of Rotational Symmetries

As we have just shown in Section 1.5, the axes of n-fold rotational symmetry which a crystal can possess are limited to values of n of 1, 2, 3, 4, or 6. These axes lie normal to a net. In principle, a crystal might conceivably be symmetric with respect to many intersecting n-fold axes. However, it turns out that the possible angular relationships between axes are severely limited. To discover these we need a method to combine the possible rotations. One possible method is to use spherical trigonometric relationships, such as the approach adopted by Euler and developed by Buerger [7, 8]. An equivalent approach is to make use of the homomorphism between unit quaternions and rotations, described and developed by Grimmer [9], Altmann [10], and Kuipers [11].

---

[10] Depending on how the motif showing fourfold symmetry is arranged with respect to the two axes of the crystal – the sides of the square in Figure 1.14b – additional symmetry elements may arise. A full discussion of this point for any of the arrangements of symmetry elements in Figure 1.14 – that is, a discussion of consistent arrangements of symmetry elements in *space* – would take us immediately into the subject of *space groups*. We defer this until much later (see Section 2.12), but the discerning reader may wish to glance at Section 2.12 before proceeding further, both to gain reassurance that this subject matter is understood and to appreciate the mass of detail which is avoided by *not* following up this question now. The question of how additional symmetry elements may arise is considered in Problem 1.14.

Combinations of successive rotations about different axes are always inextricably related in groups of three. This arises because a rotation about an axis of unit length, say $\mathbf{n}_A$, of an amount α followed by a rotation about another axis of unit length, say $\mathbf{n}_B$, of amount β can always be expressed as a single rotation about some third axis of unit length, $\mathbf{n}_C$, of amount γ'.[11]

In an orthonormal coordinate system (one in which the axes are of equal length and at 90° to one another) the rotation matrix $\mathbf{R}$ describing a rotation of an amount θ (in radians) about an axis $\mathbf{n}$ with direction cosines n_1, n_2 and n_3 takes the form (Section A1.4):

$$\mathbf{R} = \begin{bmatrix} \cos\theta + n_1^2(1-\cos\theta) & n_1n_2(1-\cos\theta) - n_3\sin\theta & n_3n_1(1-\cos\theta) + n_2\sin\theta \\ n_1n_2(1-\cos\theta) + n_3\sin\theta & \cos\theta + n_2^2(1-\cos\theta) & n_2n_3(1-\cos\theta) - n_1\sin\theta \\ n_3n_1(1-\cos\theta) - n_2\sin\theta & n_2n_3(1-\cos\theta) + n_1\sin\theta & \cos\theta + n_3^2(1-\cos\theta) \end{bmatrix}$$

$$(1.23)$$

If we first apply a rotation of an amount α about an axis $\mathbf{n}_A$, described by a rotation matrix $\mathbf{R}_A$, after which we apply a rotation of an amount β about an axis $\mathbf{n}_B$, described by a rotation matrix $\mathbf{R}_B$, then in terms of matrix algebra the overall rotation is:

$$\mathbf{R}_B\mathbf{R}_A = \mathbf{R}_C \tag{1.24}$$

where $\mathbf{R}_C$ is the rotation matrix corresponding to the equivalent single rotation of an amount γ' about an axis $\mathbf{n}_C$. It is evident that one way of deriving γ' and the direction cosines n_{1C}, n_{2C} and n_{3C} is to work through the algebra suggested by Eq. (1.24) and to use the properties of the rotation matrix evident from Eq. (1.23):

$$\mathbf{R}_{ii} = 1 + 2\cos\theta \tag{1.25}$$

$$\varepsilon_{ijk}\mathbf{R}_{jk} = -2n_i\sin\theta \tag{1.26}$$

using the Einstein summation convention (Section A1.4).

An equivalent, more elegant, way is to use quaternion algebra. The rotation matrix described by Eq. (1.23) is homomorphic (i.e. exactly equivalent) to the unit quaternion:

$$\mathbf{q} = \{q_0, q_1, q_2, q_3\} = \left\{ \cos\frac{1}{2}\theta, n_1\sin\frac{1}{2}\theta, n_2\sin\frac{1}{2}\theta, n_3\sin\frac{1}{2}\theta \right\} \tag{1.27}$$

satisfying:

$$q_0^2 + q_1^2 + q_2^2 + q_3^2 = 1 \tag{1.28}$$

In quaternion algebra, the equation equivalent to Eq. (1.24) is one in which two quaternions $\mathbf{q}_B$ and $\mathbf{q}_A$ are multiplied together. The multiplication law for two quaternions $\mathbf{p}$ and $\mathbf{q}$ is (Section A1.4):

$$\begin{aligned} \mathbf{p}.\mathbf{q} = \{ & p_0q_0 - p_1q_1 - p_2q_2 - p_3q_3, \\ & p_0q_1 + p_1q_0 + p_2q_3 - p_3q_2, \\ & p_0q_2 - p_1q_3 + p_2q_0 + p_3q_1, \\ & p_0q_3 + p_1q_2 - p_2q_1 + p_3q_0 \} \end{aligned} \tag{1.29}$$

---

[11] This conforms to the notation used by Buerger [7].

The quaternion $\mathbf{p} \cdot \mathbf{q}$ is also a unit quaternion, from which the angle and the direction cosines of the axis of rotation of this unit quaternion can be extracted using Eq. (1.27). Therefore, if:

$$
\begin{aligned}
\mathbf{p} &= \{p_0, p_1, p_2, p_3\} = \left\{ \cos \frac{1}{2}\beta, n_{1B} \sin \frac{1}{2}\beta, n_{2B} \sin \frac{1}{2}\beta, n_{3B} \sin \frac{1}{2}\beta \right\} \\
\mathbf{q} &= \{q_0, q_1, q_2, q_3\} = \left\{ \cos \frac{1}{2}\alpha, n_{1A} \sin \frac{1}{2}\alpha, n_{2A} \sin \frac{1}{2}\alpha, n_{3A} \sin \frac{1}{2}\alpha \right\}
\end{aligned}
\tag{1.30}
$$

the angle γ' satisfies the equation:

$$
\cos \frac{1}{2}\gamma' = \cos \frac{1}{2}\alpha \cos \frac{1}{2}\beta - (n_{1A}n_{1B} + n_{2A}n_{2B} + n_{3A}n_{3B}) \sin \frac{1}{2}\alpha \sin \frac{1}{2}\beta
\tag{1.31}
$$

The term $(n_{1A}\,n_{1B} + n_{2A}\,n_{2B} + n_{3A}\,n_{3B})$ is simply the cosine of the angle between $\mathbf{n}_A$ and $\mathbf{n}_B$, and therefore $\mathbf{n}_A \cdot \mathbf{n}_B$, since $\mathbf{n}_A$ and $\mathbf{n}_B$ are of unit length. Defining $\gamma = 2\pi - \gamma'$, so that γ and γ' are the same angle measured in opposite directions, and rearranging the equation, it is evident that:

$$
\mathbf{n}_A \cdot \mathbf{n}_B = \frac{\cos \frac{1}{2}\gamma + \cos \frac{1}{2}\alpha \cos \frac{1}{2}\beta}{\sin \frac{1}{2}\alpha \sin \frac{1}{2}\beta}
\tag{1.32}
$$

Permitted values of γ are 60°, 90°, 120° and 180°, as shown in Table 1.1, and so possible values of $\cos \gamma/2$ are:

$$
\frac{\sqrt{3}}{2}, \frac{1}{\sqrt{2}}, \frac{1}{2} \text{ and } 0
\tag{1.33}
$$

To apply these results to crystals, let us assume that the rotation about $\mathbf{n}_A$ is a tetrad, so that $\alpha = 90°$ and $\alpha/2 = 45°$. Suppose the rotation about $\mathbf{n}_B$ is a diad, so that $\beta = 180°$ and $\beta/2 = 90°$. Then, in Eq. (1.32):

$$
\mathbf{n}_A \cdot \mathbf{n}_B = \sqrt{2} \cos \frac{1}{2}\gamma
\tag{1.34}
$$

Since $\mathbf{n}_A \cdot \mathbf{n}_B$ has to be less than one for non-trivial solutions of Eq. (1.34), the possible solutions of Eq. (1.34) are when $\mathbf{n}_A$ and $\mathbf{n}_B$ make an angle of (i) 90° or (ii) 45° with one another, corresponding to values of γ of 180° and 120°, respectively. From Eq. (1.29) the direction cosines n_{1C}, n_{2C} and n_{3C} of the axis of rotation, $\mathbf{n}_C$, are given by the expressions:

$$
\begin{aligned}
n_{1C} \sin \frac{1}{2}\gamma &= n_{1A} \sin \frac{1}{2}\alpha \cos \frac{1}{2}\beta + n_{1B} \sin \frac{1}{2}\beta \cos \frac{1}{2}\alpha + (n_{2B}n_{3A} - n_{3B}n_{2A}) \sin \frac{1}{2}\alpha \sin \frac{1}{2}\beta \\
n_{2C} \sin \frac{1}{2}\gamma &= n_{2A} \sin \frac{1}{2}\alpha \cos \frac{1}{2}\beta + n_{2B} \sin \frac{1}{2}\beta \cos \frac{1}{2}\alpha + (n_{3B}n_{1A} - n_{1B}n_{3A}) \sin \frac{1}{2}\alpha \sin \frac{1}{2}\beta \\
n_{3C} \sin \frac{1}{2}\gamma &= n_{3A} \sin \frac{1}{2}\alpha \cos \frac{1}{2}\beta + n_{3B} \sin \frac{1}{2}\beta \cos \frac{1}{2}\alpha + (n_{1B}n_{2A} - n_{2B}n_{1A}) \sin \frac{1}{2}\alpha \sin \frac{1}{2}\beta
\end{aligned}
\tag{1.35}
$$

because $\sin \frac{1}{2}\gamma = \sin \frac{1}{2}\gamma'$. If we choose $\mathbf{n}_A$ to be the unit vector [001] in the reference orthonormal coordinate system then for case (i) we can choose $\mathbf{n}_B$ to be a vector

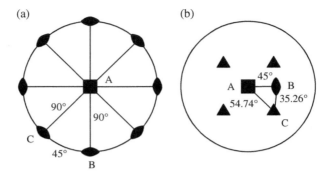

(a)   (b)

Figure 1.17 Examples of possible allowed combinations of rotational symmetries in crystals. In (a) a tetrad, A, is perpendicular to two diads, B and C, at 45° to one another, while in (b) the tetrad, A, is 45° away from a diad, B, and 54.74° from a triad, C, with B and C 35.26° apart. In (a) other diad axes which must be present are also indicated, and in (b) other triad axes (but not other tetrad and diad axes) which must be present are also indicated

such as [100]. Hence in Eq. (1.30):

$$\mathbf{p} = \{p_0, p_1, p_2, p_3\} = \{0, 1, 0, 0\}$$

$$\mathbf{q} = \{q_0, q_1, q_2, q_3\} = \left\{ \frac{1}{\sqrt{2}}, 0, 0, \frac{1}{\sqrt{2}} \right\} \tag{1.36}$$

and so:

$$\mathbf{p} \cdot \mathbf{q} = \left\{ 0, \frac{1}{\sqrt{2}}, -\frac{1}{\sqrt{2}}, 0 \right\} \tag{1.37}$$

That is, $\gamma = 180°$, $\gamma' = 180°$, and $\mathbf{n}_C$ is a unit vector parallel to $[1\bar{1}0]$. This possibility is illustrated in Figure 1.17a, with axes A, B and C lettered according to these assumptions. The other diad axes marked in Figure 1.17a must automatically be present, since A is a tetrad.

For case (ii), we can again choose $\mathbf{n}_A$ to be [001]. $\mathbf{n}_B$ can be chosen to be a vector such as:

$$\left[0, \frac{1}{\sqrt{2}}, \frac{1}{\sqrt{2}} \right] \tag{1.38}$$

Therefore, in Eq. (1.30):

$$\mathbf{p} = \{p_0, p_1, p_2, p_3\} = \left\{ 0, 0, \frac{1}{\sqrt{2}}, \frac{1}{\sqrt{2}} \right\}$$

$$\mathbf{q} = \{q_0, q_1, q_2, q_3\} = \left\{ \frac{1}{\sqrt{2}}, 0, 0, \frac{1}{\sqrt{2}} \right\} \tag{1.39}$$

and so:

$$\mathbf{p} \cdot \mathbf{q} = \left\{ -\frac{1}{2}, \frac{1}{2}, \frac{1}{2}, \frac{1}{2} \right\} \tag{1.40}$$

Table 1.2 Permissible combinations of rotation axes in crystals

A	B	C	α	β	γ	u	v	w	System
2	2	2	180°	180°	180°	90°	90°	90°	Orthorhombic
2	2	3	180°	180°	120°	90°	90°	60°	Trigonal
2	2	4	180°	180°	90°	90°	90°	45°	Tetragonal
2	2	6	180°	180°	60°	90°	90°	30°	Hexagonal
2	3	3	180°	120°	120°	70.53°	54.74°	54.74°	Cubic
2	3	4	180°	120°	90°	54.74°	45°	35.26°	Cubic

u is the angle between $\mathbf{n}_B$ and $\mathbf{n}_C$, *v* is the angle between $\mathbf{n}_C$ and $\mathbf{n}_A$, and *w* is the angle between $\mathbf{n}_A$ and $\mathbf{n}_B$.

i.e. $\gamma = 120°$, $\gamma' = 240°$, and $\mathbf{n}_C$ is a unit vector parallel to [111], making an angle of 54.74° with the fourfold axis and 35.26° with the twofold axis. This arrangement is shown in Figure 1.17b, again with the original axes marked. It should be noted that the presence of the tetrad at A automatically requires the presence of the other triad axes (and of other diads, not shown), since the fourfold symmetry about A must be satisfied. The triad axes lie at 70.53° to one another.

As a third example, suppose that the rotation about $\mathbf{n}_A$ is a hexad, so that $\alpha = 60°$ and $\alpha/2 = 30°$, and suppose the rotation about $\mathbf{n}_B$ is a tetrad, so that $\beta = 90°$ and $\beta/2 = 45°$. Under these circumstances, Eq. (1.32) becomes:

$$\mathbf{n}_A \cdot \mathbf{n}_B = \frac{\cos\frac{1}{2}\gamma + \frac{\sqrt{3}}{2\sqrt{2}}}{\frac{1}{2\sqrt{2}}} = 2\sqrt{2}\cos\frac{1}{2}\gamma + \sqrt{3} \tag{1.41}$$

Since $\mathbf{n}_A \cdot \mathbf{n}_B$ has to be less than 1, and $\cos\frac{1}{2}\gamma \geq 0$, because from Table 1.1 permitted values of γ are 60°, 90°, 120° and 180°, it follows that there are no solutions for $\mathbf{n}_A$ and $\mathbf{n}_B$ in Eq. (1.41) for Statement (1.33) to be valid. Therefore, we have shown that a sixfold axis and a fourfold axis cannot be combined together in a crystal to produce a rotation equivalent to a single sixfold, fourfold, threefold or twofold axis.

Statement (1.33) and Eq. (1.35) can be studied to find the possible combinations of rotational axes in crystals. The resulting permissible combinations and the angles between the axes corresponding to these are listed in Table 1.2, following M.J. Buerger [7].

In deriving these possibilities from Eqs. (1.33) and (1.35), it is useful to note that $\cos^{-1}(1/\sqrt{3}) = 54.74°$, $\cos^{-1}\sqrt{(2/3)} = 35.26°$, and $\cos^{-1}(1/3) = 70.53°$. The sets of related rotations shown in Table 1.2 can always be designated by three numbers, such as 222, 233, or 234, each number indicating the appropriate rotational axis.

1.7 Crystal Systems

The permissible combinations of rotation axes, listed in Table 1.2, are each identified with a crystal system in the far right-hand column of that table. A crystal system contains all those crystals that possess certain axes of rotational symmetry. In any crystal there is a necessary connection between the possession of an axis of rotational symmetry and the geometry of

Table 1.3 The crystal systems

System	Symmetry	Conventional cell
Triclinic	No axes of symmetry	$a \neq b \neq c; \alpha \neq \beta \neq \gamma$
Monoclinic	A single diad	$a \neq b \neq c; \alpha = \gamma = 90° < \beta$
Orthorhombic	Three mutually perpendicular diads	$a \neq b \neq c; \alpha = \beta = \gamma = 90°$
Trigonal	A single triad	$\begin{cases} a = b = c; \alpha = \beta = \gamma < 120°^a \\ a = b \neq c; \alpha = \beta = 90°, \gamma = 120°^b \end{cases}$
Tetragonal	A single tetrad	$a = b \neq c; \alpha = \beta = \gamma = 90°$
Hexagonal	One hexad	$a = b \neq c; \alpha = \beta = 90°, \gamma = 120°$
Cubic	Four triads	$a = b = c; \alpha = \beta = \gamma = 90°$

[a] Rhombohedral unit cell.
[b] This is also the conventional cell of the hexagonal system.

the lattice of that crystal. We shall explore this in the next section, and we have seen some simple examples in two dimensions in Section 1.5. Because of this connection between the rotational symmetry of the crystal and its lattice, a certain convenient conventional cell can always be chosen in each crystal system. These systems are listed in Table 1.3, in which the name of the system is given, along with the rotational symmetry operation or operations which define the system and the conventional unit cell, which can always be chosen. This cell is in many cases non-primitive; that is, it contains more than one lattice point. The symbol $\neq$ means 'not necessarily equal to'. The general formula for the volume, V, of the unit cell of a crystal with cell dimensions a, b, c, α, β and γ is:

$$V = abc\sqrt{1 - \cos^2\alpha - \cos^2\beta - \cos^2\gamma + 2\cos\alpha\cos\beta\cos\gamma} \qquad (1.42)$$

(see Problem 1.17).

Particular note should be made of the trigonal crystal system in Table 1.3. Here, there are two possible conventional unit cells, one rhombohedral and one the same as the hexagonal crystal system. This is because some trigonal crystals have a crystal structure based on a rhombohedral lattice, while others have a crystal structure based on the primitive hexagonal lattice [12].

1.8 Space Lattices (Bravais[12] Lattices)

All of the symmetry operations in a crystal must be mutually consistent. There are no five-fold axes of rotational symmetry because such axes are not consistent with the translational symmetry of the lattice. In Section 1.6 we derived the possible combinations of pure rotational symmetry operations that can pass through a point. These combinations are classified into different crystal systems and we will now investigate the types of space lattice (that is, the regular arrangement of points in three dimensions as defined in Section 1.1) that are

---

[12] These are termed Bravais lattices after Auguste Bravais (1811–1863). Eight years before Bravais wrote about the space filling of lattices, Moritz Ludwig Frankenheim (1801–1869) published his book *System der Krystalle* in which he also considered the same problem [13]. Frankenheim believed he had identified 15 arrangements. However, two of these in the monoclinic system (*C* and *I*) were shown by Bravais in a footnote on page 97 of his 1850 article to be equivalent [14]. This equivalence is shown in Figure 1.22. As a consequence of Bravais showing that the monoclinic *C* and *I* lattices are identical, these space lattices are known as Bravais lattices, rather than Frankenheim lattices.

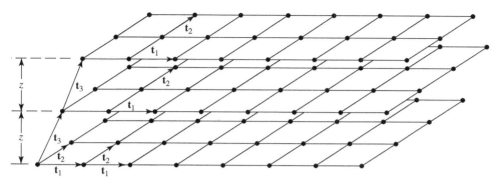

Figure 1.18 Stacking of nets to build up a space lattice. The triplet of vectors t_1, t_2, t_3 defines a unit cell of the Bravais lattice

consistent with the various combinations of rotation axes. We shall find, as we did for the two-dimensional lattice (or net) consistent with mirror symmetry (Section 1.5), that more than one arrangement of points is consistent with a given set of rotational symmetry operations. However, the number of essentially different arrangements of points is limited to 14. These are the 14 Bravais, or space, lattices. Our derivation is by no means a rigorous one because we do not show that our solutions are unique. This derivation of the Bravais lattices is introduced to provide a background for a clear understanding of the properties of imperfections studied in Part II of this book. The lattice is the most important symmetry concept for the discussion of dislocations and martensitic transformations.

We start with the planar lattices or nets illustrated in Figure 1.14. To build up a space lattice we stack these nets regularly above one another to form an infinite set of parallel sheets of spacing z. All of the sheets are in identical orientation with respect to an axis of rotation normal to their plane, so that corresponding lattice vectors t_1 and t_2 in the nets are always parallel. The stacking envisaged is shown in Figure 1.18. The vector t_3 joining lattice points in adjacent nets is held constant from net to net. The triplet of vectors t_1, t_2, t_3 defines a unit cell of the Bravais lattice.

We now consider the net based on a parallelogram (Figure 1.14a). If we stack nets of this form so that the points of intersection of twofold axes in successive nets do *not* lie vertically above one another then we destroy the twofold symmetry axes normal to the nets. We have a lattice of points showing no rotational symmetry. The unit cell is an arbitrary parallelepiped with edges a, b, c, no two of which are necessarily equal, and where the angles of the unit cell α, β, γ can take any value; the cell is shown in Figure 1.19a. By choosing a, b, c appropriately we can always ensure that the cell is primitive. Although this lattice contains no axis of rotational symmetry, the set of lattice points is of course necessarily centrosymmetric.

To preserve twofold symmetry we can proceed in one of two different ways. We can arrange parallelogram nets vertically above one another so that t_3 is normal to the plane of the sheets, as in Figure 1.20a, or we can produce the staggered arrangement shown in plan, viewed perpendicular to the nets, in Figure 1.20b.

In the first of these arrangements, in Figure 1.20a, the twofold axes at the corners of the unit parallelogram of the nets all coincide and we produce a lattice of which one unit cell

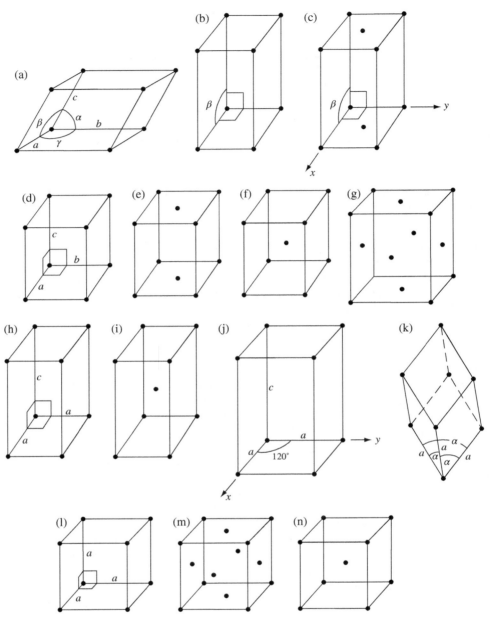

Figure 1.19 Unit cells of the 14 Bravais space lattices. (a) Primitive triclinic. (b) Primitive monoclinic. (c) Side-centred monoclinic – conventionally the twofold axis is taken parallel to *y* and the (001) face is centred (*C*-centred). (d) Primitive orthorhombic. (e) Side-centred orthorhombic – conventionally centred on (001) (*C*-centred). (f) Body-centred orthorhombic. (g) Face-centred orthorhombic. (h) Primitive tetragonal. (i) Body-centred tetragonal. (j) Primitive hexagonal. (k) Primitive rhombohedral. (l) Primitive cubic. (m) Face-centred cubic. (n) Body-centred cubic

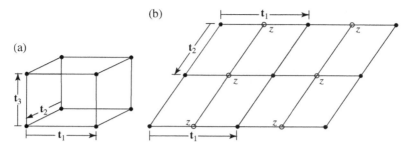

Figure 1.20 Lattice points in the net at height zero are marked as dots, those at height z with rings

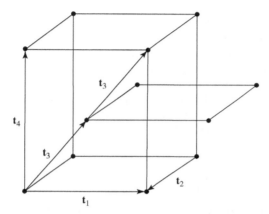

Figure 1.21 A three-dimensional view of the staggered arrangement of nets in Figure 1.20(b) in which two-fold symmetry is preserved. In this diagram, the vector $\mathbf{t}_4$ is normal to $\mathbf{t}_1$ and $\mathbf{t}_2$

is shown in Figure 1.19b. This has no two sides of the primitive cell necessarily equal, but two of the axial angles are 90°. A frequently used convention is to take α and γ as 90° so that y is normal to x and to z; β is then taken as the obtuse angle between x and z.

The staggered arrangement of the parallelogram nets in Figure 1.20b is such that the twofold axes at the corners of the unit parallelograms of the second net coincide with those at the centres of the sides of the unit parallelogram of those of the first (or zero-level) net. A lattice is then produced of which a possible unit cell is shown in Figure 1.21. This is multiply primitive, containing two lattice points per unit cell, and the vector $\mathbf{t}_4$ is normal to $\mathbf{t}_1$ and $\mathbf{t}_2$. Such a cell with lattice points at the centres of a pair of opposite faces parallel to the diad axis is also consistent with twofold symmetry. The cell centred on opposite faces shown in Figure 1.21 is chosen to denote the lattice produced from the staggered nets because it is more naturally related to the twofold symmetry than a primitive unit cell would be for this case.

The staggered arrangement of nets shown in Figures 1.20b and 1.21 could also have shown diad symmetry if we had arranged that the corners of the net at height z had lain not above the midpoints of the side containing $\mathbf{t}_1$ in Figure 1.20b but vertically above either the centre of the unit parallelogram of the first net or above the centre of the side containing

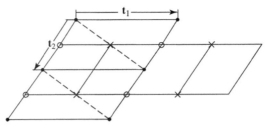

Figure 1.22 Lattice points in the net at height zero are marked with dots. The rings and crosses indicate alternative positions of the lattice points in staggered nets at height z, arranged so as to preserve twofold symmetry. The dotted lines show an alternative choice of unit cell

t_2 in Figure 1.20b. These two staggered arrangements are not essentially different from the first one, since, as is apparent from Figure 1.22, a new choice of axes in the plane of the nets is all that is needed to make them completely equivalent.

There are then two lattices consistent with monoclinic symmetry: the primitive one with the unit cell shown in Figure 1.19b and a lattice made up from staggered nets of which the conventional unit cell is centred on a pair of opposite faces. The centred faces are conventionally taken as the faces parallel to the x- and y-axes; that is, (001), with the diad parallel to y (see Figure 1.19c). This lattice is called the monoclinic C lattice. The two lattices in the monoclinic system can be designated P and C, respectively.

The two tetragonal lattices can be rapidly developed. The square net in Figure 1.14b has fourfold symmetry axes arranged at the corners of the squares and also at the centres. This fourfold symmetry may be preserved by placing the second net with a corner of the square at $00z$ with respect to the first (t_3 normal to t_1 and t_2) or with a corner of the square at $\left(\frac{1}{2}, \frac{1}{2}, z\right)$ with respect to the first. The unit cells of the lattices produced by these two different arrangements are shown in Figures 1.19h and 1.19i, respectively. They can be designated P and I. The symbol I indicates a lattice with an additional lattice point at the centre of the unit cell (German: *innenzentrierte*). In the tetragonal system the tetrad axis is usually taken parallel to c, so a and b are necessarily equal and all of the axial angles are 90°.

The nets shown in Figures 1.14d and 1.14e are each consistent with the symmetry of a diad axis lying at the intersection of two perpendicular mirror planes. It is shown in Section 2.1 that a mirror plane is completely equivalent to what is called an inverse diad axis: a diad axis involving the operation of rotation plus inversion. This inverse diad axis, given the symbol $\bar{2}$, lies normal to the mirror plane. The symmetry of a diad axis at the intersection of two perpendicular mirror planes could therefore be described as $2\bar{2}\bar{2}$, indicating the existence of three orthogonal axes: one diad and two inverse diads. The lattice consistent with this set of symmetry operations will also be consistent with the arrangement 222 in the orthorhombic crystal system (Table 1.3).[13] To develop the lattices consistent with orthorhombic symmetry, therefore, the two relevant nets are the rectangular net (Figure 1.14d) and the rhombus net (Figure 1.14e). The positions of diad axes are shown on the right-hand sides of Figures 1.14d and 1.14e. The rhombus net can also be described as the centred rectangular net.

---

[13] The validity of this statement does not follow immediately at this point. Its truth is plausible if it is noted, as shown later (Section 2.1), that an inverse diad axis plus a centre of symmetry is equivalent to a diad axis normal to a mirror plane, and that the lattice points of a lattice are centres of symmetry of the lattice.

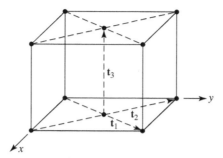

Figure 1.23   The stacking of rhombus nets vertically above one another to form an orthorhombic lattice with centring on one pair of faces

If we stack rectangular nets vertically above one another so that a corner lattice point of the second net lies vertically above a similar lattice point in the net at zero level ($\mathbf{t}_3$ normal to $\mathbf{t}_1$ and $\mathbf{t}_2$) then we produce the primitive lattice P. The unit cell is shown in Figure 1.19d. It is a rectangular parallelepiped.

If we stack rhombus nets (centred rectangles) vertically above one another, we obtain the lattice shown in Figure 1.23. This is the orthorhombic lattice with centring on one pair of faces.

We can also preserve the symmetry of a diad axis at the intersection of two mirror planes by stacking the rectangular nets in three staggered sequences. These are shown in Figures 1.24a, b and c. The lattices designated A-centred and B-centred are not essentially different since they can be transformed into one another by appropriate relabelling of the axes.[14] The staggered sequence shown in Figure 1.24c is described by the unit cell shown in Figure 1.19f. It is the orthorhombic body-centred lattice, symbol I. There is only one possibility for the staggered stacking of rhombus nets. Careful inspection of the right-hand side of Figure 1.14e shows that the only places in the net where a diad axis lies at the intersection of two perpendicular mirror planes is at points with coordinates $(0, 0)$ and $\left(\frac{1}{2}, \frac{1}{2}\right)$ of the *rhombus primitive cell*. We have already dealt with the vertical stacking of the rhombus nets. If we take the only staggered sequence possible, where the second net lies with a lattice point of the rhombus net vertically above the centre of the rhombus in the zero-level net (so that the end of $\mathbf{t}_3$ has coordinates $\frac{1}{2}, \frac{1}{2}, z$ in the rhombus net) then we produce the arrangement shown in Figure 1.25. This is most conveniently described in terms of a unit cell shown in Figure 1.19g, which is a rectangular parallelepiped with lattice points at the corners and also in the centres of all faces of the parallelepiped. This is the orthorhombic F cell. The symbol F stands for face-centred, indicating additional lattice points at the centres of all faces of the unit cell.

All of the lattices consistent with 222 – that is, orthorhombic symmetry – are shown in Figures 1.19d, e, f and g. The unit cells are all rectangular parallelepipeds, so that the crystal axes can always be taken at right angles to one another – that is, $\alpha = \beta = \gamma = 90°$ – but the cell edges a, b and c may all be different. The primitive lattice P can then be described by a unit cell with lattice points only at the corners, the body-centred lattice I by a cell with an additional lattice point at its centre and the F lattice

---

[14] A means a lattice point on the (100) face, B a lattice point on (010), and C a lattice point on (001), in all crystal systems.

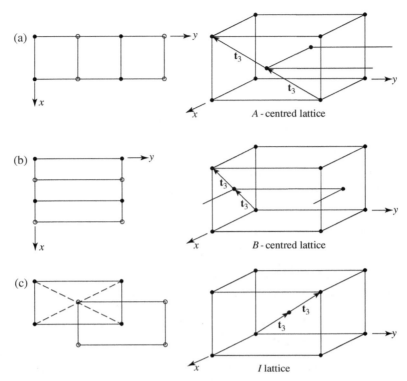

Figure 1.24 The three possible stacking sequences of rectangular nets. In the three left-hand diagrams, lattice points in the net at zero level are denoted with dots and those in the net at level *z* with open circles. The corresponding three-dimensional views of the arrangements of the nets are shown in the three right-hand diagrams

by a cell centred on all faces. The *A*-, *B*- and *C*-centred lattices, shown in Figures 1.24a, 1.24b and 1.23, respectively, are all described by choosing the axes so as to give a cell centred on the (001) face; that is, a *C*-centred cell.

We have so far described nine of the Bravais space lattices. All further lattices are based upon the stacking of triequiangular nets of points. The triequiangular net is shown in Figure 1.14c. There are sixfold axes only at the lattice points of the net. To preserve sixfold rotational symmetry in a three-dimensional lattice, such nets must be stacked vertically above one another so that t_3 is normal to t_1 and t_2. The lattice produced has the unit cell shown in Figure 1.19j. The unique hexagonal axis is taken to lie along the *z*-axis so $a = b \neq c$, $\gamma = 120°$ and α and β are both 90°. This lattice (i.e. the array of points) possesses sixfold rotational symmetry and is the only lattice to do so. However, it is also consistent with threefold rotational symmetry about an axis parallel to *z*. A crystal in which an atomic motif possessing threefold rotational symmetry was associated with each lattice point of this lattice would belong to the trigonal crystal system (Table 1.3).

A lattice consistent with a single threefold rotational axis can be produced by stacking triequiangular nets in a staggered sequence. A unit cell of the triequiangular net of points is shown outlined in Figure 1.26 by the vectors t_1, t_2 along the *x*- and *y*-axes. Axes of threefold symmetry pierce the net at the origin of the cell (0, 0) – at points such as A – and

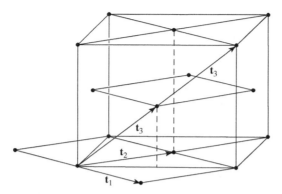

Figure 1.25 The staggered stacking of rhombus nets. This form of stacking generates the orthorhombic *F* Bravais lattice

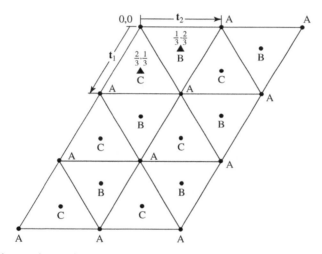

Figure 1.26 The stacking of triequiangular nets of points in a staggered sequence. Lattice points labelled A, B and C belong to nets of lattice points at different heights relative to the plane of the paper

also at two positions within the cell with coordinates $\left(\frac{1}{3}, \frac{2}{3}\right)$ and $\left(\frac{2}{3}, \frac{1}{3}\right)$, respectively, which are labelled B and C, respectively in Figure 1.26. We can preserve the threefold symmetry (while of course destroying the sixfold one) by stacking nets so that the extremity of $\mathbf{t}_3$ has coordinates of either $\left(\frac{2}{3}, \frac{1}{3}, z\right)$.or $\left(\frac{1}{3}, \frac{2}{3}, z\right)$. The two positions B and C in Figure 1.26 are equivalent to one another in the sense that the same lattice is produced whatever the order in which these two positions are used.

A plan of the lattice produced, viewed along the triad axis, is shown in Figure 1.27, and a sketch of the relationship between the triequiangular nets and the primitive cells of this lattice is shown in Figure 1.28. In Figures 1.27 and 1.28 the stacking sequence of the nets has been set as ABCABCABC … Exactly the same lattice but in a different orientation (rotated 60° clockwise looking down upon the paper in Figure 1.27) would have been produced if the sequence ACBACBACB … had been followed. The primitive cell of the trigonal lattice

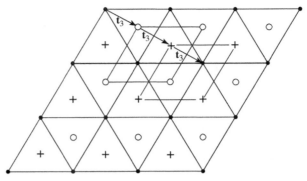

Figure 1.27   Lattice points in the net at level zero are marked with a dot, those in the net at height *z* by an open circle, and those at 2*z* by a plus sign. The projection of t_3 onto the plane of the nets is shown

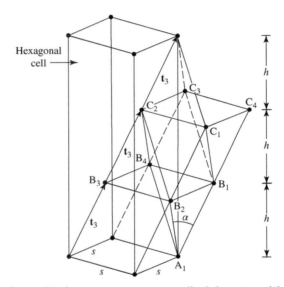

Figure 1.28   The relationship between a primitive cell of the trigonal lattice and the triply primitive hexagonal cell

in Figure 1.28 is shown in Figure 1.19k. It can be given the symbol *R*. It is a rhombohedron, the edges of the cell being of equal length, each equally inclined to the single threefold axis. To specify the cell we must state $a = b = c$ and the angle $\alpha = \beta = \gamma < 120°$.

An alternative cell is sometimes used to describe the trigonal lattice *R* because of the inconvenience in dealing with a lattice of axial angle α, which may take any value between 0 and 120°. The alternative cell is shown in Figure 1.28 and in plan viewed along the triad axis in Figure 1.29. It is a triply primitive cell, three mesh layers high, with internal lattice points at elevations of $\frac{1}{3}$ and $\frac{2}{3}$ of the repeat distance along the triad axis. The cell is of the same shape as the conventional unit cell of the hexagonal Bravais lattice and to specify it we must know $a = b \neq c$, $\alpha = \beta = 90°$ and $\gamma = 120°$.

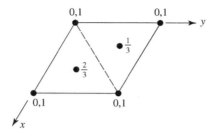

Figure 1.29 Plan view of the alternative triply primitive hexagonal unit cell used to describe the trigonal *R* lattice

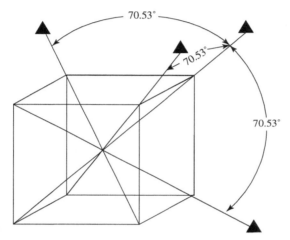

Figure 1.30 The four ⟨111⟩ three-fold axes with acute angles of 70.53° between one another which together define the minimum symmetry requirements for a crystal to belong to the cubic crystal system

Crystals belonging to the cubic system possess four threefold axes of rotational symmetry. The angles between the four threefold axes are such that these threefold axes lie along the body diagonals of a cube (Figure 1.30), with angles of 70.53° ($\cos^{-1}(1/3)$) between them. Reference to Table 1.2 and Figure 1.17b shows that these threefold axes cannot exist alone in a crystal. They must be accompanied by at least three twofold axes. To indicate how the lattices consistent with this arrangement of threefold axes arise, we start with the *R* lattice shown in Figure 1.28 and call the separation of nearest-neighbour lattice points in the triequiangular net *s* and the vertical separation of the nets along the triad axis *h*. The positions of the lattice points in the successive layers when all are projected onto the plane perpendicular to the triad axis can be designated ABCABC … as in Figures 1.26 and 1.28. In a trigonal lattice, the spacing of the nets, *h*, is unrelated to the separation of the lattice points within the nets, *s*. If we make the spacing of the nets such that $h = \sqrt{2/3}s \, (= 2s/\sqrt{6})$, the angle α in Figure 1.28 becomes 60° and triangles $A_1B_1B_2$, $A_1B_2B_4$, $A_1B_1B_4$ all become equilateral. Planes such as $A_1B_1C_1B_2$, $A_1B_2C_2B_4$, $A_1B_1C_3B_4$ all contain triequiangular nets of points. Planes parallel to each of these three planes also contain triequiangular nets of points and are also stacked so as to preserve triad symmetry along lines normal to them.

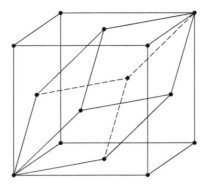

Figure 1.31   The relationship between the primitive unit cell and the conventional cell in the face-centred cubic lattice

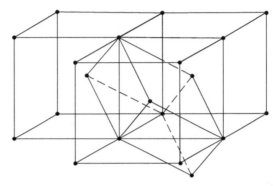

Figure 1.32   The relationship between the primitive unit cell and the conventional cell in the body-centred cubic lattice

When $\alpha = 60°$, the original trigonal lattice becomes consistent with the possession of four threefold axes. The conventional unit cell of this lattice is shown in Figure 1.19m; it is a cube centred on all faces. The relationship between this cell and the primitive one with $\alpha = 60°$ is shown in Figure 1.31.

The large non-primitive unit cell in Figures 1.19m and 1.31 is the face-centred cubic lattice, which can be designated F. It contains four lattice points. These are at the corners and centres of each of the faces.

When h in Figure 1.28 becomes equal to $s/\sqrt{6}$, the primitive unit cell of the R lattice becomes a cube with $\alpha = 90°$. This is the cubic primitive lattice P shown in Figure 1.19l, containing lattice points at the corners of the cubic unit cell.

Lastly, if in Figure 1.28 h takes the value $\frac{1}{6}\sqrt{\frac{3}{2}}s = s/(2\sqrt{6})$, the angle α is equal to $109.47°$ ($= 180° - 70.53°$) $= \cos^{-1}(-1/3)$. The lattice formed by such an array of points also contains four threefold axes of symmetry. The conventional unit cell of this lattice is shown in Figure 1.19n. It is a cube with lattice points at the cube corners and one at the centre. It can be designated I, the cubic body-centred lattice. The relationship between the doubly primitive unit cell shown in Figure 1.19n and the primitive unit cell, which is a rhombohedron with axial angles of $109.47°$, is shown in Figure 1.32.

We have just described the three lattices consistent with the possession of four threefold axes of rotational symmetry. They are shown together in Figures 1.19l, m and n. The unit cell of each can be taken as a cube with $a = b = c$; $\alpha = \beta = \gamma = 90°$. The primitive cell contains one lattice point, the face-centred cell four, and the body-centred cell two.

The unit cells of the 14 Bravais space lattices are shown together in Figure 1.19. All crystals possess one or other of these lattices, with an identical atomic motif associated with each lattice point. In some crystals a single spherically symmetric atom is associated with each lattice point. In this case the lattice itself possesses direct physical significance because the lattice and the crystal structure are identical. In other cases the lattice is a very convenient framework for describing the translational symmetry of the crystal. If the lattice is given and the arrangement of the atomic motif about a *single* lattice point is given, the crystal structure is fully described. The lattice is the most important symmetry aspect for describing the properties of imperfections in crystals.

Problems

The material in Appendix 1 may assist in some of these exercises.

1.1 (a) Select any convenient point on the plane pattern appearing on the diagram below and mark all the corresponding points, thus indicating the lattice.
(b) Outline the unit cell in several different ways, mark in the *x*- and *y*-axes in each case, and measure the cell dimensions *a*, *b* and γ in each case.
(c) Draw a line parallel to MN through any one lattice point and add all the lines of this set. Determine the indices of this set of lines for each of your different choices of unit cell.
(d) Repeat (c) for the set of lines parallel to PQ.

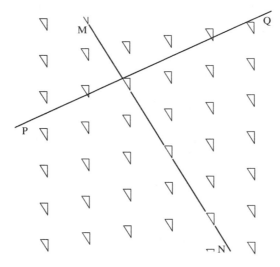

Figure for Problem 1.1

1.2 Rutile, TiO_2, has $a = b = 4.58$ Å, $c = 2.95$ Å and $\alpha = \beta = \gamma = 90°$. The atomic coordinates are:

$$Ti: 0, 0, 0; \tfrac{1}{2}, \tfrac{1}{2}, \tfrac{1}{2}$$

$$O: u, u, 0; -u, -u, 0; \tfrac{1}{2} + u, \tfrac{1}{2} - u, \tfrac{1}{2}; \tfrac{1}{2} - u, \tfrac{1}{2} + u, \tfrac{1}{2}, \text{ where } u = 0.31$$

(a) Draw an accurate projection of a unit cell on the plane containing the *x*- and *y*-axes.

(b) Determine the number of formula units per cell.

(c) Find the number of oxygen atoms surrounding each titanium atom. Calculate the interatomic distances Ti–O for the titanium atom at $\tfrac{1}{2}, \tfrac{1}{2}, \tfrac{1}{2}$.

1.3 The unit cells of several orthorhombic crystals are described below. What is the Bravais lattice of each?

(a) Two atoms of the same kind per unit cell located at:

$$0, \tfrac{1}{2}, 0; \tfrac{1}{2}, 0, \tfrac{1}{2}.$$

(b) Four atoms of the same kind per unit cell located at:

$$x, y, z; \bar{x}, \bar{y}, z; \tfrac{1}{2} + x, \tfrac{1}{2} - y, \bar{z}; \tfrac{1}{2} - x, \tfrac{1}{2} + y, \bar{z}.$$

(c) Two atoms of one kind per unit cell located at $\tfrac{1}{2}, 0, 0; 0, \tfrac{1}{2}, \tfrac{1}{2}$ and two of another kind located at $0, 0, \tfrac{1}{2}; \tfrac{1}{2}, \tfrac{1}{2}, 0$.

1.4 Show with a sketch that a face-centred tetragonal lattice is equivalent to a body-centred tetragonal lattice in a different orientation. Why are there no tetragonal *C*, monoclinic *I*, cubic *C*, and hexagonal *I* Bravais lattices?

A crystal structure is proposed for which $a = c \neq b$; $\alpha = \gamma = 90°$; $\beta > 90°$. Close examination of the proposed structure shows that it has a monoclinic *I* lattice. Specify a more conventional description of the lattice type and describe the relationship of the sides of the unit cell for this new lattice type in terms of the monoclinic *I* unit cell.

1.5 The spinel $MgAl_2O_4$ has $a = b = c = 8.11$ Å and $\alpha = \beta = \gamma = 90°$. Crystals show the faces (111), $(\bar{1}1\bar{1})$, $(1\bar{1}1)$ and $(11\bar{1})$ and the faces parallel to these, namely $(\bar{1}\bar{1}\bar{1})$, $(1\bar{1}\bar{1})$, $(\bar{1}1\bar{1})$ and $(\bar{1}\bar{1}1)$. Such crystals look like regular or distorted octahedra.

(a) Find the zone axis symbol (indices) of the zone containing (111) and $(\bar{1}\bar{1}1)$. Which of the other faces listed above also lie in this zone?

(b) Show that the zone containing (111) and $(\bar{1}\bar{1}1)$ also contains (001) and (110).

(c) Sketch the projection of the lattice on (001). Draw in the trace of the (110) lattice plane and the normal to the plane through the origin. Find the perpendicular distance of the first plane of this set from the origin in terms of the cell side.

(d) Draw the section of the lattice through the origin which contains the normal to (110) and the normal to (001). Mark on this section the traces of the (111), $(\bar{1}\bar{1}1)$, $(\bar{1}\bar{1}\bar{1})$ and $(11\bar{1})$ planes. Measure the angles between adjacent planes.

1.6 Find the angle between [111] and the normal to (111) in (a) a cubic crystal and (b) a tetragonal one with $a = 5.67$ Å, $c = 12.70$ Å.

1.7 Show that the faces (111), (231), and ($1\bar{2}4$) all lie in a zone. (a) Find the zone symbol (indices) of this zone and (b) calculate the angle between this zone and the [100] zone in the cubic system.

1.8 In a tetragonal crystal $CuFeS_2$, the angle between (111) and ($\bar{1}\,\bar{1}\,1$) is 108.67°. Calculate (a) the ratio of the axial lengths and (b) the interzone angle $[236]\wedge[001]$.

1.9 Following the method of Section 1.6, determine the angles between the rotational axes in the axial combinations of (a) two triads and a diad and (b) a tetrad, a triad and a diad. Draw a sketch to show the arrangement of the symmetry operations in each case.

1.10 Show that (a) successive combinations of rotations about two different tetrad axes are unable to be expressed as a rotation about a single triad axis in crystals, and (b) successive combinations of rotations about two different triad axes are unable to be expressed as a rotation about a single tetrad axis in crystals.

1.11 The unit cell of a two-dimensional lattice has $a = b$, $\gamma = 120°$. Dispose three atoms about each lattice point in different ways so that the two-dimensional crystal can show the following symmetries in turn: (a) a sixfold axis with mirror planes parallel to it, (b) a threefold axis, (c) a diad axis at the intersection of two mirror planes, (d) a twofold axis, and (e) a onefold axis.

1.12 CdI_2 and $CdCl_2$ both belong to the trigonal crystal system. The former has a primitive hexagonal lattice and the latter a rhombohedral lattice. The coordinates of the atoms associated with each *lattice point* are given below, using a hexagonal unit cell for both crystals:

$$CdI_2 \qquad\qquad CdCl_2$$

$$Cd: 0,0,0 \qquad Cd: 0,0,0$$

$$I: \pm\left(\frac{2}{3},\frac{1}{3},\frac{1}{4}\right) \qquad Cl: \pm\left(\frac{2}{3},\frac{1}{3},\frac{1}{12}\right)$$

Draw plans of the two structures on (001) and outline on your diagram for $CdCl_2$ the projections of the cell edges of the true rhombohedral primitive unit cell.

1.13 Show that there are three cubic lattices by following the procedure used in Section 1.8 and finding the conditions for which in Figure 1.28 triad axes lie normal to (010), (100) and (001) of the rhombohedral primitive unit cell. *Hint*: find the condition under which successive planes project along their normal so that the lattice points in one lie at the centroids of triangles of lattice points in that below it.

1.14 A two-dimensional crystal possesses fourfold rotational symmetry. Sketch the net. Position four atoms of a chemical element at the net points so that the arrangement is consistent with (a) just fourfold symmetry and (b) fourfold symmetry with mirror planes parallel to the fourfold symmetry axis. Are two different arrangements of mirror planes possible in (b)?

1.15 Specify the directions of **a***, **b*** and **c*** in a hexagonal crystal relative to the lattice vectors **a**, **b** and **c**. What is the angle between **a*** and **b***? Determine the magnitudes of **a***, **b*** and **c*** in terms of the lattice parameters a and c.

Using the formula $|r^*_{hkl}| = 1/d_{hkl}$, show that the interplanar spacing of the hkl planes in a hexagonal crystal is given by the formula:

$$\frac{1}{d_{hkl}^2} = (h^2 + k^2 + hk)\frac{4}{3a^2} + \frac{l^2}{c^2}$$

Hence show that the planes (hkl), $(h, -(h+k), l)$ and $(k, -(h+k), l)$ have the same interplanar spacing.

1.16 Show that in a triclinic crystal the normal to the plane (hkl) is parallel to the vector $[uvw]$ if:

$$\begin{pmatrix} h \\ k \\ l \end{pmatrix} = \begin{pmatrix} \mathbf{a}\cdot\mathbf{a} & \mathbf{a}\cdot\mathbf{b} & \mathbf{a}\cdot\mathbf{c} \\ \mathbf{a}\cdot\mathbf{b} & \mathbf{b}\cdot\mathbf{b} & \mathbf{b}\cdot\mathbf{c} \\ \mathbf{a}\cdot\mathbf{c} & \mathbf{b}\cdot\mathbf{c} & \mathbf{c}\cdot\mathbf{c} \end{pmatrix} \begin{pmatrix} u \\ v \\ w \end{pmatrix}$$

Hence, or otherwise, determine a suitable equation to determine the indices of the vector $[uvw]$ parallel to the normal to the plane (hkl) in a triclinic crystal. *Hint*: it is useful to define an orthonormal set of vectors with respect to which the basis vectors **a**, **b** and **c** of the triclinic crystal are defined.

1.17 Using the hint in Problem 1.16, or otherwise, confirm that the volume, V, of the unit cell of a triclinic crystal with cell dimensions a, b, c, α, β and γ is given by the formula:

$$V = abc\sqrt{1 - \cos^2\alpha - \cos^2\beta - \cos^2\gamma + 2\cos\alpha\cos\beta\cos\gamma}$$

Suggestions for Further Reading

Allen, S.M. and Thomas, E.L. (1999). *The Structure of Materials*. New York: Wiley.

Buerger, M.J. (1963). *Elementary Crystallography*. New York: Wiley.

Buerger, M.J. (1970). *Contemporary Crystallography*. New York: McGraw-Hill.

Giacovazzo, C. (ed.) (2011). *Fundamentals of Crystallography (International Union of Crystallography Texts on Crystallography – 15)*, 3rd Edition. Oxford: Oxford University Press.

Hammond, C. (2015). *The Basics of Crystallography and Diffraction (International Union of Crystallography Texts on Crystallography – 12)*, 4th Edition. Oxford: Oxford University Press.

Phillips, F.C. (1971). *Introduction to Crystallography*, 4th Edition. Edinburgh: Oliver & Boyd.

Rousseau, J.-J. (1998). *Basic Crystallography*. Chichester: Wiley.

References

[1] Miller, W.H. (1839). *A Treatise on Crystallography*. Cambridge: J. & J.J. Deighton.

[2] Whewell, W. (1825). A general method of calculating the angles made by any planes of crystals, and the laws according to which they are formed. *Phil. Trans. Roy. Soc. Lond.* 115: 87–130.

[3] Lewis, W.J. (1899). *A Treatise on Crystallography*. Cambridge: Cambridge University Press.

[4] Aroyo, M.I. (ed.) (2016). *International Tables for Crystallography*, 6th, Revised Edition, Vol. A: *Space-Group Symmetry*, published for the International Union of Crystallography. Chichester: Wiley International.

[5] Santoro, A. and Mighell, A.D. (1970). Determination of reduced cells. *Acta Crystallogr. A* 26: 124–127.

[6] Niggli, P. (1928). *Handbuch der Experimentalphysik*, vol. 7, Part 1: *Krystallographische und Strukturtheoretische Grundbegriffe*, Chapter 4, §1. Leipzig, Germany: Akademische Verlags-gesellschaft m. b. H.

[7] Buerger, M.J. (1963). *Elementary Crystallography*, pp. 35–45. New York: Wiley.

[8] Buerger, M.J. (1970). *Contemporary Crystallography*, pp. 24–27. New York: McGraw-Hill.

[9] Grimmer, H. (1974). Disorientations and coincidence rotations for cubic lattices. *Acta Crystallogr. A* 30: 685–688.

[10] Altmann, S.L. (1986). *Rotations, Quaternions and Double Groups*. Oxford: Clarendon Press.

[11] Kuipers, J.B. (1999). *Quaternions and Rotation Sequences*. Princeton: Princeton University Press.

[12] Hahn, T. and Looijenga-Vos, A. (2002). *International Tables for Crystallography* (ed. T. Hahn) 5th, Revised Edition, Vol. A: *Space-Group Symmetry*, published for the International Union of Crystallography, pp. 14–16. Dordrecht: Kluwer Academic Publishers.

[13] Frankenheim, M.L. (1842). *System der Krystalle*. Breslau: Grass, Barth & Co.

[14] Bravais, A. (1850). Les systèmes formés par des points. *J. Écol. Polytech.* 19: 1–128.

2

Point Groups and Space Groups

2.1 Macroscopic Symmetry Elements

The macroscopically measured properties of a crystal, such as electrical resistance, thermal expansion, optical properties, magnetic susceptibility and the elastic constants, show a symmetry which can be defined and understood without reference to the translational symmetry elements defined by the lattice. If the translational symmetry of the crystal is disregarded, the remaining symmetry elements (such as axes of rotational symmetry, mirror planes and any centre of inversion), themselves consistent with the translational symmetry of the lattice, can be arranged into 32 consistent groups. These are the 32 crystallographic point groups. They are so called because they are consistent with translational symmetry and because all of the symmetry elements in a group pass through a single point and the operation of these elements leaves just one point unmoved – the point through which they pass. Other point groups, such as the point group symmetry of the icosahedron, which we shall consider in Chapter 4 when discussing the point group symmetries exhibited by quasicrystals, are not compatible with translational symmetry.

The axes of rotational symmetry, the mirror plane and the centre of inversion are all called macroscopic symmetry elements because their presence or absence in a given crystal can be decided in principle by macroscopic tests, such as etching of the crystal, the arrangement of the external faces or the symmetry of the physical properties, without any reference to the atomic structure of the crystal. The macroscopic symmetry elements are of two kinds. A symmetry operation of the *first kind*, such as a pure rotation axis, when operating on a right-handed object (say) produces a right-handed object from it, and all subsequent repetitions of this object are also right-handed. A symmetry operation of the *second kind* repeats an *enantiomorphous* object from an original object. The left and right hands of the ideal external form of the human body are enantiomorphously related. The

Crystallography and Crystal Defects, Third Edition. Anthony Kelly and Kevin M. Knowles.
© 2020 John Wiley & Sons Ltd. Published 2020 by John Wiley & Sons Ltd.
Companion website: www.wiley.com/go/kelly/crystallography3e

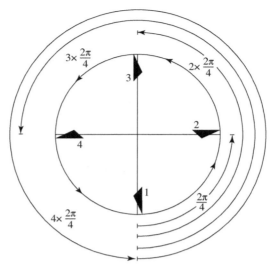

Figure 2.1 The basic operation of repetition by a rotation axis. In this example, the axis is a four-fold symmetry axis

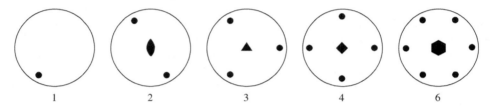

Figure 2.2 Stereograms representing the operation of one-, two-, three-, four- and sixfold axes on a single initial pole, represented as a dot

operation of reflection illustrated in Figure 1.12 is an example of a symmetry operation of the second kind since a left-handed object is repeated from an original right-handed object. Subsequent operation of the same symmetry element would produce a right-handed object again and then a left-handed object, and so forth. A symmetry operation of the second kind therefore involves a reversal of sense in the operation of repetition. Inversion through a centre is also an operation of the second kind.

In developing the 32 crystallographic point groups it is convenient to have all the macroscopic symmetry elements represented by axes and to do this we define what are called *improper rotations*. These produce repetition by a combination of a rotation and an operation of inversion. We shall use *rotoinversion* axes. These involve rotation coupled with inversion through a centre.[1] A pure rotation axis is said to produce a proper rotation.

The basic operation of repetition by a rotation axis is shown schematically in Figure 2.1. An *n*-fold axis repeats an object by successive rotations through an angle of $2\pi/n$. In the example shown in Figure 2.1, the axis is one of fourfold symmetry. The operations of monad, diad, triad, tetrad, and hexad pure rotation axes on a single initial pole are shown in Figure 2.2. These diagrams are stereograms (see Appendix 2) with the pole of the axis

---

[1] It is also possible to use rotoreflection axes [1], pp. 23–30, also termed alternating axes [2], p. 117, in developing the point groups. These repeat an object by rotation coupled with reflection in a plane normal to the axis. Onefold, twofold, threefold, fourfold, and sixfold rotoreflection axes are possible, usually denoted $\widetilde{1}$ (pronounced 'one tilde'), $\widetilde{2}, \widetilde{3}, \widetilde{4}$ and $\widetilde{6}$ respectively. $\widetilde{1}$ is clearly equivalent to a mirror plane.

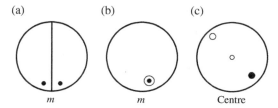

Figure 2.3 The repetition of an object by a mirror plane, (a) and (b), and by a centre of symmetry, (c)

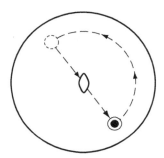

Figure 2.4 The operation of the twofold rotoinversion axis, $\bar{2}$

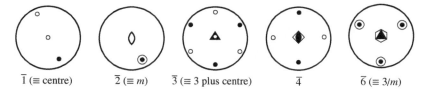

Figure 2.5 The operation of the various rotoinversion axes that can occur in crystals

at the centre of the primitive circle. Dots are used to represent poles in the northern hemisphere of projection related to one another by the rotation axis. The numbers below the stereograms give the shorthand labels for the axes 1, 2, 3, 4 and 6, indicating a onefold, twofold, threefold, fourfold, and sixfold axis, respectively. The repetition of an object by a mirror plane, symbol m, and by a centre of symmetry (or centre of inversion) is shown in Figure 2.3. In Figure 2.3a the mirror plane lies normal to the primitive circle. It is denoted by a strong vertical line | coinciding with the mirror in the stereographic projection. In Figure 2.3b the mirror coincides with the primitive; the dot representing the pole in the northern hemisphere has as its mirror image the circle shown in the southern hemisphere. In Figure 2.3c the centre of inversion is at the centre of the sphere of projection.

The onefold inversion axis is a centre of symmetry. The operations of the other rotoinversion axes are explained in Figures 2.4 and 2.5. The twofold rotation–inversion axis $\bar{2}$ shown in Figure 2.4 repeats an object by rotation through 180° (360°/2) to give the dotted circle,

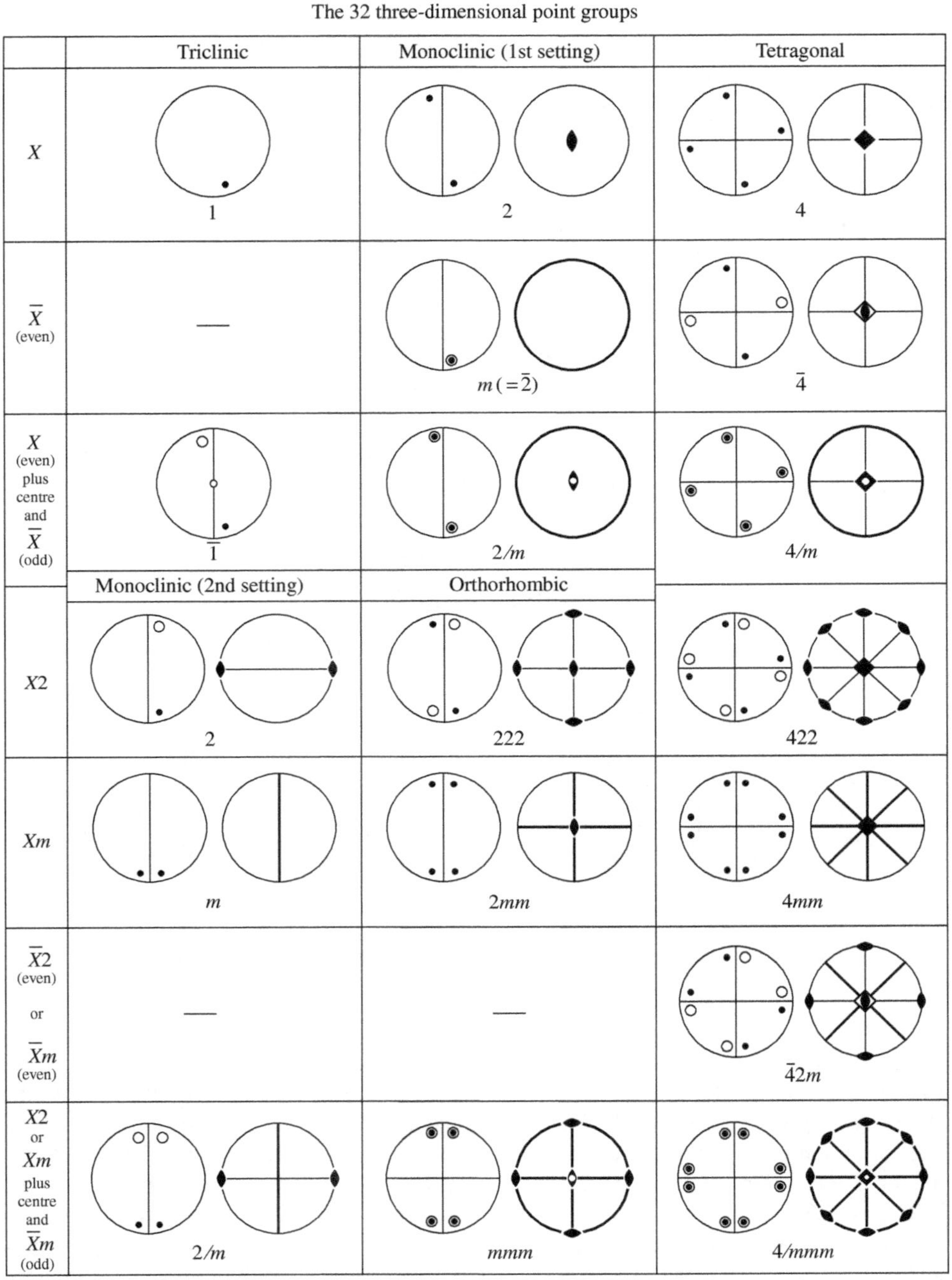

The 32 three-dimensional point groups

Figure 2.6 Stereograms of the poles of equivalent general directions and of the symmetry operations of each of the 32 crystallographic point groups. The z-axis is normal to the paper. In all the centrosymmetric classes, positions of centres of symmetry (inversion monads) lying within the primitive of the stereogram are shown. Source: Taken from the *International Tables for X-ray Crystallography*, Vol. 1 [4] and adapted to conform to current notation for the two centrosymmetric point groups in the cubic crystal system.

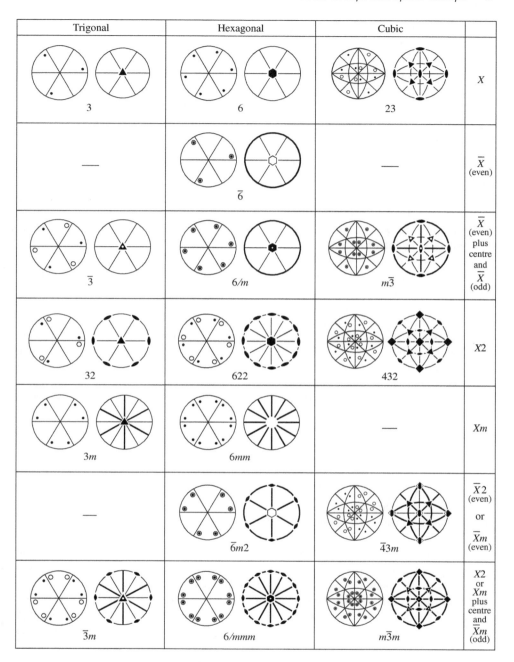

Figure 2.6   (Continued)

followed by inversion to give the full circle. Similarly, the threefold inversion axis $\bar{3}$ involves rotation through $360°/3 = 120°$ coupled with an inversion. In general, an n-fold rotoinversion axis $\bar{n}$ involves rotation through an angle of $2\pi/n$ coupled with inversion through a centre. The rotation and inversion are both part of the operation of repetition and must not be considered as separate operations. The operation of the various rotoinversion axes that can occur in crystals on a single initial pole is shown in Figure 2.5, with the pole of the rotoinversion axis at the centre of the primitive circle. The following symbols are used: inversion monad, $\bar{1}$, symbol O; inversion diad, $\bar{2}$, symbol ◊; inversion triad, $\bar{3}$, symbol ▲; inversion tetrad, $\bar{4}$, symbol ◆; inversion hexad, $\bar{6}$, symbol ◭. Inspection of Figures 2.2–2.4 shows that $\bar{1}$ is identical to a centre of symmetry, $\bar{2}$ is identical to a mirror plane normal to the inversion diad, $\bar{3}$ is identical to a triad axis plus a centre of symmetry and $\bar{6}$ is identical to a triad axis normal to a mirror plane (symbol $3/m$, the sign '$/m$' indicating a mirror plane normal to an axis of symmetry).[2] Only $\bar{4}$ is unique. The operation of repetition described by $\bar{4}$ cannot be reproduced by any combination of a proper rotation axis and a mirror plane or a centre of symmetry.

The various different combinations of 1, 2, 3, 4 and 6 pure rotation axes and $\bar{1}$, $\bar{2}$, $\bar{3}$, $\bar{4}$ and $\bar{6}$ rotoinversion axes constitute the 32 crystallographic point groups or crystal classes. These 32 classes are grouped into systems according to the presence of defining symmetry elements (see Table 1.3). Stereograms of each of the 32 crystallographic point groups or crystal classes are given in Figure 2.6, following the current conventions of the *International Tables for Crystallography* [3]. With the exception of the two triclinic point groups, each point group is depicted by two stereograms. The first stereogram shows how a single initial pole is repeated by the operations of the point group and the second stereogram shows all of the symmetry elements present. The nomenclature for describing the crystal classes is as follows. X indicates a rotation axis and $\bar{X}$ an inversion axis. X/m is a rotation axis normal to a mirror plane, Xm a rotation axis with a mirror plane parallel to it and $X2$ a rotation axis with a diad normal to it. X/mm indicates a rotation axis with a mirror plane normal to it and another parallel to it. $\bar{X}m$ is an inversion axis with a parallel plane of symmetry. A plane of symmetry is an alternative description of a mirror plane.

We shall describe each of the classes in Sections 2.2–2.8. A derivation of the 32 classes follows by noting from Sections 1.5 and 1.6 that the rotation axes consistent with translational symmetry are 1, 2, 3, 4 and 6. Individually, these give in total five crystal classes. Their consistent combinations give another six (see Table 1.2): 222, 322, 422, 622, 332 and 432, thus totalling 11. All of these 11 involve only operations of the first kind. A lattice is inherently centrosymmetric (Section 1.4) and so each of the rotation axes could be replaced by the corresponding rotoinversion axis, thus giving another five classes: $\bar{1}$, $\bar{2}$, $\bar{3}$, $\bar{4}$ and $\bar{6}$. The remaining 16 can be described as combinations of the proper and improper rotation axes. It is convenient to begin first with the three crystal systems where the angles between the axes are all 90°, then to consider the hexagonal and trigonal crystal systems, before finally turning our attention to the monoclinic and triclinic systems.

---

[2] Our symbols are identical to those used in the *International Tables for Crystallography* [3], with the exception of '◊', which in the *International Tables* is always indicated by the symbol for a mirror plane.

2.2   Orthorhombic System

A crystal in this system contains three diads, which must be at right angles to one another. In Figure 2.6 they are the crystal axes. From Table 1.3 and Figures 1.19d, e, f and g, the lattice parameters may be all unequal.

The point group containing just three diad axes at right angles to one another is shown in Figure 2.6 and is designated 222. In general, a pole in a crystal belonging to this point group is repeated four times. If the indices of the initial pole are (hkl), where there is no special relationship between h, k and l, the operation of all of the symmetry elements of the point group on this one initial pole produces three other poles. These are $(\overline{h}\,\overline{k}l)$, $(\overline{h}k\,\overline{l})$ and $(h\overline{k}\,\overline{l})$ (Figure 2.7). The assemblage of crystal faces produced by repetition of an initial crystal face with indices (hkl) is called the *form hkl* and is given the symbol $\{hkl\}$.[3] If the assemblage of faces encloses space, the form is said to be closed; otherwise it is open. In this case, $\{hkl\}$ is closed. The symbol $\{hkl\}$ with curly brackets means all faces of the form *hkl*. Here, for the point group 222, this means (hkl), $(\overline{h}\,\overline{k}l)$, $(\overline{h}k\,\overline{l})$ and $(h\overline{k}\,\overline{l})$. The form $\{hkl\}$ is said to show a *multiplicity* of four. Then $\{hkl\}$ would be said to be a *general form*; that is, a form that bears no special relationship to the symmetry elements of the point group.

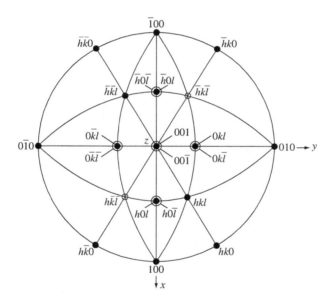

Figure 2.7   A stereogram of an orthorhombic crystal of point group 222 centred on 001. The diad axes are parallel to the *x*-, *y*- and *z*-axes

---

[3] All the individual faces of the form $\{hkl\}$ are crystallographically identical. In a similar way, if a single direction is given, say $[uvw]$, then all the directions produced if the repetition operations of the point group are carried out on the initial direction are called the family of directions of the type *uvw* and are given the symbol $\langle uvw \rangle$.

Special forms in this crystal class would be {100}, {010} and {001}; each of these forms gives just two faces: for {100}, these would be the faces (100) and ($\bar{1}$00) (Figure 2.7). These forms are all open ones. They are easily recognized as special forms since their multiplicity is less than that of the general form. Forms such as {$hk0$}, {$h0l$} and {$0kl$} which have one index zero and no special relationship between the other two would also be described as special, even though, as is evident from Figure 2.7, the multiplicity of each is four, as for the general form in this crystal class. One reason for this is that these poles lie normal to a diad axis. This is a special position with respect to this axis and has the result that if a crystal grew with faces parallel only to the planes of indices {$hk0$}, {$h0l$} and {$0kl$}, it would appear to show mirror symmetry as well as the three diad axes. A second reason is that each of the {$hk0$}, {$h0l$} and {$0kl$} forms is open, while the general {hkl} form is closed. Special forms usually correspond to poles lying normal to or on an axis of symmetry, and normal to or on mirror planes, and sometimes to poles lying midway between two axes of symmetry. However, the best definition of a special form is as follows: a form is special if the development of the complete form shows a symmetry of arrangement of the poles which is higher than the one the crystal actually possesses. Special forms in all of the crystal classes are listed later, in Table 2.1.

The orthorhombic system also contains the classes 2*mm* and *mmm* (Figure 2.6). The former contains two mirror planes, which must be at right angles. Two mirror planes at right angles automatically show diad symmetry along the line of intersection (Figure 2.6). Since $m \equiv \bar{2}$, this group could be designated $2\,\bar{2}\,\bar{2}$, and this is why it appears in the orthorhombic system, which is defined as possessing three diad axes. This crystal class could simply be designated *mm* since the diad is automatically present. However, *mm*2 is usually used for the later development of space groups (see Section 2.12). A crystal containing three diad axes can also contain mirrors normal to all of these without an axis of higher symmetry. Such a point group is designated *mmm*, or could be designated 2/*mm*. As can be seen from Figure 2.6, the multiplicity of the general form is now eight. The special forms {$hk0$}, {$h0l$} and {$0kl$} now show a lower multiplicity than the general one. The point group *mmm* shows the highest symmetry in the orthorhombic system. The point group showing the highest symmetry in a particular crystal system is said to be the *holosymmetric* class.

To plot a stereogram of an orthorhombic crystal centred on 001 if we are given the lattice parameters *a, b* and *c*, we proceed as shown in Figure 2.8. In Figure 2.8a the poles of the (001), (010) and (100) planes are immediately inserted at the centre of the primitive and where the *x*- and *y*-axes cut the primitive, respectively. A pole ($hk0$) can be inserted at an angle θ along the primitive to (100), as shown in Figure 2.8a, by noting from Figure 2.8b (which is a section of the crystal parallel to (001) showing the intersection of the plane ($hk0$) with the crystal axes) that the pole of (100) lies along *OM*, that ($hk0$) makes intercepts on the crystal axes of *a/h* along the *x*-axis and *b/k* along the *y*-axis, and that θ is the angle between the normal to ($hk0$) and the normal to (100). From Figure 2.8b we have:

$$\tan\theta = \cot(90° - \theta) = \frac{a/h}{b/k} = \frac{a\,k}{b\,h}$$

That is:

$$\tan(100) \wedge (hk0) = (a/b)(k/h) \qquad (2.1a)$$

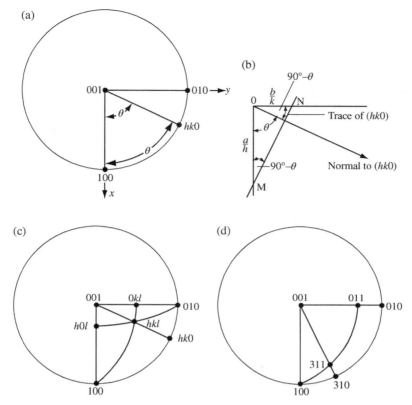

Figure 2.8 (a) General location of a *hk*0 pole on the stereogram of an orthorhombic crystal, (b) geometry to determine the angle θ between the 100 pole and the *hk*0 pole, (c) location of the *hkl* pole given the locations of the *hk*0, *h0l* and *0kl* poles, (d) the location of the 311 pole given the locations of the 100, 010, 001, 011 and 310 poles

where $(100)^{\wedge}(hk0)$ means the angle between the (100) and (*hk*0) planes; that is, the angle between the pole of (100) and the pole of (*hk*0). Similarly, if the lattice parameters are given, we can locate (*0kl*) and (*h0l*), since:

$$\tan(001) \wedge (0kl) = (c/b)(k/l) \tag{2.1b}$$

and:

$$\tan(001) \wedge (h0l) = (c/a)(h/l) \tag{2.1c}$$

These relations can be seen immediately by drawing diagrams similar to Figure 2.8b but looking along the *x*- and *y*-axes, respectively. When poles such as (*h0l*), (*0kl*) and (*hk*0) have been inserted on the stereogram, we can insert a pole such as (*hkl*) immediately by use of the zone addition rule, given in Section 1.3, Eq. (1.20).

If two poles $(h_1 k_1 l_1)$, $(h_2 k_2 l_2)$ both lie in the same zone with the indices [*uvw*] then so does the pole of the plane $(mh_1 + nh_2, mk_1 + nk_2, ml_1 + nl_2)$; that is, any plane whose indices can be formed by taking linear combinations of the indices of two planes in a given zone can provide the indices of a further plane in that same zone (Section 1.3). In general, m and n can be positive or negative, but if m and n are both positive then the pole of the plane under consideration must lie on the great circle between the poles of $(h_1k_1l_1)$ and $(h_2k_2l_2)$: this simple result can be extremely useful when locating poles on stereographic projections such as Figure 2.7.

In Figure 2.8c, after (001), (010), (100) and $(hk0)$, $(h0l)$, $(0kl)$ are plotted, then to plot, say, (hkl), we note that (hkl) must lie in the zone containing (001) and $(hk0)$, since if we multiply (001) by the number l and add the indices $(00l)$ and $(hk0)$ we obtain (hkl). It then follows that (hkl) lies somewhere on the great circle between (001) and $(hk0)$. Similarly, (hkl) lies in the zone containing $(0kl)$ and (100), since h times (100) gives $(h00)$ and this added to $(0kl)$ gives (hkl). Again, it is true that (hkl) lies somewhere on the great circle between $(0kl)$ and (100), but we can now rationalize that (hkl) must lie at the intersection of the two great circles we have considered. We then draw the great circle (or zone) containing (001) and $(hk0)$ and that containing (001) and $(0kl)$ and we know that (hkl) is situated where these intersect.

A particular example may make the procedure clear. Suppose we wish to locate (311) after plotting (001), (010) and (100) (Figure 2.8d). One way to proceed would be to locate (011) using Eq. (2.1b), setting $k = 1$ and $l = 1$ and using the known lattice parameters. We then find (310), on the primitive, by finding the angle between (100) and (310) from Eq. (2.1a), setting $h = 3$ and $k = 1$. Finally, we note that (311) lies in the zone containing (001) and (310), since (001) plus (310) yields (311). Also, (311) lies in the zone containing (100) and (011) since three times (100) plus (011) yields (311). The pole of (311) is then immediately located by drawing the great circle through (001) and (310) and that through (011) and (100); (311) is located where these great circles meet.

Using the above procedure for locating poles is usually the quickest way to draw an accurate stereogram when key poles have been located either by calculation or through the use of a computer software package. It must be strongly emphasized that, although we have chosen the orthorhombic system as an example, the use of Eq. (1.20) to locate poles applies to *any* crystal system and does not depend on the crystal axes being at any particular angle to one another. The utility of Eq. (1.20) is one of the great advantages of the Miller index for denoting crystal planes, and arises naturally from the properties of a space lattice.

Equations such as those in Eq. (2.1) can of course be used to find the ratio of the lattice parameters – the axial ratios – from measurements of the angles between poles.

2.3   Tetragonal System

The tetrad axis is always taken parallel to the z-axis. The lattice parameters a and b are equal.

The holosymmetric point group is $4/mmm$, showing three mutually perpendicular mirror planes with a tetrad normal to one of them (Figure 2.6). If a single pole is repeated according to the presence of these symmetry elements it will be found that diad axes are necessarily present normal to the mirrors and that, in addition, a second pair of diad axes also normal to mirror planes automatically arises. One of the pairs of mutually perpendicular diads is chosen as defining the directions of the x- and y-axes. The general form is $\{hkl\}$, with a multiplicity of 16. Special forms are {001}, {100}, {110}, {*hk*0}, {*h*0*l*} and {*hhl*}. The last of these, {*hhl*}, indicates a face making equal intercepts on the *x*- and *y*-axes.

The point group 422 could be specified simply as 42 since if 4 and 2 are present at right angles a second pair of diad axes arises and one of these pairs is chosen to define the *x*- and *y*-axes. The group $\bar{4}2m$ can be developed as $\bar{4}m$. It is then found that diad axes automatically arise at 45° to the two, mutually perpendicular, mirror planes. The pair of diad axes is taken to define the *x*- and *y*-axes. The four other tetragonal point groups, 4, $\bar{4}$, 4/*m* and 4*mm*, are straightforward and no diad axes arise.

The angles between poles on a stereogram and the ratio of the lattice parameters (i.e. the ratio *a* : *c* in this crystal system) are easily related by using equations such as those in Eq. (2.1) with *a* = *b*.

2.4 Cubic System

Cubic crystals possess four triad axes as a minimum symmetry requirement. These are arranged as in Figure 1.30 and are always taken to lie along the ⟨111⟩ type directions of the unit cell, which is a cube, so *a* = *b* = *c*. This is the *only* crystal system in which the direction [*uvw*] necessarily coincides with the normal to the plane (*uvw*) for all *u*, *v*, *w*.

If we put four triad axes to coincide with the ⟨111⟩ directions on a stereogram and allow these to operate on a single pole, as in Figure 2.9, we find that diad axes automatically arise along the crystal axes. The presence of the diads also follows from Table 1.2, with the row starting 233. The point group symbol used to describe the combination of a diad and two triads, shown in Table 1.2, is just 23. This is the cubic point group of lowest symmetry. The multiplicity of the general form for this point group is 12 (Figure 2.9). A feature of the symbols for describing point groups in the cubic system is that the symbol 3 or $\bar{3}$, even though it indicates the defining axis for the system, is never placed first because the triads, or inverse triads, are always at 54.74° ($\cos^{-1} (1/\sqrt{3})$) to the crystal axes. (In all other systems the defining axis comes first in the symbol – the presence of the diads or inverse diads parallel to at least two axes must be stated to distinguish orthorhombic point groups from monoclinic point groups.)

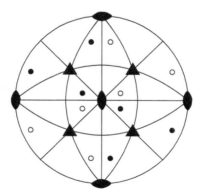

Figure 2.9   The cubic point group of lowest symmetry: 23

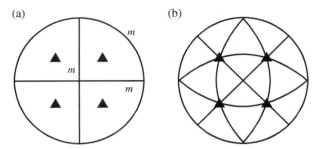

Figure 2.10   Stereograms centred on 001 of (a) mirror planes parallel to {100} planes and (b) mirror planes parallel to {110} planes in a cubic crystal. In both (a) and (b) the location of the four triads are shown; these are the minimum symmetry requirement for cubic crystals

The figure 3 (or $\bar{3}$) always occurs second in the symbol for cubic point groups, and this enables a point group in the cubic system to be distinguished from those in all other crystal systems. In a cubic crystal, mirror planes can run either parallel to {100} planes, as in Figure 2.10a, or else parallel to {110} planes, as in Figure 2.10b. The first alternative is described by putting the symbol m before $\bar{3}$ (to give $m\bar{3}$) and the second by placing m after the 3 (or $\bar{3}$) to give $X3m$ or $X\bar{3}m$, where X is an axis other than 3 or $\bar{3}$.

If we add mirror planes parallel to {100} to the class 23 we obtain $2/m3$, conventionally denoted $m\bar{3}$. (Prior to the revision of the *International Tables for Crystallography* in 1995 this point group was denoted $m3$; older textbooks will use this terminology as well.) As is evident from Figure 2.6, this contains a centre of symmetry in addition to the three diads at the intersection of three mutually perpendicular mirror planes and the four triads. The triads therefore become inversion triads, $\bar{3}$. The multiplicity of the general form $\{hkl\}$ is 24. It is worth noting from Table 2.3 (see below) that when there are no diads along $\langle 110 \rangle$ directions nor mirrors parallel to {110}, $\{hk0\}$ and $\{kh0\}$ are separate special forms. This occurs in the classes 23 and $m\bar{3}$.

Replacement of the diads in 23 by tetrads gives 43. Here we notice that diads automatically arise along the $\langle 110 \rangle$ directions, as tabulated in Table 1.2. This class is denoted 432 to indicate the diads because of later development of space groups (see Section 2.11). However, 43 is sufficient to identify it.

Replacement of 2 by $\bar{4}$ in 23 will be found to produce mirror planes automatically parallel to the {110} planes, and hence passing through the triads. Correspondingly, if mirrors parallel to {110} are added to 23 then the diad axes along the $\langle 100 \rangle$ directions become $\bar{4}$ axes. If we have mirror planes passing through the triad axes then parallel to the crystal axes we can have either $\bar{4}$ or 4. The first of these classes is $\bar{4}3m$ and the second $m\bar{3}m$. (This latter point group was known as $m3m$ conventionally prior to the 1995 revision of the *International Tables for Crystallography*.) In $\bar{4}3m$ there is no centre of symmetry and there are no additional symmetry elements, other than those indicated in the symbol. The multiplicity of the general form is 24. In $\bar{4}3m$ (as in 23), {111} and {1$\bar{1}$1} are separate special forms: each shows four planes parallel to the surfaces of a regular tetrahedron.

Class $m\bar{3}m$ has mirror planes parallel to {100} and to {110}; therefore, nine mirror planes are present in all. There are six diads, three tetrads, a centre and of course the four triads. All of these can be produced by putting mirrors parallel to both {110} and {100}, coupled

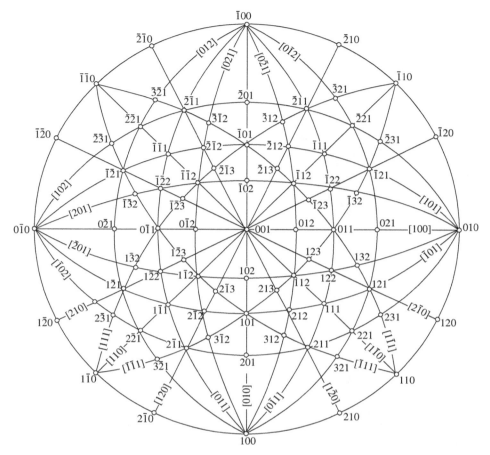

Figure 2.11 Stereogram of a cubic crystal. Source: From [5], p. 82. Copyright 1959.

with the four triads. Hence the symbol $m\bar{3}m$ is used to describe this point group, which is the cubic holosymmetric class. In full notation this would be $4/m\,\bar{3}\,2/m$, to denote the point group symmetry elements as $4/m$ along the $\langle 100 \rangle$ directions, $\bar{3}$ along the $\langle 111 \rangle$ directions and $2/m$ along the $\langle 110 \rangle$ directions. The general form has 48 faces.

A stereogram of a cubic crystal with the poles of a number of faces indicated, along with the zones in which they lie, is shown in Figure 2.11. Additional poles are easily located on a stereogram, such as that in Figure 2.11, by use of the zone addition rule explained in Section 1.3 (Eq. (1.20) and following) and used to plot poles in Section 2.2.

2.5 Hexagonal System

Crystals possessing a hexad axis have a Bravais lattice, as illustrated in Figure 1.19j in Section 1.8. The x- and y-axes are at $120°$ to one another and perpendicular to the hexagonal axis along z. The holosymmetric class of this system, $6/mmm$, possesses a hexad at the

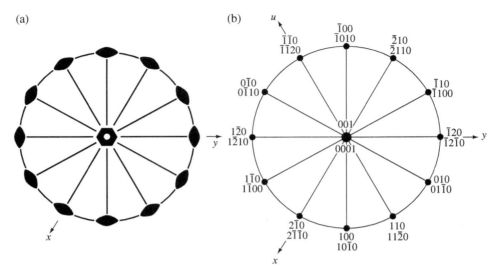

Figure 2.12   (a) Stereogram of the holosymmetric class of the hexagonal system, 6/*mmm*, (b) indices of a number of poles in both Miller–Bravais and Miller indices on this stereogram

intersection of two sets of three vertical planes of symmetry, two sets of three diad axes normal to these, a plane of symmetry normal to the hexad axis and a centre of symmetry (Figure 2.6). These symmetry elements are shown in Figure 2.12a. Diad axes at 120° to one another are taken as the crystal axes. If this is done, the indices of a number of faces are as marked in Figure 2.12b. The plane of index (100) is repeated by the hexad axis to give (010), ($\bar{1}$10), ($\bar{1}$00), (0$\bar{1}$0), (1$\bar{1}$0). All of these are identical crystallographic planes and yet their Miller indices appear different: note (010) and ($\bar{1}$10). To avoid the possibility of confusion from having planes of the same form with quite different indices, it is customary in materials science and metallurgy to employ Miller–Bravais indices for the hexagonal system.[4] To do this, we choose a third crystal axis *u* normal to the hexagonal axis and at 120° to both the *x*- and the *y*-axes.[5] The lattice repeat distance along the *u*-axis is equal to *a* (= *b*) from Figure 1.19j.

To state the Miller–Bravais indices of a plane we then take the intercepts along all three axes, *x*, *y* and *u*, express these in terms of the lattice parameters and proceed exactly as in Section 1.2. The result is that a plane always has four indices (*hkil*), where *i* is the intercept along the *u*-axis. It is obvious that there is a necessary relationship between *h*, *k* and *i* since the *u*-axis is a redundant third axis normal to the hexagonal axis. This can be deduced from Figure 2.13 and is:

$$i = -(h + k) \tag{2.2}$$

---

[4] These are termed Miller–Bravais indices (or Bravais–Miller indices, the form used in [3]), because Auguste Bravais (1811–1863) was the first to show the relationship expressed in Eq. (2.2) in Equation (13) on page 119 of his *Études Cristallographiques* [6], published posthumously in 1866, but first presented to the French Academy of Sciences on 26 February 1849.
[5] There is no 'perfect' choice of nomenclature for this redundant third axis. The choice of *u* in the term *u*-axis here should not be confused with the *u* in the nomenclature [*uvw*] introduced in Section 1.2 for vectors and is in deference to Georges Friedel who chose this nomenclature when describing Miller–Bravais indices for planes in hexagonal and trigonal materials [7]. Friedel chose a nomenclature for planes and vectors which did not use the {*hkl*} and ⟨*uvw*⟩ notations in familiar use today [3].

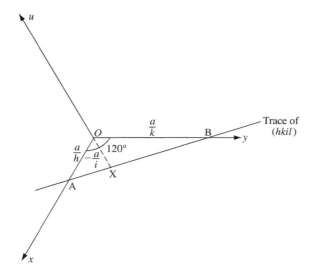

Figure 2.13 Geometry to show that in Miller–Bravais indices (*hkil*), $i = -(h + k)$

i.e. the third index is always the negative of the sum of the first two. The indices of a number of poles are given in both Miller–Bravais and Miller indices in Figure 2.12b. The hexagonal symmetry is then apparent in the former from the indices of planes of the same form.

A four-index system due to Leonhard Weber [8] is also used to specify directions in hexagonal crystals so that the index appears as [*UVTW*] instead of [*uvw*]. When this is done, a little care is needed in relating the results to those obtained using a three-index notation. To enable directions of a given family to have indices of similar appearance, the directions are specified by taking steps along all three axes and arranging it so that the step along the *u*-axis is of such length that the number of unit repeat vectors moved along this direction, *T*, is equal to the negative of (*U* + *V*).

In general, the indices [*UVTW*] and [*uvw*] are related to one another through the equations

$$U = \tfrac{1}{3}(2u - v)$$
$$V = \tfrac{1}{3}(2v - u)$$
$$T = -\tfrac{1}{3}(u + v)$$
$$W = w$$

(2.3)

and, conversely,

$$u = 2U + V = U - T$$
$$v = 2V + U = V - T$$
$$w = W$$

(2.4)

The direction corresponding to the *x*-axis in Figure 2.14 is then $[2\bar{1}\bar{1}0]$ in the four-index notation and [100] in the three-index system; the magnitude of $[2\bar{1}\bar{1}0]$ is three times that of [100], so therefore [100] in the three-index system is equivalent in magnitude to the

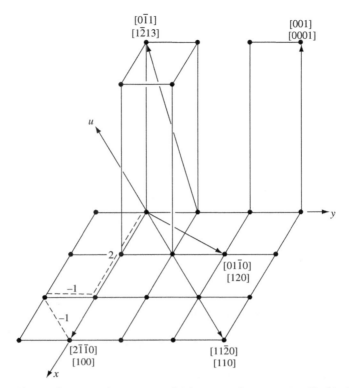

Figure 2.14 Indices of various directions in the hexagonal system specified in both the conventional three-index notation, [*uvw*], and the four-index Weber notation, [*UVTW*]

four-index vector $\frac{1}{3}[2\bar{1}\bar{1}0]$. Likewise, [010] in the three-index system is equivalent in magnitude to the four-index vector $\frac{1}{3}[\bar{1}2\bar{1}0]$ and $[\bar{1}\bar{1}0]$ to $\frac{1}{3}[\bar{1}\bar{1}20]$. Knowing the four-index vectors equivalent in magnitude and direction to the three-index vectors [100] and [010] enables more general three-index vectors [*uvw*] to be transformed into their four-index equivalent vectors relatively easily prior to the clearing of fractions to specify four-index zones or directions.

Some indices of directions specified in both ways are given in Figure 2.14. In relating planes and zone axes using Eq. (1.6), it is usually best to work entirely in the three-index notation for both planes and directions and to translate the three-index notation for a direction into the four-index system at the end of the calculation. It is also useful to note that the condition needed for a four-index vector [*UVTW*] to lie in a four-index plane (*hkil*) is:

$$hU + kV + iT + lW = 0 \qquad (2.5)$$

Other point groups besides 6/*mmm* in the hexagonal system are shown in Figure 2.6. We note that $\bar{6}$ ($\equiv 3/m$) is placed in this system because of the use of rotoinversion axes to describe symmetry operations of the second sort; 6, $\bar{6}$, 6/*m* and 6*mm* show no diad axes, just

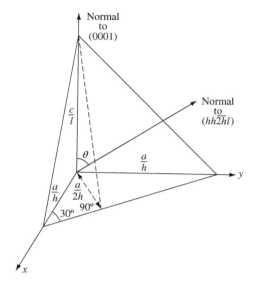

Figure 2.15 Geometry to determine the angle θ between the (0001) pole and the $(hh\overline{2h}l)$ pole

like their counterparts in the tetragonal system. The crystal axes for 6*mm* are usually chosen to be perpendicular to one set of mirrors (they then lie in the other set) and $\overline{6}m2$ could be developed as $\overline{6}m$ ($\equiv 3/mm$). The diads automatically arise and are chosen as crystallographic axes. Of course, 622 contains diads. It could be developed as 62, since the second set of diads arises automatically (see Table 1.2). The axes are chosen parallel to one set of diads. Only 6/*m* and 6/*mmm* are centrosymmetric in this system.

It is apparent from the stereogram in Figure 2.12b that stereograms with 0001 at the centre showing {*hki*0} poles are straightforward to plot. To plot more general poles on a stereogram with 0001 at the centre, it is apparent that the *c/a* ratio has to be used. Thus, for example, this has to be used to determine the angle between faces such as (0001) and $(hh\overline{2h}l)$, for example $(11\overline{2}1)$. It is convenient to choose a $(hh\overline{2h}l)$ plane because such a plane is equally inclined to the *x*- and the *y*-axes.

From Figure 2.15, the angle θ between the (0001) pole and the $(hh\overline{2h}l)$ pole is seen to be given by:

$$\tan\theta = \frac{c}{a}\frac{2h}{l} \tag{2.6}$$

Similarly, the angle θ between (0001) and $(h0\overline{h}l)$ is:

$$\tan\theta = \frac{c}{a}\frac{2h}{\sqrt{3}l} \tag{2.7}$$

An example of a stereogram centred at (0001) with poles of the forms $\{11\overline{2}1\}$, $\{10\overline{1}1\}$ and $\{12\overline{3}1\}$ indicated for a hexagonal cell is shown in Section 2.6 in connection with crystals of the trigonal system. The special forms in the various classes of the hexagonal system are listed in Table 2.1.

2.6 Trigonal System

This crystal system is defined by the possession of a single triad axis. It is closely related to the hexagonal system. The possession of a single triad axis by a crystal does not, by itself, indicate whether the *lattice* considered as a set of points is truly hexagonal, or whether it is based on the staggered stacking of triequiangular nets. When the lattice is rhombohedral, a cell of the shape of Figure 1.19k can be used. The cell in Figure 1.19k is a rhombohedron and the angle α ($< 120°$) is characteristic of the substance. When the lattice of a trigonal crystal is hexagonal, it is not appropriate to use a rhombohedral unit cell.

The symmetry elements in the holosymmetric class $\bar{3}m$ are shown in Figure 2.6 and the repetition of a single pole in accordance with this symmetry is also demonstrated in this figure. In $\bar{3}m$, three diad axes arise automatically from the presence of $\bar{3}$ and the three mirrors lying parallel to $\bar{3}$. These diad axes, which intersect in the inverse triad axis, do not lie in the mirror planes. If the rhombohedral cell is used for such a crystal then the axes cannot be chosen parallel to prominent axes of symmetry.

A stereogram of a trigonal crystal indexed according to a rhombohedral unit cell is shown in Figure 2.16. The value of α is 98°. The x-, y- and z-axes are taken to lie in the mirror planes and the inverse triad is a body diagonal of the cell, therefore lying along the direction [111], which, from the geometry of the rhombohedral unit cell, is also parallel to the normal

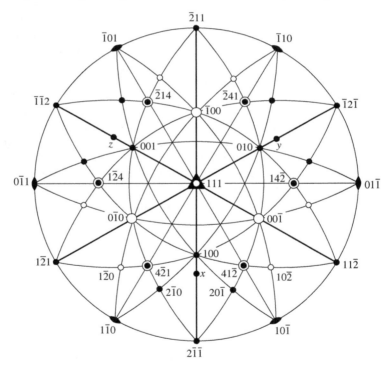

Figure 2.16 A stereogram of a trigonal crystal of class $\bar{3}m$ with a rhombohedral unit cell. When poles in the upper and lower hemispheres coincide in projection, the indices shown refer to the poles in the upper hemisphere

to the (111) plane. It is clear that the *x*-, *y*- and *z*-axes – that is, the directions [100], [010] and [001] – do not lie normal to the (100), (010) and (001) planes, respectively. However, these directions are easily located. For example, the *z*-axis, [001], is the pole of the zone containing ($\bar{1}$10), (010), (100) and (1$\bar{1}$0), shown as a great circle in Figure 2.16. Likewise, the *y*-axis is the pole of the zone containing (10$\bar{1}$), (100), (001) and ($\bar{1}$01), and the *x*-axis is the pole of the zone containing (0$\bar{1}$1), (001), (010) and (01$\bar{1}$).

In plotting a stereogram of a trigonal crystal with a rhombohedral unit cell from a given value of the angle α, it is useful to note that the angle γ between any of the crystal axes and the unique triad axis is given by:

$$\sin \gamma = \frac{2}{\sqrt{3}} \sin \alpha/2 \qquad (2.8)$$

(see Problem 2.2).

For many purposes it is more convenient to use a hexagonal cell when dealing with trigonal crystals, irrespective of whether the lattice of the trigonal crystal is primitive hexagonal or rhombohedral. The shape of the cell chosen is the same as for hexagonal crystals (Figure 1.19j) and, since it is chosen without reference to the lattice, may or may not be primitive. The value *c/a* is characteristic of the substance.

A stereogram of a trigonal crystal of the point group $\bar{3}m$ with planes indexed according to the Miller–Bravais scheme is shown in Figure 2.17. This is the same crystal as that in

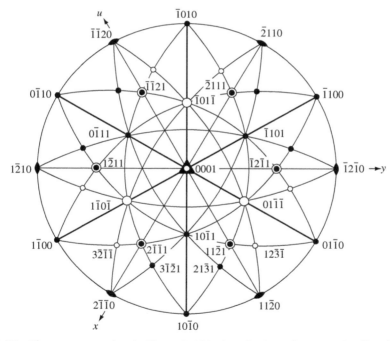

Figure 2.17 The same crystal as in Figure 2.16 indexed using a hexagonal cell. When poles in the upper and lower hemispheres superpose in projection, the indices refer to poles in the upper hemisphere

Figure 2.16 ($c/a = 1.02$). When using the hexagonal cell, the x-, y- and u-axes are chosen parallel to the diads in $\bar{3}m$.

The relationship between the indices in the two stereograms can be easily worked out using the zone addition rule and the Weiss zone law, Eq. (1.6), from the orientation relationship of the rhombohedral and hexagonal cells. In Figures 2.16 and 2.17 this is (0001) ‖ (111)[6] and (10$\bar{1}$1) ‖ (100). The plane (10$\bar{1}$0) in Figure 2.17 then has indices (2$\bar{1}$ $\bar{1}$) in Figure 2.16. The indices (2$\bar{1}$ $\bar{1}$) could be deduced by noting that the plane (2$\bar{1}$ $\bar{1}$) contains the [111] direction (and so the pole of (2$\bar{1}$ $\bar{1}$) must lie on the primitive if [111] is at the centre of the stereogram), is equally inclined to the y- and z-axes, and lies in the zone containing (111) and (100). The plotting of a stereogram and the determination of axial ratios for a trigonal crystal referred to hexagonal axes then proceed as for the hexagonal system.[7]

In the class $\bar{3}m$, special forms lie (i) normal to the triad: {0001}, (ii) parallel to the triad: {hki0}, (iii) normal to mirror planes: {$h0\bar{h}l$}, and (iv) equally inclined to two diads {$hh\overline{2h}l$}. The *six faces* in the form: {$h0\bar{h}l$}, make a rhombohedron; {10$\bar{1}$2} would be an example, consisting of the planes (10$\bar{1}$2), ($\bar{1}$102), (0$\bar{1}$12), ($\bar{1}$01$\bar{2}$), (1$\bar{1}$0$\bar{2}$), and (01$\bar{1}$$\bar{2}$). This form is similar in appearance to {$0h\bar{h}l$}, which is also a rhombohedron, rotated 60° with respect to the first one.

The relationship between {$h0\bar{h}l$} and {$0h\bar{h}l$} (or, equivalently, {$0k\bar{k}l$}) is shown in Figure 2.18. They are actually quite separate forms and each one is a special form. We therefore need to add {$0h\bar{h}l$} to the list of special forms, in addition to {$h0\bar{h}l$}. It is apparent from Figure 2.16 that when using the rhombohedral cell the two forms {10$\bar{1}$1} and {01$\bar{1}$1} have different indices, since the face above (00$\bar{1}$) in the projection in Figure 2.16 would have indices (22$\bar{1}$).

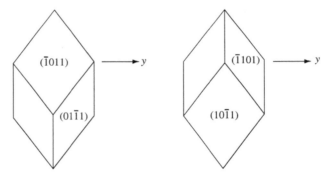

Figure 2.18 The relationship between the special forms {10$\bar{1}$1} and {01$\bar{1}$1} in the trigonal system

---

[6] The symbol ‖ means 'parallel to'.

[7] The relationship between the two cells used in Figures 2.16 and 2.17 is shown in Figure A4.2a, where general methods of transforming the indices of planes and directions, when different unit cells are chosen for the same crystal, are considered.

The other trigonal point groups are shown in Figure 2.6. In 3 and in $\bar{3}$ there are neither mirrors nor diads. The class $3m$ has three mirrors intersecting in the triad axis; the x-, y- and u-axes of the hexagonal cell are taken to lie perpendicular to the mirror planes. The point group 32 contains diads normal to 3 (consistent with the geometry for permissible combinations of rotational symmetry operations in Table 1.2), which are taken as the crystallographic axes. If three mirror planes intersect in an inversion triad, as in $\bar{3}m$, then diads automatically arise normal to the mirror planes; the diads are chosen as the x-, y- and u-axes. Finally, we note that the class $3/m$ is not specified as a point group in the trigonal system: it is placed in the hexagonal system because it is equivalent to $\bar{6}$.

2.7 Monoclinic System

Crystals in the monoclinic system possess a single twofold axis. Since a mirror plane is equivalent to an inverse diad, the class m ($\equiv \bar{2}$) is put into this system. Two settings are shown in Figure 2.6 for the monoclinic point groups, depending on whether the stereogram is centred on the twofold axis (the 1st setting) or whether the twofold axis is lying in the equatorial plane of the stereogram (the 2nd setting).

With respect to the convention for choosing x-, y- and z-axes for stereograms centred on the normal to the 001 planes for orthorhombic, tetragonal, and cubic crystals, the twofold axis is along the z-axis in the 1st setting, so that the angle γ between the x- and y-axes is obtuse (as for hexagonal crystals, where γ is fixed to be 120°), while $\alpha = \beta = 90°$. In the 2nd setting, the twofold axis is along the y-axis, so that the angle β between the x- and z-axes is obtuse, while $\alpha = \gamma = 90°$. Somewhat confusingly, both conventions are used in the scientific literature to describe the unit cells of monoclinic crystals, but it is much more common in materials science and metallurgy for the 2nd setting to be used, as we have chosen to do in Table 1.3 and Figures 1.19b and c. Irrespective of the choice of 1st or 2nd setting, the sides of the unit cells of crystals belonging to the monoclinic system are in general all unequal to one another.

The symmetry elements in the holosymmetric point group $2/m$ are shown in Figure 2.19 for the 2nd setting. The pole of (010) and the y-axis coincide on the stereogram. In this figure, [001], the z-axis, is chosen to be at the centre of the primitive and so (100) lies on the primitive 90° from (010). The x-axis, the direction [100], which makes the obtuse angle β with [001], is necessarily in the lower hemisphere. The angle between [001] and (001) is $(\beta - 90°)$, which is of course equal to the angle between the (100) pole and [100]. Poles such as $\{hk0\}$ lie around the primitive because [001] lies in $\{hk0\}$ planes (Weiss zone law, Eq. (1.6)).

A pole such as $(hk0)$ lies at angle φ to (010), given by:

$$\cot \varphi = \frac{(a/h)\sin\beta}{(b/k)} \tag{2.9a}$$

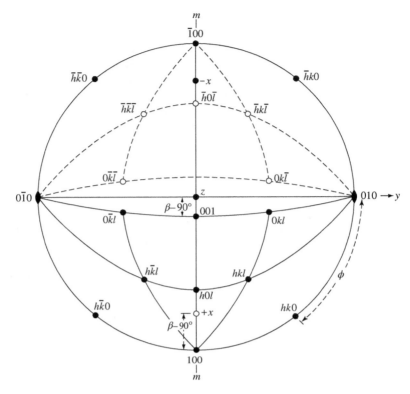

Figure 2.19 Monoclinic stereogram centred on [001] for the 2nd setting

(see Figure 2.20), or equivalently (Section A3.2):

$$\cos \varphi = \frac{\dfrac{k}{b}}{\left(\dfrac{h^2}{a^2 \sin^2 \beta} + \dfrac{k^2}{b^2} \right)^{1/2}} \tag{2.9b}$$

The factor $\sin \beta$ arises because the x-axis does not lie in the same plane as the plane perpendicular to the z-axis within which the normals to $(hk0)$ and (010) both lie. In the point group $2/m$ the only special form besides $\{010\}$ is $\{h0l\}$, with a multiplicity of two – the planes $(h0l)$ and $(\bar{h}0\bar{l})$. It is apparent that the two faces in this latter form are parallel, both lying normal to the mirror plane and parallel to the diad axis. The general form is $\{hkl\}$, with a multiplicity of four (see Figure 2.19). $\{hk0\}$ and $\{0kl\}$ are also general forms.

The remaining point groups in the monoclinic system are 2 and m. Again, $\{h0l\}$ is a special form in both and so is $\{010\}$. However, 2 does not possess a centre, and so for this point group $\{010\}$ and $\{0\bar{1}0\}$ must each be listed as separate special forms.

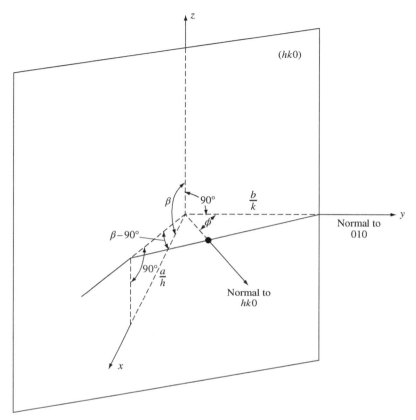

Figure 2.20 A diagram from which the angle ϕ in Figure 2.19 between 010 and $hk0$ can be derived

2.8 Triclinic System

There are no special forms in either of the point groups in this system. The unit cell is a general parallelepiped and the geometry is more complicated than even for the monoclinic system. The drawing of a stereogram whose centre is taken to be the z-axis, [001], can be carried out with the aid of Figures 2.21a and b. The two diagrams here are completely general and can be specialized to apply to any crystal system more symmetric than the triclinic by setting one or more of the axial angles to particular values and by setting two or more of the lattice parameters to be equal.

The choice of the centre of the stereogram as the z-axis in Figure 2.21a means that all planes of the general form $(hk0)$ lie on the primitive of the stereogram, just as for the monoclinic stereogram in Figure 2.19. However, for triclinic crystals there is no good reason to have the 010 pole located on the right-hand side of the horizontal axis of the stereogram, and so here it is deliberately rotated around the primitive away from this position.

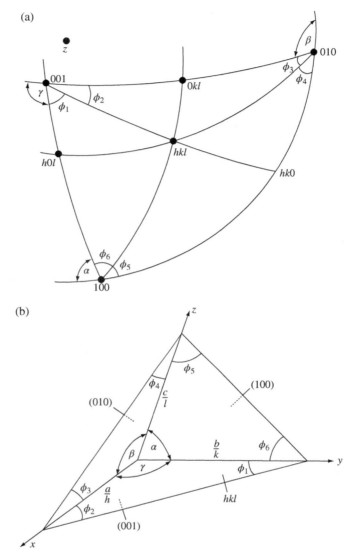

Figure 2.21 Diagrams relevant to drawing stereograms of triclinic crystals. The centre of the stereogram in (a) is the z-axis, [001]

The great circle passing through 001 and 010 is the [100] zone, containing planes of the general form (0kl). Therefore, the angle between this zone and the primitive must be the angle between [100] and [001], β, as shown in Figure 2.21a. This is because the stereographic projection is a conformal projection (see Appendix 2); that is, one for which angles are faithfully reproduced. Similarly, α and γ can be specified on Figure 2.21a from the triclinic system geometry.

Angles between poles can be determined using Eq. (A3.18). Once the position of the 010 pole (or any other pole of the form $hk0$) has been chosen, geometry determines the positions of the remaining poles on the stereogram. Thus, for example, the position of the 001 pole is fixed knowing that it lies in the [100] zone, which has to make an angle of β with the primitive, and that 001 is a given angle from 010, determined for example using Eq. (A3.18). Likewise, the position of a pole $0kl$ lying in the [100] zone can be determined, as can the positions of the poles 100, $h0l$, $hk0$ and hkl.

While stereographic projections can now be produced routinely via proprietary software packages, it is still instructive to consider further aspects of the geometry of the part of the triclinic stereogram shown in Figure 2.21a in order to gain a full appreciation of the richness of information displayed on stereograms. An (hkl) plane of a triclinic crystal is shown in Figure 2.21b. The six angles ϕ_1–ϕ_6 in Figure 2.21b are the same as those marked in Figure 2.21a.

Thus, for example, ϕ_1 in Figure 2.21b is an angle lying in the (001) plane. It is the angle between the y-axis and the vector $[\bar{k}h0]$ common to (hkl) and (001); that is, the angle between the zone containing (001), $(h0l)$ and (100) and the zone containing (001), (hkl) and $(hk0)$. We can therefore mark ϕ_1 on the stereogram. Similarly, ϕ_5 in Figure 2.21b is an angle lying in the (100) plane. It is the angle between the z-axis and the vector $[0\bar{l}k]$ common to (hkl) and (100); that is, the angle between the zone containing (100), $(hk0)$ and (010) and the zone containing (100), (hkl) and $(0kl)$. ϕ_5 can therefore also be marked on the stereogram. Proceeding in this way, we can identify all of the angles ϕ_1–ϕ_6. We have, from the geometry in Figure 2.21b:

$$\alpha + \phi_5 + \phi_6 = 180°$$

$$\beta + \phi_3 + \phi_4 = 180° \qquad (2.10)$$

$$\gamma + \phi_1 + \phi_2 = 180°$$

Furthermore, from the triangle on the (001) face in Figure 2.21b we have, using the sine rule:

$$\frac{a/h}{\sin \phi_1} = \frac{b/k}{\sin \phi_2} \qquad (2.11)$$

Therefore:

$$\frac{a/h}{b/k} = \frac{\sin \phi_1}{\sin \phi_2} \qquad (2.12a)$$

and similarly:

$$\frac{c/l}{b/k} = \frac{\sin \phi_6}{\sin \phi_5} \qquad (2.12b)$$

and:

$$\frac{a/h}{c/l} = \frac{\sin \phi_4}{\sin \phi_3} \qquad (2.12c)$$

As an aside, we note that the equations in Eq. (2.12) are also of use in finding axial ratios and axial angles from measured angles between planes on single crystal specimens of crystals belonging to the triclinic crystal system; such crystals tend to be minerals or organic materials, rather than metals, the crystal structures of which rarely tend to belong to either the monoclinic or the triclinic systems.

2.9 Special Forms in the Crystal Classes

A summary of the special forms in the various crystal classes for the various crystal systems is given in Table 2.1. The orientation of the axes is that used in this chapter. Monoclinic crystals are in the 2nd setting of Figure 2.6 and trigonal crystals are referred to hexagonal axes. There are no special forms in the triclinic system.

Table 2.1 Special forms in the crystal classes

2	32	6mm
{010}, {0$\bar{1}$0}, {h0l}	{0001}, {10$\bar{1}$0}, {2$\bar{1}\bar{1}$0}, {11$\bar{2}$0}, {hki0}, {h0$\bar{h}$l}, {0k$\bar{k}$l}, {hh$\overline{2h}$l}, {2hh$\bar{h}$l}	{0001}, {000$\bar{1}$}, {10$\bar{1}$0}, {11$\bar{2}$0}, {hki0}, {h0$\bar{h}$l}, {hh$\overline{2h}$l}
m and 2/m	4	$\bar{6}$m2
{010}, {h0l}	{001}, {00$\bar{1}$}, {hk0}	{000$\bar{1}$}, {10$\bar{1}$0}, {01$\bar{1}$0}, {11$\bar{2}$0}, {hki0}, {h0$\bar{h}$l}, {0k$\bar{k}$l}, {hh$\overline{2h}$l}
mm2	$\bar{4}$ and 4/m	622 and 6/mmm
{001}, {00$\bar{1}$}, {100}, {010}, {hk0}, {h0l}, {0kl}	{001}, {hk0}	{0001}, {10$\bar{1}$0}, {11$\bar{2}$0}, {hki0}, {h0$\bar{h}$l}, {hh$\overline{2h}$l}
222 and mmm	4mm	23
{100}, {010}, {001}, {hk0}, {h0l}, {0kl}	{001}, {00$\bar{1}$}, {100}, {110}, {hk0}, {h0l}, {hhl}	{100}, {110}, {hk0}, {kh0}, {111}, {1$\bar{1}$1}, {hll}, {h$\bar{l}$l}, {hhl}, {h$\bar{h}$l}
3	$\bar{4}$2m	m$\bar{3}$
{0001}, {000$\bar{1}$}, {hki0}	{001}, {100}, {1$\underline{1}$0}, {hk0}, {h0l}, {hhl}, {h$\bar{h}$l}	{100}, {110}, {hk0}, {kh0}, {111}, {hll}, {hhl}
$\bar{3}$	42 and 4/mmm	432
{0001}, {hki0}	{001}, {100}, {110}, {hk0}, {h0l}, {hhl}	{100}, {110}, {hk0}, {111}, {hll}, {hhl}
3m	6	$\bar{4}$3m
{0001}, {000$\bar{1}$}, {10$\bar{1}$0}, {01$\bar{1}$0}, {11$\bar{2}$0}, {hki0}, {h0$\bar{h}$l}, {0k$\bar{k}$l}, {hh$\overline{2h}$l}	{0001}, {000$\bar{1}$}, {hki0}	{100}, {110}, {hk0}, {111}, {1$\bar{1}$1}, {hll}, {h$\bar{l}$l}, {hhl}, {h$\bar{h}$l}
$\bar{3}$m	$\bar{6}$ and 6/m	m$\bar{3}$m
{0001}, {10$\bar{1}$0}, {11$\bar{2}$0}, {hki0}, {h0$\bar{h}$l}, {0k$\bar{k}$l}, {hh$\overline{2h}$l}	{0001}, {hki0}	{100}, {110}, {hk0}, {111}, {hll}, {hhl}

2.10 Enantiomorphous Crystal Classes

In some crystal classes the possibility exists of crystals being found in either a right-handed or a left-handed modification with the two not being superposable, in the sense that the right hand of the body cannot be superposed upon the left hand. This occurs if the crystal only possesses simple rotation symmetry operations; that is, no symmetry operation of the second kind. Therefore, mirror planes and rotoinversion axes of any degree do not appear in such crystal classes. There are 11 of these classes: 1, 2, 3, 4, 6, 23, 222, 32, 422, 622 and 432.

2.11 Laue Groups

Some physical methods of examination of crystals cannot determine whether or not there is a centre of symmetry present, such as when stationary single crystals are illuminated by a continuous distribution of X-ray wavelengths (known as 'white radiation') to produce Laue diffraction patterns. This is known as the Laue method. Such diffraction patterns are used for symmetry determination and for the accurate alignment of crystals; they are also useful for time-resolved macromolecular crystallography using synchrotron radiation. The Laue group is the crystal class to which a particular crystal belongs if a centre of symmetry is added to the symmetry elements already present. The 11 Laue groups are shown in Table 2.2. The loss of symmetry information in time-resolved macromolecular crystallography is more than compensated for by the ability to capture structural information over subnanosecond timescales in fast chemical reactions such as enzyme reactions.

Table 2.2 Laue groups

System	Class	Laue group
Cubic	432, $\bar{4}3m$, $m\bar{3}m$	$m\bar{3}m$
	23, $m\bar{3}$	$m\bar{3}$
Hexagonal	622, 6mm, $\bar{6}m2$, 6/mmm	6/mmm
	6, $\bar{6}$, 6/m	6/m
Tetragonal	422, 4mm, $\bar{4}m2$, 4/mmm	4/mmm
	4, $\bar{4}$, 4/m	4/m
Trigonal	32, 3m, $\bar{3}m$	$\bar{3}m$
	3, $\bar{3}$	$\bar{3}$
Orthorhombic	222, 2mm, mmm	mmm
Monoclinic	2, m, 2/m	2/m
Triclinic	1, $\bar{1}$	$\bar{1}$

2.12 Space Groups

We showed in Section 1.4 that repetition of an object (e.g. an atom or group of atoms) in a crystal is carried out by operations of rotation, reflection (or inversion) and translation. We have also described in this chapter the consistent combinations of rotations and inversions that can occur in crystals, and in Section 1.8 we described the possible translations that can occur in crystals to produce the 14 Bravais lattices, or space lattices.

However, we have so far not made any attempt to combine the operations of rotation (and inversion) with translation, except briefly in Section 1.5, where it was shown that only one-, two-, three-, four-, and sixfold rotation axes are compatible with translational symmetry. A full description of the symmetry of a crystal involves a description of the way in which all of the symmetry elements are distributed in space. This is called the *space group*. There are 230 different crystallographic space groups[8]; each one gives the fullest description of the symmetry elements present in a crystal possessing that group.

The rotation axes and rotoinversion axes possible in crystals were discussed in Section 2.1 and the possible translations in Section 1.8. An enumeration of the way these axes can be consistently combined with the translation is therefore an enumeration of the possible space groups. Space groups are most important in the solution of crystal structures. They are also very useful when establishing symmetry hierarchies in phase transitions in materials such as perovskites [10–12].

When an attempt is made to combine the operations of rotation and translation, the possibility arises naturally of what is called a *screw axis*. This involves repetition by rotation about an axis, together with translation parallel to that axis. Similarly, a repetition by reflection in a mirror plane may be combined with a translational component parallel to that plane to produce a *glide plane*. These will now each be considered in more detail.

2.12.1 Screw Axes

If, for example, a twofold rotational axis occurs in a crystal, then this means some structural unit or motif is arranged about this direction so that it is repeated by a rotation of 180° about the axis. The repetition shown in Figure 2.22a corresponds to a pure rotational diad axis. However, the rotation of 180° could be coupled with a translation of one-half of the lattice repeat distance, t, in the direction of the axis to give the screw diad axis shown in Figure 2.22b, denoted by the symbol 2_1. The translation $t/2$ will be of a length of the order of the lattice parameters of the crystal and hence of the order of a few Ångstrom units ($1 \text{ Å} = 10^{-10}$ m), and so quite undetected by the naked eye. Macroscopically, the crystal containing a 2_1 axis would then show diad symmetry about that axis in the symmetry of its external faces or of its physical properties.

Although the presence of a screw axis, say a screw diad as in Figure 2.22b, indicates the presence of identical atomic motifs arranged about it so that they are related by a rotation plus translation, the screw axis must not be thought of as translating the translation vectors of the lattice. A screw axis and a pure rotation axis of the same order n repeat a translation in the same way. It follows that the rotational components of screw axes can only be

---

[8] The 230 space groups were established in the late 1880s and early 1890s through the work of Evgraf Stepanovich Fedorov (1853–1919) and Arthur Moritz Schoenflies (1853–1928). An account of how these two scientists arrived at the definitive set of space groups is given in [9], available as a free online version at https://www.iucr.org/publ/50yearsofxraydiffraction.

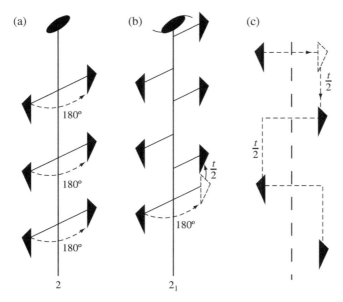

Figure 2.22   (a) A twofold rotation axis. (b) A 2_1 screw axis. (c) A glide plane normal to the surface of the paper

$2\pi/1$, $2\pi/2$, $2\pi/3$, $2\pi/4$ and $2\pi/6$, and that an n-fold screw and an n-fold pure rotation axis must have similar locations with respect to a similar set of translations. The angles between screws and between screws and rotations must therefore be the same as the permissible combinations listed in Table 1.2.

The various kinds of screw axis possible are shown in Table 2.3. An n_N-fold screw axis involves repetition by rotation through $360°/n$ with a translation of $\mathbf{t}N/n$, where $\mathbf{t}$ is a lattice repeat vector parallel to the axis. There are five types of screw hexad; for example, 6_1 involves rotation through $60°$ and translation of $\mathbf{t}/6$, while 6_5 involves rotation through $60°$ and translation of $5\mathbf{t}/6$. By drawing a diagram showing the repetition of objects by these axes, it is easily seen that 6_1 is a screw of opposite hand to 6_5, 4_1 to 4_3 and 6_2 to 6_4. A diagram to show this for 3_1 and 3_2 is given in Figure 2.23. In this figure, A and A$'$ are lattice points. The operation of 3_1 is straightforward, as shown in Figure 2.23a. When repeating an object according to 3_2, note that the object at height $\frac{4}{3}$ must also occur at height $\frac{1}{3}$, since A and A$'$ are both lattice points and the pattern is infinitely long parallel to AA$'$. It should also be noted that axes 4_2 and 6_3 include the pure rotation axes 2 and 3 respectively in the lattice. 6_2 and 6_4 also contain twofold rotation axes.

2.12.2   Glide Planes

The operation of glide plane repetition is shown in Figure 2.22c. Again, the magnitude of the translation will be of the order of the lattice parameters. Macroscopically, the glide plane in Figure 2.22c would show itself as a mirror plane.

Glide planes are described as *axial glide planes* if the translation parallel to the mirror is parallel to a single axis of the unit cell and equal to one-half of the lattice parameter in that

Table 2.3   Screw axes in crystals

Name	Symbol	Graphical symbol	Right-handed screw translation along the axis in units of the lattice parameter
Screw diad	2_1		$\dfrac{1}{2}$
Screw triads	3_1		$\dfrac{1}{3}$
	3_2		$\dfrac{2}{3}$
Screw tetrads	4_1		$\dfrac{1}{4}$
	4_2		$\dfrac{2}{4} = \dfrac{1}{2}$
	4_3		$\dfrac{3}{4}$
Screw hexads	6_1		$\dfrac{1}{6}$
	6_2		$\dfrac{2}{6} = \dfrac{1}{3}$
	6_3		$\dfrac{3}{6} = \dfrac{1}{2}$
	6_4		$\dfrac{4}{6} = \dfrac{2}{3}$
	6_5		$\dfrac{5}{6}$

direction. Such glide planes are given the symbols a, b, or c, corresponding to the directions of the glide translations. A *diagonal glide plane* involves a translation of one-half of a face diagonal or one-half of a body diagonal (the latter in the tetragonal and cubic systems), and is given the symbol n. A *diamond glide plane*, given the symbol d, can only occur in orthorhombic F, tetragonal I, and cubic I and F lattices. In orthorhombic F lattices, this involves a translation of one-quarter of a face diagonal, whereas in the tetragonal and cubic lattices this involves a translation of one-quarter of a body diagonal.

2.12.3   Combinations of Symmetry Operations to Form Space Groups

When we combine the operations of rotation with translation and look for the possible consistent combinations we can start with the point group and associate its symmetry elements with each lattice point of the lattices consistent with it. From the number of Bravais lattices and number of point groups detailed in Figures 1.19 and 2.6 it might be expected that 66 space groups could be obtained (2 triclinic, 6 monoclinic, 12 orthorhombic, 14 tetragonal, 5 trigonal with a rhombohedral lattice, 5 trigonal with a hexagonal lattice, 7 hexagonal, and 15 cubic). Seven additional ones arise because of cases in which the glide planes or screw axes *automatically* arising are different for different orientations of the point group

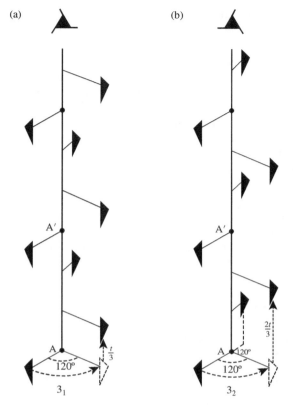

Figure 2.23 Screw axes 3_1 and 3_2: these are screw axes of opposite hand, as shown by the symbols at the top

with respect to the lattice; for example, in the tetragonal system $I\bar{4}2m$ and $I\bar{4}m2$ are distinct space groups. The 14 space groups for which this consideration is relevant are $Cmm2$, $Amm2$ ($\equiv C2mm$), $P\bar{4}2m$, $P\bar{4}m2$, $I\bar{4}2m$, $I\bar{4}m2$, $P\bar{6}m2$, $P\bar{6}2m$, $P312$, $P321$, $P3m1$, $P31m$, $P\bar{3}m1$ and $P\bar{3}1m$. Therefore, there are 73 such space groups, referred to as arithmetic crystal classes in Section 1.3.4 of the sixth edition of the *International Tables for Crystallography* [3].

When the possibilities of introducing screw axes to replace pure rotational axes and glide planes to replace mirror planes are taken into consideration, the result is to produce the total of 230 different space groups [3]. Of particular note are the space groups in the trigonal crystal system. Of the 25 trigonal space groups, only seven have a rhombohedral unit cell ($R3$, $R\bar{3}$, $R32$, $R3m$, $R3c$, $R\bar{3}m$, $R\bar{3}c$); the majority have a primitive hexagonal unit cell. Buerger [1] argues that 'to refer one crystal of the trigonal class to rhombohedral axes, and another to hexagonal axes, because the lattice types are R and P respectively, leads to confusion' (p. 106) and discourages the use of rhombohedral axes for this reason. It is certainly the case in the scientific literature that the use of rhombohedral axes for crystals where the space group is one of the seven rhombohedral space groups is markedly less prevalent than the use of hexagonal axes.

In centred cells, the possibility of a 'double' glide plane arises, given the symbol *e*. This symmetry operation is recognized officially in the fifth and sixth editions of the *International Tables for Crystallography* and incorporated into the conventional space group symbols of five orthorhombic space groups, *Aem*2, *Aea*2, *Cmce*, *Cmme*, and *Ccce*. In this operation, two glide reflections occur through the plane under consideration, with glide vectors perpendicular to one another related by a centring translation [3]. Seventeen space groups have these 'double' glide planes, recognized in the footnotes to Tables 2.1.2.2 and 2.1.2.3 of [3], but in 12 of these cases the 'double' glide planes are parallel to mirror planes. Therefore, their existence is not immediately apparent from the space group symbol because the mirror planes are stronger symmetry elements than the 'double' glide planes.

The arrangements of all of the symmetry elements in the 230 space groups are listed in the various editions of the *International Tables for Crystallography*. The numbers allocated to the various space groups are such that the space groups are in the sequence introduced by Schoenflies in 1891 [13], starting with the two triclinic space groups and finishing with the 36 cubic space groups.

Of particular interest to most materials scientists are the coordinates of general and special positions within each unit cell and diagrams showing the symmetry elements. More recent editions of the *International Tables for Crystallography* have extensive descriptions of space groups, which also include listings of maximal subgroups and minimal supergroups; these concepts will be discussed in general terms in Sections 2.14 and 2.15.

2.12.4   The Relationship Between the Space Group Symbol and the Point Group Symmetry for a Crystal

The conventional space group symbol shows that the space groups have been built up by placing a point group at each of the lattice points of the appropriate Bravais lattice. Thus, for example, $Fm\bar{3}m$ (in full, $F\,4/m\,\bar{3}\,2/m$) means the cubic face-centred lattice with the point group $m\bar{3}m$ associated with each lattice point; $P6_3/mmc$ (in full, $P\,6_3/m\,2/m\,2/c$) is a hexagonal primitive lattice derived from $P6/mmm$ by replacing the sixfold rotation axis by 6_3 and one of the mirror planes by a *c*-axis glide plane. $P6_3/mmc$ and $P6/mmm$ share the same point group symmetry. The point group symmetry of any crystal is derived immediately from the space group symbol by replacing screw axes by the appropriate rotational axes and glide planes by mirror planes in the space group symbol.

2.13   The 17 Two-Dimensional Space Groups

The essential elements of how the assignment of a crystal structure to a particular space group imposes restrictions on the possible general and special positions of atoms or ions within a particular unit cell can be illustrated by considering the 17 two-dimensional space groups (or plane groups). These space groups are shown in Figure 2.24 in a manner similar to how they are depicted in the *International Tables for Crystallography* [3]. If the plane groups are understood, the essential elements of the three-dimensional space group tables in the *International Tables* can be understood without difficulty.

Space groups are obtained by the application of point group symmetry to finite lattices, the possibility of translation symmetry being taken into account. There are five lattices or nets in two dimensions, shown in Figure 1.14. Following the convention in the *International*

Tables for Crystallography, we shall give them the symbol *p* for primitive and *c* for centred. The symbols for twofold, threefold, fourfold and sixfold axes and for the mirror plane are as in Section 2.1. In the most recent editions of the *International Tables for Crystallography*, the parallelogram net in Figure 1.14a is referred to as an oblique net, and the triequiangular net in Figure 1.14c is referred to as a hexagonal net.

The only additional symmetry element in two dimensions is the glide reflection line: symbol *g* and denoted by a dashed line in Figure 2.24. It involves reflection and a translation of one-half of the repeat distance parallel to the line.

The two-dimensional lattices and the two-dimensional point groups are combined in Table 2.4 to show the space groups that can arise, which are shown in Figure 2.24. In all the diagrams, the *x*-axis runs down the page and the *y*-axis runs across the page, the positive *y*-direction being towards the right. In each of the diagrams the left-hand one shows the equivalent general positions of the space group; that is, the complete set of positions produced by the operation of the symmetry elements of the space group upon one initial position chosen at random. The total number of general positions is the number falling within the cell, but surrounding positions are also shown to illustrate the symmetry. The right-hand diagram is that of the group of spatially distributed symmetry operators; that is, the true space (plane) group.

Below each of the diagrams in Figure 2.24, the equivalent general positions and special positions are also indicated. Special positions are positions located on at least one symmetry operator so that repetition of an initial point produces fewer equivalent positions than in the general case. The symmetry at each special position is also given.

The group *p*1 is obtained by combining the parallelogram net and a onefold axis of rotational symmetry. There are no special positions in the cell. The group *p*2 arises by combining the parallelogram net and a diad. A mirror plane *requires* the rectangular net (Section 1.5) and if this net is combined with a *single* mirror, the space group *pm*, No. 3, results. Points ○ and ⊙ are in mirror relationship to one another. If the mirror in *pm* is replaced by a glide reflection line, the space group *pg* results. In No. 4 the glide reflection line *g* runs normal to the *x*-axis. The centred rectangular lattice necessarily shows a glide reflection line as in *cm*, No. 5, but only one of these needs to be be present, corresponding to the plane point group *m*. Since, in this case, the net is multiply primitive, the motif associated with the lattice point at $(\frac{1}{2},\frac{1}{2})$ is the same as that at $(0, 0)$. Hence the coordinates of equivalent positions, if added to $(\frac{1}{2},\frac{1}{2})$, give additional equivalent positions.

If two mirror planes at right angles (point group 2*mm*) are combined with the rectangular lattice we get diads at the intersections of the mirrors as in *pmm*, No. 6. If one or both of the mirrors is replaced by *g*, the diads no longer lie at the intersections (see *pmg*, No. 7 and *pgg*, No. 8). The group *cmm*, No. 9, necessarily involves the presence of two sets of glide reflection lines, while *p*4, No. 10, denotes the square lattice and point group 4, which together necessarily involve the presence of diads. However, mirror planes are not *required*. If 4 lies at the intersection of two sets of mirrors we have *p*4*m*, No. 11, the diagonal glide reflection line necessarily being present. However, 4 can also lie at the intersection of two sets of glide reflection lines, in which case again two sets of mirrors arise but the mirrors intersect in diads, giving point group symmetry *mm* at these points (No. 12). The triequiangular net and point group 3 give the space group *p*3 (No. 13). If mirror planes are combined with the triad axis – the combination of the point group 3*m* and the triequiangular net – it is found that the mirrors can be arranged in two different ways with respect to the

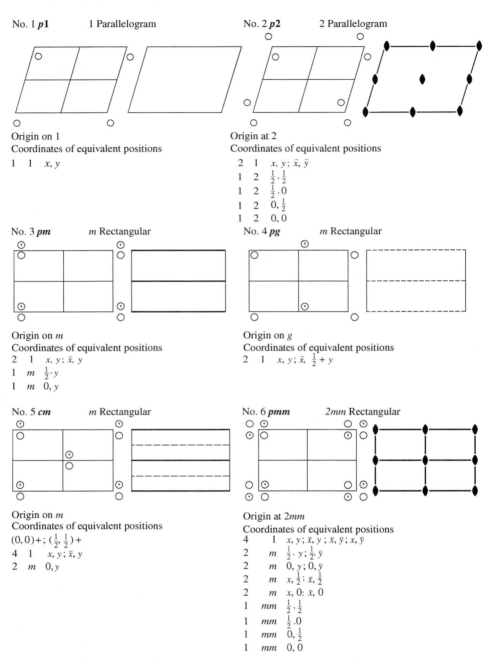

No. 1 *p*1   1 Parallelogram

Origin on 1
Coordinates of equivalent positions
1 1 x, y

No. 2 *p*2   2 Parallelogram

Origin at 2
Coordinates of equivalent positions
2 1 $x, y; \bar{x}, \bar{y}$
1 2 $\frac{1}{2}, \frac{1}{2}$
1 2 $\frac{1}{2}, 0$
1 2 $0, \frac{1}{2}$
1 2 $0, 0$

No. 3 *pm*   *m* Rectangular

Origin on *m*
Coordinates of equivalent positions
2 1 $x, y; \bar{x}, y$
1 *m* $\frac{1}{2}, y$
1 *m* $0, y$

No. 4 *pg*   *m* Rectangular

Origin on *g*
Coordinates of equivalent positions
2 1 $x, y; \bar{x}, \frac{1}{2} + y$

No. 5 *cm*   *m* Rectangular

Origin on *m*
Coordinates of equivalent positions
$(0,0)+; (\frac{1}{2}, \frac{1}{2})+$
4 1 $x, y; \bar{x}, y$
2 *m* $0, y$

No. 6 *pmm*   2*mm* Rectangular

Origin at 2*mm*
Coordinates of equivalent positions
4 1 $x, y; \bar{x}, y; \bar{x}, \bar{y}; x, \bar{y}$
2 *m* $\frac{1}{2}, y; \frac{1}{2}, \bar{y}$
2 *m* $0, y; 0, \bar{y}$
2 *m* $x, \frac{1}{2}; \bar{x}, \frac{1}{2}$
2 *m* $x, 0; \bar{x}, 0$
1 *mm* $\frac{1}{2}, \frac{1}{2}$
1 *mm* $\frac{1}{2}, 0$
1 *mm* $0, \frac{1}{2}$
1 *mm* $0, 0$

Figure 2.24 The 17 two-dimensional space groups arranged following the *International Tables for Crystallography* [3]. The headings for each figure read, from left to right, Number, Short Symbol (see Table 2.4), Point Group, Net. Below each figure the columns give the number of equivalent positions, the point group symmetry at those positions and the coordinates of the equivalent positions. The coordinates of positions *x* and *y* are expressed in units equal to the cell edge length in these two directions

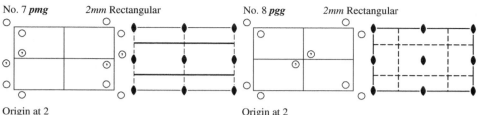

No. 7 *pmg*   2*mm* Rectangular

No. 8 *pgg*   2*mm* Rectangular

Origin at 2
Coordinates of equivalent positions

4	1	$x, y; \bar{x}, \bar{y}; \frac{1}{2}+x, \bar{y}; \frac{1}{2}-x, y$
2	m	$\frac{1}{4}, y; \frac{3}{4}, \bar{y}$
2	2	$0, \frac{1}{2}; \frac{1}{2}, \frac{1}{2}$
2	2	$0, 0; \frac{1}{2}, 0$

Origin at 2
Coordinates of equivalent positions

4	1	$x, y; \bar{x}, \bar{y}; \frac{1}{2}+x, \frac{1}{2}-y; \frac{1}{2}-x, \frac{1}{2}+y$
2	2	$\frac{1}{2}, 0; 0, \frac{1}{2}$
2	2	$0, 0; \frac{1}{2}, \frac{1}{2}$

No. 9 *cmm*   2*mm* Rectangular

No. 10 *p4*   4 Square

Origin at 2*mm*
Coordinates of equivalent positions
$(0, 0) +; (\frac{1}{2}, \frac{1}{2}) +$

8	1	$x, y; \bar{x}, y; \bar{x}, \bar{y}; x, \bar{y}$
4	m	$0, y; 0, \bar{y}$
4	m	$x, 0; \bar{x}, 0$
4	2	$\frac{1}{4}, \frac{1}{4}; \frac{1}{4}, \frac{3}{4}$
2	mm	$0, \frac{1}{2}$
2	mm	$0, 0$

Origin at 4
Coordinates of equivalent positions

4	1	$x, y; \bar{x}, \bar{y}; y, \bar{x}; \bar{y}, x$
2	2	$\frac{1}{2}, 0; 0, \frac{1}{2}$
1	4	$\frac{1}{2}, \frac{1}{2}$
1	4	$0, 0$

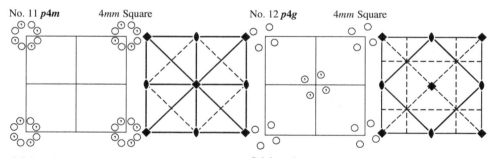

No. 11 *p4m*   4*mm* Square

No. 12 *p4g*   4*mm* Square

Origin at 4*mm*
Coordinates of equivalent positions

8	1	$x, y; \bar{x}, \bar{y}; y, \bar{x}; \bar{y}, x; \bar{x}, y; x, \bar{y}; \bar{y}, \bar{x}; y, x$
4	m	$x, x; \bar{x}, \bar{x}; \bar{x}, x; x, \bar{x}$
4	m	$x, \frac{1}{2}; \bar{x}, \frac{1}{2}; \frac{1}{2}, x; \frac{1}{2}, \bar{x}$
4	m	$x, 0; \bar{x}, 0; 0, x; 0, \bar{x}$
2	mm	$\frac{1}{2}, 0; 0, \frac{1}{2}$
1	4*mm*	$\frac{1}{2}, \frac{1}{2}$
1	4*mm*	$0, 0$

Origin at 4
Coordinates of equivalent positions

8	1	$x, y; y, \bar{x}; \frac{1}{2}-x, \frac{1}{2}+y; \frac{1}{2}-y, \frac{1}{2}-x;$ $\bar{x}, \bar{y}; \bar{y}, x; \frac{1}{2}+x, \frac{1}{2}-y; \frac{1}{2}+y, \frac{1}{2}+x$
4	m	$x, \frac{1}{2}+x; \bar{x}, \frac{1}{2}-x; \frac{1}{2}+x, \bar{x}; \frac{1}{2}-x, x$
2	mm	$\frac{1}{2}, 0; 0, \frac{1}{2}$
2	4	$0, 0; \frac{1}{2}, \frac{1}{2}$

Figure 2.24   (*Continued*)

No. 13 *p3* 3 Triequiangular No. 14 *p3m1* 3*m* Triequiangular

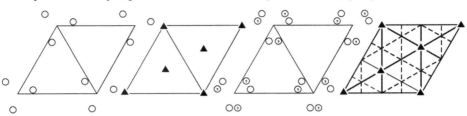

Origin at 3
Coordinates of equivalent positions

3	1	$x, y;\ \bar{y}, x - y;\ y - x, \bar{x}$
1	3	$\frac{2}{3}, \frac{1}{3}$
1	3	$\frac{1}{3}, \frac{2}{3}$
1	3	$0, 0$

Origin at 3*m*
Coordinates of equivalent positions

6	1	$x, y;\ \bar{y}, x - y;\ y - x, \bar{x};$
		$x, x - y;\ y - x, y;\ \bar{y}, \bar{x}$
3	*m*	$x, \bar{x};\ x, 2x;\ 2\bar{x}, \bar{x}$
1	3*m*	$\frac{2}{3}, \frac{1}{3}$
1	3*m*	$\frac{1}{3}, \frac{2}{3}$
1	3*m*	$0, 0$

No. 15 *p31m* 3*m* Triequiangular No. 16 *p6* 6 Triequiangular

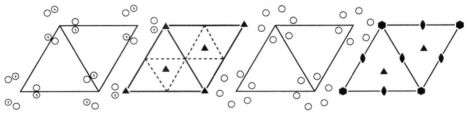

Origin at 31*m*
Coordinates of equivalent positions

6	1	$x, y;\ \bar{y}, x - y;\ y - x, \bar{x};$
		$y, x;\ \bar{x}, y - x;\ x - y, \bar{y}$
3	*m*	$x, 0;\ 0, x;\ \bar{x}, \bar{x}$
2	3	$\frac{1}{3}, \frac{2}{3};\ \frac{2}{3}, \frac{1}{3}$
1	3*m*	$0, 0$

Origin at 6
Coordinates of equivalent positions

6	1	$x, y;\ \bar{y}, x - y;\ y - x, \bar{x};$
		$\bar{x}, \bar{y};\ y, y - x;\ x - y, x$
3	2	$\frac{1}{2}, 0;\ 0, \frac{1}{2};\ \frac{1}{2}, \frac{1}{2}$
2	3	$\frac{1}{3}, \frac{2}{3};\ \frac{2}{3}, \frac{1}{3}$
1	6	$0, 0$

No. 17 *p6m* 6*mm* Triequiangular

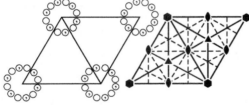

Origin at 6*mm*
Coordinates of equivalent positions

12	1	$x, y;\ y, x - y;\ y - x, \bar{x};\ y, x;\ \bar{x}, y - x;$
		$x - y, \bar{y};\ \bar{x}, \bar{y};\ y, y - x;\ x - y, x;$
		$\bar{y}, \bar{x};\ x, x - y;\ y - x, y$
6	*m*	$x, \bar{x};\ x, 2x;\ 2\bar{x}, \bar{x};\ \bar{x}, x;\ \bar{x}, 2\bar{x};\ 2x, x$
6	*m*	$x, 0;\ 0, x;\ \bar{x}, \bar{x};\ \bar{x}, 0;\ 0, \bar{x};\ x, x$
3	*mm*	$\frac{1}{2}, 0;\ 0, \frac{1}{2};\ \frac{1}{2}, \frac{1}{2}$
2	3*m*	$\frac{1}{3}, \frac{2}{3};\ \frac{2}{3}, \frac{1}{3}$
1	6*mm*	$0, 0$

Figure 2.24 (*Continued*)

Table 2.4 Two-dimensional lattices, point groups and space groups

System and lattice symbol	Point group	Space group symbols		Space group number
		Full	Short	
Parallelogram (oblique) *p* (primitive)	1	*p*1	*p*1	1
	2	*p*211	*p*2	2
Rectangular *p* and *c* (centred)	*m*	*p*1*m*1	*pm*	3
		*p*1*g*1	*pg*	4
		*c*1*m*1	*cm*	5
	2*mm*	*p*2*mm*	*pmm*	6
		*p*2*mg*	*pmg*	7
		*p*2*gg*	*pgg*	8
		*c*2*mm*	*cmm*	9
Square *p*	4	*p*4	*p*4	10
	4*mm*	*p*4*mm*	*p*4*m*	11
		*p*4*gm*	*p*4*g*	12
Triequiangular (hexagonal) *p*	3	*p*3	*p*3	13
	3*m*	*p*3*m*1	*p*3*m*1	14
		*p*31*m*	*p*31*m*	15
	6	*p*6	*p*6	16
	6*mm*	*p*6*mm*	*p*6*m*	17

Note: The two distinct space groups *p*3*m*1 and *p*31*m* correspond to different orientations of the point group relative to the lattice. This does not lead to distinct groups in any other case.

points of the net, yielding *p*31*m* and *p*3*m*1 (Nos. 14 and 15). With the hexagonal point group 6*mm*, which necessarily has two sets of mirrors, this duality does not arise and the two space groups are *p*6 and *p*6*m* (Nos. 16 and 17).

Subtle differences in the nomenclature used here for the two-dimensional space groups and that used in the various editions of the *International Tables for Crystallography,* such as choosing to describe space group No. 7 as *p*2*mg* rather than *pmg*, reflect the fact that for this two-dimensional space group the presence of a mirror plane at right angle to a glide plane generates the diad. Similarly, in space group No. 17, the presence of one set of mirrors and the sixfold axes generates the second set of mirrors, and so the space group can be described either as *p*6*m* or *p*6*mm* (Table 2.4).

2.14   Nomenclature for Point Groups and Space Groups

The nomenclature we have introduced in this chapter to describe point groups and space groups conforms to a notation known as the Hermann–Mauguin notation arising from the work of Carl Hermann [14] and Charles-Victor Mauguin [15] and used in the *Internationale Tabellen zur Bestimmung von Kristallstrukturen*, edited by Hermann and published in 1935. However, another notation for describing point group and space group symmetry is also in common use. This second notation, known as the Schoenflies notation, is named after Arthur Moritz Schoenflies [13]. This latter notation is used in particular in physical chemistry in connection with molecular symmetry and molecular spectroscopy

[16], and it is also useful when considering group theoretical aspects of point groups and space groups [17]. The equivalence between the Hermann–Mauguin notation and the Schoenflies notation is discussed in detail by Hans Burzlaff and Helmuth Zimmermann in chapter 3.3 of [3]. The reason why the Hermann–Mauguin notation might be preferred over the Schoenflies notation is neatly summarized by Phillips [2]: '*each* space group symbol [in the Hermann–Mauguin notation] conveys all the essential information, whereas we must always have a catalogue at hand to follow an arbitrary serial numbering such as that of Schoenflies' (p. 340).

2.15   Groups, Subgroups, and Supergroups

It is evident from the discussion in this chapter on point groups and space groups that each point group and space group consists of a self-consistent finite set of symmetry operations. Therefore, point groups and space groups can be described in the formalism of group theory [17–19]. A group G is a set of elements $g_1, g_2, g_3 \cdots g_n$ which conform to the following four conditions [19]:

1. The product $g_i g_j$ of any two group elements must be another element within the group.
2. Group multiplication is associative: $(g_i g_j) g_k = g_i (g_j g_k)$.
3. There is a unique group element I called the identity element belonging to G so that its operation on a member of the group g is such that $I g_i = g_i I = g_i$.
4. Each element within the group has a unique inverse; that is, for each g_i there is a unique element g_i^{-1} so that $g_i g_i^{-1} = g_i^{-1} g_i = I$.

Thus, for example, following the example of the point group *mmm* used by de Jong [5], p. 20, the eight group elements within this point group are: the identity, I; three 180° rotations about the *x*-, *y*- and *z*-axes; three reflections in the (100), (010) and (001) planes; and a centre of symmetry. A subgroup of these symmetry elements is found in *2mm*: the identity, a 180° rotation about the *x*-axis and two reflections in the (010) and (001) planes. Other subgroups are the point groups 222, 2/*m*, *m*, 2, $\bar{1}$, and 1. Conversely, the point group *mmm* is clearly a supergroup for *2mm* and 222.

Similar principles apply to subgroups and supergroups of space groups. Thus, for example, following Hans Wondratschek in chapter 8 in Volume A of the revised, fifth edition of the *International Tables for Crystallography* [20], the space group $P\bar{6}2m$ is one member of the group whose minimal nonisomorphic supergroup is $P6_3/mcm$; it has as maximal nonisomorphous subgroups $P\bar{6}$, $P321$, $P31m$ and $Pm2m$. (Groups are termed isomorphic if they have an equivalent group multiplication structure, but differ in the nature of the elements constituting them [19]. Thus, for example, the point groups 2/*m*, 222 and *2mm* are isomorphic: all three have four elements within the group and the same multiplication structure within them.)

2.16 An Example of a Three-Dimensional Space Group

To illustrate the principles we have outlined in Sections 2.12–2.14, an example of an entry from in Volume A of the revised, fifth edition of the *International Tables for Crystallography* [20] is shown in Figure 2.25. In the more recent sixth, revised edition of these *Tables*, the data tabulated in the revised, fifth edition on subgroups and supergroups of the two- and three-dimensional space groups and shown in Figure 2.25 has been moved to Volume A1.

It is apparent from the Hermann–Mauguin notation of this space group, *Pnma*, No. 62, that crystals with this space group symmetry belong to the orthorhombic crystal system, point group *mmm*. The Schoenflies notation D_{2h}^{16} indicates that the space group is dihedral (or two-sided; from the '*D*'), that it has a twofold rotation axis (from the '2'), that there is in addition a mirror plane perpendicular to this twofold axis (from the '*h*' subscript) and that it is the 16th of 28 such space groups $D_{2h}^1 \ldots D_{2h}^{28}$ (space groups 47–74 inclusive), all of which are space groups derived from the *mmm* point group.

The full Hermann–Mauguin notation for this space group is $P2_1/n\ 2_1/m\ 2_1/a$, showing that there are three mutually perpendicular sets of screw diads in addition to the mirror, *m*, and the diagonal, *n*, and axial, *a*, glide planes. The diagrams of the symmetry elements of the group follow the principles in Figure 2.24, with the use of standard graphical symbols for *n* (parallel to (100)) and *a* (parallel to (001)). The projections in the top left-hand corner and the bottom right-hand corner are both down [001], the one in the top right-hand corner down [010] and the one in the bottom left-hand corner down [100].

The asymmetric unit is 'a (simply) connected smallest part of space from which, by application of all symmetry operations of the space group, the whole of space is filled exactly' [20]. The listing under the heading 'Symmetry Operations' in Figure 2.25 is a summary of the various geometric descriptions of each of the eight symmetry operations in the space group, numbered (1)–(8). Thus, for example, here the second of these, $2\left(0, 0, \frac{1}{2}\right) \frac{1}{4}, 0, z$, defines the screw diad parallel to [001] intersecting the *x*–*y* plane at ($\frac{1}{4}$, 0). The 'Generators Selected' define the order in which the coordinates of the symmetrically equivalent positions are produced; for example, the coordinates related by symmetry to the general position *x*, *y*, *z* in the space group. (1) is the identity operation, showing that in this case integral lattice translations produce the equivalent point in adjacent unit cells. After the 'Positions', which follow the protocol described in Section 2.12, and the 'Symmetry of Special Projections', which are self-explanatory, a summary is given of subgroups and supergroups related to the space group under consideration, following the methodology set out in chapter 2 of [20]. Finally, the reflection conditions listed under 'Positions' specify the systematic absences which occur under kinematical diffraction conditions, such as X-ray diffraction.

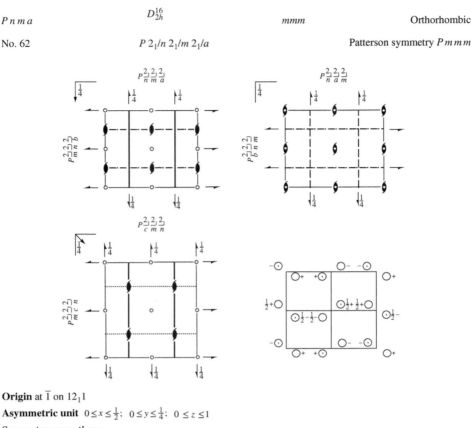

Pnma D_{2h}^{16} *mmm*   Orthorhombic

No. 62 $P\,2_1/n\,2_1/m\,2_1/a$ Patterson symmetry $P\,m\,m\,m$

Origin at $\bar{1}$ on 12_11

Asymmetric unit $0 \le x \le \frac{1}{2}$; $0 \le y \le \frac{1}{4}$; $0 \le z \le 1$

Symmetry operations

(1) 1 (2) $2(0,0,\frac{1}{2})$ $\frac{1}{4},0,z$ (3) $2(0,\frac{1}{2},0)$ $0,y,0$ (4) $2(\frac{1}{2},0,0)$ $x,\frac{1}{4},\frac{1}{4}$

(5) $\bar{1}$ $0,0,0$ (6) a $x,y,\frac{1}{4}$ (7) m $x,\frac{1}{4},z$ (8) $n(0,\frac{1}{2},\frac{1}{2})$ $\frac{1}{4},y,z$

Figure 2.25 An example of a space group. Source: Taken from [20].

2.17 Frequency of Space Groups in Inorganic Crystals and Minerals

Although there are 230 space groups, analyses of the crystal structures of organic compounds and inorganic crystals and minerals show that these structures clearly favour some space groups over others. Thus, for example, almost $\frac{2}{3}$ of inorganic crystal structures belong to one of 24 space groups [21]. For organic compounds just two space groups, $P2_1/c$ and $P\bar{1}$, account for over half the crystal structures solved by 1990 [22], while for monomeric proteins, the space group $P2_12_12_1$ accounts for almost one third of crystal structures solved by 1995 [23]. Packing and entropic considerations of atoms and molecules, discussed in Chapters 3 and 4, both help to rationalize these observations [21–23]. The consequence of these observations and considerations is that some space groups are markedly less important than others for crystal structure determination [21]. In this context, the choice of *Pnma*, No. 62, as the three-dimensional space group to highlight in Section 2.16 is deliberate – this

CONTINUED   No. 62 $Pnma$

Generators selected (1); $t(1,0,0)$; $t(0,1,0)$; $t(0,0,1)$; (2); (3); (5)

Positions

Multiplicity, Wyckoff letter, Site symmetry	Coordinates				Reflection conditions

General:

8	d	1	$(1)\,x,y,z$	$(2)\,\bar{x}+\frac{1}{2},\bar{y},z+\frac{1}{2}$	$(3)\,\bar{x},y+\frac{1}{2},\bar{z}$	$(4)\,x+\frac{1}{2},\bar{y}+\frac{1}{2},\bar{z}+\frac{1}{2}$

$0kl$: $k+l=2n$
$hk0$: $h=2n$

$(5)\,\bar{x},\bar{y},\bar{z}$ $(6)\,x+\frac{1}{2},y,\bar{z}+\frac{1}{2}$ $(7)\,x,\bar{y}+\frac{1}{2},z$ $(8)\,\bar{x}+\frac{1}{2},y+\frac{1}{2},z+\frac{1}{2}$

$h00$: $h=2n$
$0k0$: $k=2n$
$00l$: $l=2n$

Special: as above, plus

4	c	m	$x,\frac{1}{4},z$	$\bar{x}+\frac{1}{2},\frac{3}{4},z+\frac{1}{2}$	$\bar{x},\frac{3}{4},\bar{z}$	$x+\frac{1}{2},\frac{1}{4},\bar{z}+\frac{1}{2}$

no extra conditions

4	b	$\bar{1}$	$0,0,\frac{1}{2}$	$\frac{1}{2},0,0$	$0,\frac{1}{2},\frac{1}{2}$	$\frac{1}{2},\frac{1}{2},0$

hkl : $h+l,k=2n$

4	a	$\bar{1}$	$0,0,0$	$\frac{1}{2},0,\frac{1}{2}$	$0,\frac{1}{2},0$	$\frac{1}{2},\frac{1}{2},\frac{1}{2}$

hkl : $h+l,k=2n$

Symmetry of special projections

Along [001] $p2gm$ Along [100] $c2mm$ Along [010] $p2gg$
$\mathbf{a'}=\frac{1}{2}\mathbf{a}$ $\mathbf{b'}=\mathbf{b}$ $\mathbf{a'}=\mathbf{b}$ $\mathbf{b'}=\mathbf{c}$ $\mathbf{a'}=\mathbf{c}$ $\mathbf{b'}=\mathbf{a}$
Origin at $0,0,z$ Origin at $x,\frac{1}{4},\frac{1}{4}$ Origin at $0,y,0$

Maximal non-isomorphic subgroups

I   [2] $Pn2_1a$ $(Pna2_1, 33)$ 1; 3; 6; 8
      [2] $Pnm2_1$ $(Pmn2_1, 31)$ 1; 2; 7; 8
      [2] $P2_1ma$ $(Pmc2_1, 26)$ 1; 4; 6; 7
      [2] $P2_12_12_1$ (19)   1; 2; 3; 4
      [2] $P112_1/a$ $(P2_1/c, 14)$ 1; 2; 5; 6
      [2] $P2_1/n11$ $(P2_1/c, 14)$ 1; 4; 5; 8
      [2] $P12_1/m1$ $(P2_1/m, 11)$ 1; 3; 5; 7

IIa   none
IIb   none

Maximal isomorphic subgroups of lowest index
IIc   [3] $Pnma$ ($\mathbf{a'}=3\mathbf{a}$) (62); [3] $Pnma$ ($\mathbf{b'}=3\mathbf{b}$) (62) ;[3] $Pnma$ ($\mathbf{c'}=3\mathbf{c}$) (62)

Minimal non-isomorphic supergroups
I   none
II   [2] $Amma$ $(Cmcm, 63)$; [2] $Bbmm$ $(Cmcm, 63)$; [2] $Ccme$ $(Cmce, 64)$; [2] $Imma$ (74); [2] $Pcma$ ($\mathbf{b'}=\frac{1}{2}\mathbf{b}$) $(Pbam, 55)$;

      [2] $Pbma$ ($\mathbf{c'}=\frac{1}{2}\mathbf{c}$) $(Pbcm, 57)$; [2] $Pnmm$ ($\mathbf{a'}=\frac{1}{2}\mathbf{a}$) $(Pmmn, 59)$

Figure 2.25   *(Continued)*

space group is the most frequent space group found in inorganic crystals and minerals, accounting for just over 7% of all crystal structures [21], and it is relatively easy to represent in two dimensions in terms of information in the form of diagrams and symmetry data.

2.18 Magnetic Groups

These point groups and space groups are an extension of the 10 two-dimensional and 32 three-dimensional point groups and the 17 two-dimensional space groups and 230 three-dimensional space groups. They recognize the effect of an *antisymmetry* operation on these groups.

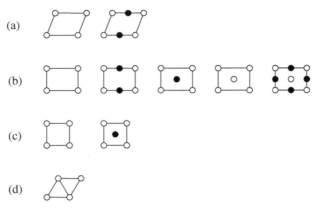

Figure 2.26 The 10 black-and-white plane lattices: (a) parallelogram (oblique), (b) rectangular, (c) square, (d) equiangular (hexagonal) (after [24])

Antisymmetry defined by Alan Mackay is 'the correspondence of faces, points or other crystallographic objects having some property denoted by a positive sign to other faces, points or objects symmetrically related in position, but having the same property denoted with the opposite sign' [24].

An example of an antisymmetry operation is magnetic spin, so that each atom site can have one of two possible values: spin 'up' or spin 'down' when the magnetic moments are not randomly arranged, but are instead aligned. A further example is where one atom is coloured one colour, say black, and the other identical atom symmetrically related in position is coloured a different colour, say white.

These groups are known by a variety of names:

- magnetic groups,
- antisymmetry groups,
- Shubnikov groups (after Aleksei Vasilevich Shubnikov, 1887–1970),
- Heesch groups (after Heinrich Heesch, 1906–1995, who introduced the concept of antisymmetry in his 1930 paper [25]),
- dichromatic groups,
- black-and-white groups.

Groups where equal and opposite black and white objects are exactly superposed are termed grey groups.

The principle behind the extension of the 'conventional' point groups and space groups can be illustrated in two dimensions by considering the five two-dimensional Bravais lattices (Figure 1.14). If these five two-dimensional lattices are centred by (lattice) points of a different colour, the number of permissible Bravais lattices increases from 5 to 10, as shown below in Figure 2.26 [24].

Here, the five 'conventional' two-dimensional Bravais lattices consistent with translational symmetry have lattice points shown as circles. The five new Bravais lattices which arise from the centring are shown as lattices with the same numbers of circles and dots per unit cell.

In two dimensions there are in total 46 black-and-white point groups, to which can be added 17 'ordinary' plane groups and 17 grey groups, giving 80 plane groups in total. In three dimensions there are 1651 Shubnikov space groups: 230 'ordinary' space groups, 230

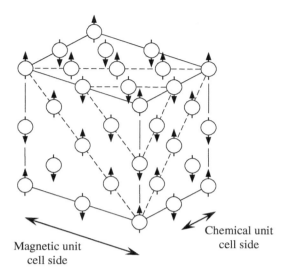

Figure 2.27 The effect of antiferromagnetic coupling on the size of the unit cell in MnO. In alternate {111} planes, shown as dotted lines, Mn atoms are either spin 'up' or spin 'down'. As a result, the magnetic unit cell has a cell size twice that of the chemical unit cell. For ease of representation, only Mn atoms are shown in this diagram

grey groups, 674 groups where antisymmetry alone generates further groups and 517 where antitranslations and antisymmetry generate further groups [26].

An example of a material where these considerations are relevant is MnO below its Curie temperature of 120 °C. Above 120 °C MnO has the NaCl crystal structure (Section 3.5.1, Figure 3.16). As is shown in Figure 2.27, the effect of antisymmetry, in this case antiferromagnetic coupling, has the effect of doubling the actual magnetic unit cell relative to the chemical unit cell [27].

An extended introduction to magnetic groups is given in chapter 3.6 of [3]. A complete discussion and descriptions of the magnetic groups is given in the recent book by Litvin [28]. The book by Shubnikov and Kopstik has, in addition, a discussion of coloured symmetry groups – groups relating objects of more than two colours, together with a full set of references to prior work in this area [26].

Problems

Material in Appendices 1–3 may be useful; in Problem 2.12 the material in Appendix 4 may also be useful.

2.1 Using Figure 2.13, confirm that in the four-index notation for a plane with indices (*hkil*):

$$i = -(h + k)$$

(Eq. (2.2).) Confirm also that the condition for a four-index vector [*UVTW*] to lie in a four-index plane (*hkil*) is:

$$hU + kV + iT + lW = 0$$

(Eq. (2.5).)

2.2 Show that the angle γ between any of the crystal axes and the unique triad axis in a trigonal crystal is given by:

$$\sin \gamma = \frac{2}{\sqrt{3}} \sin \alpha/2$$

2.3 The corners of a wooden model of a cube are cut off to make equal faces of the form $\{111\}$. The first four new faces are cut in positions (111), $(11\bar{1})$, $(\bar{1}11)$ and $(\bar{1}1\bar{1})$, in that order. Draw sketch stereograms to show the symmetry elements shown by the model after each new face is formed. What is the crystal system of the model in each case?

2.4 (a) Sketch stereograms showing the operation of the undermentioned symmetry elements on *two* distinct poles, each in a general position, and insert the symmetry elements themselves, except the centre of symmetry:

centre of symmetry
vertical plane of symmetry
horizontal plane of symmetry
vertical axes 2, 3, 4, 6
vertical inversion axes $\bar{2}, \bar{3}, \bar{4}, \bar{6}$
horizontal axes 2, 4
horizontal inversion axis $\bar{4}$.

   (b) The symbol $\bar{X}/m$ is not used in conventional point group notation. Draw sketch stereograms of the poles of one general form for each possible value of X and hence derive the conventional symbols that are actually used.

2.5 (a) In which of the trigonal classes 3, $\bar{3}$, 3*m*, 32, and $\bar{3}m$ does a rhombohedron occur as a special form? In what class is the rhombohedron the general form?
   (b) Which of the tetragonal classes 4, $\bar{4}$, 4/*m*, 422, 4*mm*, $\bar{4}2m$, and 4/*mmm* have closed forms as their general forms?
   (c) To which of the cubic classes 23, *m*$\bar{3}$, 432, $\bar{4}3m$ and *m*$\bar{3}m$ could a crystal belong which had the shape of (i) an octahedron, (ii) a tetrahedron or (iii) an icositetrahedron? (The icositetrahedron is the name given to the $\{112\}$ form with 24 faces.)

2.6 In a holosymmetric cubic crystal the angle between (110) and a face P which lies in the zone $[010]$ is $53.97°$. Find the indices of P and calculate the angle between P and the face P' related to P by the mirror plane parallel to $(1\bar{1}0)$. Determine the indices $[uvw]$ of the zone containing P and P' and calculate the interzone angle $[uvw]^\wedge[1\bar{1}0]$.

2.7 (a) Calculate the angle between $[0001]$ and $[11\bar{2}3]$ in beryllium (hexagonal, $a = 2.28\,\text{Å}$, $c = 3.57\,\text{Å}$). What face lies at the intersection of the zones $[11\bar{2}3]$ and $[\bar{2}113]$?
   (b) In a crystal of calcite (trigonal; $c/a = 0.8543$), a face lies in the zone between $(10\bar{1}1)$ and $(\bar{1}101)$ at an angle of $16.50°$ from $(10\bar{1}1)$. Determine the indices of this face.

2.8 In an orthorhombic crystal (topaz) with axial ratios $a : b : c = 0.529 : 1 : 0.477$, the following angles were measured to the face P of the general form: $(100)^\wedge P = 67.85°$, $(010)^\wedge P = 66.5°$. Determine the indices of P.

2.9 Calculate the axial ratios $a : b : c$ and the axial angle β for a monoclinic crystal (gypsum) given that $(110)\wedge(1\bar{1}0) = 68.5°$, $(001)\wedge(1\bar{1}0) = 82.3°$ and $(001)\wedge(\bar{1}01) = 33.1°$.

2.10 (a) In a hexagonal zeolite crystal the angle $(10\bar{1}0)\wedge(10\bar{1}1)$ is found to be $37.2°$. Calculate the axial ratio c/a and specify the indices of the zone containing both faces in (i) conventional three-dimensional indices and (ii) the Weber 4-index notation.

   (b) Show that $(0001)\wedge(11\bar{2}2) = 48.8°$. Hence, construct a stereogram centred on 0001 showing all the poles of the form $\{10\bar{1}0\}$, $\{10\bar{1}1\}$, $\{11\bar{2}0\}$ and $\{11\bar{2}2\}$ for this zeolite.

2.11 α-sulfur forms orthorhombic holosymmetric crystals with $a = 10.48\,\text{Å}$, $b = 12.92\,\text{Å}$ and $c = 24.55\,\text{Å}$.

   (a) Draw a sketch stereogram showing the symmetry elements shown by α-sulphur.

   (b) Calculate the angles $(001)\wedge(011)$ and $(100)\wedge(110)$. Insert these poles on a stereogram.

   (c) Hence, draw an accurate stereogram of sulphur showing all the faces of the forms $\{100\}$, $\{010\}$, $\{001\}$, $\{101\}$, $\{110\}$, $\{111\}$, $\{011\}$ and $\{113\}$. Index all the faces in the upper hemisphere. Which of these are general and which are special forms?

2.12 In dealing with some imperfect crystals with a cubic face-centred lattice it is convenient to use a hexagonal unit cell. The cubic face-centred lattice can be referred to a hexagonal cell where the z-axis is parallel to [111] and of magnitude $\sqrt{3}a$, where a is the lattice parameter of the conventional cubic unit cell and the x- and y-axes are parallel to the $\langle 101 \rangle$ directions perpendicular to [111] and of magnitude $a/\sqrt{2}$.

   (a) Draw a diagram showing the relation between the two unit cells.

   (b) Write the hexagonal lattice vectors in terms of the cubic lattice vectors. Hence, derive the matrix for transforming the indices of lattice planes.

   (c) Find the ratio of the volumes of the two unit cells by the matrix method and check by direct calculation. How many lattice points does each contain?

   (d) Obtain the hexagonal indices of the planes with indices (112), (100) and $(1\bar{1}0)$ referred to the conventional cubic cell.

Suggestions for Further Reading

See the suggestions for Chapter 1 as well as the following:

Janssen, T. (1973). *Crystallographic Groups*. North-Holland, Amsterdam.

Wooster, W.A. (1973). *Tensors and Group Theory for the Physical Properties of Crystals*. Oxford: Clarendon Press.

References

[1] Buerger, M.J. (1963). *Elementary Crystallography*. New York: Wiley.

[2] Phillips, F.C. (1971). *Introduction to Crystallography*, 4th Edition. Edinburgh: Oliver & Boyd.

[3] Aroyo, M.I. (ed.) (2016). *International Tables for Crystallography*, 6th, Revised Edition, Vol. A: *Space-Group Symmetry*, published for the International Union of Crystallography. Chichester: Wiley International.

[4] Henry, N.F.M. and Lonsdale, K. (eds.) (1952). *International Tables for X-Ray Crystallography, Volume I, International Union for Crystallography*. Birmingham, England: Kynoch Press.

[5] de Jong, W.F. (1959). *General Crystallography: A Brief Compendium*. New York: W.H. Freeman.

[6] Bravais, A. (1866). *Études Cristallographiques*, p. 119. Paris: Gauthier-Villars.

[7] Friedel, G. (1926). *Leçons de Cristallographie*, p. 80. Nancy-Paris-Strasbourg: Berger-Levrault.

[8] Weber, L. (1922). Das viergliedrige Zonensymbol des hexagonalen Systems. *Z. Kristallogr.* 57: 200–203.

[9] Ewald, P.P. (ed.) (1962). *Fifty Years of X-Ray Diffraction*. Utrecht: N.V.A. Oosthoek.

[10] Carpenter, M.A. and Howard, C.J. (2009). Symmetry rules and strain/order-parameter relationships for coupling between octahedral tilting and cooperative Jahn–Teller transitions in ABX_3 perovskites. I. Theory. *Acta Crystallogr. B* 65: 134–146.

[11] Carpenter, M.A. and Howard, C.J. (2009). Symmetry rules and strain/order-parameter relationships for coupling between octahedral tilting and cooperative Jahn–Teller transitions in ABX_3 perovskites. II. Application. *Acta Crystallogr. B* 65: 147–159.

[12] Howard, C.J. and Carpenter, M.A. (2010). Octahedral tilting in cation-ordered Jahn–Teller distorted perovskites – a group-theoretical analysis. *Acta Crystallogr. B* 66: 40–50.

[13] Schoenflies, A. (1891). *Krystallsysteme und Krystallstructur*. Leipzig: B.G. Teubner.

[14] Hermann, C. (1928). Zur systematischen Strukturtheorie I. Eine neue Raumgruppensymbolik. *Z. Kristallogr.* 68: 257–287.

[15] Mauguin, C. (1931). Sur le symbolisme des groupes de répétition ou de symétrie des assemblages cristallins. *Z. Kristallogr.* 76: 542–558.

[16] Atkins, P.W., De Paula, J. and Keeler, J. (2017). *Atkins' Physical Chemistry*, 11th Edition. Oxford: Oxford University Press.

[17] Burns, G. and Glazer, A.M. (1990). *Space Groups for Solid State Scientists*, 2nd Edition. New York: Academic Press.

[18] Mathews, J. and Walker, R.L. (1970). *Mathematical Methods of Physics*, 2nd Edition. Reading, MA: Addison-Wesley.

[19] Riley, K.F., Hobson, M.P. and Bence, S.J. (2006). *Mathematical Methods for Physics and Engineering*, 3rd Edition. Cambridge: Cambridge University Press.

[20] Hahn, T. (ed.) (2002). *International Tables for Crystallography*, 5th, Revised Edition, Vol. A: *Space-Group Symmetry*, published for the International Union of Crystallography. Dordrecht: Kluwer Academic Publishers.

[21] Urusov, V.S. and Nadezhina, T.N. (2009). Frequency distribution and selection of space groups in inorganic crystal chemistry. *J. Struct. Chem.* 50: S22–S37.

[22] Padmaja, N., Ramakumar, S. and Viswamitra, M.A. (1990). Space-group frequencies of proteins and of organic compounds with more than one formula unit in the asymmetric unit. *Acta Crystallogr. A* 46: 725–730.

[23] Wukovitz, S.W. and Yeates, T.O. (2005). Why protein crystals favour some space-groups over others. *Nat. Struct. Biol.* 2: 1062–1067.

[24] Mackay, A.L. (1957). Extensions of space-group theory. *Acta Crystallogr.* 10: 543–548.

[25] Heesch, H. (1930). Über die vierdimensionalen Gruppen des dreidimensionalen Raumes. *Z. Kristallogr.* 73: 325–345.

[26] Shubnikov, A.V. and Kopstik, V.A. (1974). *Symmetry in Science and Art*. London: Plenum Press.

[27] Shull, C.G., Strauser, W.A. and Wollan, E.O. (1951). Neutron diffraction by paramagnetic and antiferromagnetic substances. *Phys. Rev.* 83: 333–345.

[28] Litvin, D.B. (2013). *Magnetic Group Tables, 1-, 2- and 3-Dimensional Magnetic Subperiodic Groups and Magnetic Space Groups*. Chester: International Union of Crystallography. Available free online from http://www.iucr.org/publ/978-0-9553602-2-0.

3

Crystal Structures

3.1 Introduction

The relative positions of the atoms in crystals can be deduced from measurements of the relative intensities of the diffraction spectra using X-rays, electrons, or neutrons. X-rays are most useful in this regard because the intensities of the diffracted waves are not as affected by crystalline perfection as are those obtained with electrons. Neutron diffraction is a more specialized technique than X-ray diffraction or electron diffraction, but has distinct advantages when determining the crystal structures of low-atomic-number materials and when locating the atomic positions of elements such as hydrogen in a material.

Careful measurements of the spectra enable the electron density to be found throughout the unit cell of a crystal. Since only the outermost electrons are involved in binding the atoms together in a crystal, most of the electrons in the crystal reside in orbits that are the same as those in an isolated atom. The crystal structure can therefore be described by stating the relative positions of *atoms* within the unit cell. In many crystals of both elements and compounds the charge distribution of the outermost electrons, which are not involved in the chemical binding, are more or less spherically symmetric and therefore the crystal structure is sometimes regarded as being composed of spheres of the same size (if the crystal structure of an element is being considered) or of different sizes (corresponding to the different atoms in a chemical compound) packed together.

The crystal structures of the stable elements and of most simple compounds are now well established. These are listed in the Suggestions for Further Reading given at the end of this chapter. We shall now describe a number of the structures, emphasizing the geometrical features of the relative arrangement of the atoms. The imperfections in these crystals, such as point defects, dislocations, and grain boundaries, will be described in Part II. When

Crystallography and Crystal Defects, Third Edition. Anthony Kelly and Kevin M. Knowles.
© 2020 John Wiley & Sons Ltd. Published 2020 by John Wiley & Sons Ltd.
Companion website: www.wiley.com/go/kelly/crystallography3e

describing each crystal structure we shall state the space group in the style given in the *International Tables for Crystallography* [1] because from this symbol the Bravais lattice and the point group of the crystal are immediately known (see Section 2.12). Other ways of describing crystal structures are described in Section 3.10.

We shall first describe the structures of some of the elements and then those of simple compounds. In many cases a given element or compound possesses more than one crystal structure, each being the thermodynamically stable form in a given regime of temperature and pressure. When this is the case, an element is said to show *allotropy* and a compound is spoken of as *polymorphous*. In describing the crystal structures of some compounds we shall sometimes refer to the chemical formulae as being of the type *MX*, *M* denoting a metal ion or atom and *X* an electro-negative element, such as O, F, Cl and so on. Crystal structures can be described by stating the lattice and the coordinates of the atoms in terms of the cell sides as units of length, referred to the lattice point at the origin. In multiply primitive unit cells the same arrangement of atoms must, of course, occur around every lattice point.

3.2   Common Metallic Structures

The vast majority of the elements are metallic. With notable exceptions such as manganese (Mn), gallium (Ga), indium (In), tin (Sn), mercury (Hg), polonium (Po), protactinium (Pa), uranium (U) and plutonium (Pu) (see Appendix 7), the metallic elements all possess one of four structures: either the cubic close-packed (c.c.p.), hexagonal close-packed (h.c.p.), double hexagonal close-packed (double h.c.p.) or body-centred cubic (b.c.c.). The c.c.p. structure is also frequently referred to as face-centred cubic (f.c.c.); however, we will use the designation c.c.p. here to distinguish clearly between the face-centred cubic[1] Bravais lattice and the c.c.p. crystal structure.

3.2.1   Cubic Close-Packed (*Fm$\bar{3}$m*)

The noble metals (Cu, Ag, Au), the metals of higher valence (Al and Pb), the later transition metals (Ni, Rh, Pd, Ir, Pt) and the inert gases (Ne, Ar, Kr, Xe) when in the solid state all possess this structure (see Appendix 7), as does Fe between 912 and 1394 °C and Co above 420 °C. The lattice is face-centred cubic and there is one atom associated with each lattice point. A conventional unit cell of the structure with lattice parameter *a* is shown in Figure 3.1a. The conventional unit cell has four lattice points, and therefore four atoms, per unit cell. The relationship of this unit cell to the primitive rhombohedral unit cell for this crystal structure with $\alpha = 60°$ is shown in Figure 3.1b.

The coordinates of the atoms in the conventional unit cell are therefore $(0, 0, 0)$, $\left(\frac{1}{2}, \frac{1}{2}, 0\right)$, $\left(\frac{1}{2}, 0, \frac{1}{2}\right)$, and $\left(0, \frac{1}{2}, \frac{1}{2}\right)$. Each atom possesses 12 nearest neighbours at a distance of $a/\sqrt{2}$. The *coordination number* is therefore 12. This structure is the one obtained if equal spheres are placed in contact at the lattice points of a face-centred cubic lattice. A diagram illustrating this is shown in Figure 3.2a. An alternative packing arrangement of equal spheres which produces the hexagonal close-packed structure is shown in Figure 3.2b (Section 3.2.2). If the radius of the spheres is *R*, it follows that $R = a/(2\sqrt{2})$. The proportion of space filled

---

[1] Also known as a cubic face-centred lattice; the two terms are identical.

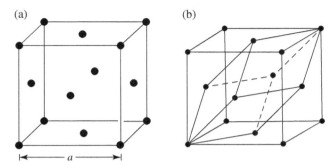

Figure 3.1 (a) The conventional unit cell of the c.c.p. crystal structure. (b) The primitive unit cell of the c.c.p. crystal structure shown within the conventional unit cell

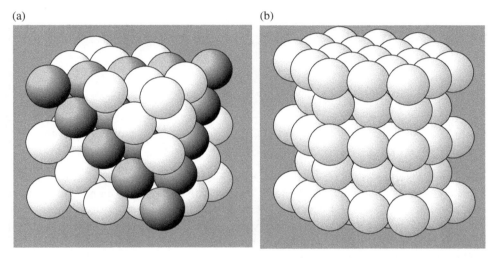

Figure 3.2 Close packing of equal spheres. (a) Cubic close-packed (c.c.p.). The close-packed layers are parallel to the darker balls. (b) Hexagonal close-packed (h.c.p.). The close-packed layers are horizontal. (After Bragg and Claringbull [2].)

by the spheres in the c.c.p. crystal structure is $\pi/(3\sqrt{2})$ or 74%. This is described by saying that the packing fraction is 0.74. A packing fraction of 0.74 is the closest packing of equal spheres that can be achieved [3–9], and is the closest packing that can be achieved with equal spheres at the lattice points of a Bravais lattice (see Section 3.6).

Rows of spheres are in contact along the line joining their centres. These lines are $\langle 110 \rangle$ directions in the lattice. Such directions are the closest packed and are termed *close-packed directions*. There are 12 such directions in all, if account is taken of change in sign. Taking the atomic centres to be at the lattice points, it is apparent from Figure 1.27 that in the $\{111\}$ planes, which lie normal to inverse triad axes, the atom centres form a triequiangular net of points. If the atoms are spheres of radius $a/(2\sqrt{2})$, the appearance of a $\{111\}$ plane is shown by the full circles in Figure 3.3. Each sphere is in contact with six equidistant spheres with centres in the plane. Since the arrangement shown in Figure 3.3 is the closest packing of

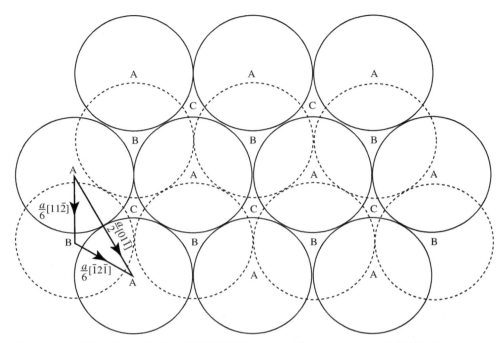

Figure 3.3 Plan view of the ABCABCABC... stacking sequence of (111) planes in a c.c.p. crystal

circles in a plane, the {111} planes are spoken of as being closest packed, or *close-packed*. There are eight such planes in the lattice if distinction is drawn between parallel normals of opposite sense (so that, for example, $(1\bar{1}1)$ and $(\bar{1}1\bar{1})$ are counted separately) and each contains six close-packed directions. The spacing of the {111} planes is $a/\sqrt{3} = 2\sqrt{2/3}R$ and so is equal to $\sqrt{2/3}$ of the atomic spacing in the {111} planes.

The centres of the atoms in a particular (111) plane occupy points such as A in Figure 3.3. If the positions of the centres of the atoms in adjacent (111) planes are projected on to this (111) plane, they occupy positions such as those marked B or C in Figure 3.3. The projections of the spheres centred on points B are shown dotted in the figure. A given crystal can be considered to be made up by stacking, one above the other, successive planar rafts of closest-packed spheres so that proceeding in the [111] direction the centres of the atoms in adjacent rafts follow the sequence ABCABCABC... (Figure 3.4a). The same crystal structure, but in a different orientation, would be described by the sequence of rafts ACBACBACB... Any atom in a c.c.p. crystal has 12 nearest neighbours along ⟨110⟩ directions at a distance of $a/\sqrt{2}$ (= 2R), 6 second-nearest along ⟨100⟩ directions at a (= $2\sqrt{2}R$), 24 third-nearest along ⟨211⟩ directions at $\sqrt{3/2}\,a$ (= $2\sqrt{3}R$) and 12 fourth-nearest along ⟨110⟩ directions at $\sqrt{2}a$ (= $2\sqrt{4}R$).

When the c.c.p. crystal structure is regarded as being made up of spheres in contact, the size of the interstices between spheres is important, because many other crystal structures contain at least one set of atoms in either a c.c.p. arrangement or a very close approximation to it. The largest interstice occurs at positions in the unit cell with coordinates $\left(\frac{1}{2}, \frac{1}{2}, \frac{1}{2}\right)$

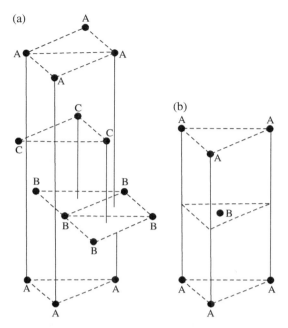

Figure 3.4 The stacking of closest-packed planes in (a) the c.c.p. structure and (b) the h.c.p. structure

and equivalent positions, i.e. $\left(\frac{1}{2},0,0\right)$, $\left(0,\frac{1}{2},0\right)$, and $\left(0,0,\frac{1}{2}\right)$. There are four of these interstices per unit cell and, hence, one per lattice point; one is illustrated in Figure 3.5. The largest sphere which can be placed in this position without disturbing the arrangement of spheres at the lattice points has radius $r = (\sqrt{2}-1)R = 0.414R$. This sphere would have octahedral coordination with six nearest neighbours (Figure 3.5); for this reason these interstices are commonly referred to as *octahedral interstices*. The sites of these octahedral interstices by themselves form a face-centred cubic lattice.

The second largest interstice occurs at points with coordinates $\left(\frac{1}{4},\frac{1}{4},\frac{1}{4}\right)$ and equivalent positions (Figure 3.6). The largest sphere which can be placed here has radius $r = (\sqrt{3/2}-1)R = 0.225R$ and possesses tetrahedral coordination; for this reason these interstices are commonly referred to as *tetrahedral interstices*. There are eight such points in the unit cell and hence two per lattice point. If the atoms in the c.c.p. crystal structure are ignored, the sites of the tetrahedral interstices can be seen to lie on a primitive cubic lattice.

3.2.2 Hexagonal Close-Packed (*P6₃/mmc*)

This structure is exhibited by the early transition metals Sc, Ti, Y and Zr, by the divalent metals Be, Mg, Zn and Cd, and by a number of the rare earth metals (see Appendix 7). The hexagonal primitive unit cell contains two atoms conventionally described as having coordinates of (0, 0, 0) and $\left(\frac{2}{3},\frac{1}{3},\frac{1}{2}\right)$ (Figure 3.7). Hence, there are two atoms associated with each lattice point. There is no pure rotational hexagonal axis in this crystal structure,

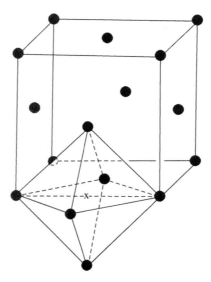

Figure 3.5 The largest interstice in the c.c.p. structure: the octahedral interstice (x)

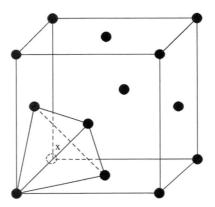

Figure 3.6 The second largest interstice in the c.c.p. structure: the tetrahedral interstice (x)

but instead 6_3 axes located at $\frac{1}{3}, \frac{2}{3}, z$ running parallel to the z-axis. This structure can be produced by packing together equal spheres, as shown in Figure 3.2b. If each sphere has 12 nearest neighbours equally far away from it, the axial ratio c/a must equal $\sqrt{8/3} = 1.633$. The packing fraction is then 0.74, as in the c.c.p. structure.

Values of the axial ratio for a number of metals with this structure are given in Table 3.1. Cobalt has an axial ratio very close to the ideal. It is noteworthy that there are significant departures from the ideal axial ratio: Ti and Zn, for example. If the atomic centres are projected on to a plane parallel to (0001), the basal plane, the structure can be regarded as being made up by stacking rafts of spheres in the sequence ABABAB or ACACAC (Figures 3.3 and 3.4b). If c/a is equal to or greater than the ideal value for sphere packing, there are six directions of the type $\langle 11\bar{2}0 \rangle$ in the basal plane along which atoms touch, and

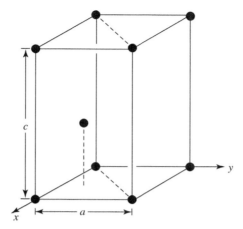

Figure 3.7   The unit cell of the h.c.p. structure

Table 3.1   Axial ratios at room temperature of some metals with a h.c.p. crystal structure.

Metal	c/a	Metal	c/a
Cd	1.886	Sc	1.597
Zn	1.857	Zr	1.593
Co	1.625	Ti	1.587
Mg	1.624	Y	1.571
Re	1.615	Be	1.568

these are the only close-packed directions. If c/a is less than the ideal value, spheres will not touch along these directions, but will be sufficiently close to one another that they can to all intents be regarded as close-packed directions.

Octahedral interstices in the conventional description of the h.c.p. crystal structure have coordinates $\left(\frac{1}{3}, \frac{2}{3}, \frac{1}{4}\right)$ and $\left(\frac{1}{3}, \frac{2}{3}, \frac{3}{4}\right)$ (Figure 3.8a), i.e. at two specific values of z along the screw hexad axes for $0 < z < 1$. There are evidently two such interstices per unit cell and hence one per atom. If $c/a = \sqrt{8/3}$ then the largest sphere that can be inserted without disturbing the spheres of radius R packed in contact has radius $r = 0.414R$, just as in the c.c.p. crystal structure. The second-largest interstices, the tetrahedral interstices, lie at $\left(0, 0, \frac{3}{8}\right)$, $\left(0, 0, \frac{5}{8}\right)$, $\left(\frac{2}{3}, \frac{1}{3}, \frac{1}{8}\right)$, and $\left(\frac{2}{3}, \frac{1}{3}, \frac{7}{8}\right)$ (Figure 3.8b). There are four such interstices per cell and hence two per atom. When $c/a = \sqrt{8/3}$ these have $r/R = 0.225$, again, just as in the c.c.p. crystal structure.

In the *International Tables for Crystallography* [1], the origin for the description of the space group for the h.c.p. crystal structure, $P6_3/mmc$ (No. 194), is chosen at an octahedral interstice position in the h.c.p. crystal structure, because these positions have $\bar{3}m$ (or, equivalently, $\bar{3}m1$) point symmetry. Under these circumstances, the positions of atoms within the unit cell can be described either as being at $\left(\frac{1}{3}, \frac{2}{3}, \frac{3}{4}\right)$ and $\left(\frac{2}{3}, \frac{1}{3}, \frac{1}{4}\right)$, or at $\left(\frac{1}{3}, \frac{2}{3}, \frac{1}{4}\right)$ and

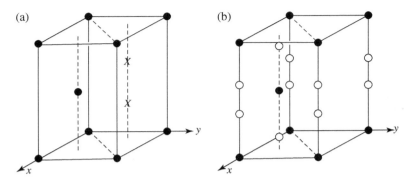

Figure 3.8   Interstices in the h.c.p. arrangement. The octahedral interstices in (a) are denoted by *X* and the tetrahedral interstices in (b) by open circles

$\left(\frac{2}{3}, \frac{1}{3}, \frac{3}{4}\right)$, depending upon which octahedral interstice is defined as the origin, with respect to which the atom positions are then specified.

3.2.3   Double Hexagonal Close-Packed (*P6₃/mmc*)

In the eight rare earth metals La, Pr, Nd, Pm, Am, Cm, Bk and Cf, the hexagonal close packing is a four-layer repeat, ABAC. This crystal structure is often referred to as the double h.c.p. structure. The axial ratios characteristic of these metals vary between 3.192 for Pm and 3.262 for Cf, so that in each case they are just below the ideal value of $2\sqrt{8/3} = 3.266$. There are four atoms per unit cell, one octahedral interstice per atom and two tetrahedral interstices per atom.

3.2.4   Body-Centred Cubic (*Im3̄m*)

This structure is shown by the alkali metals Li, Na, K, Rb and Cs, and by the transition metals V, Cr, Fe, Nb, Mo, Ta and W. It is also shown by Ti and Zr at high temperature. The lattice is b.c.c. with one atom at each lattice point, so the atomic coordinates are (0, 0, 0) and $\left(\frac{1}{2}, \frac{1}{2}, \frac{1}{2}\right)$ (Figure 3.9). Each atom has eight nearest neighbours along ⟨111⟩ directions

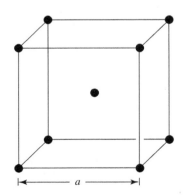

Figure 3.9   The b.c.c crystal structure

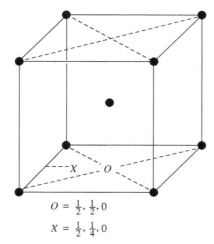

$$O = \tfrac{1}{2}, \tfrac{1}{2}, 0$$
$$X = \tfrac{1}{2}, \tfrac{1}{4}, 0$$

Figure 3.10   Tetrahedral (X) and octahedral (O) interstices in the b.c.c crystal structure

at a separation of $\sqrt{3}a/2$, where a is the lattice parameter. These eight $\langle 111 \rangle$ directions are close-packed directions.

Unlike in the c.c.p. and h.c.p. structures, none of the nearest neighbours of one particular atom are nearest neighbours of each other. If the b.c.c structure is presumed to be made up with spheres of equal size, these each have a radius $R = \sqrt{3}a/4$. The packing fraction is $\pi\sqrt{3}/8 = 0.68$. The second-nearest neighbours of an atom are closer than in the c.c.p. structure. There are six second-nearest neighbours along $\langle 100 \rangle$ directions at a distance $4R/\sqrt{3} = 2.309R$ (in c.c.p., second-nearest neighbours are at $2\sqrt{2}R = 2.828R$) and 12 third-nearest along $\langle 110 \rangle$ directions at $4\sqrt{2}R/\sqrt{3}$. There are no closest-packed planes of atoms in this structure.

This structure has its largest interstices at coordinates $\left(\tfrac{1}{2}, \tfrac{1}{4}, 0\right)$ and equivalent positions (Figure 3.10). There are 12 such positions per cell: 6 per lattice point, and therefore 6 per atom. The largest sphere fitting in such an interstice has radius $r = ((\sqrt{5/3}) - 1)R = 0.291R$. This interstice is therefore significantly smaller than the largest one in c.c.p. It has four nearest neighbours equidistant from it, and so it is a tetrahedral interstice, but the tetrahedron that these neighbours form is not regular. The second-largest interstice is at $\left(\tfrac{1}{2}, \tfrac{1}{2}, 0\right)$ and equivalent positions (Figure 3.10). There are six such sites per cell ($6 \times \tfrac{1}{2}$ at the centres of cube faces and $12 \times \tfrac{1}{4}$ at the midpoints of cube edges), and so three per lattice point. Each can accommodate a sphere of radius $r = ((2/\sqrt{3}) - 1)R = 0.155R$; such a sphere is at the centre of a distorted octahedron, and so these interstices are distorted octahedral interstices.

3.3   Related Metallic Structures

3.3.1   Indium (*I4/mmm*)

The crystal structure of indium is very similar to the c.c.p. structure. It is body-centred tetragonal with $c/a = 1.521$. This axial ratio is such that, if referred to a face-centred

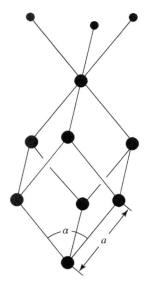

Figure 3.11   The crystal structure of mercury, showing the primitive rhombohedral unit cell

tetragonal lattice instead of the conventional body-centred one, the axial ratio $c/a = 1.076$. Therefore, the structure can be described as face-centred tetragonal ($F4/mmm$), with one atom at each lattice point and an axial ratio of 1.076.

3.3.2   Mercury ($R\bar{3}m$)

Mercury also has a structure that can be described as distorted c.c.p. The primitive unit cell has a rhombohedral lattice with one atom at each lattice point (Figure 3.11). The axial angle α is 70.74° and $a = 2.992$ Å at 78 K. The primitive unit cell of a c.c.p. crystal has $\alpha = 60°$ (Section 3.2.1 and Figure 3.1b). Thus the mercury structure can be derived from the c.c.p. crystal structure by compressing it along a body diagonal of the cell. The atoms in the (111) planes of mercury are therefore arranged to form a triequiangular net but the spacing of these planes is too small to allow closest packing of spherical atoms. The nearest neighbours of any one atom are in adjacent (111) planes. The closest-packed directions are $\langle 100 \rangle$ of the primitive rhombohedral cell of side a (Figure 3.11). There are six nearest neighbours at a distance a. Mercury can, of course, also be referred to a rhombohedral face-centred cell containing four atoms to emphasize the relationship to the conventional cell of a c.c.p. crystal. This larger cell for Hg has $a' = 4.577$ Å and axial angle $\alpha' = 98.36°$ at 78 K (see Problem 3.4).

3.3.3 β-Sn ($I4_1/amd$)

β-Sn, also known as white tin, is the form of tin that is stable at room temperature. It has a body-centred tetragonal lattice with two atoms associated with each lattice point: one at $(0, 0, 0)$ and the other at $\left(0, \frac{1}{2}, \frac{1}{4}\right)$. Both atom sites have the same point symmetry: $\bar{4}m2$. The value of c/a at room temperature is 0.546. Each atom has four nearest neighbours at the vertices of an irregular tetrahedron and two more at a slightly greater distance (see Problem 3.5). The structure is a distorted form of the diamond structure (Section 3.4.), which is taken by the form of tin stable below room temperature known as α-Sn or grey tin.

3.4   Other Elements and Related Compounds

3.4.1   Diamond ($Fd\bar{3}m$)

The elements Si, Ge, α-Sn (the low-temperature allotrope of tin) and crystalline carbon stable at high temperature and pressure (diamond) all have the structure shown in Figure 3.12. The Bravais lattice is face-centred cubic and there are two atoms associated with each lattice point. The coordinates of these atoms can be taken to be $(0, 0, 0)$ and $\left(\frac{1}{4}, \frac{1}{4}, \frac{1}{4}\right)$, or, if the origin is taken to be at a centre of symmetry, $\pm\left(\frac{1}{8}, \frac{1}{8}, \frac{1}{8}\right)$. The coordination number is 4, with the nearest neighbours at a distance $\sqrt{3}a/4$ arranged at the corners of a regular tetrahedron, outlined in Figure 3.12.

The atom centres lie at the corners of triequiangular nets in {111} planes. If these are projected on to (111) the stacking sequence of successive (111) planes can be described as CA AB BC CA AB BC. Successive planes are not equally separated from one another (Figure 3.13). The structure is very loosely packed. If equal spheres in contact have their centres at the atom centres, the packing fraction is only $\pi\sqrt{3}/16 = 0.34$. The diamond crystal structure can be regarded as a special case of the sphalerite crystal structure (Section 3.5.3), in which the atoms at $(0, 0, 0)$ and $\left(\frac{1}{4}, \frac{1}{4}, \frac{1}{4}\right)$ are of different types. For this reason, Figure 3.13 can also be used to show the stacking sequence of the sphalerite crystal structure. In sphalerite, there are no longer centres of symmetry, so the space group symmetry is $F\bar{4}3m$ (Section 3.5.3).

3.4.2   Graphite ($P6_3/mmc$)

The thermodynamically stable crystal structure of carbon at room temperature is graphite, shown in Figure 3.14a. The lattice is hexagonal with four atoms per unit cell, with

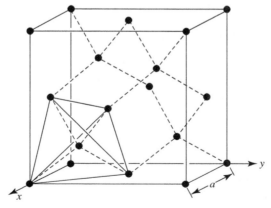

Figure 3.12   The crystal structure of diamond

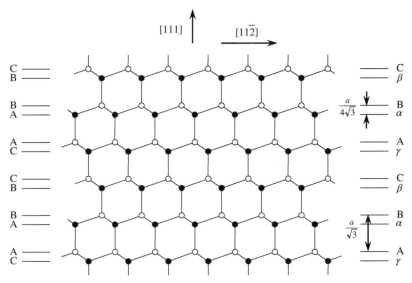

Figure 3.13 The stacking of (111) planes in the diamond and sphalerite structures. The atomic positions are projected onto ($1\bar{1}0$). In diamond, the filled and open circles represent the same type of atom. The sequence of triequiangular nets is described for diamond on the left-hand side and for sphalerite (zinc blende) on the right-hand side

coordinates (0, 0, 0), $\left(0, 0, \frac{1}{2}\right)$, $\left(\frac{1}{3}, \frac{2}{3}, 0\right)$, and $\left(\frac{2}{3}, \frac{1}{3}, \frac{1}{2}\right)$. At room temperature $a = 2.46$ Å and $c = 6.71$ Å; therefore, $c/a = 2.72$ and is large. The atoms in one (0001) plane are arranged at the corners of regular hexagons. The x- and y-axes of the unit cell are shown in Figure 1.1. The structure can be built up by stacking successive hexagonal sheets of atoms one above the other along the z-axis so that the hexagons are in the same orientation, but with half the corners of the hexagons in one plane lying in the centres of the hexagons in adjacent planes (Figure 3.14a). If the atomic centres are all projected on to (0001), the stacking sequence can be described as ABABAB or ACACAC, where it must be remembered that the letters A, B and C refer to sheets of atoms arranged at the corners of hexagons, rather than at the vertices of equilateral triangles, as in the h.c.p. crystal structure, for example.

Each atom has three nearest neighbours at a separation of $a/\sqrt{3}$ in the basal plane. Half of the atoms in any one hexagonal layer have atoms directly above and below them at a distance of $c/2$. The separation of the nearest neighbours in the (0001) planes is only 1.42 Å (compared with 1.54 Å in diamond) and this is much smaller than the separation of the hexagonal sheets – 3.35 Å. Graphite is therefore said to possess a layer structure because the atoms are strongly bonded within a sheet, and the sheets are only weakly bound to one another through van der Waals forces, also known as London dispersion forces (Section 3.9). It is this layer structure which helps to make graphite an excellent solid state lubricant.

If graphite crystals are ground or otherwise severely deformed at low temperature, another form of graphite can be detected in which the atomic hexagons are arranged in the sequence ABCABCABC or alternatively CBACBACBA. This crystal structure has a rhombohedral Bravais lattice (space group $R\bar{3}m$). The primitive rhombohedral cell has

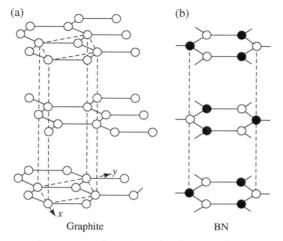

(a)   (b)

Graphite   BN

Figure 3.14   (a) The crystal structure of graphite. (b) The crystal structure of hexagonal boron nitride

$a = 3.635\,\text{Å}$ and $\alpha = 39.5°$ at room temperature. There are two atoms per unit cell with coordinates $\pm\,(u, u, u)$ where $u = \frac{1}{6}\,(=0.167)$.

3.4.3   Hexagonal Boron Nitride ($P6_3/mmc$)

The structure of hexagonal boron nitride thermodynamically stable at room temperature is closely related to that of graphite (Figure 3.14b). The atoms occur in hexagonal sheets, but the sheets are stacked directly above one another along [0001] so that the stacking sequence is described as AAAAA ... , with unlike atoms above one another in consecutive layers (see Figure 3.14b). In any one sheet there are equal numbers of B and N atoms arranged so that B and N alternate around any one atomic hexagon. At room temperature the B–N separation in the sheets is 1.45 Å and the separation of the sheets is 3.33 Å. Like graphite, this form of boron nitride is also an excellent solid state lubricant because of the van der Waals bonding between the layers.

3.4.4   Arsenic, Antimony, and Bismuth ($R\bar{3}m$)

Like the rhombohedral form of graphite described in Section 3.4.2, the crystal structures of As, Sb and Bi are also based on a primitive rhombohedral Bravais lattice, with two atoms associated with each lattice point. The structure is shown in Figure 3.15. The coordinates of the atoms in the rhombohedral unit cell are $\pm\,(u, u, u)$; u is a little less than $\frac{1}{4}$ and α somewhat less than 60° (see Table 3.2). The structure is easily visualized as being made up of sheets of atoms lying perpendicular to [111]. Within each sheet the atoms are arranged at the corners of triequiangular nets. Referring to Figure 3.15 and noting the values of u, the stacking sequence of these nets along [111] is BA CB AC BA ... If u were $\frac{1}{4}$, the spacing of adjacent nets would be regular. In fact, each atom in any one sheet has three nearest neighbours in the closest adjacent sheet at d_1 (see Table 3.2) and three at a slightly greater distance in the sheet on the other side at d_2. The possession of three nearest neighbours fits

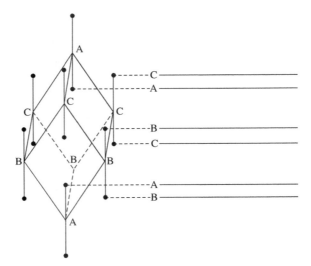

Figure 3.15 The crystal structure of As, Sb and Bi

Table 3.2 Cell dimensions of As, Sb and Bi.

	a (Å)	α (deg)	u	d_1 (Å)	d_2 (Å)
As	4.132	54.13	0.227	2.517	3.120
Sb	4.507	57.11	0.233	2.908	3.355
Bi	4.746	57.23	0.234	3.071	3.529

with the positions of As, Sb and Bi in the periodic table, since each atom is expected to form three covalent bonds. If $\alpha = 60°$ and $u = \frac{1}{4}$ in this structure, each atom has six closest neighbours and the arrangement of the atoms is at the points of a simple cubic lattice.

3.5 Simple *MX* and *MX*$_2$ Compounds

3.5.1 Sodium Chloride, NaCl (*Fm$\bar{3}$m*)

One-third of all compounds of the type *MX* crystallize in the sodium chloride (halite) struc-ture (*Fm$\bar{3}$m*), shown in Figure 3.16. In such compounds the metal atom or ion *M* is usually smaller than the electronegative element, *X*. The lattice is face-centred cubic with two dif-ferent atoms associated with each lattice point: one, say *X*, at (0, 0, 0) and the other, *M*, at $\left(0, 0, \frac{1}{2}\right)$. Each of the two types of atom, considered in isolation, lies on a face-centred cubic lattice, and each lies at the largest interstice of the other's face-centred cubic lattice. From the diagram in Figure 3.5 and from inspection of Figure 3.16, it is clear that each atom has a coordination number of 6, the neighbours being at the vertices of a regular octa-hedron. There are four formula units per conventional unit cell, but there is clearly no trace of the formation of a molecule of *MX* in this structure.

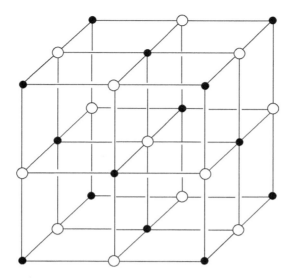

Figure 3.16   The crystal structure of sodium chloride, NaCl

Each {111} *lattice plane* specifies a double sheet of atoms, each of the sheets consisting entirely of atoms of one kind arranged at the points of a triequiangular net. If we denote sheets of atoms of one kind with a Greek letter and those of the other with a Roman letter, then the stacking sequence along a [111] direction can be described as A γ B α C β A γ B α C β … The spacing of the sheets of atoms denoted by letters from the same alphabet is $a/\sqrt{3}$, where a is the lattice parameter.

All of the alkali halides, with the exception of CsCl, CsBr and CsI, crystallize with this crystal structure, as do many of the sulfides, selenides, and tellurides of Mg, Ca, Sr, Ba, Pb and Mn, as well as the oxides of the formula *MO* of Mg, Ca, Sr, Ba, Cd, Ti, Zr, Mn, Fe, Co, Ni and U and some transition metal carbides and nitrides such as TiC, TiN, TaC, ZrC, ZrN, UN and UC (see Section A7.2 in Appendix 7).

When the metallic ion is of variable valence, crystals with this structure often form with some ion positions unoccupied. For example, FeO crystals would be perfect if they contained only divalent iron, Fe^{2+}. If some Fe^{3+} ions are present, then for every two Fe^{3+} ions, one of the sites on the iron sublattice must be empty. Such a structure is called a *defect structure* (see Section 10.1).

3.5.2   Caesium Chloride, CsCl (*Pm$\bar{3}$m*)

CsCl, CsBr, CsI, as well as many intermetallic compounds such as CuBe, CuZn, AgCd, AgMg and FeAl (see Section A.7.3 in Appendix 7), show the caesium chloride structure (*Pm$\bar{3}$m*), which has a primitive cubic lattice with one atom of each kind associated with each lattice point, one, say X, at (0, 0, 0) and the other, M, at $\left(\frac{1}{2}, \frac{1}{2}, \frac{1}{2}\right)$ (Figure 3.17). The coordination number is 8 for each atom, the nearest neighbours being at $\sqrt{3}a/2$, where a is the lattice parameter.

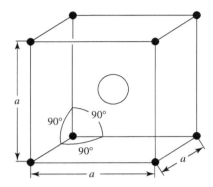

Figure 3.17 The crystal structure of caesium chloride, CsCl

3.5.3 Sphalerite, α-ZnS ($F\bar{4}3m$)

The cubic form of ZnS, designated α-ZnS, shows the sphalerite structure[2] ($F\bar{4}3m$) shown in Figure 3.18. This is also exhibited by the sulfides, selenides and tellurides of Be, Zn, Cd and Hg, the halides of Cu and AgI: see Section A.7.4 of Appendix 7 for a tabulation of selected crystals with the sphalerite structure. The lattice is face-centred cubic with one atom of each kind associated with each lattice point, one at $(0, 0, 0)$ and the other at $\left(\frac{1}{4}, \frac{1}{4}, \frac{1}{4}\right)$, so the relationship to the structure of diamond (Section 3.4) is very close (Figure 3.13). Each type of atom considered in isolation from the other lies at the lattice points of a face-centred cubic lattice and each lies in the second largest interstice of the close-packed crystal structure of the other. The coordination number is 4, the nearest neighbours being atoms of the other kind at a distance $\sqrt{3}a/4$ arranged at the vertices of a regular tetrahedron. The sequence of (111) planes is stacked at the same intervals as in diamond (Figure 3.12), but alternate planes are occupied by atoms of different chemical species. Thus, following the same nomenclature as for NaCl, the stacking sequence is γ A α B β C γ A ... (Figure 3.13).

3.5.4 Wurtzite, β-ZnS ($P6_3mc$)

The sphalerite structure is derived from the c.c.p. structure by placing atoms of a different kind to those at the lattice points at every other tetrahedrally coordinated interstice. A related structure, also shown by ZnS, is obtained by filling alternate tetrahedrally coordinated interstices in the h.c.p. structure. This is the β-ZnS or wurtzite structure ($P6_3mc$). The lattice is hexagonal with atoms of one kind at $(0, 0, 0)$ and $\left(\frac{2}{3}, \frac{1}{3}, \frac{1}{2}\right)$, and those of the other kind at $(0, 0, u)$ and $\left(\frac{2}{3}, \frac{1}{3}, \frac{1}{2} + u\right)$ (Figure 3.18b). The value of u is very close to 0.375, i.e. $\frac{3}{8}$; however, it is not constrained by symmetry to be exactly 0.375. From Figure 3.8b the relationship to the h.c.p. structure is obvious. Each atom is, of course, tetrahedrally coordinated with four of the opposite kind. The stacking of planes of atoms parallel to (0001) using the same nomenclature as for NaCl and α-ZnS is then A α B β A α B ... or equivalently A α C γ A α C γ A ... This structure is shown by a number of compounds (Table 3.3). The axial ratios

---

[2] Sometimes called zinc blende.

(a)   (b)

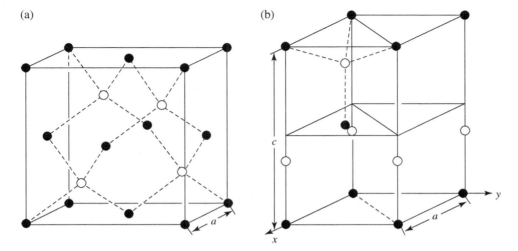

Figure 3.18 The crystal structures of (a) sphalerite (α-ZnS) and (b) wurtzite (β-ZnS)

Table 3.3 Axial ratios at room temperature of some materials with the wurtzite crystal structure.

Material	c/a	Material	c/a
NH$_4$F	1.61	β-CdS	1.62
α-AgI	1.63	CdSe	1.63
BeO	1.63	AlN	1.60
ZnO	1.60	SiC	1.65
β-ZnS	1.63	TaN	1.62

shown in Table 3.3 and Section A.7.5 of Appendix 7 are quite close to the ideal, 1.633, for hexagonal close packing of one type of ion with the other in the tetrahedral interstices.

3.5.5 Nickel Arsenide, NiAs ($P6_3/mmc$)

We described the sodium chloride structure as being derived from the c.c.p. structure by placing a second set of atoms in the octahedrally coordinated largest interstices of the structure. It is not surprising therefore that a structure exists in which atoms of a different kind are placed in the octahedral interstices of the h.c.p. structure. This is the nickel arsenide (nickeline)[3] structure ($P6_3/mmc$) (Figure 3.19). The lattice is hexagonal, with atoms of one kind at (0,0,0) and $\left(0, 0, \frac{1}{2}\right)$, and those of the other kind at $\left(\frac{2}{3}, \frac{1}{3}, \frac{1}{4}\right)$ and $\left(\frac{1}{3}, \frac{2}{3}, \frac{3}{4}\right)$. For both kinds of atom the coordination number is 6, but while the sites occupied by one kind of atom, those denoted by open circles in Figure 3.19, lie in positions corresponding to a hexagonal close-packed arrangement (As), the others (Ni) lie at the lattice points of a primitive hexagonal lattice (if the As atoms are ignored), with a repeat distance in the c direction

---

[3] Also known as niccolite.

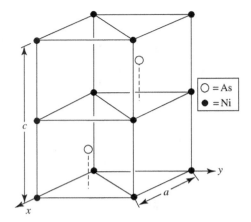

Figure 3.19 The crystal structure of nickel arsenide, NiAs

Table 3.4 Axial ratios at room temperature of some materials with the nickel arsenide structure.

Material	c/a	Material	c/a
NiAs	1.40	CoS	1.52
CrS	1.64	NiS	1.55
FeS	1.66		

half that of the lattice parameter of the NiAs crystal structure. Hence, the Ni atoms are placed in the octahedral interstices of what would otherwise be an As h.c.p. crystal structure, with the origin of the unit cell of NiAs shown in Figure 3.19 moved to one of the two octahedral interstices in this h.c.p. crystal structure.

The difference in stacking of the M and X atoms is obvious from Figure 3.8a and also by stating the stacking sequence of the planes of atoms along [0001] following the procedure for NaCl. The stacking sequence is A β A γ A β A γ … The atoms denoted by Greek letters lie in the centre of a trigonal prism of atoms denoted by Roman letters, while the atoms denoted by Roman letters are octahedrally coordinated. The axial ratios of some crystals of sulfides with this structure are listed in Table 3.4; the selenides, tellurides and antimonides of the metals listed often have the same structure: see Section A.7.6 of Appendix 7. The values of c/a at room temperature depart from the ideal for hexagonal close packing of one type of ion. Usually the metal atom is situated in the type of site occupied by Ni in NiAs, but in PtB the 'anti-NiAs' structure is shown.

3.5.6 Calcium Fluoride, CaF_2 ($Fm\bar{3}m$)

If all of the tetrahedral interstices in the c.c.p. crystal structure are filled with atoms of a different kind from those in the c.c.p. crystal structure, we obtain the calcium fluoride (CaF_2) or *fluorite* structure ($Fm\bar{3}m$) (Figure 3.20). The lattice is face-centred cubic with atoms of one kind, for example Ca, at (0, 0, 0) and equivalent positions, and with F atoms

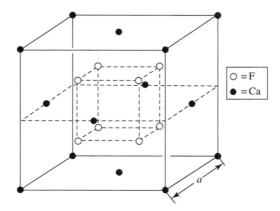

Figure 3.20 The crystal structure of calcium fluoride, CaF_2

at $\left(\frac{1}{4},\frac{1}{4},\frac{1}{4}\right)$ and $\left(\frac{1}{4},\frac{3}{4},\frac{1}{4}\right)$ and equivalent positions. There are clearly four formula units per unit cell. The coordination number of Ca is 8, with the nearest neighbours being atoms of F arranged at the corners of a cube. If the calcium atoms are ignored, it is apparent that the fluorine atoms lie at the lattice points of a primitive cubic lattice. Each F atom has tetrahedral coordination with four Ca atoms.

The centres of atoms of one kind all form triequiangular nets of points parallel to $\{111\}$ lattice planes. The separation of the atoms in these planes is the same for the two types of atom. The stacking sequence of these planes along [111] can therefore be described as … $\alpha B\gamma$ $\beta C\alpha$ $\gamma A\beta$ $\alpha B\gamma$ $\beta C\alpha$ … The planes of atoms denoted by successive Roman letters are regularly spaced from one another; the same is true of the planes of atoms denoted by Greek letters. This structure is shown by the fluorides of Ca, Sr and Ba and by the oxides of Zr, Th, Hf and U. Oxides and sulfides of alkali metals also show this structure; in these cases this crystal structure is more usually termed the *antifluorite* structure because the positions of the cations and anions are reversed relative to CaF_2. A tabulation of crystals with the fluorite structure is given in Section A.7.7 of Appendix 7.

3.5.7 Rutile, TiO_2 ($P4_2/mnm$)

A common structure of compounds with the formula AX_2, besides that of fluorite, is the rutile (TiO_2) structure ($P4_2/mnm$) (Figure 3.21a). It is also sometimes called the cassiterite (SnO_2) structure. A number of dioxides and fluorides have this crystal structure, such as those tabulated in Section A.7.8 of Appendix 7. The lattice is primitive tetragonal with c/a about 0.65. There are titanium atoms at (0, 0, 0) and $\left(\frac{1}{2},\frac{1}{2},\frac{1}{2}\right)$ and oxygen atoms at $\pm$ $(u, u, 0)$ and $\pm\left(u+\frac{1}{2},\frac{1}{2}-u,\frac{1}{2}\right)$ where u is close to 0.30 in all examples of the structure. Therefore, there are two formula units per cell. A plan of the structure is shown in Figure 3.21b, projected down [001]. Each Ti atom is surrounded by six O atoms, but the octahedron of O atoms is not a regular octahedron. It is apparent that the crystal structure of rutile can be described in terms of how these Ti octahedra pack together. This leads conceptually to the more general geometry of the packing of polyhedra, which will be discussed in Section 4.4 in the context of the packing of atoms or ions within very complicated crystal structures.

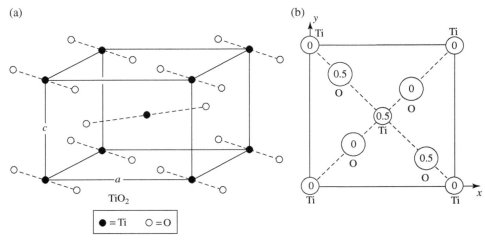

Figure 3.21 (a) The crystal structure of rutile, TiO_2. (b) The rutile crystal structure projected down [001]. The numbers indicate the z-coordinates of the atoms; those at height 0 could have been omitted to conform to standard practice (Section 1.1)

3.6 Other Inorganic Compounds

As we have seen in the descriptions of simple crystal structures in Sections 3.2–3.5, the concept of the packing of spheres into arrangements that are close-packed is of great practical use. It is also useful for circumstances where there are departures from the most close-packed ideal arrangement shown by c.c.p. crystal structures or h.c.p. crystal structures where $c/a = \sqrt{8/3}$, so that a 'close-packed' arrangement of one type of atom or ion (not necessarily in contact) occurs, with the interstices filled or partly filled with smaller atoms (or ions). Simple examples of such structures have already been considered in Section 3.5. In this section we consider other examples where the chemistry is less simple, or where the packing departs significantly from the ideal close-packed arrangement.

3.6.1 Perovskite ($Pm\bar{3}m$)

This primitive cubic crystal structure is a very common structure of crystals of compounds of the type $MM'X_3$. It is named after the mineral perovskite, $CaTiO_3$, which has this crystal structure above 900 °C; at room temperature, $CaTiO_3$ has an orthorhombic crystal structure belonging to the space group *Pnma* (see Section 2.15).[4] In this crystal structure, the calcium (M) and oxygen (X) ions taken together form a c.c.p. arrangement with the titanium (M') ions in the octahedral voids (Figure 3.22). The coordination number of Ca is 12, that of Ti is 6 and that of O is 2. The atomic coordinates in the primitive unit cell are Ti $\left(\frac{1}{2}, \frac{1}{2}, \frac{1}{2}\right)$, Ca $(0, 0, 0)$ and O $\left(\frac{1}{2}, \frac{1}{2}, 0\right), \left(0, \frac{1}{2}, \frac{1}{2}\right), \left(\frac{1}{2}, 0, \frac{1}{2}\right)$. The description of the perovskite crystal structure in terms of the packing of titanium octahedra will be deferred to Section 4.4.

---

[4] $SrTiO_3$ has this primitive cubic crystal structure at room temperature.

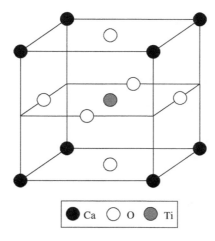

Figure 3.22   The crystal structure of perovskite, $Pm\overline{3}m$

There are many slightly distorted forms of the perovskite structure that are important in solid-state devices. At room temperature, barium titanate, $BaTiO_3$ ($P4mm$), has a crystal structure that is a tetragonally distorted form of the $Pm\overline{3}m$ high-temperature structure. At lower temperatures, further distortions convert this into an orthorhombic crystal structure ($Amm2$) and a rhombohedral crystal structure ($R3m$) [10]. $Pb(Zr_{1-x}Ti_x)O_3$, in which the Zr and Ti ions both occupy the M' position, has two different rhombohedral ($R3m$ and $R3c$) crystal structures [11] and a tetragonal ($P4mm$) form at room temperature, dependent on the value of x. The high-temperature superconducting oxide $YBa_2Cu_3O_{7-\delta}$, where δ is a small fraction $\ll 1$ indicating an oxygen deficiency, has a variety of crystal structures all based on a three-layered distorted perovskite structure [12]. Thus, for example, the super-conducting phase for values of x of the order of 0.1 or less has an orthorhombic $Pmmm$ crystal structure. The sheets of Cu and O atoms parallel to (001) in this structure are critical to the superconducting behaviour of this oxide.

3.6.2 α-Al_2O_3 ($R\overline{3}c$), $FeTiO_3$ ($R\overline{3}$) and $LiNbO_3$ ($R3c$)

Sapphire (α-Al_2O_3 or corundum) possesses a large unit cell with the oxygen ions in an h.c.p. arrangement. Aluminium ions occupy the octahedral interstices. Since the formula is Al_2O_3, only two-thirds of these are filled, as shown in Figure 3.23. The structure can be easily described by locating the 'missing' aluminium ions. If all the atomic positions are projected on to (0001) of the hexagonal cell and Roman letters denote the oxygen ion positions and Greek letters the positions of missing aluminium ions, the stacking sequence is $A\gamma_1\,B\gamma_2\,A\gamma_3\,B\gamma_1\,A\gamma_2\,B\gamma_3\,A\gamma_1\,\ldots$, where the positions $\gamma_1, \gamma_2, \gamma_3$ are indicated by their suffixes in Figure 3.23.

The hexagonal unit cell is then six oxygen layers high, and contains six formula units of Al_2O_3. At room temperature, the lattice parameters of the hexagonal cell are $a = 4.759$ Å, $c = 12.993$ Å and so $c/a = 2.731$. If c/a were to equal $2\sqrt{2} = 2.828$, the oxygen ions would be in an ideal h.c.p. arrangement (the factor of $\sqrt{3}$ greater than $\sqrt{8/3}$ arises from the 30° rotation of the x- and y-axes of the rhombohedral unit cell relative to the x- and y-axes

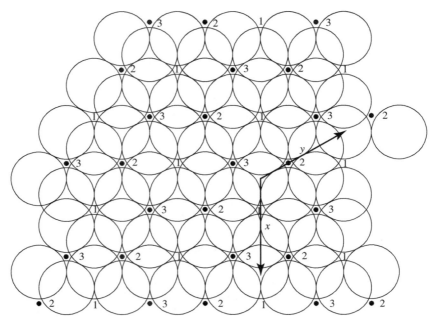

Figure 3.23 The structure of sapphire (α-Al_2O_3 or corundum). The large circles represent oxygen ions in two adjacent sheets. Between the sheets shown, aluminium ions lie in the octahedral interstices γ_2 and γ_3, while the positions γ_1 are empty. In the next layer of aluminium ions, positions γ_1 and γ_3 would be filled and γ_2 empty

defining the ideal h.c.p. arrangement). The lattice translations in (0001) of the triply primitive hexagonal cell are marked in Figure 3.23. The rhombohedral primitive cell contains two formula units of Al_2O_3 and has $a = 5.128$ Å and $\alpha = 55.27°$. If the oxygen ions were in the ideal h.c.p. arrangement, the value of α would be 53.78°.

FeTiO$_3$ (ilmenite, $R\bar{3}$) has the same structure as α-Al_2O_3 except that the Fe and Ti atoms are distributed in alternating octahedral sheets. In LiNbO$_3$ ($R3c$), the Li and Nb atoms are distributed in the same octahedral sheet so that one species occupies the γ_2 position while the other occupies the γ_3 position. The crystal structure of LiNbO$_3$ can also be described as a distorted form of the ideal perovskite structure.

3.6.3 Spinel ($Fd\bar{3}m$), Inverse Spinel and Related Structures

The *spinel* structure, shown by $MgAl_2O_4$ and by other mixed oxides of di- and trivalent metals, has a unit cell containing 32 oxygen ions in almost perfect cubic close packing. Since there would be four oxygen ions per unit cell if only the oxygen ions were present, it is apparent that the edge of the unit cell for spinel is twice that which would be expected for the oxygen ions alone in cubic close packing. Eight of the 64 tetrahedral interstices in the unit cell are filled by the divalent Mg ions and 16 of the 32 octahedral interstices are filled with the trivalent Al ions. The Mg ions considered as a group of ions form a structure of the diamond type (Figure 3.24). One-eighth of the face-centred cubic unit cell is shown in this figure.

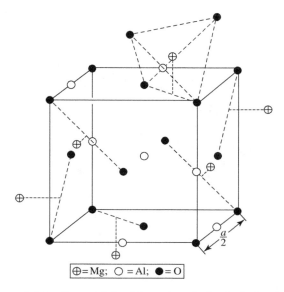

$\oplus$ = Mg; $\bigcirc$ = Al; $\bullet$ = O

Figure 3.24   One-eighth of the unit cell of spinel, $MgAl_2O_4$

Some oxides of composition MM'_2O_4 show a structure called '*inverse*' *spinel* to distinguish it from the 'normal' spinel which we have just described. $MgFe_2O_4$ is an example of an inverse spinel. In this oxide, the oxygen ions are arranged in cubic closest packing and the same two types of interstice are involved. However, the cations are arranged differently. Ideally, of the 16 iron ions per unit cell, eight occupy the eight tetrahedral interstices. The 16 octahedral interstices are occupied by the remaining eight iron ions and by the eight magnesium ions. The Mg and Fe ions can occur at random amongst the occupied octahedrally coordinated sites. To emphasize the difference from a normal spinel, the formula of this inverse spinel is sometimes written $Fe(MgFe)O_4$, or generally $M'(MM')O_4$. In fact, the structure of $MgFe_2O_4$ deviates somewhat from this ideal; the number of iron atoms in tetrahedral sites is not exactly equal to the number in octahedral sites.

Chrysoberyl, $BeO.Al_2O_3$ (*Pnma*), is isomorphous with the olivine $(Mg_{1-x}Fe_x)_2SiO_4$ group of minerals. In this case the oxygen ions are arranged in a slightly distorted h.c.p. structure [13]. As for the spinels, the metal ions in chrysoberyl are distributed amongst the tetrahedral and the octahedral sites; in the case of chrysoberyl, half of the octahedral sites are occupied by Al ions and one-eighth of the tetrahedral sites by Be.

3.6.4   Garnet ($Ia\bar{3}d$)

The concept of the occupation of the interstices in a structure by different types of ion is useful in describing the garnet structures. While these are perhaps best known as gemstones, garnets are very important in many solid-state physics devices. Various garnets can be ferrimagnetic (e.g. YIG, see below) and can form excellent laser hosts (e.g. YAG).

The garnets occur naturally as the silicates of various di- and trivalent metals. The archetypal garnet mineral has a formula $3MO.M'_2O_3.3SiO_2$ or $M_3M'_2Si_3O_{12}$, where M is a divalent metal ion and M' a trivalent ion. The technologically important examples of

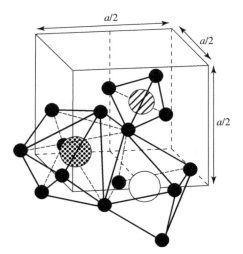

Figure 3.25 Coordination about the oxygen ions (solid black circles) in a garnet structure showing the $\{C_3\}$ tetrahedral site (striped circle), the $[A_2]$ octahedral site (solid white circle) and the (D_3) dodecahedral site (spotted circle). (After Figure 1 in [15])

yttrium iron garnet, $Y_3Fe_5O_{12}$ (YIG), and yttrium aluminium garnet, $Y_3Al_5O_{12}$ (YAG), do not contain silicon. The substitution is possible because in a typical natural garnet such as grossular,[5] $3CaO.Al_2O_3.3SiO_2$, it is possible to substitute 'YAl' for 'CaSi'. Replacing all the Ca and Si in this way produces $3YAlO_3.Al_2O_3$; that is, $Y_3Al_5O_{12}$. Because these substitutions are possible, a general formula $\{C_3\}[A_2](D_3)O_{12}$ is often used for garnets, where O denotes the oxygen ion or atom and C, A and D denote cations.

The space group is $Ia\bar{3}d$ and so the crystal structure clearly has a b.c.c. lattice. There are eight formula units per unit cell. There are 96 so-called h sites, which are occupied by oxygen. $\{C_3\}$ denotes an ion in a tetrahedral site; that is, a site surrounded by four oxygen ions. In grossular, $3CaO.Al_2O_3.3SiO_2$, these sites would be occupied by silicon. There are 24 of these sites per unit cell. $[A_2]$ represents an octahedral site, which is surrounded by six oxygen ions; there are 16 of these per unit cell. These would be occupied by aluminium in grossular. (D_3) denotes the so-called dodecahedral site, which is surrounded by eight oxygen ions; this would be occupied by Ca in grossular. This site is variously described as a triangular dodecahedron, hence the name, or as a distorted cube. It is illustrated in Figure 3.25. A triangular dodecahedron is a polyhedron with 12 faces, each of which is a triangle.

The substitution of an enormous number of cations for one another is possible within the garnet structure. For example, trivalent rare earth ions can be introduced. If strict chemical rules of valence applied, {C} sites would be occupied by four valent cations, such as silicon, by trivalent ions, such as Al, rare earth, Y or Fe^{3+} ions and so on, and (D) by divalent ions, such as Ca, Mg, Fe^{2+} or others. The ability to substitute two cations to balance the charges of two other cations gives rises to a whole host of possibilities for substitution. Hence, a single simple valence rule cannot be followed. Furthermore, since the substitution of large

---

[5] Often termed grossularite in scientific papers written before the 1971 report by the International Mineralogical Association, in which the use of the term grossular was recommended instead [14].

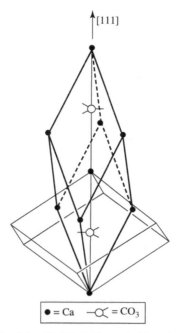

$\uparrow$[111]

| • = Ca | $-\propto$ = CO_3 |

Figure 3.26 The structure of calcite (CaCO_3). The primitive rhombohedral unit cell is shown in bold. This contains two formula units of CaCO_3. The other cell outlined is the smallest 'cleavage rhombohedron' and contains four units of CaCO_3. (After Figure 140 in [16], p. 292.)

ions such as the rare earths is possible, the oxygen ions are pushed apart to accommodate the cations, while retaining the cubic *I* lattice.

3.6.5 Calcite, CaCO_3 ($R\bar{3}c$)

Many complicated structures are most easily described as distorted forms of the simpler ones. For instance, the structure of calcite (one form of CaCO_3) can be derived from that of sodium chloride by identifying the sodium ions with calcium ones and the carbonate radical CO_3 with Cl. If the sodium chloride structure is imagined to be compressed along a [111] body diagonal until the angle between the axes, originally 90°, becomes 101.89°, the doubly primitive cell of calcite containing four formula units is produced. Cleavage occurs parallel to {100} of this cell. The CO_3 group is triangular, with the C in the centre of the triangle and the plane of the triangle normal to the direction [111] of this rhombohedral cell. This cell is not a true unit cell – it is a subcell of a true multiply primitive rhombohedral unit cell in which the repeat direction along [111] needs to be doubled.

The primitive unit cell of calcite contains just two formula units of CaCO_3. It is shown in Figure 3.26, where it is compared with the cleavage rhombohedral pseudo unit cell. Calcium ions are at (0, 0, 0) and $\left(\frac{1}{2}, \frac{1}{2}, \frac{1}{2}\right)$. The centre of the CO_3 radical is at $\pm(u, u, u)$. Note that u is close to $\frac{1}{4}$ in all examples of this structure; in calcite it is 0.259.

The primitive cell of calcite has $a = 6.375$ Å and $\alpha = 46.07°$; the cleavage pseudo unit cell has $a = 6.424$ Å and $\alpha = 101.89°$, while the true multiply primitive cleavage unit cell has $a = 12.828$ Å and $\alpha = 101.89°$. The hexagonal unit cell describing the calcite crystal structure has $a = 4.989$ Å and $c = 17.062$ Å.

3.7 Interatomic Distances

Interatomic distances can be derived from the measured lattice parameters of simple structures with the same accuracy as the parameters of the unit cell. For instance, in copper, the separation of the centres of adjacent atoms is simply $1/\sqrt{2}$ times the cell edge. Values of the interatomic distances for most metals and for some other elements deduced in this way are given in Section A.7.1 of Appendix 7. They are often useful in considering 'atomic radii', if the crystal structure is viewed as being made up of spheres in contact. A difference is found between the values of the atomic radii deduced from different crystal structures of the same element when the element shows allotropic forms. The mineralogist Victor M. Goldschmidt showed that contractions of about 3% and 12% occur when a given element alters its structure from one of coordination 12 (e.g. c.c.p.) to one of coordination 8 and 4, respectively.

In crystals of compounds, interatomic distances can again be deduced from measured lattice parameters. They are useful in considering structures of compounds in terms of hard spheres in contact. However, there is a problem in dividing up the distance between two unlike atoms or ions so as to give each its own characteristic radius. This problem can only be solved by making a theoretical estimate of the size of at least one ion. Consequently, all values of ionic radii are part experimental, part theoretical in origin. Some values of ionic radii are also given in Section A.7.1 of Appendix 7.

In crystals of compounds, the state of ionization of an atom may be quite different from that in the crystal of the element, and its size will differ accordingly. The value of the ionic radius of a metal is usually less than that of the atomic radius, defined as half the distance of closest approach of atoms in the element. This is because metals form positive ions in which the electrons are drawn inwards by the excess positive charge on the nucleus. Conversely, the ionic radius of an electronegative element is usually much greater than the atomic radius.

3.8 Solid Solutions

Many pure metals dissolve large quantities of other elements to form solid solutions. If the solute element is also metallic, the solute atom merely substitutes for the solvent atom in the crystal structure, as shown schematically in Figure 3.27a. Such a solution is called a *substitutional solid solution*. Another type is the *interstitial solid solution*, in which the solute element resides between the atoms of the solvent (Figure 3.27b). Other elements besides metals, and also inorganic compounds, form solid solutions, but, since the great majority of the elements are metallic and many of them have similar chemical properties, the formation of solid solutions is most important for metals.

Solid solutions, rather than chemical compounds, are more likely to form the more similar are the chemical properties of the components. Gold dissolves silver in all proportions and NaCl dissolves KCl. Interstitial solid solutions are often formed when elements that are

(a)   (b)

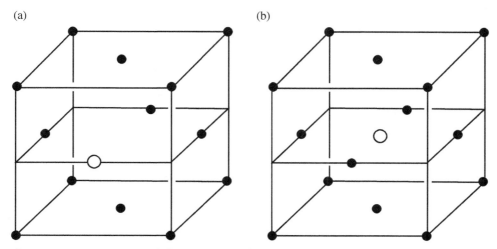

Figure 3.27   A c.c.p. crystal showing (a) a substitutional solid solution and (b) an interstitial solid solution. The solute element in (a) and (b) is shown as a circle

expected to form small atoms or ions (e.g. H, C, B, O, N) dissolve in a crystal. The two types of solution may in all cases be distinguished by density measurements and measurements of the volume of the unit cell of the solid solution. The density ρ of the crystal is given by:

$$\rho = \frac{vM}{V}$$

where M is the molecular weight (in Mg), V is the volume of the unit cell (in m^3) and v is the number of formula units per unit cell. In the pure material, v is an integer. In a substitutional solid solution, v is the same integer, but M alters to the average molecular weight $\overline{M}$ given by the chemical composition of the solution.[6] In an interstitial solution, v is again the same for the solute but the density is increased. In a *binary* interstitial solution (two components), the value of ρ is:

$$\rho = \frac{v}{V}\left(M + \frac{n_{si}}{n_s}M_{si}\right)$$

where n_{si}/n_s is the ratio of the mole fraction of solute interstitial to that of solvent and M_{si} is the molecular weight of the solute interstitial. An example of the determination of the type of solid solution is given in Problem 3.9.

The various component atoms of a solid solution are usually randomly distributed among the sites available for them, but, in some, below a certain temperature, the distribution ceases to be random and what is called ordering occurs. Ordering is most easily described for a metallic solid solution. Part of the (111) plane of a disordered alloy of copper, shown as circles, containing 25 at% of gold, shown as shaded spheres, is shown in Figure 3.28a. Above a temperature of about 390 °C, the copper and gold atoms occur in any of the positions at the lattice points of a c.c.p. lattice. There is no preferred position for gold or copper. Therefore,

---

[6] $\overline{M} = \frac{n_1M_1}{n} + \frac{n_2M_2}{n} + \cdots$ where n_1 is the number of moles of component 1 of molecular weight M_1 in the solution and $n = n_1 + n_2 + n_3 + \cdots$.

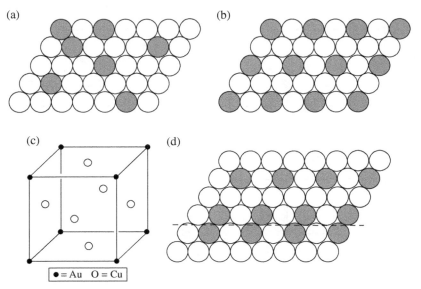

Figure 3.28 (a) A (111) plane of the disordered form of Cu_3Au. (b) A (111) plane of the ordered form of Cu_3Au. (c) The unit cell of the ordered form of Cu_3Au. (d) Schematic of an antiphase domain boundary (shown dashed) on a (111) plane of ordered Cu_3Au

while on average a (111) plane will have 25 at% gold and 75 at% copper, small regions of the plane chosen at random may have more, or less, than 25 at% gold. The part of the plane of the disordered form of Cu_3Au chosen at random in Figure 3.28a happens to be deficient in Au and richer in Cu.

In equilibrium below 375 °C, a (111) plane would appear as in Figure 3.28b – the gold atoms are all surrounded by copper atoms and there is a regular arrangement of both the gold and the copper atoms. Such a structure is called an *ordered solid solution*. It is apparent from Figure 3.28c that in the fully ordered state for Cu_3Au a conventional unit cell can be chosen with the gold atoms at cube corners (0, 0, 0) and the copper atoms at the midpoints of all of the faces. This cell is the primitive unit cell of a simple cubic *superlattice*. In this particular case, the superlattice is known as the $L1_2$ superlattice. The space group symmetry of the crystal structure changes from $Fm\bar{3}m$ in the high-temperature disordered phase to $Pm\bar{3}m$ in the low-temperature ordered phase. In this example, the conventional unit cells of the high-temperature disordered phase and low-temperature ordered phase are identical in volume, but the primitive unit cell of the ordered phase has a volume four times that of the primitive unit cell of the disordered phase.

Order–disorder changes occur in many solid solutions. The fully ordered state is always of lower symmetry than the disordered one, and usually possesses a lattice with larger cell dimensions, which is called a *superlattice*.

The essential grouping in the ordered state of this alloy, Cu_3Au, is one with all the gold atoms surrounded by copper atoms, as shown in Figure 3.28c. When ordering starts in a large crystal it may be 'out of step' in the various parts of the crystal. If this occurs, the ordering may be perfect within various regions of the crystal (as in Figures 3.28b and c). These regions are called *domains*. Where the domains are in contact, the requirement

Table 3.5 Some examples of superlattice types.

Superlattice	Examples
$B2$ (CuZn)[a]	AgMg, AuCd, CoAl, CuZn, FeAl, FeCo, NiAl, NiZn, Ti_2AlZr, $AlFe_xNi_{1-x}$
DO_3 (Fe_3Al)	Cu_3Al, Cu_3Sb, Fe_3Al, Fe_3Si, Mg_3Li
$L1_2$ (Cu_3Au)	Au_3Cd, Co_3V, Cu_3Au, Ni_3Al, Ni_3Fe, Ni_3Mn, Pt_3Fe, Zn_3Ti
DO_{19} (Mg_3Cd)	Ag_3In, Cd_3Mg, Co_3Mo, Co_3W, Fe_3Sn, Mg_3Cd, Ni_3In, Ni_3Sn, Ti_3Al, Ti_3Sn

[a] The designation $B2$, DO_3, etc.. arises because in the literature the designation of structure type used in the early volumes of *Strukturbericht* is often followed. The term $L2_0$ is still occasionally used when referring to ternary alloys with the $B2$ crystal structure. (See the Suggestions for Further Reading at the end of the chapter.)

that gold atoms are surrounded by copper atoms is not met, as is shown in Figure 3.28d. The dotted line in Figure 3.28d indicates the trace in (111) of what are called *antiphase domain* boundaries. In three dimensions, the boundaries are walls separating neighbouring domains. Since the neighbouring atoms are not fully ordered at the domain boundaries, the boundaries represent a source of extra energy in an ordered crystal. Heating a crystal for a long time near to the ordering temperature can lead to their complete removal.

The order we have just described is called *long-range order* because within any domain one type of lattice site is preferred for a particular atom. Many solid solutions, while not showing long-range order, do not show a truly random distribution of the atoms; unlike atoms occur more frequently as near neighbours than they would by pure chance. Such a state of affairs is called *short-range order*. This is very common and is shown by some ordered solutions when they are heated above their ordering temperatures.

The structures of some ordered solid solutions are illustrated in Figure 3.29. Examples of materials showing these structures are given in Table 3.5. The $B2$ (or $L2_0$) order– disorder change is characterized by a b.c.c. structure in the disordered state which changes to the caesium chloride structure on ordering. This occurs in CuZn. The perfectly ordered structure has the composition AB (Figure 3.29a). The superlattice is then simple cubic.

The DO_3 superlattice type with perfectly ordered composition AB_3 also has a b.c.c. structure in the disordered state. This superlattice type occurs in Fe_3Al. The ordered state is shown in Figure 3.29b. It is most easily described by saying that the superlattice is composed of four equal interpenetrating face-centred cubic lattices with the origins at $(0, 0, 0)$ for lattice 1, $\left(\frac{1}{2}, 0, 0\right)$ for lattice 3, $\left(\frac{1}{4}, \frac{1}{4}, \frac{1}{4}\right)$ for lattice 4 and $\left(\frac{3}{4}, \frac{1}{4}, \frac{1}{4}\right)$ for lattice 2. The ordered state consists of B atoms occupying the sites of lattices 2, 3 and 4 with A atoms at type 1 lattice sites.

The $L1_2$ superlattice has already been described (Figure 3.28c). The fully ordered condition requires the composition AB_3, as in Cu_3Au. A related superlattice type, also of ideal composition AB_3, is called DO_{19} and is typified by Mg_3Cd. The disordered state is the hexagonal close-packed structure. The ordered structure (Figure 3.29c) can be described as four interpenetrating h.c.p. structures in parallel orientation, with the c-axis the same length as in the ordered alloy but the lattice parameter a_{os} twice that of the corresponding disordered alloy. The origins of the sublattices in the ordered state are at $(0, 0, 0)$ for sublattice 1, $\left(\frac{1}{2}, \frac{1}{2}, 0\right)$ for sublattice 2, $\left(\frac{1}{2}, 0, 0\right)$ for sublattice 3 and $\left(0, \frac{1}{2}, 0\right)$ for sublattice 4. B atoms occupy the sites of sublattices 2, 3 and 4 and A atoms are found at the points of sublattice 1.

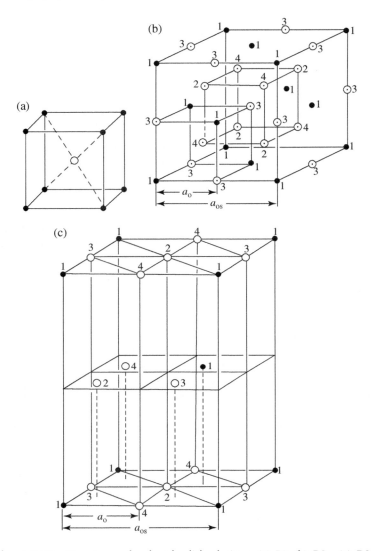

Figure 3.29   Structures of ordered solid solutions: (a) $B2$, (b) $D0_3$, (c) $D0_{19}$

3.9   Polymers

Many of the unique physical properties associated with polymers result from their ability to crystallize. In general, polymers are synthesized by the repeated addition to the growing chain of one or more small chemical units called *monomers*. Typically, polymers can have molecular weights between several hundred and several million. The predominant bond between atoms in a polymer molecule is the covalent bond, with a dissociation energy of the order of 2 eV. The same type of strong bond is found in diamond. Many polymers have a zigzag backbone consisting of covalently bonded carbon atoms. The angle between successive single bonds is usually a few degrees larger than the tetrahedral angle of $\cos^{-1}(-1/3) = 109.47°$. A change in shape of the molecule may occur by rotation about these bonds in preference to either elongation or bending of the bonds.

If the monomers are polymerized according to a regular pattern of side groups or branches the polymer is known as a *tactic* polymer and is usually able to crystallize. If the monomers are polymerized in an irregular pattern, the resulting *atactic* polymer usually cannot crystallize. The class of tactic polymers is divided into further subgroups depending upon the arrangement of side branches. For example, if an *isotactic* polymer is stretched out and viewed along its backbone, all of the branches lie along a single line, whereas in a *syndiotactic* polymer they alternate between one side of the backbone and the other.

Instantaneously induced London dispersion forces[7] hold together molecules or portions of molecules not connected by covalent (or 'primary') bonds. These forces vary as the inverse sixth power of the separation of the atoms. The inert gases are bound by similar forces. These forces are at least two orders of magnitude weaker than the covalent bond and are part of the family known as 'secondary bonds' or 'van der Waals forces'.[8] Another particularly important secondary bond for determining the crystal structure of some polymers is the hydrogen bond, since its strength is greater than that of the other secondary bonds.

The attractive force between portions of polymer molecules that are in a crystalline lattice is greater than that between portions outside the crystal, so the density of a crystalline polymer is as much as 15% greater than that of the super-cooled melt at the same temperature. Polymer crystals tend to be lamellar in external form. One, and sometimes two, dimensions are of the order of 100 Å, whereas the other dimensions can be two or more orders of magnitude larger.

The periodic placement of atoms in a tactic polymer molecule in a crystal is expressed in terms of the *repeat distance* or *repeat unit*. The repeat unit is the simplest arrangement of atoms by which the operation of linear translation alone without any rotation will generate the structure of the extended molecule. It therefore defines the conformation of the molecule in the crystal. The repeat unit usually contains one or more chemical repeat units. Some repeat units for polyethylene, $(-CH_2-)_n$, and polytetrafluoroethylene, $(-CF_2-)_n$, are shown in Figure 3.30.

The packing of the molecules is expressed in terms of the unit cell with the familiar axes **a, b** and **c** and angles α, β and γ, as in Chapter 1 of this book, and contains one or more repeat units. An example of the unit cell of polyethylene is shown in Figure 3.31. This illustrates two features of the unit cells of the crystalline forms of polymers: the unit cells tend to be of low symmetry, and they are relatively large in comparison with unit cells in metals, ceramics and minerals. It is apparent from Figure 3.31 that in polyethylene, the crystal forms with continuous covalent bonding along the c-axis and with secondary bonding in the form of London dispersion forces along the a- and b-axes. The crystal structures of some common polymers are listed in Table 3.6.

Polymer crystal imperfections may result from such factors as terminal groups (ends of molecules), branches and entanglements of the molecules, as well as modifications of the numerous types of defect found in crystals composed of small molecules. The same polymer

---

[7] Named after Fritz London (1900–1954).
[8] Named after Johannes van der Waals (1837–1923).

(a)

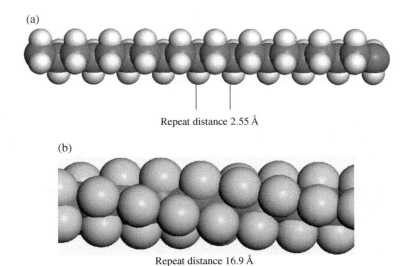

Repeat distance 2.55 Å

(b)

Repeat distance 16.9 Å

Figure 3.30   Molecular models of (a) polyethylene ($-CH_2-$)$_n$ and (b) polytetrafluoroethylene ($-CF_2-$)$_n$, where the C atoms are contained in the chain (dark grey) and the hydrogen and fluorine atoms are represented as lighter-coloured spheres. Note that in (b) the molecule forms a helix, whereas in (a) the conformation is a planar zigzag

chain may run many times through the same crystal and through amorphous regions into neighbouring crystals, as illustrated in Figure 3.32.

3.10   Additional Crystal Structures and their Designation

The crystal structures which we have considered in this chapter are important in the context of metallic, ceramic and polymeric materials within materials science, but they constitute a very small fraction of the crystal structures solved to date, some 900 000 or so, listed in the various resources given under Suggestions for Further Reading. Nevertheless, they are an important fraction, because they are likely to be candidates for engineering applications of materials. Furthermore, an appreciation of these relatively simple crystal structures is invaluable in understanding how such materials react to external factors such as temperature and stress, as we will demonstrate in Part II of this book.

The description of crystal structures which we have just given is based on stating the space group and the positions of the individual atoms. This conventional description is useful for structures in which the number of atomic positions is limited – say, less than 100 or so. When more complicated crystal structures are to be considered, descriptions in terms of clusters of atoms may be more suitable; we give examples of these for metallic and intermetallic materials in Sections 4.5 and 4.6.

During the historical development of the methods of solving and of listing the results of the elucidation of crystal structures, now more than a century old, various empirical notations for describing crystal structures were developed. It is very useful to be familiar with a small number of these when reading the literature. Two important notations are the

(a)

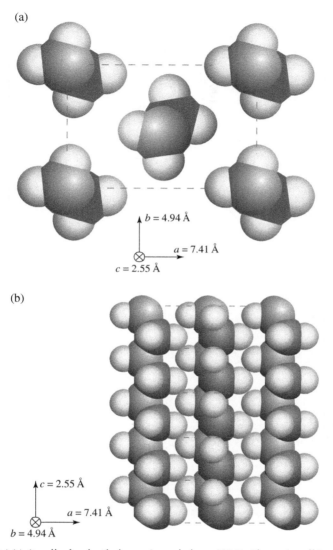

$b = 4.94$ Å

$a = 7.41$ Å

⊗

$c = 2.55$ Å

(b)

$c = 2.55$ Å

$a = 7.41$ Å

⊗

$b = 4.94$ Å

Figure 3.31   (a) Unit cell of polyethylene viewed along [001]. The unit cell is orthorhombic. At the cell centre the planar zigzag backbone is rotated by almost 100° about [001] with respect to the chains at the cell corner where the planar zigzag backbone is parallel to (110). As shown in (b) viewed along [010], the strongly bound polymer chain passes between many unit cells in the same crystal; it may also pass unbroken and in approximately the same orientation into nearby crystals (see Figure 3.32)

Table 3.6 Crystal structures of some common polymers.

Polymer and reference	Chemical formula	Type	Crystal system and space group	a, b, c (Å)	α, β, γ (°)
Poly(ethylene) (PE) [17, 18]	$-[CH_2]_n-$	I	Orthorhombic $Pnam$	7.418, 4.946, 2.546	90, 90, 90
Poly(ethylene) (PE) [19]	$-[CH_2]_n-$	II	Monoclinic $C2/m$	8.09, 2.53, 4.79	90, 107.9, 90
Poly(ethylene terephthalate) (PET) [20]	$-[O-(CH_2)_2-O-CO-C_6H_4-CO]_n-$		Triclinic $P\bar{1}$	4.56, 5.94, 10.75	98.5, 118, 112
Poly(vinylchloride) (PVC) [21]	$-[CH_2CHCl]_n-$		Orthorhombic $Pcam$	10.24, 5.24, 5.08	90, 90, 90
Poly(propylene) (PP) [22]	$-[CH(CH_3)]_n-$	synd.	Orthorhombic $C222_1$	14.4, 5.6, 7.4	90, 90, 90
Poly(propylene) (PP) [22]	$-[CH(CH_3)]_n-$	α iso.	Monoclinic $P2_1/c$	6.65, 20.96, 6.5	90, 99.80, 90
Poly(styrene) (PS) [23]	$-[CH_2CH(C_6H_5)]_n-$	iso.	Trigonal $R\bar{3}c$	21.9, 21.9, 6.65	90, 90, 120
Poly(hexamethylene adipamide) (Nylon 66) [24]	$-[NH-(CH_2)_6-NH-CO-(CH_2)_4-CO]_n-$	I	Triclinic $P\bar{1}$	4.9, 5.4, 17.2	48.5, 77, 63.5

Note: It is usual for polymer crystal structures to choose the z-direction so that it is along the axis of the polymer chain (synd., syndiotactic; iso., isotactic).

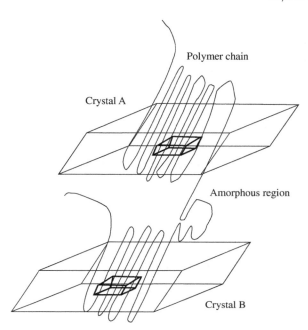

Polymer chain

Crystal A

Amorphous region

Crystal B

Figure 3.32 Illustration of a continuous polymer chain running through neighbouring crystals A and B. The small parallelepipeds indicate unit cells in the two crystals

Pearson symbol and the Strukturbericht designation. Reference to these is given in the Suggestions for Further Reading. It is usually most simple and quick to Google either 'Pearson symbol' or 'Strukturbericht designation'. It is also straightforward to obtain a representation of a structure given a specific name, such as Hägg phase, topologically close-packed structure or Laves phase using Google. A comprehensive account of methods of describing very complicated inorganic compounds is given in [25]; this has a useful appendix of prototype crystal structures with their corresponding Strukturbericht designations and Pearson symbols.

Problems

3.1 Show that if a crystal has the h.c.p. structure and the atomic sites are occupied by equal-sized hard spheres in contact, the axial ratio c/a will be equal to $\sqrt{8/3}$.

3.2 What point group describes the symmetry of the interatomic forces acting on (a) a carbon atom in the diamond structure, (b) Cs in the CsCl structure, (c) Zn in the sphalerite structure, (d) Zn in the wurtzite structure, (e) Ni in the NiAs structure, (f) As in the NiAs structure? *Hint*: draw a stereogram showing the arrangement of the neighbours at various (close) distances from a given atom.

3.3 In the fluorite structure, CaF_2, find the spacing along [111] of successive planes of (a) Ca atoms and (b) F atoms, in terms of the conventional cell side a.

3.4 Mercury has a distorted c.c.p. crystal structure with a conventional rhombohedral unit cell with $a = 2.992$ Å and $\alpha = 70.74°$ at 78 K (Section 3.3). Confirm that a face-centred rhombohedral unit cell for mercury would have lattice parameters $a' = 4.577$ Å and axial angle $\alpha' = 98.36°$ at 78 K.

3.5 Sketch the structure of white tin. Take the c/a ratio as 0.545 and show that the ratio of the separation of nearest neighbours to that of second-nearest neighbours is 0.950. Find the position of the 4_1 axis and of the axial glide planes in the unit cell of white tin.

3.6 Show with a diagram that if an element possessed the structure of bismuth with a value of α (see Table 3.2) of 60°, the lattice points in Figure 3.15 would form a c.c.p. lattice. If, in addition, $u = 0.25$, show that the atomic positions would form a simple cubic lattice.

3.7 Molybdenum disulfide MoS_2 has a hexagonal primitive lattice with atoms in the following positions:

$$\text{Mo at } (0,0,0) \text{ and } \left(\tfrac{2}{3},\tfrac{1}{3},\tfrac{1}{2}\right)$$

$$\text{S at } \left(0,0,\tfrac{1}{2}+z\right), \left(0,0,\tfrac{1}{2}-z\right), \left(\tfrac{2}{3},\tfrac{1}{3},z\right), \left(\tfrac{2}{3},\tfrac{1}{3},\bar{z}\right)$$

(a) Draw a plan on (0001) of a number of unit cells.
(b) Describe the stacking of the planes along [0001] in the A, B, ..., α, β notation and compare with that in CaF_2.
(c) Locate the centres of symmetry.
(d) Re-express the coordinates of the atoms with respect to a new origin at a centre of symmetry having positive coordinates with respect to the old origin. Write the coordinates in the short form $\pm (x, y, z)$.

3.8 The intermetallic compound $CaCu_5$ has a hexagonal P lattice with $a = 5.092$ Å and $c = 4.086$ Å. The atomic coordinates are:

$$\text{Ca at } (0,0,0)$$

$$\text{Cu (1) at } \left(\tfrac{1}{3},\tfrac{2}{3},0\right) \text{ and } \left(\tfrac{2}{3},\tfrac{1}{3},0\right)$$

$$\text{Cu (2) at } \left(\tfrac{1}{2},0,\tfrac{1}{2}\right), \left(0,\tfrac{1}{2},\tfrac{1}{2}\right) \text{ and } \left(\tfrac{1}{2},\tfrac{1}{2},\tfrac{1}{2}\right)$$

Draw a projection of the structure on (0001).
(a) How many formula units are there per unit cell?
(b) Find the interatomic distances: Ca–Cu (1), Ca–Cu (2), Cu (1)–Cu (2), Cu (2)–Cu (2).
(c) Calculate the density of $CaCu_5$.

At. wts. : Ca $= 40.08$, Cu $= 63.54$. Mass of H atom $= 1.66 \times 10^{-27}$ kg

3.9 A solution of carbon in c.c.p. iron has a density of $8.142\,Mg\,m^{-3}$ and a lattice parameter of $3.583\,Å$. The solution contains 0.8% by weight of carbon. Is it an interstitial or a substitutional solid solution?

At. wts.: $Fe = 55.85, C = 12.01$. Mass of H atom $= 1.66 \times 10^{-27}\,kg$

3.10 A crystal of wüstite (approximate composition FeO) has the sodium chloride structure and contains 76.08% by weight of iron. The density is found to be $5.613\,Mg\,m^{-3}$ and the lattice parameter is $4.2816\,Å$. Does the crystal contain iron ion vacancies or interstitial oxygen ions? How many vacancies or interstitials are there per m^3?

At. wts.: $Fe = 55.85, O = 16$. Mass of H atom $= 1.66 \times 10^{-27}\,kg$

3.11 Packing arrangements that are periodic, but in which the packing density is relatively low, are relevant for certain microporous crystal structures such as zeolites. A two-dimensional example is shown here. To what space group does this belong?

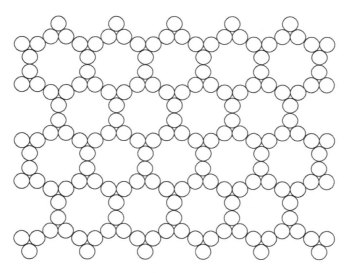

Figure for Problem 3.11

Choosing a unit cell so that the origin is at one of the points with the highest symmetry in this pattern, show that the coordinates of equivalent positions of the circles can be described as $x, \bar{x}; x, 2x; \overline{2x}, \bar{x}; \bar{x}, x; \bar{x}, \overline{2x}; 2x, x$, where $x = 1 - (1/\sqrt{3})$. Hence show that the packing fraction is $\pi(7\sqrt{3} - 12) = 0.391$.

Suggestions for Further Reading

Allen, S.M. and Thomas, E.L. (1999). *The Structure of Materials*. New York: Wiley.
Bragg, W.L. and Claringbull, G.F. (1965). *Crystal Structures of Minerals*. London: G. Bell and Sons.
Buchanan, R.C. and Park, T. (1997). *Materials Crystal Chemistry*. New York: Marcel Dekker Inc.
de Jong, W.F. (1959). *General Crystallography: A Brief Compendium*. New York: W.H. Freeman.
Elias, H.-G. (1997). *An Introduction to Polymer Science*. Weinheim: VCH.

Gaines, R.V., Skinner, H.C.W., Foord, E.E., Mason, B. and Rosenzweig, A. (1997). *Dana's New Mineralogy*, 8th Edition. New York: Wiley.

Geil, P.H. (1963). *Polymer Single Crystals*. New York: Wiley.

Hyde, B.G., Thompson, J.G. and Withers, R.L. (1998). Crystal structures of principal ceramic materials, M.V. Swain (ed.). In: *Structure and Properties of Ceramics, Materials Science and Technology*, vol. 11 (eds. R.W. Cahn, P. Haasen and E.J. Kramer). Chichester and Weinheim: Wiley–VCH Chapter 1.

Pearson, W.B. (1967). *A Handbook of Lattice Spacings and Structures of Metals and Alloys*, vol. 2. Oxford: Pergamon Press.

Rousseau, J.-J. (1998). *Basic Crystallography*. Chichester: Wiley.

Villars, P. and Calvert, L.D. (1991). *Pearson's Handbook of Crystallographic Data for Intermetallic Phases*, 2nd Edition. Metals Park, OH: American Society for Metals.

Wunderlich, B. (1973). *Macromolecular Physics, Volume 1: Crystal Structure, Morphology, Defects*. New York: Academic Press.

Wyckoff, R.W.G. (1963). *Crystal Structures*, 2nd Edition. New York: Wiley Interscience.

In addition to these books, there are a number of scientific journals where inorganic, metal–organic, organic and mineral crystal structures are published, for example,: *Acta Crystallographica Section B: Structural Science, Crystal Engineering and Materials, Acta Crystallographica Section C: Structural Chemistry, Acta Crystallographica Section D: Structural Biology, Acta Crystallographica Section E: Crystallographic Communications, Acta Crystallographica Section F: Structural Biology Communications, American Mineralogist, Journal of Solid State Chemistry* and *Zeitschrift für Kristallographie*. The International Centre for Diffraction Data maintains a powder diffraction database for the identification of crystalline materials, www.icdd.com. Other Web-based databases include the Inorganic Crystal Structure Database (ICSD), www2.fiz-karlsruhe.de/icsd_home.html, the Cambridge Structural Database (CSD), www.ccdc.cam.ac.uk, and the American Mineralogist Crystal Structure Database at www.minsocam.org.

References

[1] Aroyo, M.I. (ed.) (2016). *International Tables for Crystallography*, 6th, Revised Edition, Vol. A: *Space-Group Symmetry*, published for the International Union of Crystallography. Chichester: Wiley International.

[2] Bragg, W.L. and Claringbull, G.F. (1965). *Crystal Structures of Minerals*. London: G. Bell and Sons.

[3] Hales, T.C. (2005). A proof of the Kepler conjecture. *Ann. Math.* 162: 1065–1185.

[4] Hales, T.C. (2006). Historical overview of the Kepler conjecture. *Discret. Comput. Geom.* 36: 5–20.

[5] Hales, T.C. and Ferguson, S.P. (2006). A formulation of the Kepler conjecture. *Discret. Comput. Geom.* 36: 21–69.

[6] Hales, T.C. (2006). Sphere packing. III. Extremal cases. *Discret. Comput. Geom.* 36: 71–110.

[7] Hales, T.C. (2006). Sphere packing. IV. Detailed bounds. *Discret. Comput. Geom.* 36: 111–166.

[8] Ferguson, S.P. (2006). Sphere packing. V. Pentahedral prisms. *Discret. Comput. Geom.* 36: 167–204.

[9] Hales, T.C. (2006). Sphere packing. VI. Graphs and linear programs. *Discret. Comput. Geom.* 36: 205–265.

[10] Kwei, G.H., Lawson, A.C., Billinge, S.J.L. and Cheong, S.-W. (1993). Structures of the ferroelectric phases of barium titanate. *J. Phys. Chem.* 97: 2368–2377.

[11] Corker, D.L., Glazer, A.M., Whatmore, R.W., Stallard, A. and Fauth, F. (1998). A neutron diffraction investigation into the rhombohedral phases of the perovskite series $PbZr_{1-x}Ti_xO_3$. *J. Phys. Condens. Matter* 10: 6251–6269.

[12] Skakle, J.M.S. (1998). Crystal chemistry substitutions and doping of $YBa_2Cu_3O_x$ and related superconductors. *Mater. Sci. Eng. R* 23: 1–40.

[13] Farrell, E.F., Fang, J.H. and Newnham, R.E. (1963). Refinement of the chrysoberyl structure. *Am. Mineral.* 48: 804–810.

[14] I.M.A. Nomenclature Commission (1971). International Mineralogical Association: Commission on new minerals and mineral names. *Mineral. Mag.* 38: 102–105.

[15] Abrahams, S.C. and Geller, S. (1958). Refinement of the structure of a grossularite garnet. *Acta Crystallogr.* 11: 437–441.

[16] Ewald, P.P. and Hermann, C. (1931). *Strukturbericht 1913–1928*. Leipzig: Akademische Verlagsgesellschaft M.B.H.

[17] Bunn, C.W. (1939). The crystal structure of long-chain normal paraffin hydrocarbons. The 'shape' of the –CH_2– group. *Trans. Faraday Soc.* 35: 482–491.

[18] Zugenmaier, P. and Cantow, H.-J. (1969). Ein einfaches Verfahren zur Indizierung von Röntgenweitwinkel-Reflexen von Substanzen mit orthorhombischer Elementarzelle. *Kolloid-Zeitschrift und Zeitschrift für Polymere* 230: 229–236.

[19] Seto, T., Hara, T. and Tanaka, K. (1968). Phase transformation and deformation processes in oriented polyethylene. *Jpn. J. Appl. Phys.* 7: 31–42.

[20] Daubeny, R. de P., Bunn, C.W. and Brown, C.J. (1954). The crystal structure of polyethylene terephthalate. *Proc. R. Soc. London, Ser. A* 226: 531–542.

[21] Wilkes, C.E., Folt, V.L. and Krimm, S. (1973). Crystal structure of poly(vinyl chloride) single crystals. *Macromolecules* 6: 235–237.

[22] Lotz, B., Wittmann, J.C. and Lovinger, A.J. (1996). Structure and morphology of poly(propylenes): a molecular analysis. *Polymer* 37: 4979–4992.

[23] Natta, G., Corradini, P. and Bassi, I.W. (1960). Crystal structure of isotactic polystyrene. *Nuovo Cimento Suppl.* 15: 68–82.

[24] Bunn, C.W. and Garner, E.V. (1947). The crystal structures of two polyamides ('nylons'). *Proc. R. Soc. London, Ser. A* 189: 39–68.

[25] Ferro, R. and Saccone, A. (2008). *Intermetallic Chemistry*, Pergamon Materials Series, vol. 13. Oxford: Elsevier.

4

Amorphous Materials and Special Types of Crystal–Solid Aggregates

4.1 Introduction

So far in this book, we have been almost exclusively concerned with the description of the crystalline state where the atoms are regularly arranged in a periodic pattern in three dimensions and show translational symmetry. In the discussion of polymers in Section 3.9, reference has been made to the difficulty in crystallizing atactic polymers, without specifying fully the implications of what this would mean for the polymer under consideration.

There are many examples of solid materials where atoms are not regularly arranged in a periodic pattern in three dimensions, thereby implying that there is a degree of disorder in the constitution of the material. A familiar example is glass: this is an inorganic material in which there is no periodic pattern in three dimensions at all, but where it is known from the results of X-ray diffraction experiments that there are well-defined local structures within the material; in this case the local structures are structural units composed of silica tetrahedra in which each silicon atom is bonded to four adjacent oxygen atoms.

In this chapter we will describe a number of three-dimensional examples in which there is no longer a completely regular arrangement of atoms in a periodic pattern in three dimensions. Despite this lack of regular arrangement, there are still meaningful ways of describing the ensemble of atoms and the imperfection in these arrangements, whether it is in a higher-dimensional space, through the use of radial distribution functions or by approximating a non-periodic arrangement, such as a foam, as being periodic.

Crystallography and Crystal Defects, Third Edition. Anthony Kelly and Kevin M. Knowles.
© 2020 John Wiley & Sons Ltd. Published 2020 by John Wiley & Sons Ltd.
Companion website: www.wiley.com/go/kelly/crystallography3e

4.2 Amorphous Materials

Solids are said to be amorphous when they are homogenous and uniform on a scale of more than a few atomic or molecular units, provided that the 'average environment' is under consideration. This average is invariant against an arbitrary translation, rotation, or reflection. Examples of amorphous materials are liquids and glasses. A glass possesses the highest possible symmetry group of three-dimensional space, the Euclidean group $E(3)$, in which all translations, rotations, and reflections are possible.

A glass is formed when, on cooling, a liquid fails to crystallize and instead uniformly and continually congeals into a solid. The glass has then the essential structure of a liquid, but this is combined with the resistance to shear deformation of a solid. Non-crystalline or amorphous metallic alloys can be obtained by a number of methods, of which the most popular is rapid quenching, although there are now examples of bulk metallic glasses (BMGs) which can be obtained by cooling as slowly as $1 \, \text{K s}^{-1}$. BMGs are multicomponent systems such as $Ti_{40}Cu_{36}Pd_{14}Zr_{10}$. The principle in producing BMGs is to have a liquid containing so many different metallic species that even with such modest cooling rates the constituent atoms in the liquid do not have enough time to rearrange themselves into the equilibrium crystalline phase(s).

It is worth noting that the amorphous and glassy states of a particular substance are not necessarily identical. Silicon, like water, expands on freezing: the liquid has a higher density ($2.57 \, \text{mg m}^{-3}$) than the solid ($2.52 \, \text{mg m}^{-3}$) at the melting point. Liquid silicon is metallic, whereas crystalline silicon has tetrahedral covalent bonding and is a semiconductor. Amorphous silicon, which can be made by deposition onto a substrate directly from the vapour, or by solidification of liquid silicon, also has local tetrahedral coordination with covalent bonding. Solid amorphous silicon is therefore quite distinct from the liquid and should not be regarded as a glass, which, if formed, would be denser and metallic.

For an idealized glass containing only one atomic species in which the bonding is non-directional, such as in a metal, the structure of the glass approximates that of a random packing of close-packed hard spheres, with a density high enough that there are no cavities large enough to incorporate an additional sphere. The resultant assembly of atoms is called the random closed-packed sphere model (RCPS) and is due principally to Bernal [1]. From Bernal's experiments with ball bearings, and more recently, computer modelling of RCPS assemblies, it is found that the average number of contacts between a given sphere and its neighbours is 8.5. The packing fraction is close to 0.64; this compares with the value of $\pi/(3\sqrt{2})$ or 0.74 for rigid spheres in a close-packing arrangement where the average number of contacts between a given sphere and its neighbours is 12 (Section 3.2). The ratio of these two packing fractions is 0.86_4, which is only roughly equal to the ratio of the density of liquid copper to solid copper at the melting point (which we find to be 0.94_6).

Larger packing ratios for amorphous metals require that the size of the spheres varies, so that smaller atoms can fit into the interstices between larger atoms. With a carefully controlled hierarchy of atom sizes, it is possible to obtain packing fractions close to 0.95 [2]. Using such a sphere packing model to describe an amorphous solid or glass involves stating statistical quantities such as the average packing fraction, the average number of nearest neighbours and the average distance between nearest neighbours for each atom species in the solid.

Another way in which to describe the structure of a glass (or of a liquid), which is useful for amorphous materials generally, is by construction of Voronoi polyhedra [3]. For an amorphous material consisting of a three-dimensional arrangement of atoms, this construction process consists of first drawing straight lines between the centre of an arbitrarily chosen atom and all other nearby atom centres. Then, planes bisecting these lines are constructed. The set of planes nearest to the atom chosen as origin forms a convex polyhedron about the chosen central atom and is called the Voronoi polyhedron. Instead of atoms, molecules could be the 'elementary' species around which such a polyhedron is constructed when this principle is applied to organic materials. The process of constructing a Voronoi polyhedron is readily visualized for a set of spheres, as in Bernal's RCPS model. A two-dimensional example is shown in Figure 4.1a.

The analysis of the structure then consists of characterizing the dimensions and geometry of the polyhedra statistically. Consideration of the two-dimensional illustration in Figure 4.1a shows that, within each polyhedron, all points are closer to the centre of the central circle than to any other. The same will be true of the polyhedra found in three dimensions (e.g. Figure 4.1b), so that the average packing fraction, f_v, can be determined as:

$$f_v = \frac{\frac{4}{3}\pi R^3}{V_{cell}} \tag{4.1}$$

where R is the radius of the atom and V_{cell} the average volume of a Voronoi polyhedron. The average number of faces of the polyhedra gives the average number of nearest neighbours.

The benefit of the Voronoi construction is that it changes the problem of characterizing the amorphous structure from one only involving a set of intermolecular distances into one involving the distribution of other geometric parameters – polyhedron volume, number of faces, number of edges per face and so on. These numbers can be generated by computer. The geometry of polyhedra is considered further in Section 4.4.

4.3 Liquid Crystals

Liquid crystals can be thought of as a structure of matter intermediate between that of a solid and that of a liquid. Examples of liquid crystals are *para*-azoxyanisole (PAA) (Figure 4.2) and polyparaphenylene teraphthalamide (PPTA), more commonly known by its trade name, Kevlar.

Liquid crystals can be made either from a solution of a polymer in a solvent, such as the polymer α-helical poly-γ-benzyl-α,L-glutamate (PBLG) in the solvent dimethylformamide (DMF) [4], in which case they are called *lyotropic*, or by cooling from the melt of a pure polymer, in which case they are termed *thermotropic*.

Liquid crystals were discovered in 1888 by Friedrich Reinitzer, who noted that cholesteryl benzoate, $C_{27}H_{45}.C_7H_5O_2$, appeared to show two melting points, in between which the material looked like a cloudy liquid [5]. Above the higher of the two melting points the liquid was transparent, as expected. The term 'liquid crystal' was given to the state of matter exhibited by cholesteryl benzoate by Otto Lehmann in 1889 [6]. Sometime later, Georges Friedel coined the term 'mesophase' (meso meaning intermediate) to describe a liquid crystal, to reflect the fact that it is between a solid and a liquid [7]. The key structural requirement for the occurrence of liquid crystallinity is that the

(a)

(b)

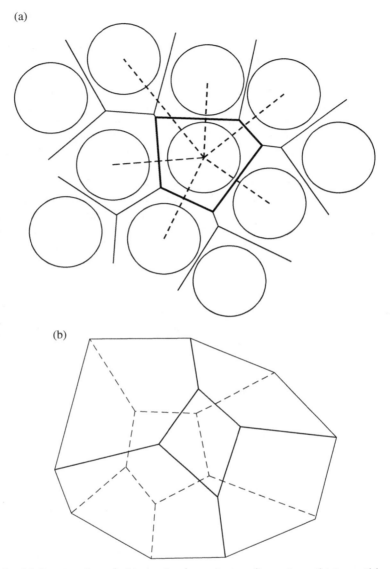

Figure 4.1 (a) Construction of a Voronoi polygon in two dimensions. (b) A possible shape for a Voronoi polyhedron in three dimensions

molecule possesses an anisotropic shape. Mesogenic molecules – those that induce liquid crystallinity – are stiff and inflexible. They either have a large length-to-diameter ratio (*l/d*) in a needle- or thread-like shape, such as PAA, or have a large diameter-to-thickness ratio (*d/h*) and are disc-like, such as liquid crystals formed by a set of linked aromatic rings in a plane. Friedel identified three types of mesogenic molecule: (i) rod-like, such as PAA, which forms a *nematic* (Greek: νεμα – thread) liquid crystal; (ii) *cholesteric*; and (iii) *smectic* (Greek: σμεγμα – soapy). These three types are illustrated in Figure 4.3.

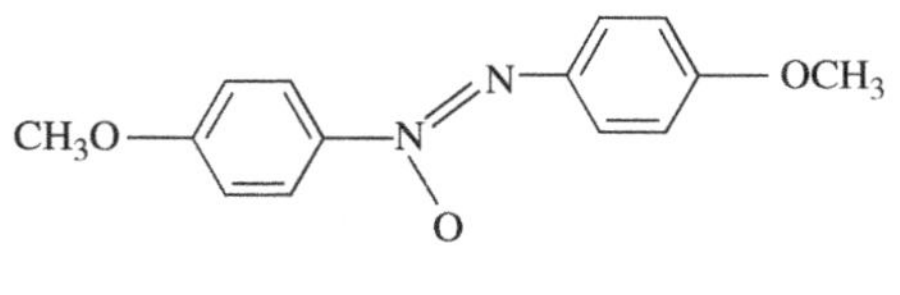

Figure 4.2   *Para*-azoxyanisole (PAA)

It is apparent from Figure 4.3 that the rod-like molecules in a nematic are roughly aligned, with their long axes more or less parallel to the vector **n**, which is known as the *director*. The quality of alignment can be defined by an order parameter *S*. **n** and − **n** are physically equivalent, and so **n** is a non-polar vector. The cholesteric type of liquid crystal illustrated in Figure 4.3b is equivalent to a nematic phase that has been twisted periodically about an axis perpendicular to the direction of the director. This twist in cholesteric crystals is due to the chiral (left- or right-hand) nature of this type of mesogenic molecule. Smectic liquid crystals all possess a basic layer structure, which gives them their soapy feel, because the individual molecules are *amphiphilic*. This word indicates that the molecule has a water-seeking end, which is hydrophilic, and a water-avoiding end, which is hydrophobic. In solution, the molecules pack so that the hydrophilic ends come together, screening one another from the water.

Since Friedel's original classification, it has been realized that liquid crystalline polymers can be constructed by the incorporation of mesogenic units (mers) into long-chain molecules, either by hanging the mesogenic units off the side chain of a long-chain molecule, or by connecting the mesogenic units at the ends; that is, end-to-end.

Cholesteric crystals are thermotropic; here, the anisotropic structure, which is cloudy in appearance and optically birefringent, is formed by the cooling of the isotropic (optically clear in this case) liquid, hence the two melting points and an explanation for the initially puzzling observations made by Reinitzer [5]. A lyotropic liquid crystal is formed by increasing the concentration of mesogenic molecules in a solution at constant temperature. Hence, thermotropic liquid crystals exist over a range of temperature, while lyotropic liquid crystals exist over a range of concentration. The ultrastrong and stiff fibre Kevlar is made by spinning a lyotropic solution of PPTA in sulfuric acid.

Thermotropic and lyotropic liquid crystals can be mixed. Pure soap (e.g. sodium stearate) forms a thermotropic smectic liquid crystal phase. When water is added, the layers swell, because the polar water molecules enter and interact with the polar headgroups of the long-chain molecules. If, instead of water, oil is added, swelling occurs through the interaction of the oil with the non-polar hydrocarbon regions of the chain. Thus, in suitable concentrations of a solvent (either water or oil), sodium stearate forms ordered structures within the solvent; that is, a lyotropic smectic liquid crystalline phase.

For further details on these three classes of liquid crystal, it is convenient to consider each one briefly in turn.

4.3.1   Nematic Phases

These show long-range orientational order: the mesogenic units may be rod-like (as in Figure 4.3a) or disc-shaped. The symmetry group consists of an infinity-fold rotation axis parallel to the director **n**, accompanied by an infinite number of mirror planes, also parallel

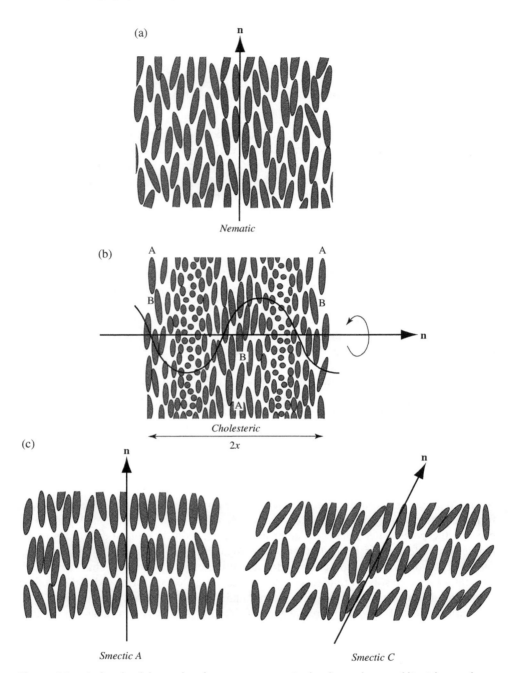

Figure 4.3   A sketch of the molecular arrangements in the three classes of liquid crystal recognized by Friedel: (a) nematic, (b) cholesteric and (c) smectic. In (b), opposite ends of a liquid crystal are labelled A and B to show the pitch, $2x$, within the cholesteric phase

to **n**, and an infinity of diad axes perpendicular to the director (since **n** and − **n** are related by symmetry). The point group describing this is ∞/*mmm*.

For rod-like nematics, **n** is parallel to the axis of the molecules, while for disc-like mesogens, it is parallel to the normal to the disc. The average of the molecular axes at a position vector **r** determines the average of director lines **n**(**r**). This quantity may vary continuously through a sample, except at imperfections – a common one of which is the *disclination* where singularities occur, at which there are rapid local changes in the direction of **n**(**r**) [8].

If forces are applied to a nematic liquid crystal, the director field may show three distinct types of distortion. Following F.C. Frank [9], these are named splay, twist, and bend. They are illustrated schematically in Figure 4.4. The distortions are associated with a gradual change in the orientation of the director with position.

4.3.2   Cholesteric Phases

These are twisted nematic phases. The components of the director **n** are:

$$n_x = n \cos\left(\frac{2\pi z}{\lambda} + \theta\right); n_y = n \sin\left(\frac{2\pi z}{\lambda} + \theta\right); n_z = 0 \tag{4.2}$$

This represents a helix of pitch λ; θ is set by the (arbitrary) choice of origin of the orientation of the director. The director is non-polar and the point group symmetry of the phase is 222. Most twisted nematic phases are based on chiral molecules in which the molecule cannot be superimposed on its mirror image – there are 'right-handed' forms and 'left-handed forms' of the same molecule. A simple example of a chiral molecule is bromochlorofluoromethane, $CHBrClF$, but this is not a liquid crystal. A second example is the rod-like molecule cholesterol. This, in common with many of these phases, has a pitch comparable with the wavelength of visible light. This produces a systematic variation in the refractive index and can cause white light to be dispersed into various colours, so that the cholesteric phase appears cloudy. For some materials, the pitch is a sensitive function of temperature, a fact which enables twisted nematic crystals to act as sensitive colour thermometers. A change of temperature as small as 0.01 °C may be detected.

4.3.3   Smectic Phases

There are many types of these layered phases, depending on the arrangement found within the layers. The two most common are Smectic A and Smectic C. The molecules in Smectic A are, on average, oriented with their long axes parallel to the normal to the layers, while in Smectic C the molecular axes are tilted by an average angle with respect to the layer normal (as in Figure 4.3c). Smectic phases tend to have more order than nematic or cholesteric phases, and so smectic liquid crystalline behaviour generally occurs at lower temperatures and higher concentrations in solvents than either of the other two.

4.4   Geometry of Polyhedra

A familiarity with the geometry of polyhedra is very useful when discussing the packing of atoms within very complicated inorganic crystal structures, and also for explaining the concept of quasicrystals; these often arise in similar types of material containing a large

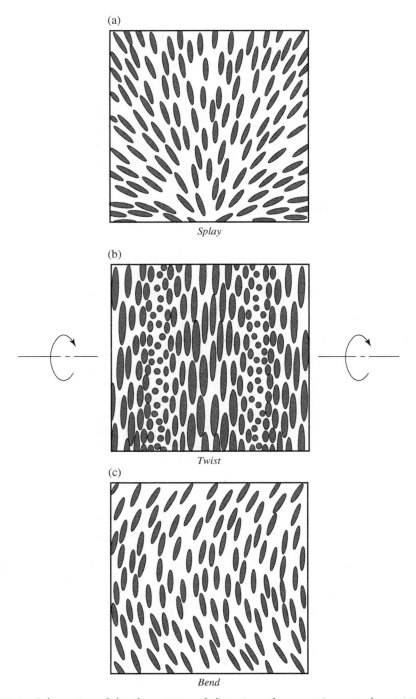

Figure 4.4   Schematics of the three types of distortion of a nematic mesophase: (a) splay, (b) twist and (c) bend

number of atoms. Furthermore, the geometry of polyhedra has application in the modern theory of packing in liquids, as we have seen. It will also be useful for considering the properties of foams and cellular structures (Section 4.8).

We begin with the basics. A convex polyhedron may be defined as a solid of which all of the faces are convex polygons; a convex polygon is a region of a plane enclosed by lines, in the sense that the interior is entirely on the same side of each line.

There is a set of very regular polyhedra called the Platonic solids [10]. These are illustrated in Figure 4.5. They are such that for a given Platonic solid, all the vertices are identical to one another and all the faces are regular polygons which are identical in shape and size, with all their angles equal and all their sides equal. The Platonic solids are shown in three forms in Figure 4.5 – the left-hand column illustrates the body, the second column shows the body with all of the faces illustrated in plan, and the third column shows the *Schlegel diagram* of each.

A Schlegel diagram [11] is a drawing of the body seen in perspective from a position just outside the centre of one face; this face appears as a large polygon with the remaining faces filling its interior. It is a very useful representation since it shows at a glance which vertices belong to which faces and edges; the faces being, of course, the polygons inside the figure together with the enclosing polygon. The *Schläfli symbol* of each solid is also shown in the figure: this lists the number of edges of any face and the number of faces meeting at any vertex; thus {3, 3} represents the simplest Platonic solid, the tetrahedron where three equilateral triangular faces meet at each vertex. {3, 5} represents the icosahedron where five equilateral triangular faces meet at each vertex.

It is useful to draw a stereogram of each of these solids with the primitive parallel to the chosen plane of the Schlegel diagram and to insert on this all of the symmetry elements (Problem 4.4).

A simple proof that there are five, and only five, such solid bodies was first given in the *13th Book of Euclid's Elements*. Each face is a regular polygon; hence the angles of the faces at any vertex must be equal and together less than 360°, and must be three or more in number. Each angle in an equilateral triangle is 60°; hence solids can be formed with three, four or five regular triangles as faces – the tetrahedron, the octahedron and the icosahedron, respectively. The angle in a square is 90°; hence three will form the solid angle of a cube, but four will not. The angle in a regular pentagon is 108°; hence three will form a solid angle and four will not, yielding the (regular) dodecahedron.

A modern proof rests on a theorem of very general utility when dealing with polyhedra and with the packing and arranging of many shapes of solids (e.g. in architecture); this proof stems from Euler's theorem and is both sophisticated and useful. To prove this theorem, we use the Schlegel diagram, remembering that we are standing outside the centre of one face, the one appearing as a large polygon, with the remaining faces filling its interior. A polyhedron capable of representation by a Schlegel diagram is said to be simply connected, or Eulerian, and its properties must satisfy the relation:

$$V + F - E = 2 \tag{4.3}$$

where V, E and F are the number of vertices, faces, and edges, respectively.

The proof of this relation is given by Coxeter [10]; he says it is based on Euler's proof. Suppose that a new edge is inserted into a Schlegel diagram. It would either connect a vertex

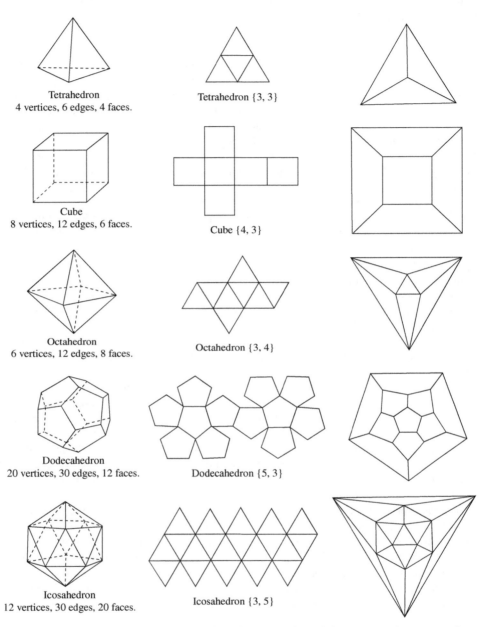

Tetrahedron
4 vertices, 6 edges, 4 faces.

Tetrahedron {3, 3}

Cube
8 vertices, 12 edges, 6 faces.

Cube {4, 3}

Octahedron
6 vertices, 12 edges, 8 faces.

Octahedron {3, 4}

Dodecahedron
20 vertices, 30 edges, 12 faces.

Dodecahedron {5, 3}

Icosahedron
12 vertices, 30 edges, 20 faces.

Icosahedron {3, 5}

Figure 4.5 The five Platonic solids. Left-hand column: the solids. Centre column: templates of the solids showing all their faces in a plane. Right-hand column: the corresponding Schlegel diagrams. The Schläfli symbols are underneath the corresponding templates of the solids. (After Figure 10.2a, Coxeter [10])

already present with a new vertex, or else connect two vertices already in the diagram. In the first situation, the number of vertices is increased by one, and also the number of edges, but no new face is made, so the increase in the sum $(V + F - E)$ is zero. In the second situation, no new vertex is made, but one new face is made, so again the increase in the sum $(V + F - E)$ is zero. Euler argued that the combination $(V + F - E)$ is unchanged throughout the construction of the Schlegel figure. At the very start of the construction of such a figure, there would be only one face (the one outside which we are placed) and one vertex (the one from which we draw a line), so it follows that:

$$V + F - E = 1 + 1 - 0 = 2 \tag{4.4}$$

This value of 2 must remain invariant throughout the construction of the Schlegel diagram.

We now use this result to show that there are just five Platonic solids. Suppose that in a particular Platonic solid, each face is an n-sided regular polygon, and that m edges meet at each vertex. Each edge belongs to two faces, so it follows that:

$$nF = 2E \tag{4.5}$$

Each edge connects two vertices, so it also follows for this solid that:

$$mV = 2E \tag{4.6}$$

We can now use Eq. (4.4) to eliminate V and F, so that we obtain an expression for E:

$$\left(\frac{2}{m} + \frac{2}{n} - 1 \right) E = 2 \tag{4.7}$$

Now, n must be ≥ 3, since each face is a polygon. Also, at least three edges must meet at a vertex, and so $m \geq 3$. However, we have the further constraint that E is a positive integer, so that we need to satisfy the inequality:

$$\left(\frac{2}{m} + \frac{2}{n} - 1 \right) > 0 \tag{4.8}$$

Hence, it follows that if $m = n$, they can both only take the value 3; integers 4 or greater would not satisfy this inequality.

If $m = 3$, possible values of n which satisfy Eq. (4.7) for integer E are 3, 4 and 5, giving $E = 6$, 12, or 30: the tetrahedron, the cube and the dodecahedron, respectively. If $n = 3$, solutions for m are 3, 4, or 5, giving the tetrahedron, the octahedron and the icosahedron, respectively. Many examples where use was made of the first three Platonic solids are discussed in Chapter 3 in order to describe a crystal structure; for example, the tetrahedron was used for the diamond structure and the cube for the sodium chloride structure.

The description of crystal structures given in Chapter 3 is useful, but its formality, based as it is on the Bravais lattice, often fails to indicate the way in which the solid crystal is bound by the forces between the component atoms. For this reason, it is often useful to describe crystal structures in terms of an assemblage of polyhedra linked in various ways. This approach can help to simplify the description of the structures and bring out the crystal chemistry or explain why a crystal possesses a desirable physical property.

Thus, for example, the rutile structure, MX_2 (Section 3.5.7), can be described either as a complex assemblage of oxygen ions, with the titanium ions occupying one-half of the octahedral cavities, or alternatively, and more powerfully, as an assemblage of slightly distorted TiO_6 octahedra linked by their corners and edges. The perovskite structure, $MM'X_3$ (Section 3.6.1), can be regarded as a framework of $M'X_6$ octahedra linked by their corners, each X being shared by two octahedra. The result is a three-dimensional assemblage of octahedra whose tetragonal axes coincide with those of the unit cell. When the relative dimensions of the octahedra of M' ions are incompatible with this configuration, the structure becomes slightly distorted, causing a lowering in symmetry.

The regular dodecahedron, the fourth Platonic solid, has $m\bar{3}\,\bar{5}$ point group symmetry. It has 12 fivefold axes, 20 triads and 30 diads and mirror planes. The 12 fivefold axes are at an angle to one another of $\tan^{-1} 2 = 63.43°$. Seen in projection looking down one of the fivefold axes, the other five adjacent fivefold axes are all at $72°$ to one another. Sets of rods (or fibres) aligned normal to the faces of a regular dodecahedron can produce an elastically isotropic solid; this requires that the cross-sections of the rods have specific shapes and that these be arranged to meet each other in specified ways [12].

4.5   Icosahedral Packing

Regular dodecahedral packing has not so far been shown to be of great use in the description and understanding of complicated crystal structures, in contrast to the icosahedron. This also has $m\bar{3}\,\bar{5}$ point group symmetry, with 12 fivefold rotation axes, 20 triads and 30 diads and mirror planes. The importance of the icosahedron arises because of its close relationship with cubic close packing.

In the cubic close-packed (c.c.p) assembly of equal spheres, each sphere is surrounded by a first coordination polyhedron of 12 identical spheres (Section 3.2). Each neighbouring sphere is situated at a vertex of a cubo-octahedron. This figure with 12 vertices, 14 faces and 24 edges has eight equilateral triangular faces and six square faces. It can be formed by suitably truncating a cube with planar slices at the eight vertices to form the eight equilateral triangular faces. At each new vertex produced, exactly halfway along the edges of the original cube, two equilateral triangular faces and two square faces meet.

A second layer of spheres packed similarly over the first layer will be found to require 42 spheres to complete it. In general, it may be shown that if the structure of spheres is continued, the nth layer will consist of $(10n^2 + 2)$ spheres [13].

The cubo-octahedron can be transformed into a regular icosahedron by compression of some 5% along the line joining the vertices to the centre and allowing the square faces to become rhombs, and each to split into two triangular faces. This movement is easily illustrated on a stereogram (Problem 4.5).

When an icosahedron of 12 spheres about a central sphere is surrounded by a second icosahedral sphere, the second shell will also contain 42 spheres. The second shell will lie over the first, so that spheres will be in contact along the six fivefold axes of symmetry. These spheres will not be close-packed. The packing density of such an icosahedral shell packing is less than that of the c.c.p. structure, 0.74, but nonetheless can be quite high; close to 0.69, which is a higher packing fraction than is found in body-centred cubic (b.c.c.) (0.68) [13]. The packing fraction in this icosahedron shell packing can be calculated by noting

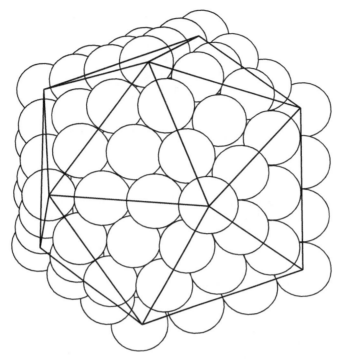

Figure 4.6   Packing of equal-sized spheres into an icosahedral arrangement. The diagram shows the third layer of spheres. (After Mackay [13].)

that within the solid icosahedron the unit that is repeated is a distorted tetrahedron; 20 of these make up the solid icosahedron. An icosahedral packing of equal spheres is shown in Figure 4.6. On each triangular face, the layers of spheres succeed each other, just as in the c.c.p. sequence. Each of these spheres touches just six neighbours, three above and three below, and is separated by a distance of 5% of its radius from its six neighbours in the plane of the icosahedron.

Icosahedral packing is extremely useful for describing structures containing many types of atom of very different sizes. The polyhedra within a given structure can have various shapes in order to accommodate atoms or ions of different sizes. They are often referred to as topologically close-packed (TCP) structures or Frank–Kasper phases [14, 15]. TCP structures enable a closer packing of atoms to arise than is found in the geometrically close-packed structures (c.c.p. and hexagonal close-packed, h.c.p.). There are many examples in the literature of metallic alloys with TCP structures containing three or more types of atom.

In general, TCP structures are made up of clusters of polyhedra, with the symmetry of the whole crystal being less than that of the individual clusters. Icosahedral coordination frequently occurs in TCP structures, as do clusters containing 55 or more atoms. Examples of icosahedral coordination can be found in compounds such as $MoAl_{12}$ (Figure 4.7), WAl_{12} and $Mn_{0.5}Cr_{0.5}Al_{12}$ [16]. The space group of these compounds is now designated $Im\bar{3}$. Some

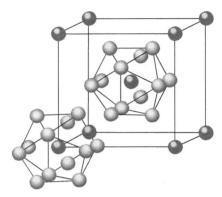

Figure 4.7 The crystal structure of $MoAl_{12}$. This figure is derived from Figure 1 of [16]. Molybdenum (Mo) atoms (heavily shaded) are found at the corners and centre of the unit cell. Each Mo is surrounded by 12 aluminium atoms (lightly shaded) arranged at the 12 vertices of a regular icosahedron

of the triads and diads from the icosahedral arrangement of the Al atoms are retained in the crystal, but the fivefold rotational axes of the individual $MoAl_{12}$ units are lost.

Icosahedral coordination is also found in individual C_{60} molecules in buckminsterfullerene. At room temperature the C_{60} molecules have a high degree of rotational disorder. The room temperature crystal structure of buckminsterfullerene has a face-centred Bravais lattice and belongs to the space group $Fm\overline{3}$ [17].

4.6 Quasicrystals

The perovskite example is one where, in order to maintain the essential packing of the ions, the deviation from ideal symmetry is towards a lower symmetry (Sections 3.6 and 4.4). In some cases, particularly where the packing is based on clumps of atoms with nearly close packing, the point symmetry may be greatly increased, so that it exceeds that compatible with repetition at the points of a Bravais lattice.

The famous first report of fivefold rotational symmetry in a solid occurred in 1984, in a paper whose title amply describes the effect: 'Metallic phase with long-range orientational order and no translational symmetry' [18]. For this discovery, the first author of the paper, Daniel Shechtman, was awarded the 2011 Nobel Prize for Chemistry. The compositions of the alloys investigated were relatively simple: Al with 10–14 wt% of Mn, Fe or Cr. Since these initial observations, many examples have been found in related alloy systems of metallic phases without translational symmetry in two or three dimensions. These can show a number of 'forbidden' rotational symmetries (e.g. 5-, 8-, 10- or 12-fold symmetry), which are completely incompatible with the possession of a Bravais lattice as defined so far in this book. Such structures are best described in terms of regular or slightly irregular (distorted) polyhedra, often icosahedra.

Phases with icosahedral symmetry are often referred to as i-phases. Useful reviews of quasicrystals can be found in [19–21]. The first quasicrystals were produced by rapid cooling from the melt. Although this is still a common way of producing quasicrystalline phases, quasicrystalline phases of suitable compositions can now be produced in centimetre sizes by conventional crystal growth methods, such as in i-Al–Mn–Pd.

It is common for i-phases to be formed over a range of compositions, such as in i-Mg–Al–Zn alloys [19]. There is often a quasicrystalline phase and a crystalline phase of the same composition in the same alloy. Three typical alloys readily showing both crystal and quasicrystal behaviour are α-Al–Mn–Si, $Mg_{32}(Al,Zn)_{49}$ and $Al_{5.6}Li_{2.9}Cu$. In such alloys, it is usual for the crystalline phase to have a relatively large unit cell, and to be referred to as an *approximant* (of the quasicrystal).

A notable feature of quasicrystals is that they can be almost free of defects: this can be deduced directly from the contrast seen in transmission electron microscopy of individual icosahedral grains, but also from the extreme sharpness of the X-ray diffraction peaks – which are often as sharp as those found from very carefully prepared silicon crystals.

4.6.1 A Little Recent History and a New Definition

As we have noted above, the history of quasicrystals began in 1984 with their discovery by Shechtman and his co-authors. Since then, hundreds of quasicrystals have been discovered. Relatively recently, evidence has been produced that quasicrystals form naturally in Al–Cu–Fe minerals, producing the $Al_{65}Cu_{24}Fe_{13}$ and $Al_{65}Cu_{24}Fe_{13}$ i-phases and the $Al_{71}Cu_{24}Fe_5$ decagonal phase [22–24]. Following these many discoveries, the International Union of Crystallography has amended its definition of a crystal so as to include ordering of the atoms in either a periodic or an aperiodic manner. Nowadays those symmetries compatible with translational symmetry are defined as *crystallographic*. This leaves room for '*noncrystallographic symmetries*'. Hence crystals can be divided into two main classes: those with crystallographic point group symmetry, to which all of the structures described in this book other than in this section belong, and those with noncrystallographic symmetry, to which the quasicrystals described in this section and the incommensurate structures described in Section 4.7 belong. These latter materials are commonly referred to as *aperiodic* crystals.

4.7 Incommensurate Structures

In addition to quasicrystals, there are other types of aperiodic structures. The first *incommensurate* phases were recognized as such in the 1970s [25]. A good example of an incommensurate phase is a magnetic system in which the magnetization has a helical structure with a pitch that does not have a rational relation to the underlying lattice [25, 26]. Expressed mathematically, two periodicities q_1 and q_2 in the same direction in a lattice are incommensurate if their ratio cannot be expressed as a ratio of two integers M and N. That is:

$$\frac{q_1}{q_2} \neq \frac{M}{N} \quad M, N = 1, 2, 3 \ldots$$

Many materials are now known which display two or more periodicities that are incommensurate with each other. A well-known example of an incommensurate structure is the modulated structure which appears over a narrow 1.3 K temperature range in the high $\leftrightarrow$ low phase transition in quartz [27]. A second example can be found in the mineral calaverite,

AuTe$_2$ [28]. The particular interatomic interactions responsible for different types of incommensurability in molecular and inorganic crystals are the subject of a recent review [29]. The formalism of the higher dimensional crystallography, or *superspace crystallography*, required to describe aperiodic crystals is discussed in [30].

4.8 Foams, Porous Materials, and Cellular Materials

While the emphasis in this book is on the arrangements of atoms in single crystals, there are a number of sections, such as those concerned with crystalline textures and interfaces, in which we note that most materials are usually arranged in polycrystalline aggregates. The interfaces between the individual crystals, known as grain boundaries or interphase boundaries (Chapters 13 and 14), are usually just a few atoms wide. These boundaries form a continuous skeleton within which the individual crystals are contained. If we imagine a solid metal or ceramic consisting only of these intercrystalline regions, with the material of the individual crystals removed, we should have a material with the geometry of a *foam*. An example of the microstructure of a solid foam is shown in Figure 4.8 [31, 32].

The foam is an important type of porous solid, which itself is a class of *cellular solids* [33]. Polymer foams such as those of polyurethane and polystyrene have been known for many years, and there are natural cellular materials, such as zeolites and sponges – the term 'sponge' usually denotes a very flexible foam. Recent technical advances in processes for foaming metals and ceramics have led to intense research into such materials.

The nomenclature for describing the architecture of foams and other forms of porous solids is not completely consistent. The terms 'three-dimensional meshes' and 'lattice structures' are also used to describe foamed materials. The International Union of Pure and Applied Chemistry (IUPAC) recommendations for the characterization of porous solids [34] describe pore diameters of less than 2 nm as micro; between 2 and 50 nm as meso and greater than 50 nm as macro, but these definitions are not consistently maintained. These authors of the IUPAC recommendations had in mind catalysts, absorbents, and membranes primarily. They define materials as constitutively porous if their final structure depends mainly on the original structure, such as pores in zeolites or amorphous carbon. Materials made porous by a subtractive processing route have material removed from them: chemically etched porous silicon is an example of such a material.

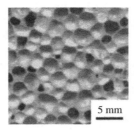

5 mm

Figure 4.8 The microstructure of a foam made from an Al – 9 wt% Si alloy containing 10 vol% of silicon carbide particulates [31, 32]. Pretreated titanium dihydride, TiH$_2$, was used as the foaming agent

The volume fraction of porosity is the fraction of the total volume occupied by pores; that is, V_p/V, where V_p is the total volume of pores and V the total volume of specimen. Pores are open if accessible by the particular probe used, and closed if they are not: the definition depends on the type of probe employed. Open pores are interconnected; the same range of sizes micro, meso, and macro may be used to define the diameter of the connecting channels. It is also possible to have mesopores with micro-connectivity or mesopores with meso-connectivity. The term *porous metal* or *porous ceramic* usually suggests a porosity range of 15–75%. This range of porosity is shown by metals used in filtration, fluid flow control, self-lubricating bearings and battery electrodes.

True foams have high porosities, such as 80–90%, and it is these that provide a distinct advantage for energy and sound absorption, vibration suppression and thermal management. A special form of closed-cell foam is a syntactic foam consisting of small spheres, which are stuck together (Greek: σύνταξις – arrangement). If the pores are hollow and the interstices are completely filled, the maximum porosity is about 65%. To a first approximation, most porous structures can be regarded as equiaxed, although it is possible to obtain porous metals and ceramics with elongated structures, the so-called lotus structures; anodized aluminium is a particularly good example. The properties of porous materials are governed by three main factors:

1. The properties of the solid forming the walls of the foam.
2. The topology (connectivity) and shape of the faces and edges of the individual cells.
3. The relative density ρ_f/ρ_s of the foam, where ρ_f is the overall density of the foam, and ρ_s that of the solid material within it.

When estimating the mechanical properties of cellular materials, reference is often made to *lattice structures*, which in this case refers to a connected set of struts, rather than the term used elsewhere in this book. In this context, a *lattice*, *truss*, or *space frame* means an array of struts pin-jointed or rigidly bonded at their connections. Foams are like the trusses and frames of space structures, and so are made up of a connected array of struts or plates. They are characterized by a typical 'unit cell' with symmetry elements, even though most foams do not show translational symmetry. However, these lattices, cellular materials and foams differ from the lattices of the engineer in one very important regard: that of scale of size. The unit cell of a cellular material is one on the scale of micrometres–millimetres, or sometimes less, so allowing it to be viewed both as a structure and as a material.

Although a foam may not show translational symmetry, the identification of a typical (or average) 'unit cell' is very important for describing the overall symmetry of the physical properties, such as elastic properties. This same procedure is used with composite materials – the best known of which, and presently the most useful, are the fibre composites, consisting of stiff fibres arranged in a relatively soft and yielding matrix. Here, the 'unit cell' is termed the *representative volume element*. We shall revisit similar structures in Chapter 13.

Problems

4.1 Determine the shape of the polyhedron which represents the Voronoi polyhedron for a c.c.p. arrangement of spheres. What is the name of this polyhedron?

4.2 Construct the Schlegel figure for the rhombic dodecahedron.

4.3 Describe a solid having five vertices and six triangular faces.

4.4 Draw a stereogram of each of the Platonic solids and insert all of the symmetry elements; make the primitive of the stereogram parallel to the chosen plane of the relevant Schlegel diagram.

4.5 By placing a triad axis of the cubo-octahedron at the centre of the primitive, draw a stereogram to illustrate the distortion of a cubo-octahedron, so as to form a regular icosahedron.

4.6 Show that in c.c.p., the nth layer packed over the layers below is the surface of a cubo-octahedron on which there are $(n+1)$ spheres along each of its edges. Hence, show that the nth layer consists of $(10n^2+2)$ spheres. Explain why in an icosahedrally-packed assembly of equal spheres the nth layer packed over the layers must also consist of $(10n^2+2)$ spheres.

4.7 Giving your reasons, explain whether $MoAl_{12}$ has crystallographic point group symmetry or noncrystallographic point group symmetry.

4.8 Give a logical reason why the porosity limit of a syntactic foam is very close to Bernal's packing fraction for a liquid.

Suggestions for Further Reading

Allen, S.M. and Thomas, E.L. (1999). _The Structure of Materials_. New York: Wiley.
Donald, A.M., Windle, A.H. and Hanna, S. (2006). _Liquid Crystalline Polymers_, 2nd Edition. Cambridge: Cambridge University Press.
Gibson, L.J. and Ashby, M.F. (1997). _Cellular Solids: Structure and Properties_, 2nd Edition. Cambridge: Cambridge University Press.
Janot, C. (1994). _Quasicrystals: A Primer_, 2nd Edition. Oxford: Clarendon Press.
Nabarro, F.R.N. (ed.) (1992). _Dislocations in Solids_, vol. 9. Dislocations and Disclinations. Amsterdam: North Holland.
van Smaalen, S. (2012). _Incommensurate Crystallography_. Oxford: Oxford University Press.

References

[1] Bernal, J.D. (1959). A geometrical approach to the structure of liquids. _Nature_ 183: 141–147.
[2] Elliott, J.A., Kelly, A. and Windle, A.H. (2002). Recursive packing of dense particle mixtures. _J. Mater. Sci. Lett._ 21: 1249–1251.
[3] Voronoï, G.F. (1908). Nouvelles applications des paramètres continus à la théorie des formes quadratiques. Deuxième mémoire. Recherches sur les parallélloèdres primitifs. _J. Reine Angew. Math. (Crelle's J.)_ 134: 198–287.
[4] Valenti, B. and Sartirana, M.L. (1984). Mesophase formation in lyotropic polymers. _Il Nuovo Cimento_ 3: 104–120.
[5] Reinitzer, F. (1888). Beiträge zur Kenntniss des Cholesterins. _Monatsh. Chem._ 9: 421–441.
[6] Lehmann, O. (1889). Über fliessende Krystalle. _Z. Phys. Chem._ 4: 462–472.
[7] Friedel, G. (1922). Les états mésomorphes de la matière. _Ann. Phys._ 18: 273–474.
[8] Nabarro, F.R.N. (1987). _Theory of Crystal Dislocations_. New York: Dover Publications.
[9] Frank, F.C. (1958). On the theory of liquid crystals. _Disc. Faraday Soc._ 25: 19–28.

[10] Coxeter, H.S.M. (1961). *Introduction to Geometry*. New York: Wiley.

[11] Schlegel, V. (1883). Theorie der homogen zusammengesetzten Raumgebilde. *Nova Acta Academiae Caesareae Leopoldino-Carolinae Germanicae Naturae Curiosorum* 44: 343–459.

[12] Parkhouse, J.G. and Kelly, A. (1998). The regular packing of fibres in three dimensions. *Proc. Roy. Soc. Lond. A* 454: 1889–1909.

[13] Mackay, A.L. (1962). A dense non-crystallographic packing of equal spheres. *Acta Crystallogr.* 15: 916–918.

[14] Frank, F.C. and Kasper, J.S. (1958). Complex alloy structures regarded as sphere packings. I. Definitions and basic principles. *Acta Crystallogr.* 11: 184–190.

[15] Frank, F.C. and Kasper, J.S. (1959). Complex alloy structures regarded as sphere packings. II. Analysis and classification of representative structures. *Acta Crystallogr.* 12: 483–499.

[16] Adam, J. and Rich, J.B. (1954). The crystal structure of WAl_{12}, $MoAl_{12}$ and $(Mn, Cr)Al_{12}$. *Acta Crystallogr.* 7: 813–816.

[17] Fleming, R.M., Siegrist, T., Marsh, P.M., Hessen, B., Kortan, A.R., Murphy, D.W., Haddon, R.C., Tycko, R., Dabbagh, G., Mujsce, A.M., Kaplan, M.L. and Zahurak, S.M. (1991). Diffraction symmetry in crystalline, close-packed C_{60}. *Mater. Res. Soc. Symp. Proc.* 206: 691–695.

[18] Shechtman, D., Blech, I., Gratias, D. and Cahn, J.W. (1984). Metallic phase with long-range orientational order and no translational symmetry. *Phys. Rev. Lett.* 53: 1951–1953.

[19] Goldman, A.I. and Kelton, K.F. (1993). Quasicrystals and crystalline approximants. *Rev. Mod. Phys.* 65: 213–230. Erratum 579.

[20] Steurer, W. (2004). Twenty years of structure research on quasicrystals. Part I. Pentagonal, octagonal, decagonal and dodecagonal quasicrystals. *Z. Kristall.* 219: 391–446.

[21] Steurer, W. (2018). Quasicrystals: what do we know? What do we want to know? What can we know? *Acta Crystallogr. A* 74: 1–11.

[22] Bindi, L., Steinhardt, P.J., Yao, N. and Lu, P.J. (2009). Natural quasicrystals. *Science* 324: 1306–1309.

[23] Bindi, L., Yao, N., Lin, C., Hollister, L.S., Andronicos, C.L., Distler, V.V., Eddy, M.P., Kostin, A., Kryachko, V., MacPherson, G.J., Steinhardt, W.M., Yudovskaya, M. and Steinhardt, P.J. (2015). Natural quasicrystal with decagonal symmetry. *Sci. Rep.* 5 9111-1–9111-5.

[24] Bindi, L., Lin, C., Ma, C. and Steinhardt, P.J. (2016). Collisions in outer space produced an icosahedral phase in the Khatyrka meteorite never observed previously in the laboratory. *Sci. Rep.* 6 38117-1–38117-6.

[25] Bak, P. (1982). Commensurate phases, incommensurate phases and the devil's staircase. *Rep. Prog. Phys.* 45: 587–629.

[26] Habenschuss, M., Stassis, C., Sinha, S.K., Deckman, H.W. and Spedding, F.H. (1974). Neutron diffraction study of the magnetic structure of erbium. *Phys. Rev. B* 10: 1020–1026.

[27] Putnis, A. (1992). *Introduction to Mineral Sciences*, Section 12.3. Cambridge: Cambridge University Press.

[28] Schutte, W.J. and de Boer, J.L. (1988). Valence fluctuations in the incommensurately modulated structure of calaverite $AuTe_2$. *Acta Crystallogr. B* 44: 486–494.

[29] Pinheiro, C.B. and Abakumov, A.M. (2015). Superspace crystallography: a key to the chemistry and properties. *IUCrJ* 2: 137–154.

[30] Janssen, T. and Janner, A. (2014). Aperiodic crystals and superspace concepts. *Acta Crystallogr. B* 70: 617–651.

[31] Gergely, V., Curran, D.C. and Clyne, T.W. (2003). The FOAMCARP process: foaming of aluminium MMCs by the chalk-aluminium reaction in precursors. *Compos. Sci. Technol.* 63: 2301–2310.

[32] Gergely, V. and Clyne, T.W. (2000). The FORMGRIP process: foaming of reinforced metals by gas release in precursors. *Adv. Eng. Mater.* 2: 175–178.

[33] Gibson, L.J. and Ashby, M.F. (1997). *Cellular Solids: Structure and Properties*, 2nd Edition. Cambridge: Cambridge University Press.

[34] Rouquerol, J., Avnir, D., Fairbridge, C.W., Everett, D.H., Haynes, J.H., Pernicone, N., Ramsay, J.D.F., Sing, K.S.W. and Unger, K.K. (1994). Recommendations for the characterization of porous solids. *Pure & Appl. Chem* 66: 1739–1758.

5

Tensors

5.1 Nature of a Tensor

There are a number of ways to introduce tensors. We shall begin by considering tensors of zero, first and second rank, before extending our discussion to tensors of third and fourth rank.

Scalar quantities are examples of tensors of zero rank; these simply have magnitude. Vectors are tensors of the first rank. These have both direction and magnitude and represent a definite physical quantity. Tensors of the second rank are quantities that relate two vectors.

Suppose we wish to know the relationship between the electric field in a crystal, represented by the vector $\mathbf{E}$, and the current density (i.e. current per unit area of cross-section perpendicular to the current), represented by the vector $\mathbf{J}$.[1] In general, in a crystal the components of $\mathbf{J}$ referred to three mutually perpendicular axes (Ox_1, Ox_2, Ox_3), which we can call J_1, J_2 and J_3, will be related to the components of $\mathbf{E}$, referred to the same set of axes in such a way that they each depend linearly on *all three of the components* E_1, E_2 and E_3. It is usual to write this in the following way:

$$J_1 = \sigma_{11}E_1 + \sigma_{12}E_2 + \sigma_{13}E_3$$
$$J_2 = \sigma_{21}E_1 + \sigma_{22}E_2 + \sigma_{23}E_3 \qquad (5.1)$$
$$J_3 = \sigma_{31}E_1 + \sigma_{32}E_2 + \sigma_{33}E_3$$

---

[1] It is conventional to denote current density by $\mathbf{j}$, as in Section A1.3 and in [1]. However, $\mathbf{j}$ is used in this chapter to denote a unit vector along Ox_2.

Crystallography and Crystal Defects, Third Edition. Anthony Kelly and Kevin M. Knowles.
© 2020 John Wiley & Sons Ltd. Published 2020 by John Wiley & Sons Ltd.
Companion website: www.wiley.com/go/kelly/crystallography3e

Table 5.1 Properties represented by second-rank tensors.

Tensor	Vectors related	
Electrical conductivity	Electric field	Current density
Thermal conductivity	Thermal gradient (negative)	Thermal current density
Diffusivity	Concentration gradient (negative)	Flux of atoms
Permittivity	Electric field	Dielectric displacement
Dielectric susceptibility	Electric field	Dielectric polarization
Permeability	Magnetic field	Magnetic induction
Magnetic susceptibility	Magnetic field	Intensity of magnetization

The nine quantities σ_{11}, σ_{12}, σ_{13}, σ_{21}, σ_{22}, σ_{23}, σ_{31}, σ_{32}, σ_{33} are called the components of the conductivity tensor. The electrical conductivity tensor relates the vectors **J** and **E**. If we write all of the relations (5.1) in the shorthand form:

$$\mathbf{J} = \sigma\mathbf{E} \tag{5.2}$$

we see that σ is a quantity that multiplies the vector **E** in order to obtain the vector **J**. When a tensor relates two vectors in this way it is called a tensor of the second rank or second order.

Many physical properties are represented by tensors like the electrical conductivity tensor. Such tensors are called *matter tensors*. Some examples are given in Table 5.1.

In addition there are *field tensors*, of which two second-rank tensors are very important, namely stress and strain. The stress tensor relates the vector traction (force per unit area) and the orientation of an element of area in a stressed body. The strain tensor relates the displacement of a point in a strained body and the position of the point (see Chapter 6).

5.2   Transformation of Components of a Vector

If we know the components of a vector, say **p**, referred to a set of orthogonal axes (Ox_1, Ox_2, Ox_3), each of unit length – that is, orthonormal axes (Appendix 1) – it is often necessary to know the components of the same vector referred to a different set of axes (Ox'_1, Ox'_2, Ox'_3), which are again orthonormal and have the same origin as Ox_1, Ox_2 and Ox_3 (Figure 5.1). We must first define how the two sets of axes are related to one another.

We do this by setting down a table of the cosines of the angles between each new axis and the three of the old set. The table will appear as:

$$
\begin{array}{c c}
 & \text{Old} \\
 & \begin{array}{ccc} x_1 & x_2 & x_3 \end{array} \\
\text{New}\begin{array}{c} x'_1 \\ x'_2 \\ x'_3 \end{array} & \left|\begin{array}{ccc} a_{11} & a_{12} & a_{13} \\ a_{21} & a_{22} & a_{23} \\ a_{31} & a_{32} & a_{33} \end{array}\right.
\end{array} \tag{5.3}
$$

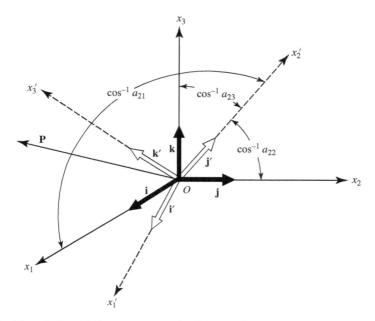

Figure 5.1 The relationship between a set of orthonormal axes (Ox_1, Ox_2, Ox_3) and a different set of orthonormal axes (Ox'_1, Ox'_2, Ox'_3) with a common origin O to enable the components of a vector **p** to be specified relative to both sets of axes.

Here, a_{32}, for example, is the cosine of the angle between the new axis 3 and the old axis 2; that is, the angle x'_3Ox_2 in Figure 5.1. Similarly, a_{11} is the cosine of the angle between Ox'_1 and Ox_1, and so on. In an array of the type of Eq. (5.3) it should be noted that the sum of the squares of any row or column of the array is equal to one, because both sets of axes are orthonormal. Therefore, for example, $a^2_{11} + a^2_{21} + a^2_{31} = 1$ and $a^2_{21} + a^2_{22} + a^2_{23} = 1$, and so on. It is evident that the elements a_{ij} constitute the elements of a rotation matrix (Section A1.4). If we call this matrix **A**, the inverse of this matrix, $\mathbf{A}^{-1}$, is equal to its transpose, $\tilde{\mathbf{A}}$ (Section A1.4).

Now let **p** have components p_1, p_2, p_3 along the old axes so that:

$$\mathbf{p} = p_1\mathbf{i} + p_2\mathbf{j} + p_3\mathbf{k} \tag{5.4}$$

where **i**, **j**, **k** are unit vectors along Ox_1, Ox_2, Ox_3, respectively. The component of **p** along the new axis Ox'_1 can be found by resolving the three old components, $p_1\mathbf{i}$, $p_2\mathbf{j}$ and $p_3\mathbf{k}$, along the new axis Ox'_1 and adding up the projections, thus obtaining:

$$p'_1 = a_{11}p_1 + a_{12}p_2 + a_{13}p_3 \tag{5.5a}$$

Similarly, we find the new components of **p** along Ox'_2 and Ox'_3, p'_2 and p'_3 respectively, to be:

$$p'_2 = a_{21}p_1 + a_{22}p_2 + a_{23}p_3 \tag{5.5b}$$

and

$$p'_3 = a_{31}p_1 + a_{32}p_2 + a_{33}p_3 \tag{5.5c}$$

The vector **p** has the same magnitude and direction referred to any set of axes. If $\mathbf{i}'$, $\mathbf{j}'$, $\mathbf{k}'$ are unit vectors along Ox_1', Ox_2', Ox_3' then:

$$p_1\mathbf{i} + p_2\mathbf{j} + p_3\mathbf{k} = p_1'\mathbf{i}' + p_2'\mathbf{j}' + p_3'\mathbf{k}' \tag{5.6}$$

From Eq. (5.3) we can also write:

$$\mathbf{i}' = a_{11}\mathbf{i} + a_{12}\mathbf{j} + a_{13}\mathbf{k}$$
$$\mathbf{j}' = a_{21}\mathbf{i} + a_{22}\mathbf{j} + a_{23}\mathbf{k} \tag{5.7}$$
$$\mathbf{k}' = a_{31}\mathbf{i} + a_{32}\mathbf{j} + a_{32}\mathbf{k}$$

and, conversely, since the elements a_{ij} constitute the elements of a rotation matrix:

$$\mathbf{i} = a_{11}\mathbf{i}' + a_{21}\mathbf{j}' + a_{31}\mathbf{k}'$$
$$\mathbf{j} = a_{12}\mathbf{i}' + a_{22}\mathbf{j}' + a_{32}\mathbf{k}' \tag{5.8}$$
$$\mathbf{k} = a_{13}\mathbf{i}' + a_{23}\mathbf{j}' + a_{33}\mathbf{k}'$$

Equations (5.5a)–(5.5c) could also be deduced by using Eq. (5.8) to express **i**, **j** and **k** each in terms of $\mathbf{i}'$, $\mathbf{j}'$ and $\mathbf{k}'$ and then substituting in Eq. (5.4), noting that p_1' is the coefficient of the unit vector $\mathbf{i}'$, and so on. It is a useful exercise to write out the substitution in full.

Proceeding in the same way, we can also obtain expressions for the old components of the vector **p**, $[p_1, p_2, p_3]$, in terms of the new components by substituting for $\mathbf{i}'$, $\mathbf{j}'$, $\mathbf{k}'$ in terms of **i**, **j**, **k** (using Eq. (5.7)) on the right-hand side of Eq. (5.6) and subsequently comparing terms on the two sides of the equation. Doing this, we find:

$$p_1 = a_{11}p_1' + a_{21}p_2' + a_{31}p_3'$$
$$p_2 = a_{12}p_1' + a_{22}p_2' + a_{32}p_3' \tag{5.9}$$
$$p_3 = a_{13}p_1' + a_{23}p_2' + a_{33}p_3'$$

Relations (5.9) are the converse of Eqs. (5.5a)–(5.5c). We are now in a position to relate these results on the transformation of the components of a vector to a relation such as Eq. (5.2).

A tensor of the second rank is a quantity that multiplies one vector to give another, generally non-parallel, vector of a different magnitude. If we return to our example of electrical conductivity in Section 5.1, we note that we could state the equations in Eq. (5.1) in the following way:

$$J_1\mathbf{i} = (\sigma_{11}E_1 + \sigma_{12}E_2 + \sigma_{13}E_3)\mathbf{i}$$
$$J_2\mathbf{j} = (\sigma_{21}E_1 + \sigma_{22}E_2 + \sigma_{23}E_3)\mathbf{j} \tag{5.10}$$
$$J_3\mathbf{k} = (\sigma_{31}E_1 + \sigma_{32}E_2 + \sigma_{33}E_3)\mathbf{k}$$

Having done this, let us add up these three equations and express the right-hand side in a new form so that the equation reads:

$$\begin{aligned} J_1\mathbf{i} + J_2\mathbf{j} + J_3\mathbf{k} = [\sigma_{11}\mathbf{ii} + \sigma_{12}\mathbf{ij} + \sigma_{13}\mathbf{ik} + \sigma_{21}\mathbf{ji} + \sigma_{22}\mathbf{jj} + \sigma_{23}\mathbf{jk} \\ + \sigma_{31}\mathbf{ki} + \sigma_{32}\mathbf{kj} + \sigma_{33}\mathbf{kk}].(E_1\mathbf{i} + E_2\mathbf{j} + E_3\mathbf{k}) \end{aligned} \tag{5.11}$$

The quantity in square brackets is to be regarded as an operator, which operates on vector **E** in order to obtain the vector **J** from **E**. The form in which the operator is written here is the one that is appropriate to the orthogonal axes defined by the unit vectors **i**, **j** and **k**.[2] We let the operator inside the square brackets 'multiply' the vector **E** ($= E_1\mathbf{i} + E_2\mathbf{j} + E_3\mathbf{k}$), as if we were forming the scalar (or dot) product of vector multiplication (Appendix 1).

To understand Eq. (5.11), it is best to write out the right-hand side term by term. The first term is $\sigma_{11}\mathbf{ii} \cdot E_1\mathbf{i} = \sigma_{11}E_1\mathbf{ii} \cdot \mathbf{i} = \sigma_{11}E_1\mathbf{i}$, since $\mathbf{i} \cdot \mathbf{i} = 1$. The next term is $\sigma_{11}\mathbf{ii} \cdot E_2\mathbf{j} = 0$, since $\mathbf{i} \cdot \mathbf{j} = 0$ because the **i** and **j** axes are orthogonal. Similarly, $\sigma_{11}\mathbf{ii} \cdot E_3\mathbf{k} = 0$ and $\sigma_{12}\mathbf{ij} \cdot E_1\mathbf{i} = 0$, but the fifth term is $\sigma_{12}\mathbf{ij} \cdot E_2\mathbf{j} = \sigma_{12}E_2\mathbf{i}$, since $\mathbf{j} \cdot \mathbf{j} = 1$. If all terms are written out and similar ones collected, we finally obtain just Eq. (5.10), which can also be represented as in Eq. (5.1). Equation (5.11) is the expanded form of Eq. (5.2), or, put another way, the most succinct expression of Eq. (5.11) is Eq. (5.2).

Equation (5.11) contains the components of the conductivity tensor σ_{11}, σ_{12}, σ_{13}, ..., σ_{33} referred to the axes (Ox_1, Ox_2, Ox_3), defined by the unit vectors **i**, **j** and **k**. We have seen how to transform the components of a vector, so we can also transform the components of this second-rank tensor.

If we take new axes represented by the unit vectors $\mathbf{i}'$, $\mathbf{j}'$, $\mathbf{k}'$, the components of the conductivity tensor σ will be different and we can find what they are by substituting for **i**, **j**, **k** in Eq. (5.11) the values of $\mathbf{i}'$, $\mathbf{j}'$, $\mathbf{k}'$ referred to the new axes, using Eq. (5.7). We would also have to substitute for E_1, E_2, E_3 the components of **E** along the new axes – E_1', E_2', E_3' – so that the new components of the tensor σ_{11}', σ_{12}', σ_{13}', ... σ_{33}', multiplied by the new components of **E** in a form such as Eq. (5.9), yield the components of the vector **J**, also referred to the new set of axes. This is very lengthy to write out and so we shall first introduce a convenient short notation.

5.3 Dummy Suffix Notation

The tensor T relates the vector **p** with components p_1, p_2 and p_3 to the vector **q** with components q_1, q_2 and q_3 so that:

$$\begin{aligned} p_1 &= T_{11}q_1 + T_{12}q_2 + T_{13}q_3 \\ p_2 &= T_{21}q_1 + T_{22}q_2 + T_{23}q_3 \\ p_3 &= T_{31}q_1 + T_{32}q_2 + T_{33}q_3 \end{aligned} \tag{5.12}$$

These three equations can be written as:

$$p_1 = \sum_{j=1}^{j=3} T_{1j}q_j, \qquad p_2 = \sum_{j=1}^{j=3} T_{2j}q_j, \qquad p_3 = \sum_{j=1}^{j=3} T_{3j}q_j \tag{5.13}$$

or, even more succinctly, as:

$$p_i = \sum_{j=1}^{j=3} T_{ij}q_j \qquad (i = 1, 2, 3) \tag{5.14}$$

---

[2] Note that inside the square brackets in Eq. (5.11) we do not have $\mathbf{i} \cdot \mathbf{i}$ (which equals one), but **ii**.

We now leave out the summation sign and introduce the Einstein summation convention [2]: if a suffix occurs twice in the same term, summation with respect to that suffix is automatically implied. In what follows, the summation will always run over the values 1, 2 and 3 for both i and j. Equation (5.12) can then be written as:

$$p_i = T_{ij}q_j \tag{5.15}$$

Here j is called a *dummy suffix*, because it does not matter which letter except i is taken to represent it. Equation (5.12) could equally well read:

$$p_i = T_{ik}q_k$$

using k as the dummy suffix. If we apply this notation to Eqs. (5.5a)–(5.5c), all three are contained in the single expression:

$$p_i' = a_{ij}p_j \tag{5.16}$$

Note that the dummy suffix occurs in neighbouring places in Eq. (5.16). If we apply the same notation to Eq. (5.9) we obtain:

$$p_i = (a^{-1})_{ij}p_j' = (\tilde{a})_{ij}p_j' = a_{ji}p_j' \tag{5.17}$$

so that when writing the 'old' components in terms of the 'new', the dummy suffixes are separated because, as we have noted in Section 5.2, the elements a_{ij} constitute the elements of a rotation matrix, and so the elements of the inverse of this matrix, $(a^{-1})_{ij}$, are simply the transpose of this matrix.

When first learning this notation it is best to write out the sums term by term. Thus $p_i' = a_{ij}p_j$ gives, on expansion:

$$p_1' = a_{1j}p_j$$
$$p_2' = a_{2j}p_j \tag{5.18}$$
$$p_3' = a_{3j}p_j$$

These three sums can then be expanded without fear of making an error.

5.4   Transformation of Components of a Second-Rank Tensor

Let two vectors $\mathbf{p} = [p_1\ p_2\ p_3]$ and $\mathbf{q} = [q_1\ q_2\ q_3]$ be related by the tensor T_{ij} for a specific choice of the axes $\mathbf{i}, \mathbf{j}, \mathbf{k}$. The components of T_{ij} relate the components of $\mathbf{q}$ to those of $\mathbf{p}$ according to Eq. (5.12), or, more succinctly, according to Eq. (5.15):

$$p_i = T_{ij}q_j$$

Now, the *components* of the two vectors depend upon the choice of axes, because this choice determines the values $[p_1\ p_2\ p_3]$ and $[q_1\ q_2\ q_3]$. The vectors $\mathbf{p}$ and $\mathbf{q}$ themselves do not change. When the axes are changed, and hence the components of **p** and **q** change, the components T_{ij} will also change.

If now we choose new axes $\mathbf{i}', \mathbf{j}', \mathbf{k}'$ so that:

$$p'_i = T'_{ij} q'_j \tag{5.19}$$

we will wish to find the relation between the nine components T'_{ij} and the nine components T_{ij}. We can find these directly by writing Eq. (5.15) in the operational form of Eq. (5.11):

$$
\begin{aligned}
p_1\mathbf{i} + p_2\mathbf{j} + p_3\mathbf{k} = [T_{11}\mathbf{ii} + T_{12}\mathbf{ij} + T_{13}\mathbf{ik} + T_{21}\mathbf{ji} + T_{22}\mathbf{jj} + T_{23}\mathbf{jk} \\
+ T_{31}\mathbf{ki} + T_{32}\mathbf{kj} + T_{33}\mathbf{kk}] \cdot (q_1\mathbf{i} + q_2\mathbf{j} + q_3\mathbf{k})
\end{aligned}
\tag{5.20}
$$

We can now substitute for $\mathbf{i}, \mathbf{j}$ and $\mathbf{k}$ in this equation the values of these quantities in terms of $\mathbf{i}', \mathbf{j}'$ and $\mathbf{k}'$ according to Eq. (5.8).

The substitution in the left-hand side yields $p'_1\mathbf{i}' + p'_2\mathbf{j}' + p'_3\mathbf{k}'$, directly from Eq. (5.6). Likewise, the term in round brackets on the right-hand side of Eq. (5.20) becomes $q'_1\mathbf{i}' + q'_2\mathbf{j}' + q'_3\mathbf{k}'$. The tensor operator in square brackets is treated in the same way and substitution is made for $\mathbf{i}, \mathbf{j}$ and $\mathbf{k}$ in terms of $\mathbf{i}', \mathbf{j}'$ and $\mathbf{k}'$ from Eq. (5.8). There will be 81 terms in total in this square bracket, but these can be grouped together in nine separate new terms. If all those containing (say) the vectors $\mathbf{j}'\mathbf{k}'$ are collected, there will be nine of these and the coefficient of $\mathbf{j}'\mathbf{k}'$ will be the new component T'_{23}. If we write out the expression for T'_{23}, it is:

$$
\begin{aligned}
T'_{23} = (a_{21}a_{31}T_{11} + a_{21}a_{32}T_{12} + a_{21}a_{33}T_{13} + a_{22}a_{31}T_{21} + a_{22}a_{32}T_{22} \\
+ a_{22}a_{33}T_{23} + a_{23}a_{31}T_{31} + a_{23}a_{32}T_{32} + a_{23}a_{33}T_{33})
\end{aligned}
\tag{5.21}
$$

Using the summation convention, this can be written succinctly in the form:

$$T'_{23} = a_{2k}a_{3l}T_{kl} \tag{5.22}$$

Similarly, the other eight components of T'_{ij} may be found by collecting the terms in $\mathbf{i}'\mathbf{i}'$, $\mathbf{i}'\mathbf{j}', \mathbf{i}'\mathbf{k}', \mathbf{j}'\mathbf{i}', \mathbf{j}'\mathbf{j}', \mathbf{k}'\mathbf{i}', \mathbf{k}'\mathbf{j}'$ and $\mathbf{k}'\mathbf{k}'$, respectively. Using the summation convention, the full relationship can be written in the form:

$$T'_{ij} = a_{ik}a_{jl}T_{kl} \tag{5.23}$$

For each value of i and j, the sum is to be taken over both k and l. To confirm Eq. (5.23) is correct, it is helpful to write out the expansion summing over each of these separately. For example, if $i = 2$ and $j = 3$, Eq. (5.21) is the expanded version of Eq. (5.23).

An alternative, and more elegant, method to establish the relationship between T'_{ij} and T_{ij} is to perform the following mathematical operations with suitable dummy suffixes to determine the nine components of T'_{ij}:

1. Write p' in terms of p: $p'_i = a_{ik}p_k$ (using Eq. (5.16)).
2. Write p in terms of q: $p_k = T_{kl}q_l$ (using Eq. (5.15)).
3. Write q in terms of q': $q_l = a_{jl}q'_j$ (using Eq. (5.17)).

When we combine these three operations we have:

$$p'_i = a_{ik}p_k = a_{ik}T_{kl}q_l = a_{ik}T_{kl}a_{jl}q'_j \tag{5.24}$$

or:

$$p'_i = a_{ik}T_{kl}a_{jl}q'_j \tag{5.25}$$

This is of the form of Eq. (5.19). Therefore, by comparing Eqs. (5.19) and (5.26), we have the important result:

$$T'_{ij} = a_{ik}T_{kl}a_{jl} \tag{5.26}$$

which is Eq. (5.23), because the order in which a product is written on the right-hand side of this equation does not matter when the dummy suffix notation is used.

The relationship between T'_{ij} and T_{ij} is most easy to remember in the form:

$$T'_{ij} = a_{ik}a_{jl}T_{kl} \tag{5.27}$$

The reverse transformation to Eq. (5.27) – that is, the transformation giving the components of the tensor T_{ij} in terms of those of the new components T'_{ij} – can be accomplished in exactly the same way. This will lead to:

$$T_{ij} = a_{ki}a_{lj}T'_{kl} \tag{5.28}$$

It is a useful exercise to prove this.

5.5   Definition of a Tensor of the Second Rank

If the operator T relates two vectors **p** and **q**, so that T may be written in the form of the operator given in Eq. (5.20), then T is known as a dyadic or tensor of the second rank. Alternatively, we can define a tensor of the second rank as a physical quantity which, with respect to a set of axes x_i, has nine components which transform according to Eq. (5.27) when the axes of reference are changed.[3]

A tensor of the second rank, T_{ij}, is said to be symmetric if $T_{ij} = T_{ji}$, and to be skew symmetric or antisymmetric if $T_{ij} = -T_{ji}$. Thus the tensor with components:

$$\begin{bmatrix} 9 & -4 & 1 \\ -4 & 7 & 2 \\ 1 & 2 & 6 \end{bmatrix} \tag{5.29}$$

is symmetric. An antisymmetric tensor necessarily has all the diagonal terms T_{ij} equal to zero, such as the tensor:

$$\begin{bmatrix} 0 & -\gamma & \beta \\ \gamma & 0 & -\alpha \\ -\beta & \alpha & 0 \end{bmatrix} \tag{5.30}$$

---

[3] Generalizing this definition, a scalar is a tensor of zero rank, because it transforms according to the law $\rho' = \rho$ when the axes are changed. A vector is a tensor of the first rank, because when the axes are changed, its components transform according to the rule $p'_j = a_{ij}p_j$ (Eq. (5.16)). Sometimes, the word 'order' is used instead of rank to describe the number of suffixes appropriate to a tensor quantity.

Whether or not a tensor is symmetric or antisymmetric is independent of the choice of axes. The condition for two tensors to be equal is that each component of one be equal to the corresponding component of the other.

Any second-rank tensor may be expressed as the sum of a symmetric tensor and of an antisymmetric tensor. This is because any component T_{ij} can always be written as:

$$T_{ij} = \tfrac{1}{2}(T_{ij} + T_{ji}) + \tfrac{1}{2}(T_{ij} - T_{ji}) \tag{5.31}$$

The first term then gives the component of a symmetric second-rank tensor and the second term that of an antisymmetric second-rank tensor.

Any *symmetric* tensor S_{ij} can be transformed by a suitable choice of axes so that it takes on the simple form:

$$\begin{bmatrix} S_{11} & 0 & 0 \\ 0 & S_{22} & 0 \\ 0 & 0 & S_{33} \end{bmatrix} \tag{5.32}$$

That is, all $S_{ij} = 0$ unless $i = j$. Such a tensor when expressed in this form is said to be referred to its *principal axes*. When referred to its principal axes, the components S_{11}, S_{22}, S_{33} are called the *principal components* and are often written simply as S_1, S_2, S_3, respectively. This second-rank tensor can be written in the form:

$$S_1\mathbf{ii} + S_2\mathbf{jj} + S_3\mathbf{kk} \tag{5.33}$$

We shall not prove any of the above statements. Proofs can be found in the books listed at the end of this chapter.

5.6 Tensor of the Second Rank Referred to Principal Axes

When referred to its principal axes, the symmetric tensor S_{ij} relating vectors **p** and **q** only has its diagonal components S_{11}, S_{22} and S_{33} as possible non-zero components. Under these circumstances, the equations:

$$p_i = S_{ij}q_j \tag{5.34}$$

reduce to:

$$p_1 = S_{11}q_1, \qquad p_2 = S_{22}q_2, \qquad p_3 = S_{33}q_3 \tag{5.35}$$

Now let us return to the simple example of electrical conductivity. The conductivity tensor σ_{ij} is symmetric. When referred to its principal axes, all the σ_{ij} are zero except σ_{11}, σ_{22} and σ_{33}. If the crystal under consideration is orthorhombic, monoclinic or triclinic, there is no symmetry requirement that two or more of σ_{11}, σ_{22} and σ_{33} have to be equal. For our purposes here we shall assume that none of σ_{11}, σ_{22} and σ_{33} are equal.

Suppose we apply an electrical field **E** in the x_1–x_2 plane with components $[E_1, E_2, 0]$ along these principal axes. Then:

$$J_1 = \sigma_{11}E_1 + \sigma_{12}E_2 + \sigma_{13}E_3 = \sigma_{11}E_1 \tag{5.36}$$

because σ_{12} and σ_{13} are both zero.

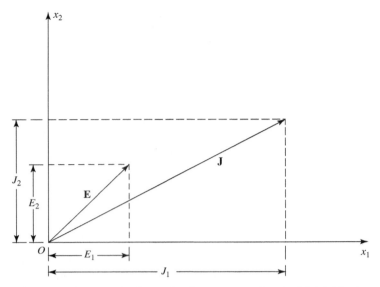

Figure 5.2   The general tensorial relationship between an electrical field **E** in the x_1–x_2 plane and the resultant current density **J**, also in the x_1–x_2 plane.

Similarly, $J_2 = \sigma_{22}E_2$. However, $J_3 = 0$ because $E_3 = 0$. We can represent this relationship between **J** and **E** on a diagram such as Figure 5.2. This diagram can be constructed by drawing **E**, finding E_1 and E_2, and multiplying E_1 by σ_{11} to obtain J_1 and E_2 by σ_{22} to obtain J_2. We then construct **J** from its components along the axes; that is, J_1 and J_2. It should be noted carefully that, in general, **J** and **E** are not parallel. If **E** were directed along Ox_1, we would have $J_1 = \sigma_{11}E_1$ because then both J_2 and J_3 would be zero. Likewise, if **E** were directed along Ox_2, **J** would be parallel to **E**; **J** will only be parallel to **E** if **E** is directed along a principal axis.

When we speak of the conductivity in a particular direction, what we actually mean in practice is that if **E** is applied in that direction and the current density is measured *in the same direction*, to give a value $J_\parallel$, the conductivity in this direction is $J_\parallel$ divided by the magnitude of **E**; that is, $J_\parallel/|E|$. We can find an expression for this by resolving **J** parallel to **E**.

Suppose **E** is applied in a direction so that its direction cosines with respect to the principal axes of the conductivity tensor are $\cos\alpha$, $\cos\beta$ and $\cos\gamma$. Then we have:

$$J_1 = \sigma_{11}E_1 = \sigma_{11}E\cos\alpha$$
$$J_2 = \sigma_{22}E_2 = \sigma_{22}E\cos\beta \qquad (5.37)$$
$$J_3 = \sigma_{33}E_3 = \sigma_{33}E\cos\gamma$$

where E is the magnitude of **E** (i.e. |**E**|). Then, a unit vector $\hat{\mathbf{n}}$ parallel to **E** is simply the vector $[\cos\alpha, \cos\beta, \cos\gamma]$, since $\cos^2\alpha + \cos^2\beta + \cos^2\gamma = 1$. Resolving **J** parallel to **E** produces a vector parallel to **E** of magnitude $J_\parallel = \mathbf{J}.\hat{\mathbf{n}}$. Hence:

$$J_\parallel = J_1\cos\alpha + J_2\cos\beta + J_3\cos\gamma = E(\sigma_{11}\cos^2\alpha + \sigma_{22}\cos^2\beta + \sigma_{33}\cos^2\gamma) \qquad (5.38)$$

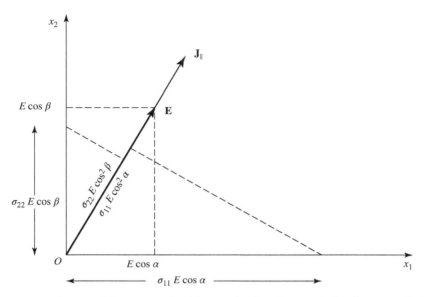

Figure 5.3 Derivation of the magnitude of the conductivity in a particular direction. The figure is drawn for $\sigma_{11} = 250$ and $\sigma_{22} = 75\,\Omega^{-1}\,\mathrm{m}^{-1}$, so that $\sigma_{22} = 0.3\sigma_{11}$.

Therefore the conductivity in the direction parallel to **E** is:

$$\sigma = J_{\|}/E = \sigma_{11}\cos^2\alpha + \sigma_{22}\cos^2\beta + \sigma_{33}\cos^2\gamma \tag{5.39}$$

The steps in deriving Eq. (5.39) are illustrated diagrammatically in Figure 5.3 for the simple case where **E** is normal to the principal axis Ox_3 of the conductivity tensor, so that $\cos\gamma = 0$.

It is instructive to derive the result given in Eq. (5.39) in a different way. Suppose we consider the meaning of the component σ'_{11} of the conductivity tensor irrespective of whether it is referred to principal axes. The component σ'_{11} relates the electric field along axis Ox'_1 to the component of the current along the same axis Ox'_1. If, therefore, we wish to find the value of the conductivity in a particular direction, having been given the components of the conductivity tensor referred to its principal axes, we can proceed as follows:

Choose a new set of axes such that Ox'_1 is along the direction of interest. Then the component σ'_{11} of the conductivity tensor referred to this new set of axes gives us the conductivity along this particular direction – the $J_{\|}/E$. We are only interested in σ'_{11}, so in writing out the array of the a_{ij} for this transformation we only need to know the values of the cosines of the angles between Ox'_1 and the principal axes of the conductivity tensor, Ox_1, Ox_2 and Ox_3. Following the scheme of the expressions in Eq. (5.3), we therefore only need to know a_{11}, a_{12} and a_{13}. These quantities are $\cos\alpha$, $\cos\beta$ and $\cos\gamma$, respectively. Using the transformation formula in Eq. (5.27), we have:

$$\sigma'_{ij} = a_{ik}a_{jl}\sigma_{kl} \tag{5.40}$$

and so when $i = 1$ and $j = 1$:

$$\sigma'_{11} = a_{1k}a_{1l}\sigma_{kl} \tag{5.41}$$

Since σ_{ij} is defined relative to its principal axes, the only non-zero terms within this conductivity tensor are σ_{11}, σ_{22} and σ_{33}. Therefore:

$$\sigma'_{11} = a_{11}a_{11}\sigma_{11} + a_{12}a_{12}\sigma_{22} + a_{13}a_{13}\sigma_{33} \tag{5.42}$$

Substituting for a_{11}, a_{12}, a_{13}, we obtain:

$$\sigma'_{11} = \sigma_{11}\cos^2\alpha + \sigma_{22}\cos^2\beta + \sigma_{33}\cos^2\gamma \tag{5.43}$$

in agreement with Eq. (5.39).

It is clear that we could have proceeded in exactly the same way to find the conductivity in a particular direction even if the values of the components of σ_{ij}, the conductivity tensor, had not been given to us referred to principal axes. In that case there would have been, in general, nine terms in the expansion of Eq. (5.40). This transformation formula clearly holds irrespective of whether the σ_{ij} are referred to principal axes. Therefore, we can state that to find the value of a property of a crystal in a particular direction we proceed as follows:

Let the components of the tensor representing this property be given as T_{ij} referred to axes (Ox_1, Ox_2, Ox_3). Choose an axis along the direction of interest and call this Ox'_1. Let this axis have direction cosines referred to (Ox_1, Ox_2, Ox_3) of a_{11}, a_{12}, a_{13}, respectively. Then the value of the property T'_{11} in the direction we are interested in is given by:

$$T'_{11} = a_{1i}a_{1j}T_{ij} \tag{5.44}$$

This relation holds for all second-rank tensors, whether or not they are symmetrical. This equation can be written in a convenient shorthand notation by removing the dash and the '1' subscripts in Eq. (5.44), so that the property T along the direction of interest is:

$$T = l_i l_j T_{ij} \tag{5.45}$$

where the relevant direction cosines are now defined as l_1, l_2 and l_3 to conform to the notation used by Nye [1].

5.7 Limitations Imposed by Crystal Symmetry for Second-Rank Tensors

The discussion in this section applies to tensors used to represent *physical properties* of crystals, and strictly only to perfect single crystals; that is, ones without defects.

Suppose two vector quantities **p** and **q** are related by the tensor T_{ij} in a *crystal* so that this tensor represents a property of the crystal. The crystal then determines the relationship between components of **p** and those of **q** and we choose to represent this relationship by means of the tensor T_{ij}. We state the components of T_{ij} with respect to certain axes. If we now take a different set of axes in the crystal, the components of **p** and of **q** along these new axes will in general be different and, furthermore, the relationship between the components of **p** and those of **q** along the new axes will alter because the tensor T'_{ij} referred to these new axes will in general be different from T_{ij}.

However, if we select new axes in the crystal which are related to the old ones by some symmetry operation, the values of the components of **p** and of **q** with respect to the new set of axes will be different from the values of those referred to the old ones, but the *relationship* between the components of **p** and those of **q** will be the same, so that:

$$T'_{ij} = T_{ij}$$

This is because the properties of the crystal are the same along the new axes as along the old ones. In order for this to be true for particular symmetry operations, limitations are imposed upon the components of T_{ij} when representing a property of a crystal. The same general type of argument will also impose restrictions upon the components of tensors of the third and higher orders; examples occur in Sections 6.4 and 6.5.

The fact that physical properties of crystals should remain the same when the system of coordinates to which the properties are referred are rotated to a new set of coordinates by a symmetry operation was first appreciated by Franz Neumann, who applied this principle to elastic coefficients of crystals in a course in elasticity at the University of Königsberg in 1873/4 [3, 4]. While this is one way [5] of defining what has now come to be known as *Neumann's principle*, it is nowadays more usual to state the principle in the form:

The symmetry operators of a physical property of a crystal must include the symmetry operators of the point group of the crystal [1] (p. 20), [6, 7].[4]

It is important to appreciate that the principle does not state that the symmetry of the physical property is the same as that of the point group: the symmetry of physical properties is often higher than that of the point group.

Physical properties characterized by a second-order tensor are necessarily invariant with respect to the operation of a centre of symmetry. This is implicit in the linear relations:

$$p_i = T_{ij}q_j$$

because if we substitute $-p_i$ for p_i and $-q_j$ for q_j (i.e. we reverse the directions of **p** and of **q**), the relation is still satisfied by the same values of T_{ij}. In terms of applying Neumann's principle, this means that physical properties represented by second-order tensors for all crystals must include the symmetry operation of $\bar{1}$; this will be true even for crystals belonging to non-centrosymmetric point groups.

It will assist in understanding what follows if we look at the centre-of-symmetry statement in another way. Suppose that for one set of axes (Ox_1, Ox_2, Ox_3) the relation between the p_i and the q_j is given by T_{ij}. If we now reverse the axes of reference, leaving **p** and **q** the same as before, this corresponds to choosing a new set of axes such that the array of the a_{ij} relating the axes is:

$$
\begin{array}{c}
\text{Old} \\
\begin{array}{c|ccc}
 & x_1 & x_2 & x_3 \\
\hline
\end{array}
\end{array}
$$

	x_1	x_2	x_3
x'_1	-1	0	0
x'_2	0	-1	0
x'_3	0	0	-1

(New)   (5.46)

---

[4] Here, to conform to International Union of Crystallography current nomenclature, the word 'operators' has been substituted for the word 'elements' in the form used in [1] (see the footnote to Section 1.4).

Here, all a_{ij} are equal to zero unless $i = j$. We now apply the transformation formula, Eq. (5.27):

$$T'_{ij} = a_{ik}a_{jl}T_{kl}$$

Therefore:

$$T'_{ij} = a_{ii}a_{jj}T_{ij} = T_{ij} \quad \text{since } a_{11} = a_{22} = a_{33} = -1 \tag{5.47}$$

What we have done here is to leave the measured quantities **p** and **q** the same and to imagine the crystal inverted through a centre-of-symmetry operation. We have obtained the same result as if we had reversed the directions of **p** and **q**.

Now suppose the crystal contains a diad axis of symmetry. If we measure a certain property along a certain direction with respect to this axis, and then rotate the crystal 180° about this axis and re-measure the property, we will get the same value as before by Neumann's principle. This imposes restrictions on the values of the components of any *symmetric* second-rank tensor S_{ij} which represents this property. In this context, it is relevant that the tensors of the second rank in which we will be interested will all be symmetric.

To see what these restrictions are, take axes (Ox_1, Ox_2, Ox_3) and suppose there is a diad axis along Ox_2, such as might arise in a material with monoclinic symmetry. Initially, we assume the tensor S_{ij} is symmetric, so that it has six independent components. Now, if we take new axes related to the old by a rotation of 180° about Ox_2, the physical property must remain the same as before. The new axes are related to the old by the array of a_{ij} as:

$$
\begin{array}{c}
\text{Old} \\
\begin{array}{cccc}
 & x_1 & x_2 & x_3 \\
x'_1 & -1 & 0 & 0 \\
\text{New } x'_2 & 0 & 1 & 0 \\
x'_3 & 0 & 0 & -1
\end{array}
\end{array}
\tag{5.48}
$$

The components of the tensor with respect to the new axes, S'_{ij}, are given in terms of the old components, S_{ij}, by:

$$S'_{ij} = a_{ik}a_{jl}S_{kl} \tag{5.49}$$

and we must have $S'_{ij} = S_{ij}$ for all i and j. If we work out the components S'_{ij} one by one, we find:

$$S'_{11} = a_{11}a_{11}S_{11} = S_{11}, \qquad S'_{22} = a_{22}a_{22}S_{22} = S_{22},$$
$$S'_{33} = a_{33}a_{33}S_{33} = S_{33}, \qquad S'_{13} = a_{11}a_{33}S_{13} = S_{13} \tag{5.50}$$

but:

$$S'_{23} = a_{22}a_{33}S_{23} = -S_{23} \quad \text{and} \quad S'_{12} = a_{11}a_{22}S_{12} = -S_{12} \tag{5.51}$$

However, we must also have $S'_{23} = S_{23}$ and $S'_{12} = S_{12}$. Therefore, $S_{23} = S_{12} = 0$. Hence, a symmetrical second-rank tensor representing a physical property of a crystal with a twofold axis of rotational symmetry must have the components S_{23} and S_{12} equal to zero when referred to axes so that Ox_2 corresponds to the diad axis.

The limitations on the components of a symmetrical second-rank tensor representing a physical property of a crystal are summarized in Table 5.2 for each of the seven crystal

Table 5.2 Number of independent components of physical properties represented by second-rank (order) tensors.

Crystal system	Orientation of principal axes with respect to the crystal axes	Form of tensor	Number of independent components
Cubic	Any; representation quadric is a sphere	$\begin{bmatrix} S & 0 & 0 \\ 0 & S & 0 \\ 0 & 0 & S \end{bmatrix}$	1
Tetragonal Hexagonal Trigonal[a]	x_3 parallel to 4, 6, 3, or $\bar{3}$	$\begin{bmatrix} S_1 & 0 & 0 \\ 0 & S_1 & 0 \\ 0 & 0 & S_3 \end{bmatrix}$	2
Orthorhombic	x_1, x_2, x_3 parallel to the diads along the x-, y- and z-axes	$\begin{bmatrix} S_1 & 0 & 0 \\ 0 & S_2 & 0 \\ 0 & 0 & S_3 \end{bmatrix}$	3
Monoclinic	x_2 parallel to the diad along the y-axis	$\begin{bmatrix} S_{11} & 0 & S_{13} \\ 0 & S_{22} & 0 \\ S_{13} & 0 & S_{33} \end{bmatrix}$	4
Triclinic	Not fixed	$\begin{bmatrix} S_{11} & S_{12} & S_{13} \\ S_{12} & S_{22} & S_{23} \\ S_{13} & S_{23} & S_{33} \end{bmatrix}$	6

[a] A hexagonal cell is used here for trigonal crystals.

systems. The derivation of this table is dealt with in Section 5.8 by a rapid method. However, we could deduce part of it using the procedure we have just outlined for a diad axis and applying this to the defining symmetry operations of each of the crystal classes. The constraints listed in Table 5.2 apply to any of the physical properties of a crystal listed in Table 5.1 (see Footnote 5 in this chapter).

5.8 Representation Quadric

The discussion in Section 5.6 demonstrated that the measured value of a property that can be represented by a second-order tensor will vary with the direction of measurement and that this variation can be found by the procedure of that section, leading to Eq. (5.45), which is:

$$T = l_i l_j T_{ij}$$

If we expand this equation, we obtain:

$$T = l_i(l_1 T_{i1} + l_2 T_{i2} + l_3 T_{i3})$$
$$= l_1^2 T_{11} + l_1 l_2 T_{12} + l_1 l_3 T_{13} + l_2 l_1 T_{21} + l_2^2 T_{22} + l_2 l_3 T_{23} + l_3 l_1 T_{31} + l_3 l_2 T_{32} + l_3^2 T_{33}$$
$$(5.52)$$

Here, l_1, l_2 and l_3 are the direction cosines of the direction we are considering with respect to the same axes to which T_{ij} is referred. We shall now restrict our discussion to symmetrical second-rank tensors, so that $T_{ij} = T_{ji}$. When second-order tensors are used to represent physical properties, in nearly every case the tensor can be shown to be symmetrical (a proof always involves thermodynamic considerations).[5] We shall show in Chapter 6 that when second-rank tensors are used to represent stresses and small strains, they too are always symmetric. When the tensor is symmetric, terms such as $a_1 a_2 T_{12}$ and $a_2 a_1 T_{21}$ are clearly equal. If we denote the tensor now by S_{ij}, S being chosen to remind ourselves that it is symmetric, we have, from Eq. (5.52):

$$S = l_1^2 S_{11} + l_2^2 S_{22} + l_3^2 S_{33} + 2l_1 l_2 S_{12} + 2l_1 l_3 S_{13} + 2l_2 l_3 S_{23} \tag{5.53}$$

If S_{ij} were referred to principal axes as well as being symmetric, Eq. (5.52) would reduce to:

$$S = S_1 l_1^2 + S_2 l_2^2 + S_3 l_3^2 \tag{5.54}$$

because $S_{12} = S_{13} = S_{23} = 0$. In Eq. (5.54) we have also chosen to use an abbreviated notation, so that the principal axis coefficients are written S_1, S_2 and S_3, rather than S_{11}, S_{22} and S_{33}.

Equation (5.53) is of the same form as the equation describing the general surface of the second degree (called a quadric) written in polar coordinates and referred to its centre as the origin. The general equation of a quadric is:

$$\frac{1}{r^2} = A\cos^2\alpha + B\cos^2\beta + C\cos^2\gamma + 2D\cos\alpha\cos\beta + 2E\cos\alpha\cos\gamma$$
$$+ 2F\cos\beta\cos\gamma \tag{5.55}$$

where r is the radius vector and $\cos\alpha$, $\cos\beta$ and $\cos\gamma$ are the direction cosines of r with respect to a set of orthogonal axes. When the general surface of the second degree is referred to its principal axes, it takes the form:

$$\frac{1}{r^2} = A\cos^2\alpha + B\cos^2\beta + C\cos^2\gamma \tag{5.56}$$

which is of the same form as Eq. (5.54). When the axes, to which a general surface of the second degree is referred, are altered, the coefficients in Eq. (5.55) transform in the same way as do the components of a symmetrical second-rank tensor. Because of this, the variation of a given *property* of a crystal with direction, given by Eqs. (5.53) and (5.54), can be represented by a figure in three-dimensional space, which gives the variation of the property S with direction.

The general surface of the second degree can be an ellipsoid or a hyperboloid of one or two sheets. To explain the procedure for representing the variation of the property S with direction in a crystal, we shall confine ourselves to the case where the values of S_1, S_2 and S_3 in Eq. (5.54) and A, B and C in Eq. (5.56) are all positive. In this case, the second-degree surface, which we shall call the representation quadric, is an ellipsoid.[6]

---

[5] All of the properties listed in Table 5.1 can be shown to be represented by *symmetric* second-rank tensors [1]. However, in the case of the transport properties in materials, such as electrical or thermal conductivity and diffusion, appeal to experiment is necessary to complete the proof. In some cases, the tensor is definitely not symmetric; examples are the tracer mixing tensor and the antisymmetric stirring tensor, both of which are used in models of ocean circulation behaviour [8].

[6] If two of the coefficients S_1, S_2, S_3 are positive and the other one is negative, the quadric is a hyperboloid of one sheet; if two are negative and the other is positive, it is a hyperboloid of two sheets. If all three values are negative, the quadric is an imaginary ellipsoid [1].

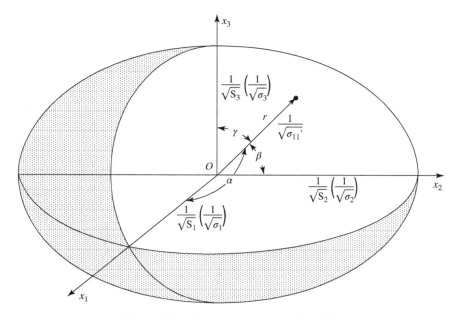

Figure 5.4 The representation ellipsoid.

Fixing attention on Eqs. (5.54) and (5.56), we see that if we construct an ellipsoid of semiprincipal axes $1/\sqrt{S_1}$, $1/\sqrt{S_2}$, $1/\sqrt{S_3}$, as in Figure 5.4, so that a general point on the ellipsoid satisfies the equation:

$$S_1 x_1^2 + S_2 x_2^2 + S_3 x_3^2 = 1 \qquad (5.57)$$

then the length r of any radius vector of the ellipsoid (representation quadric) is equal to the reciprocal of the square root of the magnitude of the property S in that direction. Therefore, if we return to the example of electrical conductivity, and we know the values σ_1, σ_2 and σ_3 of the components of the electrical conductivity referred to principal axes, we can then construct the conductivity quadric (which, if the σ_{ii} are all positive, will be an ellipsoid). If now a field **E** is applied in any direction, the magnitude of the conductivity in that direction can be found by drawing a radius vector r in the direction of **E**, measuring the value of r and taking the reciprocal of the square root of r to find the conductivity in that direction.

We are now in a position to deal rapidly with the limitations imposed by crystal symmetry on any physical property of a crystal which can be described by a *symmetric* second-rank tensor. The number of independent coefficients in the equation of the representation surface (Eq. (5.55)) equals the number of independent components of the tensor representing the physical property (Eq. (5.53)). If we look for the symmetry of the representation quadric of a property of a crystal belonging to each of the Laue groups,[7] we find that the symmetry of the representation quadric for a symmetrical second-rank tensor is governed only by the symmetry of the crystal system. The results are collected in Table 5.2.

---

[7] The Laue groups are important here because we showed in Section 5.7 that all properties represented by a second-rank tensor are necessarily invariant upon the operation of a centre of symmetry.

A *cubic* crystal must possess four triad axes. The representation quadric is then a sphere, and so the axes, to which the symmetrical second-rank tensor representing any of the properties shown in Table 5.1 is referred, are then of no consequence. The crystal is *isotropic* with regard to this property. *Hexagonal, trigonal and tetragonal* crystals have *two* independent components for the properties indicated in Table 5.1. This is because the representation quadric must be a surface of revolution about the hexad, triad or tetrad, respectively.

A general quadric shows three mutually perpendicular diad axes. In the *orthorhombic* system these must lie along the crystal diads. There are then just *three* independent components of the tensor necessary to describe the physical property. In the monoclinic system, one of the diad axes of the quadric must lie parallel to the crystallographic diad. There are *four* independent components of the symmetrical second-rank tensor describing a property. These can be thought of as three necessary to specify the length of the semi-axes of the quadric and one to fix the angle in the plane, normal to the crystal diad, between a crystal axis and a principal axis of the quadric. In the *triclinic system* there are six independent components possible for any property that may be represented by a symmetrical second-rank tensor – a symmetrical second-rank tensor has to be invariant upon a centrosymmetric operation and the holosymmetric class in this system, $\bar{1}$, possesses just a centre of symmetry.

5.9 Radius–Normal Property of the Representation Quadric

The representation ellipsoid has a further very useful property known as the *radius–normal* property. It can be stated as follows: if S_{ij} are the components of a symmetrical second-rank tensor relating the vectors **p** and **q** so that $p_i = S_{ij}q_j$ then the direction of **p** for a given **q** can be found by drawing a radius vector OQ of the representation quadric parallel to **q** and finding the normal to the quadric at Q. This is shown in two dimensions in Figure 5.5.

A proof of this property can be undertaken straightforwardly using the principles of vector calculus applied to surfaces [9]. From Eq. (5.57), the equation of the ellipsoid with respect to the principal axes of S_{ij} takes the form:

$$S_1 x_1^2 + S_2 x_2^2 + S_3 x_3^2 = 1,$$

Suppose a unit vector parallel to **q** has direction cosines l_1, l_2 and l_3 with respect to the three mutually perpendicular principal axes (Ox_1, Ox_2, Ox_3). The vector **p** in Figure 5.5 will be parallel to the vector $[S_1 l_1, S_2 l_2, S_3 l_3]$. We wish to determine the normal to the representation ellipsoid at a point such as Q, so that $\overrightarrow{OQ}$ is in the direction of **q** and is the vector $\mathbf{y} = [x_1, x_2, x_3] = [rl_1, rl_2, rl_3]$, where the reciprocal of the square root of r is the property of the tensor S_{ij} in that direction (Section 5.8).

In general from Eq. (5.57) we can solve for x_3 so that:

$$x_3 = \sqrt{\frac{1 - S_1 x_1^2 - S_2 x_2^2}{S_3}} \tag{5.58}$$

and:

$$\mathbf{y} = \left[x_1, x_2, \sqrt{\frac{1 - S_1 x_1^2 - S_2 x_2^2}{S_3}} \right] \tag{5.59}$$

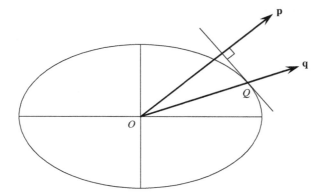

Figure 5.5   The radius–normal property of a representation ellipsoid.

We can now determine the gradients $\partial y/\partial x_1$ and $\partial y/\partial x_2$. Since the surface of the representation ellipsoid is smooth, these two gradients define the tangent plane at the point of interest [9]. A vector normal to this tangent plane will be parallel to the vector:

$$\mathbf{n} = \frac{\partial \mathbf{y}}{\partial x_1} \times \frac{\partial \mathbf{y}}{\partial x_2} \tag{5.60}$$

It is evident from Eq. (5.59) that:

$$\frac{\partial \mathbf{y}}{\partial x_1} = \left[1, 0, -\frac{S_1 x_1}{\sqrt{S_3}\sqrt{1 - S_1 x_1^2 - S_2 x_2^2}} \right] = \left[1, 0, -\frac{S_1 x_1}{S_3 x_3} \right]$$

$$\frac{\partial \mathbf{y}}{\partial x_2} = \left[0, 1, -\frac{S_2 x_2}{\sqrt{S_3}\sqrt{1 - S_1 x_1^2 - S_2 x_2^2}} \right] = \left[0, 1, -\frac{S_2 x_2}{S_3 x_3} \right] \tag{5.61}$$

so that:

$$\mathbf{n} = \begin{vmatrix} \mathbf{i} & \mathbf{j} & \mathbf{k} \\ 1 & 0 & -\frac{S_1 x_1}{S_3 x_3} \\ 0 & 1 & -\frac{S_2 x_2}{S_3 x_3} \end{vmatrix} = \left[\frac{S_1 x_1}{S_3 x_3}, \frac{S_2 x_2}{S_3 x_3}, 1 \right] \tag{5.62}$$

That is, $\mathbf{n}$ is parallel to the vector:

$$\mathbf{n} = [S_1 x_1, S_2 x_2, S_3 x_3] = r\,[S_1 l_1, S_2 l_2, S_3 l_3] \tag{5.63}$$

Hence, $\mathbf{n}$ is parallel to the vector $\mathbf{p}$. Finally, we note that we could have solved Eq. (5.57) for either x_1 or x_2, rather than x_3, and then defined $\mathbf{n}$ by either $\mathbf{n} = \frac{\partial \mathbf{y}}{\partial x_2} \times \frac{\partial \mathbf{y}}{\partial x_3}$ or $\mathbf{n} = \frac{\partial \mathbf{y}}{\partial x_3} \times \frac{\partial \mathbf{y}}{\partial x_1}$ as appropriate; we would still have obtained the same result for $\mathbf{n}$.

5.10   Third- and Fourth-Rank Tensors

Just as a second-rank tensor relates two vectors, so a third-rank tensor relates a second-rank tensor and a vector, and a fourth-rank tensor relates two second-rank tensors, and so on. Thus, for a tensor of the third rank, T_{ijk}, we can envisage a relationship between a second-rank tensor s_{jk} and a vector p_i so that, for example:

$$p_i = T_{ijk}s_{jk} \tag{5.64}$$

Equally, relationships could be of the general form:

$$s_{ij} = T_{ijk}p_k \tag{5.65}$$

where we require the tensor of the third rank to transform as:

$$T'_{ijk} = a_{il}a_{jm}a_{kn}T_{lmn} \tag{5.66}$$

generalizing the result in Section 5.4 for a tensor of the second rank. Likewise, a tensor of the fourth rank, T_{ijkl}, can relate two tensors of the second rank, s_{ij} and g_{kl}, through an equation of the form:

$$s_{ij} = T_{ijkl}g_{kl} \tag{5.67}$$

where T_{ijkl} transforms as:

$$T'_{ijkl} = a_{im}a_{jn}a_{kp}a_{lq}T_{mnpq} \tag{5.68}$$

We shall consider third- and fourth-rank tensors in more detail in Chapter 6, once we have introduced the two important field tensors of stress and strain.

Problems

5.1 Define a tensor of the second rank. Write down two physical properties of crystals that can be represented by a second-rank tensor, and for each state the two physical quantities that are related by the tensor. For one of your examples, write down in full the equations relating the components of the two physical quantities and explain your notation. Explain the physical significance of the tensor component D_{12}, where D is the diffusivity tensor.

5.2 Explain why the array of the a_{ij} in Eq. (5.3) does not represent the components of a second-rank tensor.

5.3 Express the following tensor as the sum of a symmetric tensor and an antisymmetric tensor:

$$\begin{pmatrix} 12 & 6 & 0 \\ 4 & 7 & 0 \\ 0 & 0 & 3 \end{pmatrix}$$

5.4 If σ is a symmetric dyadic and **n** and **b** are vectors, show that:

$$(\sigma \cdot \mathbf{n}) \cdot \mathbf{b} = (\sigma \cdot \mathbf{b}) \cdot \mathbf{n}$$

5.5 A crystal possesses a single fourfold rotational axis of symmetry parallel to the z-axis. Find the necessary relations between the components of a second-rank tensor representing a physical property of the crystal when the tensor is referred to axes parallel to the crystal axes. How can this result be reconciled with the entry for tetragonal crystals in Table 5.2?

5.6 Prove that whether a tensor is symmetric or antisymmetric is independent of the choice of axes. *Hint*: show that if $T_{ij} = T_{ji}$ then $T'_{ij} = T'_{ji}$.

5.7 The electrical conductivity of an orthorhombic crystal, with lattice parameters $a = 10.2\,\text{Å}$, $b = 7.8\,\text{Å}$ and $c = 6.3\,\text{Å}$, is found in general to be anisotropic. However, in a $(hk0)$ plane whose normal is $52°$ to the $[010]$ axis, the electrical conductivity is isotropic. In a separate experiment on the same crystal, it was found that the magnitude of the electrical conductivity in the $[001]$ direction was $4 \times 10^7\,\Omega^{-1}\,\text{m}^{-1}$ and that the electrical conductivity in the $[010]$ direction was three times that in the $[100]$ direction.
   (a) Determine the electrical conductivity tensor of the crystal with respect to the crystallographic axes of the orthorhombic crystal.
   (b) Determine the electrical conductivity in the $[110]$ direction.

5.8 (a) The electrical conductivity tensor of a crystal has the components:
$$\sigma_{ij} = \begin{pmatrix} 18.25 & -\sqrt{3} \times 2.25 & 0 \\ -\sqrt{3} \times 2.25 & 22.75 & 0 \\ 0 & 0 & 9 \end{pmatrix} \times 10^8\,\Omega^{-1}\text{m}^{-1}$$
Take new axes rotated $60°$ about x_3 in a clockwise direction looking along negative x_3 and write down a table of the direction cosines between the new and old axes as in Section 5.2. Check that the sum of the squares of the a_{ij} in each row and column equals 1.
   (b) Write down the components of the conductivity tensor referred to the new axes. Check that σ_{ij} has not altered. The tensor is now referred to principal axes.
   (c) Draw a section of the representation quadric (i.e. the conductivity ellipsoid in this case) in the plane $x'_3 = 0$. Draw radius vectors of the resulting ellipse at angles of $30°$ and $60°$ to x'_1 and hence find the conductivity in these directions.
   (d) Check your results in (c) by direct calculations.
   (e) Suppose an electric field **E** of $100\,\text{Vm}^{-1}$ is established in a direction at $60°$ to x'_1 in the plane $x'_3 = 0$. Write down the components of this electric field along x'_1 and x'_2 and hence find the electric current densities in these two directions. Finally, find the component of the resultant current density along **E** and hence the electrical conductivity in this direction.
   (f) Find the direction of the resultant current vector **J** in (e).
   (g) Draw the direction of the resultant current density vector on your diagram prepared in (c) and confirm that **J** is parallel to the normal to the surface of the representation quadric at the point where a vector from the origin parallel to **E** meets the representation quadric.

Suggestions for Further Reading

Lovett, D.R. (1999). *Tensor Properties of Crystals*, 2nd Edition. Bristol and Philadelphia: Adam Hilger.

Nye, J.F. (1985). *Physical Properties of Crystals*. Oxford: Clarendon Press.

Sands, D.E. (1982). *Vectors and Tensors in Crystallography*. Reading, Massachusetts and London: Addison-Wesley.

Wooster, W.A. (1973). *Tensors and Group Theory for the Physical Properties of Crystals*. Oxford: Clarendon Press.

References

[1] Nye, J.F. (1985). *Physical Properties of Crystals*. Oxford: Clarendon Press.

[2] Einstein, A. (1916). Die Grundlage der allgemeinen Relativitätstheorie. *Ann. Phys.* 49: 769–822.

[3] Neumann, F. (1885). Elasticität krystallinischer Stoffe, Chapter 12. In: *Vorlesungen über die Theorie der Elasticität der festen Körper und des Lichtäthers* (ed. O.E. Meyer), pp. 164–202. Leipzig: B.G. Teubner.

[4] Katzir, S. (2004). The emergence of the principle of symmetry in physics. *Hist. Stud. Phys. Biol. Sci.* 35: 35–65.

[5] Wooster, W.A. (1973). *Tensors and Group Theory for the Physical Properties of Crystals*, p. 15. Oxford: Clarendon Press.

[6] Gay, P. (1972). *The Crystalline State*, p. 259. Edinburgh: Oliver and Boyd.

[7] Lovett, D.R. (1999). *Tensor Properties of Crystals*, 2nd Edition, p. 31. Bristol and Philadelphia: Adam Hilger.

[8] Griffies, S.M. (1998). The Gent–McWilliams skew flux. *J. Phys. Oceanogr.* 28: 831–841.

[9] Riley, K.F., Hobson, M.P. and Bence, S.J. (2006). *Mathematical Methods for Physics and Engineering*, 3rd Edition, pp. 345–347. Cambridge: Cambridge University Press.

6

Strain, Stress, Piezoelectricity and Elasticity

6.1 Strain: Introduction

When forces are applied to crystals, or when imperfections are formed inside them (discussed in Part II of this book), the atoms change their *relative* positions. This change in the relative positions is called strain, and in this section we describe a way of specifying it.

The basic ideas of strain are those of *extension* and *shear*. The first is easily understood. Consider a very thin rod of length l (Figure 6.1a). Let the rod be stretched so that l' is its length in the strained state. Then the extension, e, is defined as:

$$e = \frac{l' - l}{l} \tag{6.1}$$

that is:

$$l' = l\,(1 + e)$$

Therefore, e is the ratio of the change in length to the original length. It is positive in tension and negative in compression. Shear is a measure of the change in angle between two lines in a body when the body is distorted. It is defined as follows (Figure 6.1b). If OP, OR are two perpendicular straight lines in the unstrained state and $O'P'$, $O'R'$ are the positions of the corresponding lines in the strained state then the (engineering) shear strain γ associated with these two directions at the point O is defined as:

$$\gamma = \tan\left(\frac{\pi}{2} - \theta\right) \tag{6.2}$$

where θ is the angle between $O'P'$ and $O'R'$ in the strained state.

Crystallography and Crystal Defects, Third Edition. Anthony Kelly and Kevin M. Knowles.
© 2020 John Wiley & Sons Ltd. Published 2020 by John Wiley & Sons Ltd.
Companion website: www.wiley.com/go/kelly/crystallography3e

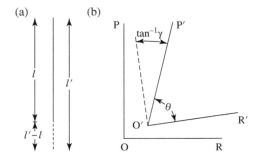

Figure 6.1 (a) Definition of extension. (b) Definition of shear

The description of general distortions, involving both extensions and shears, can be very complicated. We shall consider two important special cases:

1. *Infinitesimal strain.* Here, it is assumed that the quantities e and γ are so small that their squares and products can be neglected. For elastic strains within a crystal, this assumption is a good one, except at points which are only a very few atomic distances from a defect.
2. *Homogeneous strain.* Here, every part of the strained region suffers the same distortion. Large homogeneous strains will be encountered in Chapters 7, 11 and 12.

In this chapter we will restrict our attention to infinitesimal strains, because these are relevant for a description of how elastic strains affect perfect crystals when considering piezo-electricity and elasticity. Homogeneous strains are considered in Appendix 6. The discussion of homogeneous strains in the context of slip, twinning and martensitic transformations will be deferred to the relevant chapters later in the book.

6.2 Infinitesimal Strain

The distortion of a body can be described by giving the displacement of each point from its location in the undistorted state. Any displacements which do *not* correspond to a translation or rotation of the body as a whole will produce a strain.

We now confine ourselves to two dimensions and choose an origin fixed in space such as O in Figure 6.2. Let P be a point with coordinates (x_1, x_2) in the unstrained state, which after distortion of the body moves to the point P'. The displacement of the point P is the vector PP'. Let the coordinates of P' be $(x_1 + u_1, x_2 + u_2)$.

Now consider a point Q with coordinates $(x_1 + dx_1, x_2 + dx_2)$ lying infinitesimally close to P in the unstrained state. After deformation, Q moves to Q'. In a strained body, the displacement of Q will not be exactly the same as that of P. The displacement of Q to Q' has components $(u_1 + du_1, u_2 + du_2)$. We can write:

$$du_1 = \frac{\partial u_1}{\partial x_1}dx_1 + \frac{\partial u_1}{\partial x_2}dx_2 \tag{6.3}$$

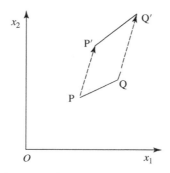

Figure 6.2 Distortion of a body in two dimensions. Points *P* and *Q* move to *P'* and *Q'* respectively after the distortion has taken place

and:

$$du_2 = \frac{\partial u_2}{\partial x_1} dx_1 + \frac{\partial u_2}{\partial x_2} dx_2 \tag{6.4}$$

Defining the four quantities at the point *P*:

$$e_{11} = \frac{\partial u_1}{\partial x_1}, \ e_{12} = \frac{\partial u_1}{\partial x_2}, \ e_{22} = \frac{\partial u_2}{\partial x_2}, \ e_{21} = \frac{\partial u_2}{\partial x_1}, \tag{6.5}$$

Eqs. (6.3) and (6.4) can be written compactly as:

$$du_i = e_{ij} dx_j \ (j = 1, 2) \tag{6.6}$$

Since du_i and dx_j are vectors, according to the definition given in Chapter 5, e_{ij} is a tensor, which we can term the *relative displacement tensor*. We shall now demonstrate the physical meaning of the various e_{ij} when each is very small compared to unity.

Let us take two special positions of the vector *PQ*: first parallel to Ox_1 (PQ_1) and then parallel to Ox_2 (PQ_2), and find out how a rectangular element at *P* is distorted (Figure 6.3). For PQ_1 we put $dx_2 = 0$ and obtain:

$$du_1 = \frac{\partial u_1}{\partial x_1} dx_1 = e_{11} dx_1 \tag{6.7}$$

$$du_2 = \frac{\partial u_2}{\partial x_1} dx_1 = e_{21} dx_1 \tag{6.8}$$

From Figure 6.3 it is clear that e_{11} measures the extension per unit length of PQ_1 resolved along Ox_1, while e_{21} measures the anticlockwise rotation of PQ_1, provided that e_{11} and e_{21} are small. Similarly, e_{22} measures the change in length per unit length of PQ_2, resolved along Ox_2, and e_{12} measures the small clockwise rotation of PQ_2.

This relative displacement tensor e_{ij} is not completely satisfactory as a measure of strain, because it is possible to have non-zero components of e_{ij} without there being any distortion of the body. For example, consider a rigid-body rotation through a small angle ω, illustrated in Figure 6.4. Here we have $e_{11} = e_{22} = 0$, but $e_{12} = -\omega$ and $e_{21} = \omega$ (since for small angles, $\tan \omega = \omega - O(\omega^3) \cong \omega$, where ω is measured in radians, neglecting the higher-order terms because we are considering infinitesimal relative displacements). To remove the component

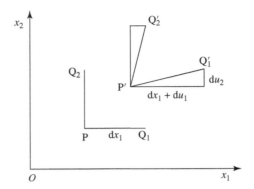

Figure 6.3 Distortion in two dimensions of a rectangular element at P. Two special positions Q_1 and Q_2 relative to P are also shown prior to deformation. After deformation, P, Q_1 and Q_2 have moved to P', Q_1' and Q_2' respectively

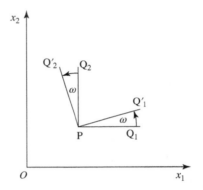

Figure 6.4 Imposition of a rigid body rotation ω to enable a relative displacement tensor e_{ij} in this two-dimensional example to be written solely as an antisymmetrical rotation tensor, ω_{ij}, without any distortion of the body taking place

of rotation from a general e_{ij} we express it as the sum of a symmetrical tensor, ε_{ij}, and an antisymmetrical tensor, ω_{ij}, so that:

$$e_{ij} = \varepsilon_{ij} + \omega_{ij} = \tfrac{1}{2}(e_{ij} + e_{ji}) + \tfrac{1}{2}(e_{ij} - e_{ji}) \tag{6.9}$$

$\varepsilon_{ij} = \tfrac{1}{2}(e_{ij} + e_{ji})$ is then defined as the *pure strain* and $\omega_{ij} = \tfrac{1}{2}(e_{ij} - e_{ji})$ measures the rotation, as shown in Figure 6.4. The geometrical interpretation of Eq. (6.9) is shown in Figure 6.5. It should be noted that the change in angle between the two lines originally at right angles in Figure 6.5 is $2\varepsilon_{12}$. Therefore, the shear component of the pure strain tensor ε_{12} is half the engineering shear strain γ defined in Eq. (6.2).

In specifying infinitesimal strain in three dimensions, the method is the same as for two dimensions. The variation of displacement $\mathbf{u}$ ($= u_1, u_2, u_3$) with variation in position

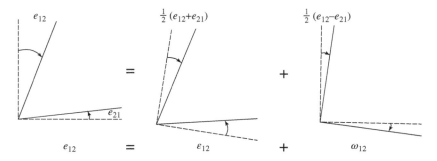

Figure 6.5 The geometrical interpretation of Eq. (6.9) for e_{12} and e_{21}

$\mathbf{x}$ $(= x_1, x_2, x_3)$ is used to define nine components of a relative displacement tensor e_{ij}:

$$e_{ij} = \frac{\partial u_i}{\partial x_j} \qquad (i,j = 1, 2, 3) \tag{6.10}$$

The strain tensor ε_{ij} is then defined as the symmetrical part of e_{ij}:

$$\varepsilon_{ij} = \tfrac{1}{2} (e_{ij} + e_{ji}) \tag{6.11}$$

In full:

$$\begin{bmatrix} \varepsilon_{11} & \varepsilon_{12} & \varepsilon_{13} \\ \varepsilon_{12} & \varepsilon_{22} & \varepsilon_{23} \\ \varepsilon_{13} & \varepsilon_{23} & \varepsilon_{33} \end{bmatrix} = \begin{bmatrix} e_{11} & \tfrac{1}{2}(e_{12} + e_{21}) & \tfrac{1}{2}(e_{13} + e_{31}) \\ \tfrac{1}{2}(e_{21} + e_{12}) & e_{22} & \tfrac{1}{2}(e_{23} + e_{32}) \\ \tfrac{1}{2}(e_{31} + e_{13}) & \tfrac{1}{2}(e_{32} + e_{23}) & e_{33} \end{bmatrix} \tag{6.12}$$

The diagonal components of ε_{ij} are the changes in length per unit length of lines parallel to the axes and are called the *tensile strains*. The off-diagonal components measure *shear strains* so that, for instance, ε_{13} is one-half the change in angle in radians between two lines originally parallel to the Ox_1 and Ox_3 axes.

Since pure strain is a symmetrical second-rank tensor, it can be referred to principal axes (Chapter 5). Under these circumstances, the shear components vanish and we have:

$$\varepsilon = \begin{bmatrix} \varepsilon_1 & 0 & 0 \\ 0 & \varepsilon_2 & 0 \\ 0 & 0 & \varepsilon_3 \end{bmatrix} \tag{6.13}$$

The geometrical meanings of the principal strains ε_1, ε_2 and ε_3 are shown in Figure 6.6, which shows that a unit cube whose edges are parallel to the principal axes is changed into a rectangular-sided figure with edges $(1 + \varepsilon_1)$, $(1 + \varepsilon_2)$ and $(1 + \varepsilon_3)$. The change in volume per unit volume is called the dilatation, and is given by:

$$\Delta = (1 + \varepsilon_1)(1 + \varepsilon_2)(1 + \varepsilon_3) - 1 = \varepsilon_1 + \varepsilon_2 + \varepsilon_3 \tag{6.14}$$

where terms such as $\varepsilon_1 \varepsilon_2$ and $\varepsilon_1 \varepsilon_2 \varepsilon_3$ have been neglected, since the strains are infinitesimal. When the strain is referred to any other axes, the dilatation is always given by the

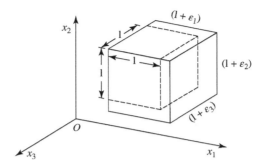

Figure 6.6 Distortion of a unit cube whose edges are parallel to the principal axes into a rectangular-sided figure with edges $(1 + \varepsilon_1)$, $(1 + \varepsilon_2)$ and $(1 + \varepsilon_3)$

invariant sum of the trace of the strain matrix, $(\varepsilon_{11} + \varepsilon_{22} + \varepsilon_{33})$. (This is a general result for symmetrical tensors of the second rank, the proof of which follows from the properties of similarity transformations and can be found in references [1–4] of Appendix 1.)

The components of either e_{ij} or ε_{ij} can be transformed to new axes by means of the standard transformation rule for second-rank tensors (Chapter 5).

In a number of books on elasticity, the strain is described by the quantities:

$$\begin{bmatrix} \varepsilon_x & \gamma_{xy} & \gamma_{zx} \\ \gamma_{xy} & \varepsilon_y & \gamma_{yz} \\ \gamma_{zx} & \gamma_{yz} & \varepsilon_z \end{bmatrix} \tag{6.15}$$

We then see that, for instance, $\gamma_{xy} \equiv 2\varepsilon_{12}$. The γ values are usually called shear strains or engineering shear strains, and it should be noted that they are equal to twice the corresponding tensor shear strains defined in this book. This array of the values of ε and γ does *not* form a tensor.

6.3 Stress

When atoms in a crystal are displaced relative to one another, forces act on them, tending to restore their normal spatial relationship. If a plane is passed through a strained region of crystal, it will be found that the atoms on one side of the plane are exerting forces upon the atoms on the other side. At any point on the plane, the force acting per unit area of plane is defined as the traction on the plane at that point. The force is customarily resolved into components normal and parallel to the plane, f_n and f_s, respectively, and corresponding normal and shear stress components, σ_n and τ, are defined as follows:

$$\sigma_n = \underset{\delta A \to 0}{\mathrm{Lt}} \frac{f_n}{\delta A} \tag{6.16}$$

$$\tau = \underset{\delta A \to 0}{\mathrm{Lt}} \frac{f_s}{\delta A} \tag{6.17}$$

The area δA is shown in Figure 6.7; it surrounds the point P, so that by finding the limit of the series of average stresses $f/\delta A$ as δA decreases, the stress at the point itself can be

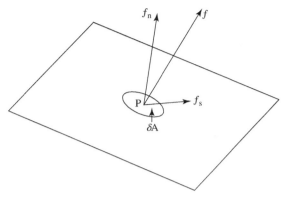

Figure 6.7 The force f acting on a small area δA in a plane surrounding a point P resolved into components normal and parallel to the plane, f_n and f_s, respectively

defined. For this definition to be meaningful, the force at P must not be changing too sharply on an atomic scale.

The procedure outlined above could be repeated for any plane passing through the point P. To define the stress at P completely, it is sufficient to specify the stresses acting on three mutually perpendicular planes passing through P. It is then possible to calculate the stress acting on any other plane by means of a force balance. The stresses acting on the faces of an infinitesimal cube located at P are shown in Figure 6.8a, using two alternative notations. In the numerical suffix notation, the stress component σ_{ij} is the component in the direction of the x_j-axis of the traction acting on the face whose normal is parallel to x_i.[1] The sign

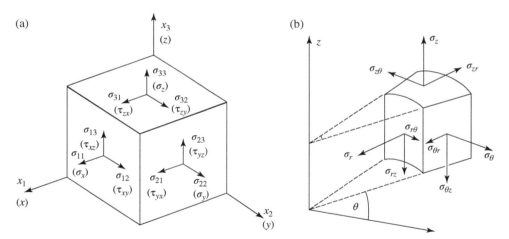

Figure 6.8 Definition of stress components (a) in Cartesian coordinates and (b) in cylindrical polar coordinates

---

[1] Nye [1] and Lovett [2] use a notation where σ_{ij} is the component in the direction of the x_i-axis of the traction acting on the face whose normal is parallel to x_j. The notation we have chosen is that used elsewhere, such as [3–6], but, irrespective of which notation is used, $\sigma_{ij} = \sigma_{ji}$ in equilibrium. There is no single 'correct' notation.

convention is that a stress is positive if the force component and the outward-going normal of the face on which it acts have the same sense, relative to the axes to which they are respectively parallel. The letter suffix notation follows the same pattern, except that shear stresses are distinguished by the letter τ. It is sometimes convenient to work with cylindrical polar coordinates instead of Cartesian coordinates; the stress components for this case are shown in Figure 6.8b.

As the cube in Figure 6.8a is supposed to be infinitesimally small, the forces acting on opposite faces must be equal and opposite. If the cube were of side δ, the shear stresses τ_{xy} and τ_{yx} would produce a couple $(\tau_{xy} - \tau_{yx})\delta^3$, tending to rotate the cube about the z-axis. Since the moment of inertia of the cube varies as δ^5, it vanishes as $\delta \to 0$ more rapidly than does the couple $(\tau_{xy} - \tau_{yx})\delta^3$, so that, to avoid an infinite angular acceleration:

$$\tau_{xy} = \tau_{yx}$$

and in general in equilibrium:

$$\sigma_{ij} = \sigma_{ji} \tag{6.18}$$

In order to calculate the stress acting upon any given plane at P, an infinitesimal body is constructed having one face, ABC, parallel to the given plane and all its other faces parallel to the cube faces (Figure 6.9). A balance of the forces acting on the tetrahedron $ABCO$ then determines the required stress. Let $\mathbf{P} = [P_1, P_2, P_3]$ be the traction exerted on the face ABC by the material outside the tetrahedron. Setting the sum of forces in the Ox_1 direction equal to zero:

$$P_1(ABC) = \sigma_{11}(BOC) + \sigma_{21}(AOC) + \sigma_{31}(AOB) \tag{6.19}$$

Now, the volume of a tetrahedron is (area of base × perpendicular height)/3. If we specify the perpendicular from O to ABC to be d, then:

$$d.(ABC) = OA.(BOC) = OB.(AOC) = OC.(AOB) \tag{6.20}$$

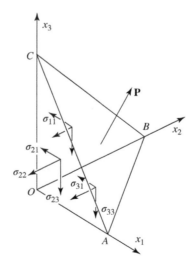

Figure 6.9   Calculation of the stress acting on the plane *ABC*

and so

$$\frac{(BOC)}{(ABC)} = \frac{d}{OA} = l_1; \qquad \frac{(AOC)}{(ABC)} = \frac{d}{OB} = l_2; \qquad \frac{(AOB)}{(ABC)} = \frac{d}{OC} = l_3 \qquad (6.21)$$

where l_1 is the cosine of the angle between the normal to ABC and Ox_1, and so on for l_2 and l_3. Hence:

$$P_1 = \sigma_{11}l_1 + \sigma_{21}l_2 + \sigma_{31}l_3 \qquad (6.22)$$

where the l_i are the direction cosines of the normal to the plane *ABC*. We obtain similar equations for P_2 and P_3. All three equations can be written compactly as:[2]

$$P_i = \sigma_{ji}l_j \qquad (6.23)$$

Resolving **P** into two components, one normal and one parallel to *ABC*, we can obtain the normal stress σ_n and shear stress τ, where σ_n is given by the sum of the components P_i, each resolved normal to *ABC*:

$$\sigma_n = P_i l_i$$

Therefore, from Eq. (6.23):

$$\sigma_n = \sigma_{ji}l_j l_i \qquad (6.24)$$

By resolving the forces:

$$\tau = [P^2 - \sigma_n^2]^{1/2} \qquad (6.25)$$

where P is the magnitude of **P**. It is apparent from Eq. (6.23) that the components of the vector **P** representing the traction on a plane are linearly related to those of the unit vector which is normal to that plane. It follows that the relating coefficients σ_{ij} form a tensor of the second rank. When the axes of reference are rotated, the components of a given stress transform according to the general transformation law:

$$\sigma_{ij}' = a_{ik}a_{jl}\sigma_{kl} \qquad (6.26)$$

where a_{ij} is the cosine of the angle between the new Ox_i-axis and the old Ox_j-axis (Chapter 5).

Because the stress tensor is symmetrical ($\sigma_{ij} = \sigma_{ji}$), it is always possible to find a set of axes, the principal axes, so that a cube with its edges parallel to them has no shear stresses acting upon its faces. Referred to the principal axes, the stress takes the form:

$$\begin{pmatrix} \sigma_1 & 0 & 0 \\ 0 & \sigma_2 & 0 \\ 0 & 0 & \sigma_3 \end{pmatrix} \qquad (6.27)$$

where σ_1, σ_2 and σ_3 are called the principal stresses.

In very special cases, a body may be under a homogeneous stress; that is, the stress may be the same at every point. If the stress is homogeneous, parallel cubes of any size and at

---

[2] It is evident that this equation is written $P_i = \sigma_{ij}l_j$ in the notation used by Nye [1] and Lovett [2].

any location have the same tractions on their faces. Such a condition occurs in the tensile test of a uniform rod, the stress at every point being:

$$\begin{pmatrix} F/A & 0 & 0 \\ 0 & 0 & 0 \\ 0 & 0 & 0 \end{pmatrix} \tag{6.28}$$

where F is the tensile force and A is the cross-sectional area of the rod, and the tensile axis is parallel to Ox_1.

More generally, the stress in a body varies from point to point. However, the different components of stress cannot vary in an entirely arbitrary way, independently of one another. The variations in the different stress components must be related in such a way that there is a balance of the forces acting on any internal body (assuming that the body as a whole is in static equilibrium). One face of a cube of side 2δ in a varying stress field is shown in Figure 6.10. By setting the sum of forces in the Ox_1 direction equal to zero, we find:

$$\frac{\partial \sigma_{11}}{\partial x_1} + \frac{\partial \sigma_{21}}{\partial x_2} + \frac{\partial \sigma_{31}}{\partial x_3} + f_1 = 0 \tag{6.29}$$

where f_1 is the component of body force per unit volume (e.g. due to gravity) acting in the Ox_1 direction. Two similar equations are obtained with respect to the Ox_2 and Ox_3 directions. All three equations of equilibrium can be written compactly as:

$$\frac{\partial \sigma_{ij}}{\partial x_i} + f_i = 0 \tag{6.30}$$

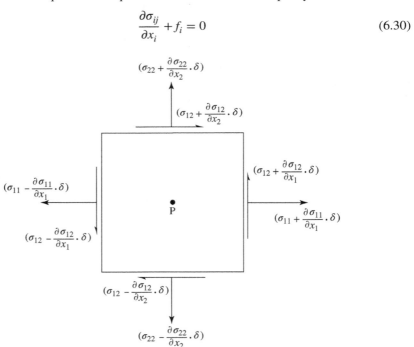

Figure 6.10   Stresses acting on one face of a cube of side 2δ in a varying stress field. P represents the point at the centre of the cube

The repetition of the suffix i in Eq. (6.30) implies the summation of the differential coefficients obtained by setting $i = 1, 2$ and 3 in turn.

We shall now mention some important types of stress. A pure shear stress is of the form $\sigma_{12} = S$, with all other σ_{ij} zero, or:

$$\begin{pmatrix} 0 & S & 0 \\ S & 0 & 0 \\ 0 & 0 & 0 \end{pmatrix} \qquad (6.31)$$

When the axes of reference are rotated through 45° about Ox_3, in an anticlockwise sense, the stress components transform to:

$$\begin{pmatrix} S & 0 & 0 \\ 0 & -S & 0 \\ 0 & 0 & 0 \end{pmatrix} \qquad (6.32)$$

The principal stresses of a pure shear are therefore equal tensile and compressive stresses acting on planes at 45° to those on which the shear stress acts alone (Figure 6.11).

In a general state of stress, shear stress components exist on all planes other than those normal to the principal axes. It can be shown that the largest shear stress occurs on the plane which bisects the angle between the planes on which the greatest and least of the principal stresses act (Figure 6.12). If $\sigma_1 < \sigma_2 < \sigma_3$, the magnitude of the largest shear stress is $\frac{1}{2}(\sigma_3 - \sigma_1)$.

A hydrostatic pressure p takes the form:

$$\begin{pmatrix} -p & 0 & 0 \\ 0 & -p & 0 \\ 0 & 0 & -p \end{pmatrix} \qquad (6.33)$$

which remains the same no matter what axes of reference are chosen. In other words, a hydrostatic pressure produces no shear stress anywhere.

It is possible to produce any state of stress by superimposing a hydrostatic pressure or tension upon three pure shear stresses. To prove this, we note that a general stress can be written as the sum of a hydrostatic component and a component called the deviatoric stress.

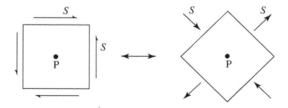

Figure 6.11   A shear stress at the point P

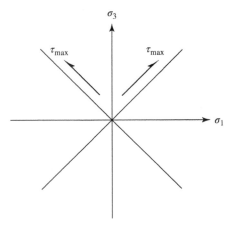

Figure 6.12 If σ_3 is the largest principal stress and σ_1 is the smallest, the largest shear stress acts in the directions shown, on planes that are normal to the plane of the figure

Referred to principal axes, the stress can be written as:

$$\begin{pmatrix} \sigma_1 & 0 & 0 \\ 0 & \sigma_2 & 0 \\ 0 & 0 & \sigma_3 \end{pmatrix} = \frac{1}{3}\begin{pmatrix} \sigma_{ii} & 0 & 0 \\ 0 & \sigma_{ii} & 0 \\ 0 & 0 & \sigma_{ii} \end{pmatrix} + \begin{pmatrix} \sigma_1 - \frac{1}{3}\sigma_{ii} & 0 & 0 \\ 0 & \sigma_2 - \frac{1}{3}\sigma_{ii} & 0 \\ 0 & 0 & \sigma_3 - \frac{1}{3}\sigma_{ii} \end{pmatrix} \quad (6.34)$$

where $\sigma_{ii} = \sigma_1 + \sigma_2 + \sigma_3$. The hydrostatic component $\frac{1}{3}\sigma_{ii}$ is invariant, because the sum $(\sigma_{11} + \sigma_{22} + \sigma_{33})$ remains the same whatever axes of reference are chosen, since, just like strain tensors (Section 6.2), stress tensors are symmetrical second-rank tensors. The deviatoric stress can be expressed as the sum of three pure shear stresses, as follows:

$$\begin{pmatrix} \sigma_1 - \frac{1}{3}\sigma_{ii} & 0 & 0 \\ 0 & \sigma_2 - \frac{1}{3}\sigma_{ii} & 0 \\ 0 & 0 & \sigma_3 - \frac{1}{3}\sigma_{ii} \end{pmatrix} = \frac{1}{3}\begin{pmatrix} \sigma_1 - \sigma_2 & 0 & 0 \\ 0 & \sigma_2 - \sigma_1 & 0 \\ 0 & 0 & 0 \end{pmatrix}$$

$$+ \frac{1}{3}\begin{pmatrix} \sigma_1 - \sigma_3 & 0 & 0 \\ 0 & 0 & 0 \\ 0 & 0 & \sigma_3 - \sigma_1 \end{pmatrix} + \frac{1}{3}\begin{pmatrix} 0 & 0 & 0 \\ 0 & \sigma_2 - \sigma_3 & 0 \\ 0 & 0 & \sigma_3 - \sigma_2 \end{pmatrix} \quad (6.35)$$

6.4   Piezoelectricity

When some single crystals are stressed in uniaxial tension, an electric polarization, $\mathbf{P}$,[3] can occur within the crystal, the magnitude of which increases linearly with the magnitude of the

---

[3] There is of course no connection between $\mathbf{P}$ used in this section to define an electrical polarization and $\mathbf{P}$ defined in Section 6.2 as the traction on a plane though a point P.

stress. This manifests itself through the development of electrical charges on the surfaces of the crystals. **P** is a vector describing the electric dipole moment per unit volume, or equivalently, the surface charge developed per unit area [1, 7]. Conversely, the application of an electric field, **E**, to these single crystals produces a strain within the crystals, and again there is a linear relationship between the magnitude of the electric field and the magnitude of the strain produced. This phenomenon is known as piezoelectricity and was first reported by Jacques and Pierre Curie in 1880 [8].

The piezoelectric tensor, d_{ijk}, is a tensor of the third rank which relates the stress applied, σ_{jk}, to the polarization produced, P_i:

$$P_i = d_{ijk}\sigma_{jk} \tag{6.36}$$

Reversing the sign of σ_{jk} reverses the sign of P_i. As we will now demonstrate, it is straightforward to show that this means that centrosymmetric crystals cannot exhibit piezoelectricity using Neumann's principle (Section 5.7). From Eq. (5.66), tensors of the third rank transform as:

$$T'_{ijk} = a_{il}a_{jm}\sigma_{kn}T_{lmn}$$

If a crystal has a centre of symmetry at the origin, the direction $[uvw]$ maps exactly on to the direction $[\overline{uvw}]$ for all u, v and w. We can examine what happens to the d_{ijk} specified relative to an 'old' orthonormal set 1, 2 and 3 if we then choose new orthonormal axes 1', 2' and 3' related to 1, 2 and 3 by the centre of symmetry (Figure 6.13).

Under these circumstances, $a_{ij} = -\delta_{ij}$, where δ_{ij} is the Kronecker delta, since the angle between the ith 'old' axis and the jth 'new' axis is either 90 or 180°. Hence, the elements of the tensor d_{ijk} produce transformed moduli so that:

$$d'_{ijk} = a_{il}a_{jm}a_{kn}d_{lmn} = -\delta_{il}\delta_{jm}\delta_{kn}d_{lmn} = -d_{ijk} \tag{6.37}$$

However, if the crystal has a centre of symmetry, and we have performed this centre-of-symmetry operation, the transformed modulus d'_{ijk} defined with respect to the 'new' axis set must be equal to the modulus d_{ijk} defined with respect to the 'old' axis set. That is, we require:

$$d'_{ijk} = d_{ijk} \tag{6.38}$$

for all i, j and k. For Eqs. (6.37) and (6.38) to be compatible, it follows that $d_{ijk} = 0$ for all i, j and k, so that we can conclude that materials with a centre of symmetry have zero

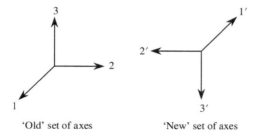

'Old' set of axes   'New' set of axes

Figure 6.13 Orthonormal 'old' and 'new' sets of axes related to one another by a centre of symmetry operation through the origin

piezoelectric moduli, and are not therefore piezoelectric. Of the 32 crystal classes, 21 are non-centrosymmetric (Chapter 2); a necessary (but not sufficient) condition for a crystal to exhibit piezoelectricity is that it belongs to a non-centrosymmetric crystal class.

In principle, we might expect the tensor d_{ijk} to have 27 independent components. However, the fact that for a shear stress, $\sigma_{kj} = \sigma_{jk}$, irrespective of body torques being present [9], means in turn that $d_{ikj} = d_{ijk}$ and so there are actually at most 18 independent coefficients of d_{ijk}.

At this stage, it becomes possible to introduce a contracted notation in which two suffixes are replaced by one, both in σ_{jk} and in d_{ijk} [1].[4] The notation for the stress σ_{ij} is contracted as follows:

$$\begin{pmatrix} \sigma_{11} & \sigma_{12} & \sigma_{13} \\ \sigma_{12} & \sigma_{22} & \sigma_{23} \\ \sigma_{13} & \sigma_{23} & \sigma_{33} \end{pmatrix} \rightarrow \begin{pmatrix} \sigma_1 & \sigma_6 & \sigma_5 \\ \sigma_6 & \sigma_2 & \sigma_4 \\ \sigma_5 & \sigma_4 & \sigma_3 \end{pmatrix} \tag{6.39}$$

A similar procedure applies to d_{ijk}, but in this case it is necessary to introduce factors of $\frac{1}{2}$ into the contracted form of d_{ijk}:

$$d_{ijk} \rightarrow d_{mn} \text{ for } n = 1, 2 \text{ and } 3 \tag{6.40a}$$

$$d_{ijk} \rightarrow \frac{1}{2} d_{mn} \text{ for } n = 4, 5 \text{ and } 6 \tag{6.40b}$$

The new d_{mn} form a 3×6 matrix of 3 rows and 6 columns, but are not tensors. It is instructive to see why the factors of $\frac{1}{2}$ arise by examining one component of Eq. (6.36). For example, when $i = 2$:

$$P_2 = d_{211}\sigma_{11} + d_{212}\sigma_{12} + d_{213}\sigma_{13} + d_{221}\sigma_{21} + d_{222}\sigma_{22} + d_{223}\sigma_{23} + d_{231}\sigma_{31} + d_{232}\sigma_{32}$$
$$+ d_{233}\sigma_{33}$$

from which it is evident that the two terms $d_{212}\sigma_{12}$ and $d_{221}\sigma_{21}$ in the full notation become together $d_{26}\sigma_6$ in the contracted notation only if the factor of $\frac{1}{2}$ is included. Similar considerations apply to the other terms for $n = 5$ and 6 in this equation.

The zero and non-zero d_{ij} for a particular crystal class can be represented conveniently using a matrix notation introduced by Van Dyke in 1950 [11]. In this notation, and the variation used by Nye [1], non-zero d_{ij} are represented by large filled dots, •, zero d_{ij} are represented by either small dots or a space, numerically equivalent d_{ij} are joined by lines, and rings denote values which are the negative of •. A complete presentation of the piezoelectric moduli in this form for each crystal class is given in Appendix E of Nye [1]. Here, we will restrict our attention to consideration of determining the non-zero d_{ij} of a few non-centrosymmetic crystal classes.

6.4.1 Class 2

In this crystal class there is a single 180° rotation axis. Let this be about x_2 in a crystal. 'Old' and 'new' axes can be defined as in Figure 6.14.

---

[4] This is often referred to as the Voigt notation after the notation introduced by Woldemar Voigt [10]. A different contracted notation is used by Wooster [7].

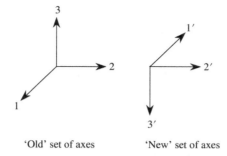

'Old' set of axes   'New' set of axes

Figure 6.14 Orthonormal 'old' and 'new' sets of axes related to one another by a 180° rotation about [010]

The only non-zero a_{ij} are $a_{11} = -1$, $a_{22} = 1$ and $a_{33} = -1$. Hence, for example:

$$d'_{113} = a_{11}a_{11}a_{33}d_{113} = -d_{113}; \quad d'_{112} = a_{11}a_{11}a_{22}d_{112} = d_{112} \quad (6.41)$$

However, because of the diad axis, d'_{113} must be equal to d_{113} and d'_{112} must be equal to d_{112}. Hence, we conclude that $d_{113} = 0$ but $d_{112} \neq 0$. Therefore, if in the transformation from d_{ijk} to d'_{ijk} the sign is $+1$, the modulus is non-zero, but if the sign is -1, the modulus is zero. Clearly, only those d_{ijk} with either a single '2' or three '2's survive. Hence the non-zero moduli are the eight in bold in the array:

$$
\begin{array}{ccccccccc}
d_{111} & \mathbf{d_{112}} & d_{113} & \mathbf{d_{211}} & d_{212} & \mathbf{d_{213}} & d_{311} & \mathbf{d_{312}} & d_{313} \\[4pt]
 & d_{122} & \mathbf{d_{123}} & & \mathbf{d_{222}} & d_{223} & & d_{322} & \mathbf{d_{323}} \\[4pt]
 & & d_{133} & & & \mathbf{d_{233}} & & & d_{333}
\end{array}
\qquad (6.42)
$$

which then become the eight dots in the corresponding matrix representation:

$$
\begin{pmatrix}
\cdot & \cdot & \cdot & \bullet & \cdot & \bullet \\
\bullet & \bullet & \bullet & \cdot & \bullet & \cdot \\
\cdot & \cdot & \cdot & \bullet & \cdot & \bullet
\end{pmatrix}
\qquad (6.43)
$$

6.4.2   Class 222

Here, the occurrence of two mutually perpendicular diads forces the third one to occur (Section 1.6). Of the eight non-zero d_{ij} in Class 2, only d_{14}, d_{25} and d_{36} are non-zero for Class 222 (Problem 6.9), so that the matrix representation is the following:

$$
\begin{pmatrix}
\cdot & \cdot & \cdot & \bullet & \cdot & \cdot \\
\cdot & \cdot & \cdot & \cdot & \bullet & \cdot \\
\cdot & \cdot & \cdot & \cdot & \cdot & \bullet
\end{pmatrix}
\qquad (6.44)
$$

6.4.3 Class 23

Crystal Class 23 has three mutually perpendicular diads and four triad axes along $\langle 111 \rangle$ and belongs to the cubic crystal system. Clearly, from the above consideration of Class 222, we know that there are at most three independent d_{ij} in Class 23 before we consider the effect of the triad axes. Suppose we examine the effect of a 120° anticlockwise rotation about [111]. In terms of the 'old' axes, this is $1 \to 2$, $2 \to 3$ and $3 \to 1$ (Figure 6.15). The only non-zero a_{ij} are $a_{12} = 1$, $a_{23} = 1$ and $a_{31} = 1$.

Therefore, for example,

$$d'_{123} = a_{11}a_{22}a_{33}d_{123} + a_{11}a_{23}a_{32}d_{132} + a_{12}a_{21}a_{33}d_{213} + a_{12}a_{23}a_{31}d_{231} + a_{13}a_{21}a_{32}d_{312}$$
$$+ a_{13}a_{22}a_{31}d_{321}$$

reduces to:

$$d'_{123} = a_{12}a_{23}a_{31}d_{231} = d_{231} \tag{6.45}$$

and likewise $d'_{231} = d_{312}$ and $d'_{312} = d_{123}$. Thus, in the d_{ij} terminology, $d_{14} = d_{25} = d_{36}$, giving the matrix representation

$$\begin{pmatrix} \cdot & \cdot & \cdot & \bullet & \cdot & \cdot \\ \cdot & \cdot & \cdot & \cdot & \bullet & \cdot \\ \cdot & \cdot & \cdot & \cdot & \cdot & \bullet \end{pmatrix} \tag{6.46}$$

Further rotations of 120° about the three other triad axes leave this result unchanged. This result is also true for class $\bar{4}3m$: the effect of a shear stress σ_{12} acting in the (001) plane of a crystal such as sphalerite is to produce a polarization, P_3, in the $[00\bar{1}]$ direction by displacing tetrahedrally coordinated zinc ions relative to the centre of mass of the four surrounding sulphur ions (Problem 6.12). (By convention, the direction of polarization of a vector from the centre of the negative charge to the centre of the positive charge determines the sense of a dipole.)

6.4.4 Class 432

Here we have 90° symmetry operations to superimpose upon the 'basic' symmetry elements of Class 23. Consider the effect of a 90° anticlockwise symmetry operation about x_3 on the

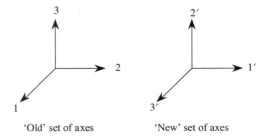

'Old' set of axes   'New' set of axes

Figure 6.15 Orthonormal 'old' and 'new' sets of axes related to one another by a 120° rotation about [111]

surviving moduli of Class 23; that is, $d_{123} = d_{213} = d_{312}$. The non-zero a_{ij} are $a_{12} = 1$, $a_{21} = -1$ and $a_{33} = 1$. Hence,

$$d'_{213} = a_{21}a_{12}a_{33}d_{123} = -d_{123},$$

$$d'_{123} = a_{12}a_{21}a_{33}d_{213} = -d_{213},$$

$$d'_{321} = a_{33}a_{21}a_{12}d_{312} = -d_{312}.$$

and similarly, $d'_{231} = -d_{132}$, $d'_{132} = -d_{231}$ and $d'_{312} = -d_{321}$. However, since we must have $d_{123} = d_{213} = d_{312}$, and since also $d_{ijk} = d_{ikj}$, it follows that $d_{123} = d_{213} = d_{312} = 0$. Hence, all the piezoelectric moduli vanish for Class 432.

Therefore, even though Class 432 is not centrosymmetric, crystals belonging to Class 432 are not piezoelectric. This is the only non-centrosymmetic class for which all the piezoelectric moduli vanish; the remaining 20 crystal classes have at least one non-vanishing piezoelectric modulus [1].

6.4.5   The Converse Effect

In the converse piezoelectric effect, predicted by Gabriel Lippman in 1881 [12] and confirmed experimentally the same year by Jacques and Pierre Curie [13], the relevant tensor equation is:

$$\varepsilon_{jk} = d_{ijk}E_i \tag{6.47}$$

where the d_{ijk} are the *same* as in Eq. (6.36) [1]. When Eq. (6.47) is expressed in the contracted form:

$$\varepsilon_j = d_{ij}E_i \tag{6.48}$$

the ε_{jk} transform as:

$$
\begin{pmatrix} \varepsilon_{11} & \varepsilon_{12} & \varepsilon_{13} \\ \varepsilon_{12} & \varepsilon_{22} & \varepsilon_{23} \\ \varepsilon_{13} & \varepsilon_{23} & \varepsilon_{33} \end{pmatrix} \rightarrow \begin{pmatrix} \varepsilon_1 & \frac{1}{2}\varepsilon_6 & \frac{1}{2}\varepsilon_5 \\ \frac{1}{2}\varepsilon_6 & \varepsilon_2 & \frac{1}{2}\varepsilon_4 \\ \frac{1}{2}\varepsilon_5 & \frac{1}{2}\varepsilon_4 & \varepsilon_3 \end{pmatrix} \tag{6.49}
$$

(Problem 6.13). Note that the contracted shear strains ε_4, ε_5 and ε_6 are engineering shear strains (γ), as defined in Section 6.2, and *not* tensor shear strains.

6.5   Elasticity of Crystals

In a strained crystal, the stress at any point is linearly related to the strain at that point, provided that the strain is very small. This relationship is known as Hooke's law. The physical basis for Hooke's law is merely that at small strains the relative displacements of the atoms are small, so that the interatomic force–separation relationship operates over such a small range as to be sensibly linear. In most solids, strains do in fact remain small (e.g. <0.001) over a quite useful range of stress. The most common family of solids for which this is not true are elastomers (elastic polymers) above their glass transition temperature where very large strains can be achieved through changes in molecular confirmation, and therefore secondary bonding, while keeping primary bonds intact (Section 3.9). Even so, for small strains, elastomers have a linear relationship between stress and strain.

We have seen that a stress and a small strain can each be represented by a symmetrical second-rank tensor, with six independent components. The most general linear dependence of stress on strain has the form:

$$\sigma_{11} = c_{1111}\varepsilon_{11} + c_{1112}\varepsilon_{12} + c_{1113}\varepsilon_{13} + c_{1121}\varepsilon_{21} + c_{1122}\varepsilon_{22} + c_{1123}\varepsilon_{23}$$
$$+ c_{1131}\varepsilon_{31} + c_{1132}\varepsilon_{32} + c_{1133}\varepsilon_{33} \tag{6.50}$$

together with similar equations for the remaining stress components. All these equations can be written in the concise form:

$$\sigma_{ij} = c_{ijkl}\varepsilon_{kl} \tag{6.51}$$

The constants c_{ijkl} are called *stiffness constants*. The existence of non-zero values of certain of the c_{ijkl} has rather surprising consequences. For example, if c_{1112} is not zero, the occurrence of a finite shear strain ε_{12} implies the existence of a proportional tensile stress σ_{11}. By symmetry arguments, it can be shown that constants of the type c_{1112} are zero in an isotropic medium but are in general not zero in a single crystal. The existence of a single strain component in a crystal may require that there be non-zero values of all the stress components.

The elastic strain produced by an applied stress is given by a set of equations similar to Eq. (6.51):

$$\varepsilon_{ij} = s_{ijkl}\sigma_{kl} \tag{6.52}$$

The constants s_{ijkl} are called *compliances*.[5] It is apparent from Eq. (6.52) that, in general, one component of stress will produce non-zero values of all of the strain components. Earlier experimenters usually determined compliances directly, by applying a stress and measuring the resulting strain. More recent determinations of the elastic properties of crystals have been based on measurements of the velocities of elastic waves in a crystal; these measurements generate stiffness constants. The compliances can be calculated if the stiffness constants are known through Eqs. (6.51) and (6.52).

Since the sets of Eqs. (6.51) and (6.52) both consist of nine equations, each containing nine terms on the right-hand side, it might appear that 81 compliances or stiffness constants must be specified. However, the number of independent constants can always be reduced from 81 to 21.

Because $\varepsilon_{12} = \varepsilon_{21}$, the constants c_{ij12} and c_{ij21} always occur together in $\sigma_{ij} = (c_{ij12} + c_{ij21})$ ε_{12} in Eq. (6.51). It is therefore permissible to set $c_{ij12} = c_{ij21}$, and, in general, $c_{ijkl} = c_{ijlk}$. Next, suppose that only the strain component ε_{11} exists. We have:

$$\sigma_{12} = c_{1121}\varepsilon_{11}$$

---

[5] As Nye notes in a footnote to Chapter 8 of his book [1], it is unfortunate that, in the English language, the initial letters of 'stiffness' and 'compliance' are the reverse of the corresponding symbols c and s. This is a historical legacy: when the theory of the mechanical behaviour of solids was developed in the nineteenth century, the c_{ijkl} terms were defined first and called 'elastic constants'. Later, Voigt [10] introduced the s_{ijkl} terms, which he called 'elastic moduli'.

and:

$$\sigma_{21} = c_{2111}\varepsilon_{11}$$

Since $\sigma_{12} = \sigma_{21}$, $c_{1211} = c_{2111}$ and in general $c_{ijkl} = c_{jikl}$. Similar arguments apply to the s_{ijkl}. The number of independent constants is now reduced to 36, and so we can use the contracted notation for stress and strain introduced in Section 6.4. A corresponding contraction is applied to c_{ijkl}, so that, for example, $c_{1233} \to c_{63}$, $c_{2323} \to c_{44}$, $c_{2332} \to c_{44}$ etc. Eq. (6.51) can now be written in the contracted form:

$$\sigma_i = c_{ij}\varepsilon_j \tag{6.53}$$

Factors of 2 and 4 must also be introduced into the definition of s_{ij}, as follows (Problem 6.13):[6]

$$s_{mn} = 2s_{ijkl} \tag{6.54}$$

when one only of either m or n is 4, 5 or 6. Or:

$$s_{mn} = 4s_{ijkl} \tag{6.55}$$

when both m and n are 4, 5, or 6. More succinctly:

$$s_{mn} = \frac{4s_{ijpq}}{(1 + \delta_{ij})(1 + \delta_{pq})} \tag{6.56}$$

For example, $s_{13} = s_{1133}$, $s_{14} = 2\,s_{1123}$ and $s_{46} = 4\,s_{2312}$. With these definitions, Eq. (6.52) can be written as:

$$\varepsilon_i = s_{ij}\sigma_j \tag{6.57}$$

The fact that the energy stored in an elastically strained crystal depends on the strain, and not on the path by which the strained state is reached, makes $c_{ij} = c_{ji}$, reducing to 21 the maximum number of non-zero c_{ij}. When the strain in a body is increased by $\mathrm{d}\varepsilon_{ij}$, the tractions acting upon a unit cube within the body do work $\mathrm{d}w$, where:

$$\mathrm{d}w = \sigma_{11}\mathrm{d}\varepsilon_{11} + \sigma_{12}\mathrm{d}\varepsilon_{12} + \sigma_{13}\mathrm{d}\varepsilon_{13} + \sigma_{21}\mathrm{d}\varepsilon_{21} + \sigma_{22}\mathrm{d}\varepsilon_{22} + \sigma_{23}\mathrm{d}\varepsilon_{23}$$
$$+ \sigma_{31}\mathrm{d}\varepsilon_{31} + \sigma_{32}\mathrm{d}\varepsilon_{32} + \sigma_{33}\mathrm{d}\varepsilon_{33} \tag{6.58}$$

or, in contracted notation:

$$\mathrm{d}w = \sigma_i\mathrm{d}\varepsilon_i \tag{6.59}$$

where $i = 1, 6$. (The quantity $\mathrm{d}w$ must be positive, otherwise the crystal would not be stable in the unstrained state.) It must be possible to integrate Eq. (6.59) to obtain the energy due to a finite strain. Suppose that a strain ε_1 is first imposed, keeping all other ε_i zero. The work done per unit volume is:

$$w_1 = \int_0^{\varepsilon_1} \sigma_1\mathrm{d}\varepsilon_1 = \int_0^{\varepsilon_1} c_{11}\varepsilon_1\mathrm{d}\varepsilon_1 = \tfrac{1}{2}c_{11}\varepsilon_1^2 \tag{6.60}$$

---

[6] As Nye notes in a footnote to Chapter 8 of his book [1], this is to conform to the Voigt notation [10]. A different notation is used by Wooster [7] which results in a different form of Eq. (6.57).

Suppose that at this level of strain of ε_1, ε_2 is increased from zero. The work is then:

$$w_2 = \int_0^{\varepsilon_2} \sigma_2 d\varepsilon_2 = \int_0^{\varepsilon_2} (c_{22}\varepsilon_2 + c_{21}\varepsilon_1) \, d\varepsilon_2 = \tfrac{1}{2}c_{22}\varepsilon_2^2 + c_{21}\varepsilon_1\varepsilon_2 \qquad (6.61)$$

The total stored energy is:

$$w = w_1 + w_2 = \tfrac{1}{2}c_{11}\varepsilon_1^2 + \tfrac{1}{2}c_{22}\varepsilon_2^2 + c_{21}\varepsilon_1\varepsilon_2 \qquad (6.62)$$

The same state of strain can be reached by imposing the strains ε_1 and ε_2 in the reverse order, giving:

$$w = \tfrac{1}{2}c_{22}\varepsilon_2^2 + \tfrac{1}{2}c_{11}\varepsilon_1^2 + c_{12}\varepsilon_2\varepsilon_1 \qquad (6.63)$$

Since w in Eqs. (6.62, 6.63) must be the same, it follows that $c_{12} = c_{21}$ and, in general:

$$c_{ij} = c_{ji} \qquad (6.64)$$

By writing Eq. (6.59) in the form:

$$dw = \sigma_i s_{ij} d\sigma_j \qquad (6.65)$$

and obtaining the energy increase in terms of the applied stress, it can be shown that:

$$s_{ij} = s_{ji} \qquad (6.66)$$

The most general stiffness and compliance constants for crystals of Class 1 can now be written in the form of matrices with 21 independent terms; in the equivalent of the matrix notation used in Section 6.4, both 6×6 stiffness matrices can be represented in the form:

$$
\begin{pmatrix}
\bullet & \bullet & \bullet & \bullet & \bullet & \bullet \\
\bullet & \bullet & \bullet & \bullet & \bullet & \bullet \\
\bullet & \bullet & \bullet & \bullet & \bullet & \bullet \\
\bullet & \bullet & \bullet & \bullet & \bullet & \bullet \\
\bullet & \bullet & \bullet & \bullet & \bullet & \bullet \\
\bullet & \bullet & \bullet & \bullet & \bullet & \bullet
\end{pmatrix}
\qquad (6.67)
$$

or, since $c_{ij} = c_{ji}$, and $s_{ij} = s_{ji}$:

$$
\begin{pmatrix}
\bullet & \bullet & \bullet & \bullet & \bullet & \bullet \\
 & \bullet & \bullet & \bullet & \bullet & \bullet \\
 & & \bullet & \bullet & \bullet & \bullet \\
 & & & \bullet & \bullet & \bullet \\
 & & & & \bullet & \bullet \\
 & & & & & \bullet
\end{pmatrix}
$$

When arbitrary axes of references are chosen, any crystal has 21 different elastic constants, although, because of crystal symmetry, there may be some relations between the constants. However, if the axes of reference are related to the crystal structure, the requirements of symmetry will ensure that some constants are zero through Neumann's principle. Just as for piezoelectricity, it is instructive to examine how this applies to a few crystal classes to appreciate the methodology of applying this principle.

6.5.1 Class $\bar{1}$

This is a centre-of-symmetry operation (Chapter 2). Under these circumstances, $a_{ij} = -\delta_{ij}$, as we have noted in Section 6.4. Hence, the elements of the tensor c_{ijkl} produce transformed moduli so that:

$$c'_{ijkl} = a_{il}a_{jm}a_{kn}a_{lp}c_{lmnp} = \delta_{il}\delta_{jm}\delta_{kn}\delta_{lp}c_{lmnp} = c_{ijkl} \tag{6.68}$$

which we know to be the case from the centre-of-symmetry operation. Therefore, in contrast to piezoelectric moduli, the presence or absence of a stiffness constant (or compliance constant) is not affected by a centre-of-symmetry operation. Hence, crystals belonging to both Class 1 and Class $\bar{1}$ each have 21 separate stiffness constants and 21 separate compliance constants. This is an example of a more general result which holds for these fourth-rank tensors: all classes within the same Laue group (Section 2.11) behave alike, in that they have the same number of separate stiffness constants and separate compliance constants. As a result, it can be deduced that there cannot be more than 11 separate forms of the matrix representations for c_{ij} and s_{ij} because there are 11 Laue groups (Table 2.2).

6.5.2 Class 2

In this crystal class there is a single 180° rotation axis. Let this be about x_2 in a crystal, as in Section 6.4. Therefore, the only non-zero a_{ij} are $a_{11} = -1$, $a_{22} = 1$ and $a_{33} = -1$. Hence, for example, as a consequence of performing this 180° symmetry operation,

$$c'_{2323} = a_{22}a_{33}a_{22}a_{33}c_{2323} = c_{2323}; \quad c'_{3312} = a_{33}a_{33}a_{11}a_{22}c_{3312} = -c_{3312} \tag{6.69}$$

However, because of the diad axis, all c'_{ijkl} must be equal to their corresponding c_{ijkl}. Hence, we deduce from Eq. (6.69) that $c_{3312} = 0$, or in compact notation, $c_{36} = 0$. The form of the equations in (6.69) show that the c_{ijkl} must be zero whenever there are either one '2' or three '2's amongst the *ijkl*. The non-zero c_{ij} (and s_{ij}) for Class 2 can then be represented in matrix form as:

$$
\begin{pmatrix}
\bullet & \bullet & \bullet & \cdot & \bullet & \cdot \\
 & \bullet & \bullet & \cdot & \bullet & \cdot \\
 & & \bullet & \cdot & \bullet & \cdot \\
 & & & \bullet & \cdot & \bullet \\
 & & & & \bullet & \cdot \\
 & & & & & \bullet
\end{pmatrix}
\tag{6.70}
$$

It is evident that the same result holds for Class m, a result which follows from the statement above with reference to Class $\bar{1}$ that all classes within the same Laue group behave alike; however, it is helpful to see why the same result holds for Class m. If (010) is a mirror plane, $1 \rightarrow 1$, $2 \rightarrow -2$ and $3 \rightarrow 3$, so the non-zero a_{ij} are $a_{11} = 1$, $a_{22} = -1$ and $a_{33} = 1$, and we arrive at the same condition that the c_{ijkl} must be zero whenever there are either one '2' or three '2's amongst the $ijkl$. This is an example of the general result that the presence or absence of a stiffness constant (or compliance constant) is not affected by an inversion operation, since m is equivalent to $\bar{2}$ (Section 2.1). Adding a centre of symmetry to give Class $2/m$ clearly does not alter the result.

Hence, we have shown that the matrix form in Eq. (6.70) is valid for all three crystal classes of the monoclinic system when x_2 is the unique axis: the three monoclinic classes each have 13 separate stiffness constants and 13 separate compliance constants.

6.5.3 Class 222

Here, the occurrence of two mutually perpendicular diads forces the third one to occur (Section 1.6). We need only look at the effect of having a diad along x_1 on the 13 surviving moduli that we know will still exist after we have looked at the consequence of a diad along x_2 for Class 2.

The result is that c_{ijkl} must be zero whenever there are either one '1' or three '1's amongst the $ijkl$. An equivalent result holds if this second 180° rotation axis is about x_3. The consequence of these considerations is that four of the 13 separate stiffness constants represented in Eq. (6.70) become zero: c_{15}, c_{25}, c_{35} and c_{46}. The nine remaining non-zero c_{ij} (and s_{ij}) for Class 222, and indeed for all three orthorhombic classes, can then be represented in matrix form as:

$$\begin{pmatrix} \bullet & \bullet & \bullet & \cdot & \cdot & \cdot \\ & \bullet & \bullet & \cdot & \cdot & \cdot \\ & & \bullet & \cdot & \cdot & \cdot \\ & & & \bullet & \cdot & \cdot \\ & & & & \bullet & \cdot \\ & & & & & \bullet \end{pmatrix} \tag{6.71}$$

6.5.4 Class 23

Again, we can repeat the process we adopted for piezoelectric moduli (Section 6.4) and examine the effect of adding four triad axes to the three mutually perpendicular diads of Class 222. If, as in Section 6.4, we examine the effect of a 120° anticlockwise rotation about [111], the only non-zero a_{ij} are $a_{12} = 1$, $a_{23} = 1$ and $a_{31} = 1$ (Figure 6.15). Hence, for example:

$$c'_{1111} = a_{12}a_{12}a_{12}a_{12}c_{2222} = c_{2222}$$

and likewise $c'_{2222} = c_{3333}$ and $c'_{3333} = c_{1111}$. We therefore deduce $c_{11} = c_{22} = c_{33}$. Similarly, $c_{12} = c_{23} = c_{31}$ and $c_{44} = c_{55} = c_{66}$. What were nine separate stiffness constants for Class 222 reduce to three separate stiffness constants for Class 23. An equivalent result holds for the compliance constants. The matrix form for both c_{ij} and s_{ij} for Class 23 is thus:

$$(6.72)$$

This result is valid for *both* Laue classes of cubic crystals: the elastic properties of a cubic crystal are completely defined by the three constants c_{11}, c_{12} and c_{44}.

The compliances are related to the stiffness constants through Eqs. (6.53) and (6.57). Clearly:

$$c_{ij}s_{jk} = \delta_{ik} \qquad (6.73)$$

for $i,j,k = 1,6$, where δ_{ik} is the Kronecker delta. For cubic crystals, this means:

$$s_{11} = \frac{c_{11} + c_{12}}{(c_{11} - c_{12})(c_{11} + 2c_{12})}$$

$$s_{12} = \frac{-c_{12}}{(c_{11} - c_{12})(c_{11} + 2c_{12})} \qquad (6.74)$$

and:

$$s_{44} = \frac{1}{c_{44}}$$

(Problem 6.19). The stiffness constants are obtained in terms of the compliances merely by interchanging c and s in Eq. (6.74). The elastic constants of various cubic crystals are listed in Table 6.1.

6.5.5 Reduction in the Number of Independent c_{ij} and s_{ij} for Other Crystal Classes

As we have already noted, symmetry relations reduce the number of independent elastic constants in all crystal systems other than the triclinic. The forms which the compliance and stiffness matrices take when a coordinate axis lies along a given symmetry axis are shown in Table 6.2. When one coordinate axis is parallel to one axis of symmetry and a second coordinate axis is parallel to another axis of symmetry, the matrix must obey the restrictions on its form given by *both* the corresponding entries in Table 6.2. With the aid of Table 6.2, the number of independent elastic constants for each class of symmetry is then readily obtained for the classes we have already considered, and also for the tetragonal, trigonal and hexagonal crystal classes.

The Laue group $4/m$ of the tetragonal system (see Table 2.1) has seven independent constants, as can be seen from the third row of Table 6.2. The second Laue group of the tetragonal system, $4/mmm$, has six independent constants. This can be seen by superposing the entry for a fourfold axis parallel to Ox_3, say, upon the entry for a mirror plane normal to Ox_2 (i.e. $\bar{2}$ parallel to Ox_2), which shows that the constant 16 must be zero. In the trigonal

Table 6.1 Elastic constants of cubic crystals at room temperature

Material	c_{11}	c_{12}	c_{44}	s_{11}	s_{12}	s_{44}	Ref.
Ag	124.0	93.4	46.1	0.0229	−0.00983	0.0217	[14]
Al	108.2	61.3	28.5	0.0157	−0.00568	0.0351	[14]
Au	186	157	42.0	0.0233	−0.01065	0.0238	[14]
Cu	168.4	121.4	75.4	0.01498	−0.00629	0.01326	[14]
Ni	246.5	147.3	124.7	0.00734	−0.00274	0.00802	[14]
Pb	49.5	42.3	14.9	0.0951	−0.0438	0.0672	[15]
Fe	228	132	116.5	0.00762	−0.00279	0.00858	[16]
Mo	460	176	110	0.0028	−0.00078	0.0091	[14]
Na	7.32	6.25	4.19	0.64	−0.295	0.239	[17]
Nb	245.6	138.7	29.3	0.0069	−0.00249	0.0342	[18]
Ta	267	161	82.5	0.00685	−0.00258	0.0121	[19]
V	228	119	42.6	0.00683	−0.00234	0.0235	[19]
W	501	198	151.4	0.00257	−0.00073	0.0066	[14]
C (diamond)	1076	125.0	575.8	0.000953	−0.000099	0.00174	[14]
Ge	128.9	48.3	67.1	0.00978	−0.00266	0.0149	[14]
Si	165.7	63.9	79.6	0.00768	−0.00214	0.01256	[14]
NaCl	48.7	12.4	12.6	0.0229	−0.00465	0.0794	[14]
LiF	111.2	42.0	62.8	0.01135	−0.0031	0.0159	[14]
MgO	289.2	87.9	154.6	0.00403	−0.00094	0.0647	[20]
TiC	500	113	175	0.00218	−0.00040	0.0572	[21]

Axes of reference parallel to cube axes: $c_{11} = c_{22} = c_{33}$, $c_{12} = c_{23} = c_{13}$, $c_{44} = c_{55} = c_{66}$. Units of c_{ij} are GPa; units of s_{ij} are (GPa)$^{-1}$.

system, the Laue group $\bar{3}$ has seven independent constants and the Laue group $\bar{3}m$ has six independent constants. Both of these results can be verified by inspecting Table 6.2.

It is apparent from Table 6.2 that a hexagonal crystal has five independent elastic constants, irrespective of its crystal class. When the x_3-axis is chosen to be parallel to the sixfold c-axis, as is conventional, the constants c_{11}, c_{33}, c_{44}, c_{12} and c_{13} must be specified. The compliances can be derived from the stiffness constants by applying the equations listed in Table 6.3. The corresponding equations for tetragonal and trigonal crystals are also given in this table. The elastic constants of some hexagonal crystals are given in Table 6.4.

Hence, overall, there are nine separate forms of the matrix representations for c_{ij} and s_{ij} because all hexagonal crystals have the same form, as do all cubic crystals. For ease of reference, these nine forms are collected together in Table 6.5. It is a useful exercise to confirm that the forms of these matrix representations are consistent with those in Table 6.2 following the principles we have discussed in this section.

6.5.6   Isotropic Media

In an isotropic medium, the elastic constants must be independent of the choice of coordinate axes. This requirement imposes an extra condition in addition to the conditions of cubic symmetry, which can be shown to be:

$$2c_{44} = c_{11} - c_{12} \tag{6.75}$$

Table 6.2 Forms of the compliance and stiffness constant arrays

Axis of:	Symmetry axis parallel to Ox_1	Symmetry axis parallel to Ox_2	Symmetry axis parallel to Ox_3
2 or $\bar{2}$			
3 or $\bar{3}$	$\alpha \quad -a \quad b$	$-c$ $\beta \quad d$	$-e$ f γ
4 or $\bar{4}$			
6 or $\bar{6}$	α	β	γ

	a	b	c	d	e	f	α	β	γ
Stiffness	c_{26}	c_{25}	c_{16}	c_{14}	c_{15}	c_{14}	$\frac{1}{2}(c_{22}-c_{23})$	$\frac{1}{2}(c_{11}-c_{13})$	$\frac{1}{2}(c_{11}-c_{12})$
Compliance	$2s_{26}$	$2s_{25}$	$2s_{16}$	$2s_{14}$	$2s_{15}$	$2s_{14}$	$2(s_{22}-s_{23})$	$2(s_{11}-s_{13})$	$2(s_{11}-s_{12})$

After Table 1 of Waterman [22], using a modified notation of Van Dyke [9].

Table 6.3 The equations giving the compliances s_{ij} in terms of the stiffness constants c_{ij} for the more symmetrical crystal systems

Crystal system	Equations
Cubic (all crystal classes)	$s_{11} = \dfrac{c_{11}+c_{12}}{(c_{11}-c_{12})(c_{11}+2c_{12})}$ $s_{12} = \dfrac{-c_{12}}{(c_{11}-c_{12})(c_{11}+2c_{12})}$ $s_{44} = \dfrac{1}{c_{44}}$
Hexagonal (all crystal classes)	$s_{11} + s_{12} = \dfrac{c_{33}}{c}$ $s_{11} - s_{12} = \dfrac{1}{c_{11}-c_{12}}$ $s_{13} = -\dfrac{c_{13}}{c}$ $s_{33} = \dfrac{c_{11}+c_{12}}{c}$ $s_{44} = \dfrac{1}{c_{44}}$ $c = c_{33}(c_{11}+c_{12}) - 2c_{13}^2$
Tetragonal (classes 422, 4mm, $\overline{4}$2m and 4/mmm)	$s_{11} + s_{12} = \dfrac{c_{33}}{c}$ $s_{11} - s_{12} = \dfrac{1}{c_{11}-c_{12}}$ $s_{13} = -\dfrac{c_{13}}{c}$ $s_{33} = \dfrac{c_{11}+c_{12}}{c}$ $s_{44} = \dfrac{1}{c_{44}}$ $s_{66} = \dfrac{1}{c_{66}}$ $c = c_{33}(c_{11}+c_{12}) - 2c_{13}^2$
Trigonal (classes 32, 3m and $\overline{3}$m)	$s_{11} + s_{12} = \dfrac{c_{33}}{c}$ $s_{11} - s_{12} = \dfrac{c_{44}}{c'}$ $s_{13} = -\dfrac{c_{13}}{c}$ $s_{14} = -\dfrac{c_{14}}{c'}$ $s_{33} = \dfrac{c_{11}+c_{12}}{c}$ $s_{44} = \dfrac{c_{11}-c_{12}}{c'}$ $c = c_{33}(c_{11}+c_{12}) - 2c_{13}^2$ $c' = c_{44}(c_{11}-c_{12}) - 2c_{14}^2$

Stiffness constants are expressed in terms of compliances by interchanging c_{ij} and s_{ij} in the equations.

Table 6.4 Elastic constants of hexagonal crystals at room temperature; x_3-axis parallel to [0001], x_1- and x_2-axes anywhere in the basal plane; $c_{11} = c_{22}$, $c_{44} = c_{55}$, $c_{13} = c_{23}$, $c_{66} = \frac{1}{2}(c_{11} - c_{12})$, $s_{66} = 2(s_{11} - s_{12})$

Material	c_{11}	c_{33}	c_{44}	c_{12}	c_{13}	s_{11}	s_{33}	s_{44}	s_{12}	s_{13}	Ref.
Be	292.3	336.4	162.5	26.7	14.0	0.00348	0.00298	0.00616	−0.00030	−0.00013	[23]
C (graphite)	1160	46.6	2.3	290	109	0.00111	0.0332	0.435	−0.00004	−0.00249	[24]
Cd	115.8	51.4	20.4	39.8	40.6	0.0124	0.0352	0.0498	−0.00076	−0.00920	[25]
Co	307	358.1	78.3	165	103	0.00472	0.00319	0.01324	−0.00231	−0.00069	[14]
Hf	181.1	196.9	55.7	77.2	66.1	0.00715	0.00613	0.018	−0.00247	−0.00157	[26]
Mg	59.7	61.7	16.4	26.2	21.7	0.022	0.0197	0.061	−0.00785	−0.0050	[14]
Re	612.5	682.7	162.5	270	206	0.00212	0.0017	0.00616	−0.00080	−0.00040	[27]
Ti	162.4	180.7	46.7	92.0	69.0	0.00958	0.00698	0.0214	−0.00462	−0.00189	[26]
Zn	161	61.0	38.3	34.2	50.1	0.00838	0.02838	0.0261	−0.00053	−0.00731	[14]
ZnO	209.7	210.9	42.5	121.1	105.1	0.00787	0.00694	0.0235	−0.00344	−0.00221	[28]
Zr	143.4	164.8	32.0	72.8	65.3	0.01013	0.00799	0.0313	−0.00404	−0.00241	[26]

Units of c_{ij} are GPa; units of s_{ij} are (GPa)$^{-1}$.

Table 6.5 Forms of the compliance and stiffness constant arrays for the different crystal classes

Triclinic: both classes (classes 1 and $\bar{1}$)	Monoclinic: all classes (classes 2, m and $2/m$). Unique axis x_2.	Orthorhombic: all classes (classes 222, $2mm$, mmm)

Trigonal: classes 3 and $\bar{3}$. Unique axis x_3.	Trigonal: classes 32, $3m$ and $\bar{3}m$. Unique axis x_3.	Tetragonal: classes 4, $\bar{4}$ and $4/m$. Unique axis x_3.

Tetragonal: classes 422, $4mm$, $\bar{4}2m$ and $4/mmm$. Unique axis x_3.	Hexagonal: all classes (classes 6, $\bar{6}$, $6/m$, 622, $6mm$, $\bar{6}2m$ and $6/mmm$). Unique axis x_3.	Cubic: all classes (classes 23, $m\bar{3}$, 432, $\bar{4}3m$ and $m\bar{3}m$)

	e	f	γ
Stiffness	c_{15}	c_{14}	$\frac{1}{2}(c_{11} - c_{12})$
Compliance	$2s_{15}$	$2s_{14}$	$2(s_{11} - s_{12})$

For ease of reference, the nomenclature used in Table 6.2 has been retained in this table.

Equivalently:

$$s_{44} = 2(s_{11} - s_{12}) \tag{6.76}$$

These conditions for isotropy can also be written down immediately from the form of the matrices for a hexagonal crystal (Table 6.2).

Isotropic materials are often encountered. For example, metals are ordinarily found in the form of an aggregate of crystals which are oriented at random, unless a preferred orientation has developed during solidification or subsequent deformation. Similar principles apply to polycrystalline ceramics and minerals, and, of course, to glasses. The elastic properties of an isotropic material are commonly specified by the two constants, Young's modulus, E, and Poisson's ratio, v. These are defined from the effect of a simple tensile stress, σ, as follows:

$$E = \frac{\sigma}{\varepsilon} \tag{6.77}$$

$$v = -\frac{\varepsilon'}{\varepsilon} \tag{6.78}$$

where ε is the tensile strain in the direction of the tensile force and ε' is the tensile strain in directions normal to this. In terms of the compliances:

$$E = \frac{1}{s_{11}} \tag{6.79}$$

$$v = -\frac{s_{12}}{s_{11}} \tag{6.80}$$

Other constants, dependent on E and v, are also used. For example, the rigidity modulus or shear modulus μ is defined as:

$$\mu = \tau/\gamma \tag{6.81}$$

where γ is the elastic engineering shear strain produced by a shear stress, τ. It follows that:

$$\mu = \frac{1}{s_{44}} = c_{44} \tag{6.82}$$

and using Eqs. (6.76), (6.79) and (6.80):

$$\mu = \frac{E}{2(1 + v)} \tag{6.83}$$

Using the constants E, v and μ, Hooke's law for an isotropic solid can be written as:

$$\varepsilon_{11} = \frac{1}{E}[\sigma_{11} - v(\sigma_{22} + \sigma_{33})]$$

$$\varepsilon_{22} = \frac{1}{E}[\sigma_{22} - v(\sigma_{33} + \sigma_{11})]$$

$$\varepsilon_{33} = \frac{1}{E}[\sigma_{33} - v(\sigma_{11} + \sigma_{22})]$$

$$\varepsilon_{12} = \frac{\sigma_{12}}{2\mu} \tag{6.84}$$

$$\varepsilon_{23} = \frac{\sigma_{23}}{2\mu}$$

$$\varepsilon_{31} = \frac{\sigma_{31}}{2\mu}$$

Two other constants in common use are the Lamé constants, λ and μ. The Lamé constant μ is identical to the shear modulus, while λ is identical to the stiffness constant c_{12} of an isotropic material:

$$\lambda = \frac{\nu E}{(1 + \nu)(1 - 2\nu)} \tag{6.85}$$

In terms of the Lamé constants, Hooke's law can be written as:

$$\sigma_{ij} = \lambda \varepsilon_{kk} \delta_{ij} + 2\mu \varepsilon_{ij} \tag{6.86}$$

where $\varepsilon_{kk} = \varepsilon_{11} + \varepsilon_{22} + \varepsilon_{33}$ and δ_{ij} is the Kronecker delta.

6.6   Elasticity of Cubic Crystals

It is evident from Section 6.5 that a cubic crystal fulfils all of the conditions for isotropy except Eq. (6.75). Its degree of anisotropy can be measured by the departure from unity of the ratio A, where:

$$A = \frac{2c_{44}}{(c_{11} - c_{12})} = \frac{2(s_{11} - s_{12})}{s_{44}} \tag{6.87}$$

[29]. A is known as the *Zener ratio* or the *Zener anisotropy factor*. It is a measure of the relative resistance of the crystal to two types of shear strain: c_{44} is a measure of the resistance to shear on the (010) plane in the [001] direction, while $\frac{1}{2}(c_{11} - c_{12})$ is the stiffness with respect to shear on (110) in the direction [1$\bar{1}$0]. Of the crystals listed in Table 6.1, only W and Al come close to being isotropic. By comparison, Cu is noticeably anisotropic, with a value of A of 3.21.

6.6.1   Transformation of the Stiffness and Compliance Tensors for Cubic Materials

Since there are only three independent non-zero elastic constants needed for cubic materials, the general formula for the transformation of tensors of the fourth rank, Eq. (5.68), from axes 1, 2 and 3 with respect to the crystal axes to new axes 1′, 2′ and 3′ can be expressed in the succinct form for the transformation of c_{ijkl} as:

$$c'_{ijkl} = c_{12}\delta_{ij}\delta_{kl} + c_{44}(\delta_{ik}\delta_{jl} + \delta_{il}\delta_{jk}) + (c_{11} - c_{12} - 2c_{44})\, a_{iu}a_{ju}a_{ku}a_{lu} \tag{6.88}$$

where the dummy suffix u takes the values 1, 2 and 3 and where δ is the Kronecker delta, and where on the right-hand side of this equation the contracted Voigt notation has been used [30, 31]. Likewise, for the same change in axes,

$$s'_{ijkl} = s_{12}\delta_{ij}\delta_{kl} + \frac{1}{4}(\delta_{ik}\delta_{jl} + \delta_{il}\delta_{jk})s_{44} + \left(s_{11} - s_{12} - \frac{1}{2}s_{44}\right) a_{iu}a_{ju}a_{ku}a_{lu} \tag{6.89}$$

[31]. These equations help to determine how Young's modulus, c'_{1111} and s'_{1111} all vary as a function of orientation in cubic crystals (Problem 6.17). It is then easy to show that cubic metals with $A > 1$ are most stiff along $\langle 111 \rangle$ directions and least stiff along $\langle 001 \rangle$ directions; the reverse is true for cubic metals with $A < 1$. Eqs. (6.88) and (6.89) are also useful for examining other more complex loading situations, such as how the modulus changes as a function of orientation in a cubic crystal subjected to an equi-biaxial elastic strain [32], and

the variation with orientation of Young's modulus in plane strain in a cubic single crystal when a uniaxial stress is applied, and where the strain is constrained to be zero along a direction perpendicular to the direction along which the stress is applied [33].

6.6.2 Shear Stiffnesses and Compliances on the (111) Plane of Cubic Crystals

As a straightforward example of the application of Eqs. (6.88) and (6.89), it is evident that

$$c'_{3232} = c_{44} + (c_{11} - c_{12} - 2c_{44})(a_{31}^2 a_{21}^2 + a_{32}^2 a_{22}^2 + a_{33}^2 a_{23}^2) \tag{6.90}$$

$$s'_{3232} = \tfrac{1}{4}s_{44} + \left(s_{11} - s_{12} - \tfrac{1}{2}s_{44}\right)(a_{31}^2 a_{21}^2 + a_{32}^2 a_{22}^2 + a_{33}^2 a_{23}^2) \tag{6.91}$$

To examine the variation of these tensors on the (111) plane, we might choose our 3' axis to be the normal to the (111) plane, leaving two axes 1' and 2' to define in this plane. Since the plane of interest is (111), it follows that $a_{31} = a_{32} = a_{33} = 1/\sqrt{3}$. Therefore,

$$c'_{3232} = c_{44} + \tfrac{1}{3}(c_{11} - c_{12} - 2c_{44})(a_{21}^2 + a_{22}^2 + a_{23}^2) = \tfrac{1}{3}(c_{11} - c_{12} + c_{44}) \tag{6.92}$$

since $(a_{21}^2 + a_{22}^2 + a_{23}^2) = 1$ irrespective of how we define our 1' and 2' axes. Therefore, we have the result that $c'_{3232} = c'_{44}$ in contracted notation is isotropic in the (111) plane. Similarly, $s'_{3232} = \tfrac{1}{4}s'_{44}$ in contracted notation is also isotropic in the (111) plane, as is the shear modulus $G = 1/s'_{44}$. A similar result holds for shear on {001} planes (Problem 6.16). On all planes other than {001} and {111} in cubic crystals the shear modulus varies as a function of shear direction on the plane under consideration [31].

Problems

6.1 Derive expressions for the displacements u_1 and u_2 parallel to fixed axes Ox_1 and Ox_2 of a point (x_1, x_2) in a body which is rotated about the origin through an angle θ. Hence, show that, when θ is very small, its magnitude is given by

$$\left|\tfrac{1}{2}(\partial u_1/\partial x_2 - \partial u_2/\partial x_1)\right|.$$

6.2 A body is subjected to an elastic pure shear strain by stretching it along the x_1-axis and compressing it along the x_2-axis. The displacements are $u_1 = ex_1$, parallel to Ox_1, and $u_2 = -ex_2$, parallel to Ox_2, where e is very small. Write down the strain tensor referred to the axes Ox_1, Ox_2 and Ox_3, and obtain the strain tensor referred to axes Ox'_1, Ox'_2 and Ox'_3, which are derived by rotating the axes Ox_1 and Ox_2 through 45° anticlockwise about Ox_3.

6.3 A body is subjected to an elastic tensile strain along the x_1-axis. The displacements are $u_1 = ex_1$, $u_2 = -vex_2$ and $u_3 = -vex_3$, where v is Poisson's ratio (a material constant). Write down the strain tensor referred to the axes Ox_1, Ox_2 and Ox_3 and obtain the tensor referred to the axes Ox'_1, Ox'_2 and Ox'_3 whose orientation is given in Problem 6.2.

6.4 When a body is subjected to a uniaxial tensile stress σ, prove that the planes on which the largest shear stress exists lie at an angle of 45° to the tensile axis, and determine the magnitude of the largest shear stress.

6.5 The cylindrical coordinates of a point (r, θ, z) are related to a set of mutually perpendicular axes Ox_1, Ox_2 and Ox_3 as follows: the z-axis is parallel to Ox_3 and the angle θ is measured anticlockwise from Ox_2 looking along the positive Ox_3 direction. Show that:

$$\sigma_{rz} = \sigma_{13} \sin \theta + \sigma_{23} \cos \theta$$

$$\sigma_{\theta z} = \sigma_{13} \cos \theta - \sigma_{23} \sin \theta$$

6.6 Show that a uniaxial compressive stress is equivalent to a hydrostatic pressure superimposed on two pure shear stresses. Write down the equation in terms of the stress tensors that demonstrate this result and illustrate it with sketches showing the stresses.

6.7 The only non-zero stress components at a point P are σ_{11}, σ_{22} and σ_{12}, referred to axes Ox_1, Ox_2 and Ox_3. Consider a plane through P, parallel to Ox_3 and making an angle of θ with the x_2-axis, measured anticlockwise from Ox_2 looking along Ox_3. Show that the normal stress σ_n acting on this plane at this point is given by:

$$\sigma_n = \sigma_{11} \cos^2 \theta + \sigma_{22} \sin^2 \theta - 2\sigma_{22} \sin \theta \cos \theta$$

and that the shear stress τ is given by:

$$\tau = \sigma_{11} \sin \theta \cos \theta - \sigma_{22} \sin \theta \cos \theta + \sigma_{12}(\cos^2 \theta - \sin^2 \theta)$$

6.8 Determine the principal stresses for the stress state:

$$\sigma = \begin{pmatrix} 300 & 0 & 100 \\ 0 & 100 & 0 \\ 100 & 0 & 300 \end{pmatrix} \text{MPa}$$

Specify the directions of the principal stresses and the maximum shear stress. Confirm that the principal directions are orthogonal and construct a suitable orthogonal matrix $\mathbf{C}$ from unit lengths of vectors parallel to these principal directions. Evaluate $\mathbf{C}^{-1}\sigma\mathbf{C}$ and confirm that the diagonal elements of this matrix are the principal stresses of σ and that the other matrix elements are all zero.

6.9 Show that Eq. (6.44) is correct. *Hint*: consider the effect of having a diad along x_1 on the eight surviving moduli from Class 2 given in Eq. (6.43).

6.10 Show for a crystal with point symmetry m where the mirror plane is (010) that there are 10 non-zero d_{ij} in the matrix representation for the piezoelectric moduli. Hence, deduce from the form of this matrix and that of the matrix representation for the non-zero d_{ij} for crystals with a diad axis parallel to [010] (Eq. (6.43)) that all the piezoelectric moduli vanish for crystals with $2/m$ point symmetry.

6.11 Show that for orthorhombic crystals where the (100) and (010) planes are mirror planes that there are five non-zero d_{ij} in the matrix representation for the piezoelectric moduli. Hence show for a crystal of 4mm point symmetry with the fourfold axis along [001] that the matrix representation for the non-zero d_{ij} is:

6.12 Draw a projection of 2×2 unit cells of sphalerite on (001), with a motif of a sulfur ion at (0, 0, 0) and a zinc ion at $\left(\frac{1}{4}, \frac{1}{4}, \frac{1}{4}\right)$ (Section 3.5.3). Identify the four sulfur ions surrounding the zinc ion at $\left(\frac{1}{4}, \frac{1}{4}, \frac{1}{4}\right)$ and determine the position of the centre of their charge. Suppose a shear stress σ_{12} is applied to these unit cells, so that they lengthen in the [110] direction and contract in the [1$\bar{1}$0] direction. What must happen to the zinc ion at $\left(\frac{1}{4}, \frac{1}{4}, \frac{1}{4}\right)$ to enable the four Zn–S distances to remain unchanged? What has happened to the position of the centre of charge of the sulfur ions surrounding this zinc atom? What has happened to the other zinc ions in the unit cell? What, therefore, must have occurred as a result of this shear stress? Why does a similar effect not occur in diamond?

6.13 Deduce that the contracted form of the converse piezoelectric effect $\epsilon_j = d_{ij}E_i$ (Eq. (6.48)) requires the relationship between the ϵ_j and the ϵ_{jk} to be as given in Eq. (6.49). Confirm that Eq. (6.53) is then valid and that the definitions of the s_{ij} conform to Eqs. (6.54)–(6.56).

6.14 An elastically isotropic body of Young's modulus E and Poisson's ratio ν is elastically strained by a tensile stress σ. Determine the dilatation of the body; that is, the change in volume per unit volume. The bulk modulus of a solid is defined as $K = -P/\Delta$, where Δ is the dilatation produced by a stress whose hydrostatic pressure component is P. Show that

$$K = \frac{E}{3(1 - 2\nu)}.$$

6.15 In a cubic crystal, the stiffness constant c_{44} is by definition the shear modulus applicable to shear on a cube plane, {001}, in a ⟨100⟩ direction. Show that the same modulus, c_{44}, applies to shear on a cube plane in *any* direction.

6.16 Show that, in a cubic crystal, the shear modulus applicable to shear on a {110} plane in a ⟨1$\bar{1}$0⟩ direction is $\frac{1}{2}(c_{11} - c_{12})$. *Hint*: consider the principal strains of a shear strain on {110} in a ⟨1$\bar{1}$0⟩ direction.

6.17 In a single crystal, Young's modulus in a particular crystallographic direction is defined as follows: when a tensile stress is applied in that direction, the ratio of the stress to the tensile strain in that direction is Young's modulus. Show that, in a cubic crystal, the reciprocal of Young's modulus is given by:

$$1/E = s_{11} - 2\left(s_{11} - s_{12} - \tfrac{1}{2}s_{44}\right)(a_{11}^2 a_{12}^2 + a_{11}^2 a_{13}^2 + a_{12}^2 a_{13}^2)$$

where a_{11}, a_{12} and a_{13} are the cosines of the angles between the direction in question and the Ox_1, Ox_2 and Ox_3 axes, respectively. Hence, deduce Eq. (6.76) for isotropic materials.

Show that Young's modulus, s'_{1111} and c'_{1111} for a cubic crystal are all isotropic in sections normal to the $\langle 111 \rangle$ directions. *Hint*: transform the compliance s_{1111} to new axes, with Ox'_1 parallel to the direction in question. Take great care in replacing the tensor compliances with the corresponding components in contracted notation. Repeat for the corresponding stiffness, c_{1111}.

6.18 Write down expressions for Young's modulus measured along [100] for a cubic crystal measured under the following conditions:

(a) Uniaxial tension along [100].

(b) Uniaxial tension along [100] with no strains allowed to occur along [010] or [001] (so that there have to be tensile stresses along [010] and [001]).

(c) Equal biaxial tension along [100] and [010] with no strain allowed to occur along [001] (so that there has also to be a tensile tress along [001]).

(d) Equal triaxial tension.

6.19 Using the fact that in contracted notation $c_{ij}s_{jk} = \delta_{ik}$ for $i,j,k = 1,6$, where δ_{ik} is the Kronecker delta Eq. (6.73), show that the equations giving the stiffness constants in terms of the compliances for a cubic crystal in this notation are:

$$c_{11} = \frac{s_{11} + s_{12}}{(s_{11} - s_{12})(s_{11} + 2s_{12})}$$

$$c_{12} = \frac{-s_{12}}{(s_{11} - s_{12})(s_{11} + 2s_{12})}$$

$$c_{44} = \frac{1}{s_{44}}$$

Suggestions for Further Reading

Huntington, H.B. (1958). The elastic constants of crystals. *Solid State Phys.* 7: 213–251.

Jaeger, J.C. (1971). *Elasticity, Fracture and Flow*, 3rd Edition. Dordrecht: Kluwer Academic Publishers.

Love, A.E.H. (1944). *The Mathematical Theory of Elasticity*. Mineola, NY: Dover Press.

Nye, J.F. (1985). *Physical Properties of Crystals*. Oxford: Clarendon Press.

Simmons, G. and Wang, H. (1971). *Single Crystal Elastic Constants and Calculated Aggregate Properties*, 2nd Edition. Cambridge, MA/London: MIT Press.

References

[1] Nye, J.F. (1985). *Physical Properties of Crystals*. Oxford: Clarendon Press.

[2] Lovett, D.R. (1999). *Tensor Properties of Crystals*, 2nd Edition. Bristol/Philadelphia: Adam Hilger.

[3] Backofen, W.A. (1972). *Deformation Processing*. Reading, MA: Addison-Wesley.

[4] Johnson, W. and Mellor, P.B. (1973). *Engineering Plasticity*. Wokingham, UK: Van Nostrand Reinhold.

[5] Benham, P.P., Crawford, R.J. and Armstrong, C.G. (1996). *Mechanics of Engineering Materials*, 2nd Edition. Harlow, Essex: Prentice-Hall/Pearson Education Limited.

[6] Hosford, W.F. and Caddell, R.M. (2011). *Metal Forming: Mechanics and Metallurgy*, Fourth Edition. Cambridge: Cambridge University Press.

[7] Wooster, W.A. (1973). *Tensors and Group Theory for the Physical Properties of Crystals*. Oxford: Clarendon Press.

[8] Curie, J. and Curie, P. (1880). Développement par compression de l'électricité polaire dans les cristaux hémièdres à faces inclinées. *Bull. Soc. Min. Fr.* 3: 90–93.

[9] Tiffen, R. and Stevenson, A.C. (1956). Elastic isotropy with body force and couple. *Quart. J. Mech. Appl. Math.* 9: 306–312.

[10] Voigt, W. (1910). *Lehrbuch der Kristallphysik*. Leipzig: B.G. Teubner.

[11] Van Dyke, K.S. (1950). Matrices of piezoelectric, elastic and dielectric constants, Program of the 39th Meeting of the Acoustical Society of America. *J. Acoust. Soc. Am.* 22: 681.

[12] Lippmann, G. (1881). Principe de la conservation de l'électricité. *Ann. Chim. Phys.* 24: 145–178.

[13] Curie, J. and Curie, P. (1881). Contractions et dilatations produites par des tensions électriques dans les cristaux hémièdres à faces inclinées. *Comptes Rendus* 93: 1137–1140.

[14] Huntington, H.B. (1958). The elastic constants of crystals. *Solid State Phys.* 7: 213–251.

[15] Waldorf, D.L. and Alers, G.A. (1962). Low-temperature elastic moduli of lead. *J. Appl. Phys.* 33: 3266–3269.

[16] Lord, A.E. and Beshers, D.N. (1965). Elastic stiffness coefficients of iron from 77° to 673°K. *J. Appl. Phys.* 36: 1620–1623.

[17] Daniels, W.B. (1960). Pressure variation of the elastic constants of sodium. *Phys. Rev.* 119: 1246–1252.

[18] Carroll, K.J. (1965). Elastic constants of niobium from 4.2° to 300°K. *J. Appl. Phys.* 36: 3689–3690.

[19] Bolef, D.I. (1961). Elastic constants of single crystals of the bcc transition elements V, Nb and Ta. *J. Appl. Phys.* 32: 100–105.

[20] Chung, D.-H. (1963). Elastic moduli of single crystal and polycrystalline MgO. *Phil. Mag.* 8: 833–841.

[21] Gilman, J.J. and Roberts, B.W. (1961). Elastic constants of TiC and TiB_2. *J. Appl. Phys.* 32: 1405.

[22] Waterman, P.C. (1959). Orientation dependence of elastic waves in single crystals. *Phys. Rev.* 113: 1240–1253.

[23] Smith, J.F. and Arbogast, C.L. (1960). Elastic constants of single crystal beryllium. *J. Appl. Phys.* 31: 99–102.

[24] Spence, G.B. (1963). Extended dislocations in the anisotropic continuum approximation. In: *Proceedings of the Fifth Conference on Carbon*, vol. 2, Oxford: Pergamon Press, pp. 531–538.

[25] Garland, C.W. and Silverman, J. (1960). Elastic constants of cadmium from 4.2°K to 300°K. *Phys. Rev.* 119: 1218–1222.

[26] Fisher, E.S. and Renken, C.J. (1964). Single-crystal elastic moduli and the hcp → bcc transformation in Ti, Zr and Hf. *Phys. Rev.* 135: A482–A494.

[27] Shepard, M.L. and Smith, J.F. (1965). Elastic constants of rhenium single crystals in the temperature range 4.2°–298°K. *J. Appl. Phys.* 36: 1447–1450.

[28] Bateman, T.B. (1962). Elastic moduli of single-crystal zinc oxide. *J. Appl. Phys.* 33: 3309–3312.

[29] Zener, C. (1948). *Elasticity and Anelasticity of Metals*. Chicago: University of Chicago Press.

[30] Thomas, T.Y. (1966). On the stress-strain relations for cubic crystals. *Proc. Natl. Acad. Sci. USA* 55: 235–239.

[31] Knowles, K.M. and Howie, P.R. (2015). The directional dependence of elastic stiffness and compliance shear coefficients and shear moduli in cubic materials. *J. Elast.* 120: 87–108.

[32] Knowles, K.M. (2016). The biaxial moduli of cubic materials subjected to an equi-biaxial elastic strain. *J. Elast.* 124: 1–25.

[33] Knowles, K.M. (2017). The plane strain Young's modulus in cubic materials. *J. Elast.* 128: 147–173.

Part II
Imperfect Crystals

7

Glide

7.1 Translation Glide

A useful distinction between a solid and a liquid is that a liquid cannot permanently resist small forces which tend to change the shape of a liquid body, but which allow its volume to remain constant. Such forces are shear forces. A solid can resist these, provided that they are small. However, if large shear forces are applied to a crystal, the crystal will either break or *yield*. If it yields, it will do so in a *plastic* manner. That is to say, if the shear stresses are large enough and are applied for a sufficient length of time, the shape of the crystal will be altered permanently, though it will remain in one piece.

All crystals will yield plastically under normal confining pressures of 1 atm or less at a sufficiently high temperature. Crystals of many metals and of the alkali halides and some other non-metals, such as graphite or the crystals of inert gases, yield plastically at very low temperatures – sometimes close to absolute zero. These temperatures are much too low for atomic diffusion to be occurring during the time of the experiment. At low temperatures, crystals yield plastically by a process called *glide*.

Crystals that do not glide normally break under the action of sufficiently large forces, as we have already noted. These are said to be *brittle*. The property of brittleness and that of its antonym, ductility, is not an absolute property, but depends upon the state of stress, and in particular, upon the hydrostatic pressure. Under a sufficiently high hydrostatic pressure, a superposed shear stress can always produce glide in a crystal. For instance, sapphire crystals can be made to glide at room temperature under a pressure of 25 000 atm. At high temperatures, permanent deformation can occur by means of diffusion in crystals.

Glide is the translation of one part of a crystal with respect to another without a change in volume. The translation usually takes place upon a specific crystallographic plane and in a particular direction in that plane.

Crystallography and Crystal Defects, Third Edition. Anthony Kelly and Kevin M. Knowles.
© 2020 John Wiley & Sons Ltd. Published 2020 by John Wiley & Sons Ltd.
Companion website: www.wiley.com/go/kelly/crystallography3e

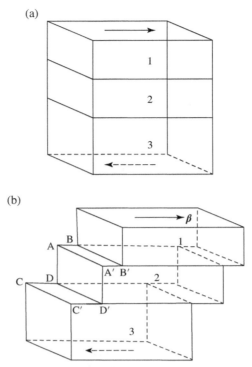

Figure 7.1 Illustration of the process of glide. Blocks 1, 2 and 3 in a crystal, (a), are sheared in the direction β to produce the appearance of the crystal shown in (b) after glide has taken place

The process of glide is illustrated schematically in Figure 7.1. When a shearing stress too small to produce glide is applied to the crystal, the crystal is elastically deformed. If the stress is uniform throughout the crystal, the strain within the crystal is homogeneous (Figure 7.1a). In discussing the change of shape due to glide we can neglect the elastic deformation. This is because the plastic deformation due to glide is able to operate over a far larger range of strain than that induced by elastic deformation. The appearance of the crystal after glide has occurred in the direction β on a particular glide plane is shown in Figure 7.1b. Comparing Figures 7.1a and b, it is apparent that the shape of the crystal is altered, but not its volume. The orientation of the lattice is unchanged. The two parts of the crystal on either side of the glide plane remain in an identical orientation.

Glide was first discovered in 1867 by Reusch [1], who recognized steps on the surface of a crystal of rock salt as a consequence of glide in compression, noting that sliding occurred on {110} planes in ⟨1$\bar{1}$0⟩ directions of the crystal. It is apparent from the schematic in Figure 7.1b that, at the surface of a crystal undergoing glide, steps will be produced: AA′B′B and CC′D′D in Figure 7.1b. With the optical microscope, the steps are usually seen as lines, and were called slip-bands by Ewing and Rosenhain, who reported the first serious investigation of them in a large number of metal crystals in 1899 and 1900 [2, 3]. Very careful measurement fails to detect any change in crystal orientation of the parts of a crystal on either side of a slip line, and there is no evidence for any change in crystal structure as the two parts of the crystal slide over one another. It is therefore clear that the translation of Part 1 of the crystal with respect to Part 2 in Figure 7.1b and that of Part 2 with respect to Part 3 must each be equal to *an integral number of lattice translation vectors*. In view of this, the process is also called *translation glide*.

Slip lines are easily seen under an optical microscope and sometimes can be observed with the naked eye. Therefore, the translation at a single step can be larger than a micrometre (10^{-6} m), corresponding to a movement of some thousands of lattice vectors. Glide in a crystal usually occurs on a well-defined crystallographic plane of low indices, which is called the *glide plane* or *slip plane*, and always in a definite crystallographic direction, the *slip direction*.

The indices of the slip plane in a crystal can be determined easily by a two-surface analysis of the direction of the trace of the glide plane in two (or more) faces of a crystal which have been cut or polished at a known angle to one another; the procedure for a two-surface analysis is given in Appendix 2. The determination of the glide direction presents a little more difficulty. It is clear from Figure 7.2 that the slip steps on the surface of a crystal will be of maximum height on a crystal face lying normal to the direction of glide, but should produce no observable effect on a smooth, clean crystal face lying

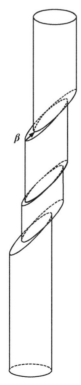

Figure 7.2   Schematic diagram of glide occurring in the direction β in a rod-shaped crystal of circular cross-section

parallel to the direction of glide. In practice, however, a crystal face usually contains dirt or corrosion products which give rise to a surface film. This surface film is torn during glide, and some effects due to glide are then observable on a face parallel to the glide direction.

One method of finding the direction of glide is to observe in which faces of a crystal the slip lines appear least well marked or in which face they disappear. The glide direction must be parallel to this plane; since it must also lie in the slip plane, it is then identifiable. Careful measurements of the change of shape of the crystal can of course reveal the direction in which glide is occurring. This method is used particularly with metallic crystals, which undergo large amounts of glide without breaking (see Section 7.4).

Only rarely are the slip steps on crystals deep enough to be seen with the naked eye. The scanning electron microscope shows that these, and those appearing under the optical microscope as single steps, are in reality composed of a complicated fine structure of smaller steps (Figure 7.3). As a measure of the amount of glide in a crystal, a macroscopic average over a volume of the crystal containing many individual slip steps is used. If s is the relative translation in the direction of glide of two planes, both parallel to the glide plane, and separated by a distance h, measured normal to the glide plane (Figure 7.4), the crystallographic glide strain α is defined as:

$$\alpha = \frac{s}{h}$$

so that $\alpha = \tan \gamma$ in Figure 7.4. When α is very small, it can be written in terms of the pure strain tensor ε_{ij}. If x_1 is in the glide direction and x_3 is normal to the glide plane:

$$\alpha = 2\varepsilon_{13} = 2\varepsilon_{31}$$

and is therefore equal to the engineering shear strain, $\gamma_{13} = \gamma_{31}$. In terms of the tensor e_{ij} (Section 6.3), $\alpha = e_{13}$, provided α is defined for a sufficiently large volume of the crystal for the strain to be considered homogeneous.

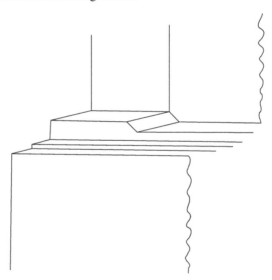

Figure 7.3   Schematic diagram of the fine structure observed in slip lines in the scanning electron microscope

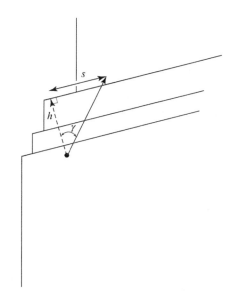

Figure 7.4   Macroscopic measurement of the amount of glide in a crystal

7.2   Glide Elements

The glide elements of a crystal are the glide direction and the glide plane. These are also named the slip direction and slip plane, respectively. A glide plane and a glide direction in that plane together constitute a *glide system* or *slip system*.

The deformation taking place during glide is a simple shear, so that the displacements relative to the centre of the crystal are, as is shown in Figure 7.5, inherently centrosymmetric. Because of this, a single glide system defined, say, by the slip plane, unit normal **n**, and a slip direction normal to **n**, say β, produces exactly the same deformation as slip

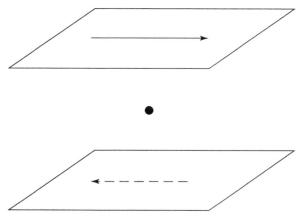

Figure 7.5   Illustration of the inherent centrosymmetric nature of the simple shear deformation which takes place during glide

on $-\mathbf{n}$ in the direction $-\beta$. The point group of the crystal governs the multiplicity of the slip planes and directions, and because of the centrosymmetric nature of the process, it is the Laue group that must be used to determine the multiplicity of glide systems.

When a slip direction and a slip plane are given to define a glide system, then in all crystals of symmetry higher than the triclinic system there is usually a multiplicity of glide systems, because the glide plane and glide direction are repeated by the point group symmetry of the crystal. All those combinations of slip planes and slip directions that arise from the point group symmetry of the crystal if one slip plane and one slip direction are given are called the *family of slip systems*.

A given glide system, say $\mathbf{n}$, β, is taken to be capable of operation in either a positive or a negative sense, so that on the plane of normal $\mathbf{n}$, slip can proceed in either the direction β or the direction $-\beta$. The deformation produced by slip in the direction β is, of course, just the reverse of that produced by slip in the direction $-\beta$. However, slip in these two directions will only be *crystallographically equivalent* if one, or more, of four conditions is fulfilled. These conditions are:

1. An even-fold axis of rotational symmetry (diad, tetrad, hexad) lies parallel to $\mathbf{n}$.
2. A mirror plane lies parallel to the slip plane; that is, a $\bar{2}$ axis lies parallel to $\mathbf{n}$.
3. β is parallel to an even-fold axis of rotational symmetry.
4. A mirror plane lies normal to β; that is, β is parallel to a $\bar{2}$ axis.

It is noteworthy from Table 2.2 that if a centre of symmetry is added to the point group, fulfilment of condition (1) implies that (2) is obeyed and, correspondingly, fulfilment of (3) implies that (4) is fulfilled. If none of the conditions (1)–(4) is fulfilled then slips in the directions β and $-\beta$ on the same slip plane are not crystallographically identical and cannot be necessarily expected to occur under the same shear stress. Some examples are considered later in this section.

As Reusch observed [1], crystals of NaCl slip on $\{110\}$ planes in $\langle 1\bar{1}0 \rangle$ directions. The crystal structure has a space group $Fm\bar{3}m$ (Section 3.5.1), and therefore has a point group $m\bar{3}m$. The normals to the glide plane – the $\{110\}$ poles – occupy special positions. There are 12 of these. Each slip plane contains two slip directions, which are of opposite sense: for example, $[1\bar{1}0]$ and $[\bar{1}10]$ in (110). There are then initially 24 glide systems in the $\{110\}$ $\langle 1\bar{1}0 \rangle$ family to consider.

Since the process of glide is centrosymmetric, these 24 glide systems yield 12 systems which produce different values of the tensor components e_{ij} defining the deformation (regarding positive and negative values of the same component as being different). In this point group, conditions (1)–(4) are all satisfied, so that reversal of the glide direction produces crystallographically equivalent effects. We say then that there are 12 physically distinct glide systems (different values of the e_{ij}) but, since reversal of the slip direction produces crystallographically equivalent effects, we regard a slip system as capable of operation in both a positive and a negative sense, and hence we speak of there being six different glide systems in the $\{110\}$ $\langle 1\bar{1}0 \rangle$ family in this case.

The glide elements of a large number of crystals are listed in Table 7.1. With few exceptions, the glide direction is parallel to the shortest lattice translation vector of the Bravais lattice. This is so even when the shortest lattice translation vector is considerably larger than the interatomic distance, as, for example, in bismuth. This rule finds a simple explanation in terms of the physical mechanism giving rise to glide. Defects known as dislocations are present in the crystal (Chapters 8 and 9). Glide always occurs by the motion of dislocations.

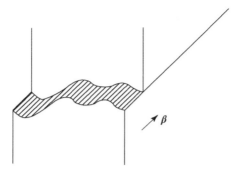

Figure 7.6   Pencil glide, also known as wavy glide. Here, the slip direction β is fixed, but the slip plane can be any plane which has the slip direction as the zone axis

As a rule, the slip planes in simple crystal structures at low temperatures are parallel to the closest-packed atomic planes in the crystal, but this rule has many exceptions. In addition, crystals often show a number of crystallographically different slip planes; for example, the h.c.p. metal magnesium glides on $\{0001\}$, $\{10\bar{1}0\}$ and $\{10\bar{1}1\}$ [7]. In many crystals, at moderate temperatures, while the slip direction remains fixed, the slip plane in a given slip line varies, so that what is called *pencil glide* or *wavy glide* occurs, with the slip plane being any plane with the slip direction as the zone axis (Figure 7.6). The slip trace then appears to be irrational on all crystal faces except those parallel to the slip direction.

Crystals can show all variations between very well-defined slip planes and wavy glide. For instance, in NaCl at temperatures less than $200\,^{\circ}$C the slip plane is accurately $\{110\}$ and slip traces on all crystal faces appear to be very straight. Above a temperature of about $250\,^{\circ}$C slip appears to occur on any plane in the $\langle 1\bar{1}0 \rangle$ type zone [31]. For crystals that show wavy glide, it is the well-defined slip plane observed at low temperatures that is listed in Table 7.1.

A number of cases in which slip on a given slip plane in a certain direction is not crystallographically equivalent to slip in the reverse direction are pointed out in Table 7.1. A common example is slip in the $\langle 111 \rangle$ directions on the $\{11\bar{2}\}$ planes in b.c.c. metals. There are 24 planes of the type $\{11\bar{2}\}$. Each $\{11\bar{2}\}$ plane contains two $\langle 111 \rangle$ directions, for example $[111]$ and $[\bar{1}\,\bar{1}\,\bar{1}]$ in $(11\bar{2})$, so initially there are 48 slip systems. The centrosymmetric nature of glide reduces this number to 24. None of conditions (1)–(4) of this section is satisfied and hence there are, strictly speaking, two crystallographically different families, each containing 12 members.

It is apparent from Table 7.1 that a knowledge of the crystal structure alone is not sufficient to deduce the slip planes and directions of a crystal. Many alkali halides with the NaCl structure glide on $\{110\}$ $\langle 1\bar{1}0 \rangle$, but in PbS and PbTe, which have this structure, this is not the dominant glide system [10, 11]. TiC also has this structure, but has the same slip systems as c.c.p. metals [8]. In all metals with the h.c.p. structure the most commonly observed slip direction is $\langle 11\bar{2}0 \rangle$, but the slip plane most commonly observed at room temperature varies with the metal (Table 7.2) and is not explained simply by variation of the c/a ratio. It is clear that glide planes, and to some extent the glide directions, do not depend solely on the geometry of the arrangement of the constituent atoms, but also upon the type of interatomic binding in the crystal [34].

Table 7.1   Glide elements of crystals (at room temperature and at atmospheric pressure except where stated)

Crystal type, examples and references	Class	Lattice	Slip direction	Slip plane	Remarks	References
c.c.p. metals and solid solutions, e.g. Al, Cu, α-Cu–Zn	$m\bar{3}m$	F	$\langle 1\bar{1}0 \rangle$	$\{111\}$		[4–6]
b.c.c. metals, e.g. Fe, Nb, Ta, W, Na, K	$m\bar{3}m$	I	$\langle 1\bar{1}1 \rangle$	$\{110\}$	Predominant system; wavy glide frequent	[5, 6]
			$\langle 1\bar{1}1 \rangle$	$\{\bar{2}\,\bar{1}1\}^a$		
			$\langle 1\bar{1}1 \rangle$	$\{\bar{1}23\}^a$	At high homologous temperatures in alkali metals	
h.c.p. metals: Zn, Cd, Mg, Be	$6/mmm$	P	$\langle 11\bar{2}0 \rangle$	(0001)	Predominant	[7]
			$\langle 11\bar{2}0 \rangle$	$\{1\bar{1}00\}$		
			$\langle 11\bar{2}0 \rangle$	$\{1\bar{1}01\}$		
			$\langle 11\bar{2}3 \rangle$	$\{11\bar{2}\,\bar{2}\}^a$		
h.c.p. metals: Ti, Zr	$6/mmm$	P	$\langle 11\bar{2}0 \rangle$	$\{1\bar{1}00\}$	Predominant	[7]
			$\langle 11\bar{2}0 \rangle$	(0001)		
			$\langle 11\bar{2}0 \rangle$	$\{1\bar{1}01\}$		
			$\langle 11\bar{2}3 \rangle$	$\{11\bar{2}\,\bar{2}\}^a$		
Sodium chloride-type, e.g., NaCl, LiF, MgO, NaF, AgCl, NH_4I, KI, UN, LiCl, LiBr, KCl, NaBr, RbCl, KBr, NaI, AgBr	$m\bar{3}m$	F	$\langle 1\bar{1}0 \rangle$	$\{110\}$		[4–6]
TiC, UC	$m\bar{3}m$	F	$\langle 1\bar{1}0 \rangle$	$\{111\}$	At high temperature	[8, 9]
PbS	$m\bar{3}m$	F	$\langle 1\bar{1}0 \rangle$	$\{001\}$	Predominant system	[10]
			$\langle 1\bar{1}0 \rangle$	$\{110\}$		
			$\langle 1\bar{1}0 \rangle$	$\{111\}$		
PbTe	$m\bar{3}m$	F	$\langle 001 \rangle$	$\{110\}$		[11]
Fluorite-type, e.g. CaF_2, UO_2	$m\bar{3}m$	F	$\langle 1\bar{1}0 \rangle$	$\{001\}$		[4, 6, 12, 13]
			$\langle 1\bar{1}0 \rangle$	$\{110\}$	At higher temperatures	
			$\langle 1\bar{1}0 \rangle$	$\{111\}$	At higher temperatures	
Diamond C, Si, Ge	$m\bar{3}m$	F	$\langle 1\bar{1}0 \rangle$	$\{111\}$	At temperatures above one-half of the melting temperature	[6]

(continued)

Table 7.1 (*continued*)

Crystal type, examples and references	Class	Lattice	Slip direction	Slip plane	Remarks	References
Al$_2$MgO$_4$ spinel	$m\bar{3}m$	F	$\langle 1\bar{1}0\rangle$	{111} {110}	In nonstoichiometric crystals with excess Al$_2$O$_3$, {110} is the predominant slip plane	[13, 14]
Caesium chloride-type, e.g. CsCl, CsBr, NH$_4$Cl, NH$_4$Br, Tl(Br,I), LiTl, MgTl, AuZn, AuCd	$m\bar{3}m$	P	$\langle 001\rangle$	{110}		[4, 6]
β-CuZn	$m\bar{3}m$	P	$\langle 1\bar{1}1\rangle$	{110}		[6]
Perovskite, SrTiO$_3$	$m\bar{3}m$	P	$\langle 1\bar{1}0\rangle$ $\langle 1\bar{1}0\rangle$	{110} {001}		[15]
Sphalerite α-ZnS, InSb, GaAs	$\bar{4}3m$	F	$\langle 1\bar{1}0\rangle$	{111}		[16–18]
Graphite	$6/mmm$	P	$\langle 11\bar{2}0\rangle$	(0001)		[19]
β-Sn	$4/mmm$	I	$\langle 001\rangle$ $\langle 010\rangle$ $\langle 1\bar{1}1\rangle$ $\langle 011\rangle$	{110} {100} {110} {100}	Other modes are also able to operate at high temperatures and/or high stresses	[20]
Rutile, TiO$_2$	$4/mmm$	P	$\langle 10\bar{1}\rangle$ $\langle 001\rangle$	{101}[a] {110}		[21, 22]
Bi	$\bar{3}m$	R	$\langle 10\bar{1}\rangle$	(111)	Rhombohedral face-centred cell	[23–25]
Hg	$\bar{3}m$	R	$\langle 011\rangle$ $\langle 1\bar{1}0\rangle$	{11$\bar{1}$} {11$\bar{1}$}	Rhombohedral face-centred cell	[26]
α-Al$_2$O$_3$	$\bar{3}m$	R	$\langle 11\bar{2}0\rangle$ $\langle 10\bar{1}0\rangle$ $\langle 11\bar{2}0\rangle$	(0001) {1$\bar{2}$10} {1$\bar{1}$02}	Above 1000 °C Above 1200 °C Above 1150 °C Indices refer to triply-primitive hexagonal cell (Section 3.6)	[13, 27]
Te	32	P	$\langle 11\bar{2}0\rangle$	{10$\bar{1}$0}		[28]
Ga	mmm	A	[010] [010] $\langle 0\bar{1}1\rangle$	(001) (100) {011}[a]		[29]
α-U	mmm	C	[100] [100] [1$\bar{1}$0]	(010) (001) (110)	above 150 °C	[30]

[a]Indicates systems where reversing the slip direction does not produce crystallographically equivalent motions.

Table 7.2 Predominant slip planes for h.c.p. metals at room temperature. In each case the predominant slip direction is $\langle 11\bar{2}0 \rangle$

Metal	c/a	Predominant slip plane	References
Cd	1.886	(0001)	[5, 7]
Zn	1.857	(0001)	[5, 7]
Co	1.625	(0001)	[5, 7]
Mg	1.624	(0001)	[5, 7]
Re	1.615	(0001)	[5]
Tl	1.598	both (0001) and (10$\bar{1}$0) occur with approximately equal ease	[32]
Zr	1.593	(10$\bar{1}$0)	[5, 7]
Ti	1.587	(10$\bar{1}$0)	[5, 7]
Hf	1.581	(10$\bar{1}$0)	[7]
Y	1.571	(10$\bar{1}$0)	[33]
Be	1.568	(0001)	[5, 7]

7.3 Independent Slip Systems

The change of shape produced by glide in a crystal is a simple shear. This is an example of a homogeneous strain in which every part of the strained region suffers the same distortion (Section 6.1). Under such circumstances, the e_{ij} in Eqs. (6.5) and (6.9) are constants, independent of position in the body. The equation:

$$du_i = \frac{\partial u_i}{\partial x_j}dx_j = e_{ij}dx_j \tag{7.1}$$

can then be integrated immediately to obtain:

$$u_i = (u_0)_i + e_{ij}x_j \qquad (i, j = 1, 2, 3) \tag{7.2}$$

Hence, in homogeneous strain, the displacements are linear functions of the position coordinates.

The constants $(u_0)_i$ represent the translation of the body as a whole, and can all be subtracted from the displacements suffered as a consequence of the homogeneous strain. We therefore obtain the displacement arising from the strain:

$$x_i' - x_i = e_{ij}x_j \tag{7.3}$$

where x_i' are the new coordinates (referred to axes that have been translated by $(u_0)_i$ but not rotated) of the point which was originally at x_i. Eq. (7.3) written out in full is:

$$x_1' = (1 + e_{11})x_1 + e_{12}x_2 + e_{13}x_3$$
$$x_2' = e_{21}x_1 + (1 + e_{22})x_2 + e_{23}x_3 \tag{7.4}$$
$$x_3' = e_{31}x_1 + e_{32}x_2 + (1 + e_{33})x_3$$

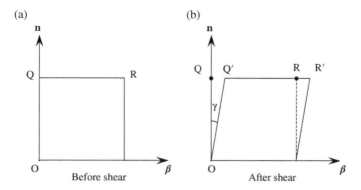

Figure 7.7 Illustration of simple shear in two dimensions: (a) before shear and (b) after shear

Suppose we first consider a two-dimensional example of simple shear of a crystal through a small angle γ on a slip plane defined by a unit vector $\mathbf{n}$ normal to it in a slip direction lying in the slip plane defined by the unit vector β, as in Figure 7.7. The vector $\mathbf{OQ}$ initially parallel to $\mathbf{n}$ in Figure 7.7 is sheared to become a vector $\mathbf{OQ'}$ after the simple shear has taken place; likewise, the vector $\mathbf{OR}$ lying in the plane containing $\mathbf{n}$ and β is sheared to a vector $\mathbf{OR'}$. From this diagram it is apparent that the displacement $\mathbf{QQ'}$ is of the same magnitude as the displacement $\mathbf{RR'}$.

Clearly, all vectors $\mathbf{r}$ in the plane containing $\mathbf{n}$ and β whose projected length along $\mathbf{n}$ is OQ suffer the same displacement. Such vectors will have a component along the unit vector $\mathbf{n}$ of magnitude $\mathbf{r} \cdot \mathbf{n}$. Also, from the diagram:

$$\frac{|QQ'|}{|OQ|} = \tan \gamma \cong \gamma \tag{7.5}$$

if γ is small (with γ measured in radians). Hence, if γ is small:

$$\mathbf{QQ'} = \gamma(\mathbf{r} \cdot \mathbf{n})\beta \tag{7.6}$$

showing that, as we expect, the magnitude of $\mathbf{QQ'}$ scales with the magnitude of $\mathbf{r}$.

For a more general three-dimensional geometry, in which glide occurs on a plane of unit normal $\mathbf{n}$ in the direction of the unit vector β, such as in Figure 7.8, it is convenient to consider a point P in a crystal so that, from an origin O, the vector $\mathbf{OP}$ is $\mathbf{r}$. Let P move to P' at $\mathbf{r'}$ from the origin as a consequence of the simple shear. Then, following the same methodology as for the two-dimensional example:

$$\mathbf{PP'} = \mathbf{r'} - \mathbf{r} = \gamma(\mathbf{r} \cdot \mathbf{n})\beta \tag{7.7}$$

where γ is the amount of crystallographic glide strain.

Provided γ is small, we can write down the components of the relative displacement tensor e_{ij} and of the pure strain tensor ε_{ij}, referred to the axes (x_1, x_2, x_3), describing the deformation. From Eq. (6.10), for example:

$$e_{11} = \frac{\partial u_1}{\partial x_1} = \frac{\partial}{\partial x_1}(\mathbf{r'} - \mathbf{r})_1 = \frac{\partial}{\partial x_1}\gamma(\mathbf{r} \cdot \mathbf{n})\beta_1 \tag{7.8}$$

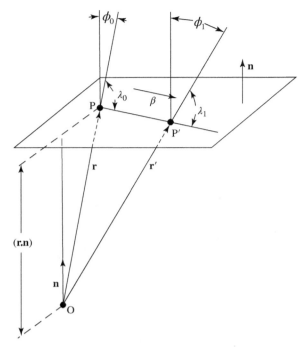

Figure 7.8 Simple shear in three dimensions. Here, a point P relative to O moves to a point P′ as a consequence of the simple shear

If we write:

$$\mathbf{r} = x_1\mathbf{i} + x_2\mathbf{j} + x_3\mathbf{k}$$

$$\mathbf{n} = n_1\mathbf{i} + n_2\mathbf{j} + n_3\mathbf{k}$$

and:

$$\beta = \beta_1\mathbf{i} + \beta_2\mathbf{j} + \beta_3\mathbf{k}$$

where $\mathbf{i}$, $\mathbf{j}$ and $\mathbf{k}$ form an orthonormal basis set (Appendix 1), it follows that:

$$e_{11} = \frac{\partial}{\partial x_1}\gamma(\mathbf{r}\cdot\mathbf{n})\beta_1 = \gamma\frac{\partial}{\partial x_1}[(x_1\mathbf{i} + x_2\mathbf{j} + x_3\mathbf{k})\cdot(n_1\mathbf{i} + n_2\mathbf{j} + n_3\mathbf{k})]\beta_1$$

$$= \gamma\frac{\partial}{\partial x_1}(x_1 n_1 + x_2 n_2 + x_3 n_3)\beta_1 \tag{7.9}$$

$$= \gamma n_1\beta_1$$

Similarly, a component such as e_{23} is given by:

$$e_{23} = \frac{\partial u_2}{\partial x_3} = \frac{\partial}{\partial x_3}\gamma(\mathbf{r}\cdot\mathbf{n})\beta_2 \tag{7.10}$$

i.e.:

$$e_{23} = \gamma n_3\beta_2 \tag{7.11}$$

The relative displacement tensor e_{ij} defining the deformation therefore has components:

$$e_{ij} = \gamma \begin{pmatrix} n_1\beta_1 & n_2\beta_1 & n_3\beta_1 \\ n_1\beta_2 & n_2\beta_2 & n_3\beta_2 \\ n_1\beta_3 & n_2\beta_3 & n_3\beta_3 \end{pmatrix} \qquad (7.12)$$

This tensor contains a pure strain tensor ε_{ij}. In addition, it also describes the rotation produced by the glide. Since glide corresponds to a simple shear, the deformation necessarily proceeds without a change of volume. The tensor in Eq. (7.12) shows this to be the case, because if $\mathbf{n}$ and $\boldsymbol{\beta}$ are orthogonal, the trace of the tensor sums to zero. That is:

$$e_{11} + e_{22} + e_{33} = \gamma n_1\beta_1 + \gamma n_2\beta_2 + \gamma n_3\beta_3 = 0 \qquad (7.13)$$

Equation (7.12) can be decomposed into the symmetric tensor describing the pure strain produced by the glide and the tensor describing the rotation. From Section 6.2, the pure strain tensor is:

$$\varepsilon_{ij} = \gamma \begin{pmatrix} n_1\beta_1 & \frac{1}{2}(n_1\beta_2 + n_2\beta_1) & \frac{1}{2}(n_1\beta_3 + n_3\beta_1) \\ \frac{1}{2}(n_1\beta_2 + n_2\beta_1) & n_2\beta_2 & \frac{1}{2}(n_3\beta_2 + n_2\beta_3) \\ \frac{1}{2}(n_1\beta_3 + n_3\beta_1) & \frac{1}{2}(n_3\beta_2 + n_2\beta_3) & n_3\beta_3 \end{pmatrix} \qquad (7.14)$$

and the antisymmetrical tensor describing the rotation is:

$$\omega_{ij} = \gamma \begin{pmatrix} 0 & \frac{1}{2}(n_2\beta_1 - n_1\beta_2) & \frac{1}{2}(n_3\beta_1 - n_1\beta_3) \\ -\frac{1}{2}(n_2\beta_1 - n_1\beta_2) & 0 & \frac{1}{2}(n_3\beta_2 - n_2\beta_3) \\ -\frac{1}{2}(n_3\beta_1 - n_1\beta_3) & -\frac{1}{2}(n_3\beta_2 - n_2\beta_3) & 0 \end{pmatrix} \qquad (7.15)$$

To see simply what Eqs. (7.14) and (7.15) mean, suppose we take axes in Figure 7.7 so that $\mathbf{n}$, the slip plane normal, is parallel to x_2 and $\boldsymbol{\beta}$ is parallel to x_1; Eqs. (7.12), (7.14) and (7.15) then reduce to:

$$e_{ij} = \begin{pmatrix} 0 & \gamma & 0 \\ 0 & 0 & 0 \\ 0 & 0 & 0 \end{pmatrix}, \quad \varepsilon_{ij} = \begin{pmatrix} 0 & \gamma/2 & 0 \\ \gamma/2 & 0 & 0 \\ 0 & 0 & 0 \end{pmatrix} \quad \text{and} \quad \omega_{ij} = \begin{pmatrix} 0 & \gamma/2 & 0 \\ -\gamma/2 & 0 & 0 \\ 0 & 0 & 0 \end{pmatrix} \qquad (7.16)$$

as in Figure 7.9.

Physically distinct slip systems of the same family may produce similar pure strains. Suppose as an example we consider glide of amount γ to occur on the $[0\bar{1}1](011)$ system of a cubic crystal, for example NaCl (Table 7.1). Then if we take axes parallel to the cubic axes of the crystal we have:

$$\boldsymbol{\beta} = 0\mathbf{i} - \frac{1}{\sqrt{2}}\mathbf{j} + \frac{1}{\sqrt{2}}\mathbf{k}$$

$$\mathbf{n} = 0\mathbf{i} + \frac{1}{\sqrt{2}}\mathbf{j} + \frac{1}{\sqrt{2}}\mathbf{k}$$

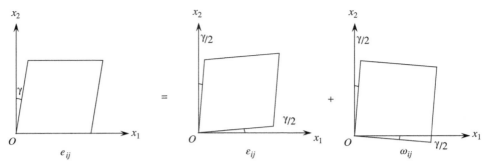

Figure 7.9 A small simple shear deformation e_{ij} is equivalent to a pure strain ε_{ij} plus a rotation ω_{ij} describing the rotation of the x_1 and x_2 axes by $\gamma/2$

since we have defined both **n** and $\boldsymbol{\beta}$ to be unit vectors. Therefore, from Eqs. (7.12), (7.14) and (7.15), we have:

$$e_{ij} = \begin{pmatrix} 0 & 0 & 0 \\ 0 & -\gamma/2 & -\gamma/2 \\ 0 & \gamma/2 & \gamma/2 \end{pmatrix}, \quad \varepsilon_{ij} = \begin{pmatrix} 0 & 0 & 0 \\ 0 & -\gamma/2 & 0 \\ 0 & 0 & \gamma/2 \end{pmatrix} \quad \text{and} \quad \omega_{ij} = \begin{pmatrix} 0 & 0 & 0 \\ 0 & 0 & -\gamma/2 \\ 0 & \gamma/2 & 0 \end{pmatrix} \quad (7.17)$$

If we consider glide of the same amount γ on the system $[011](0\bar{1}1)$, which is a second slip system in NaCl orthogonal to our first example, then we find that:

$$e_{ij} = \begin{pmatrix} 0 & 0 & 0 \\ 0 & -\gamma/2 & \gamma/2 \\ 0 & -\gamma/2 & \gamma/2 \end{pmatrix}, \quad \varepsilon_{ij} = \begin{pmatrix} 0 & 0 & 0 \\ 0 & -\gamma/2 & 0 \\ 0 & 0 & \gamma/2 \end{pmatrix} \quad \text{and} \quad \omega_{ij} = \begin{pmatrix} 0 & 0 & 0 \\ 0 & 0 & \gamma/2 \\ 0 & -\gamma/2 & 0 \end{pmatrix} \quad (7.18)$$

Therefore, the operation of each of these two physically different systems produces the same pure strain, but opposite rigid-body rotations (Figure 7.10).

Clearly, slip (also of amount γ) on the system $[01\bar{1}](011)$ – that is, just reversing the direction of glide in the first case – produces the opposite pure strain to slip on the system

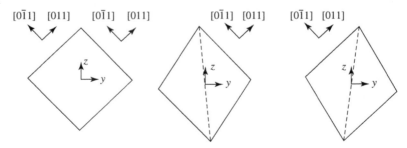

Figure 7.10 The strains produced by the physically different slip systems $[0\bar{1}1](011)$ and $[011](0\bar{1}1)$ are the same, but the rigid-body rotation of $[0\bar{1}1](011)$ (centre diagram) is in the opposite sense to that for $[011](0\bar{1}1)$ (the diagram on the right)

$[011](0\bar{1}1)$, but the *same rotation*, since for this slip system:

$$
e_{ij} = \begin{pmatrix} 0 & 0 & 0 \\ 0 & \gamma/2 & \gamma/2 \\ 0 & -\gamma/2 & -\gamma/2 \end{pmatrix}, \quad \varepsilon_{ij} = \begin{pmatrix} 0 & 0 & 0 \\ 0 & \gamma/2 & 0 \\ 0 & 0 & -\gamma/2 \end{pmatrix} \quad \text{and} \quad \omega_{ij} = \begin{pmatrix} 0 & 0 & 0 \\ 0 & 0 & \gamma/2 \\ 0 & -\gamma/2 & 0 \end{pmatrix} \quad (7.19)
$$

Since, in the absence of any constraint, glide occurs without any deformation of the *lattice* of the crystal, so that the lattice is subject to neither pure strain nor rotation, it is very important to realize that simultaneous slip *on more than one system* can produce a *pure* strain of a crystal without change in the orientation of the lattice and alternatively can rotate the crystal without changing the *orientation of the lattice*.

From Eqs. (7.17)–(7.19) it is apparent that equal amounts, γ, of slip on $[0\bar{1}1](011)$ and $[011](0\bar{1}1)$ produce a total deformation defined by the sum of the relative displacement tensors in Eqs. (7.17) and (7.18):

$$
e_{ij} = \begin{pmatrix} 0 & 0 & 0 \\ 0 & -\gamma & 0 \\ 0 & 0 & \gamma \end{pmatrix} \quad (7.20)
$$

which produces an expansion along the z-axis and a contraction along the y-axis without a change in volume, and without subjecting the lattice to either a pure strain or rotation.

In contrast to this, equal amounts, γ, of slip on $[01\bar{1}](011)$ and $[011](0\bar{1}1)$ produce a total deformation defined by the sum of the relative displacement tensors in Eqs. (7.18) and (7.19):

$$
e_{ij} = \begin{pmatrix} 0 & 0 & 0 \\ 0 & 0 & \gamma \\ 0 & -\gamma & 0 \end{pmatrix} \quad (7.21)
$$

which produces a clockwise rotation of magnitude γ about the x-axis, without any strain, and without subjecting the lattice to either a pure strain or rotation. Schematics illustrating Eqs. (7.20) and (7.21) are shown in Figure 7.11. In order for these two operations to be accomplished, slip lines corresponding to the slip on various slip systems must interpenetrate; whether or not this is possible depends on the properties of individual dislocations (Chapter 9).

Physically distinct slip systems may produce the same *pure* strain; we have just seen an example of this. Slip on $[0\bar{1}1](011)$ in a cubic crystal and slip on $[011](0\bar{1}1)$ produce the same pure strain tensor, as is apparent from Eqs. (7.17) and (7.18). Slip on $[\bar{1}10](110)$ of amount γ in a cubic crystal produces the pure strain:

$$
\begin{pmatrix} -\gamma/2 & 0 & 0 \\ 0 & \gamma/2 & 0 \\ 0 & 0 & 0 \end{pmatrix} \quad (7.22)
$$

This system then produces a pure strain that cannot be produced by slip on either of the systems $[0\bar{1}1](011)$ or $[011](0\bar{1}1)$. Slip systems that produce different pure strains are said

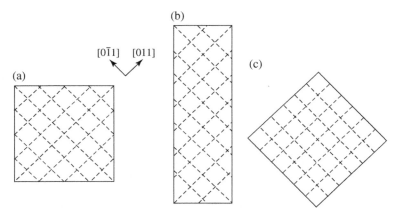

Figure 7.11 Schematic diagram to illustrate that the cubic crystal in (a) slipping by equal amounts on the glide systems $[0\bar{1}1](011)$ and $[011](0\bar{1}1)$ would produce a pure shear, (b), without any rotation of the crystal in space, or of the lattice, while if it were to slip by equal amounts on $[01\bar{1}](011)$ and $[011](0\bar{1}1)$, (c), the crystal as a body would be rotated in space without any pure strain and without any rotation of the crystal lattice

to be *independent*. A general pure strain can be produced in a crystal by glide, provided the crystal can slip on five independent systems [35]. The reason that five independent systems are needed is as follows. Plastic deformation of a crystal by glide proceeds without a change in volume, so that the pure strain tensor describing the deformation always has the property that:

$$\varepsilon_{11} + \varepsilon_{22} + \varepsilon_{33} = 0 \tag{7.23}$$

Therefore, only two of the components ε_{11}, ε_{22}, ε_{33} can be chosen arbitrarily, since the third is always fixed by the condition in Eq. (7.23). Since the pure strain tensor is symmetric, $\varepsilon_{ij} = \varepsilon_{ji}$, there are just five numbers that can be selected arbitrarily to be the strain components of the general pure strain tensor when the volume is conserved.

Glide on any one slip system can alter just one component of the pure strain tensor independently of its effect on the others, so five independent glide systems must be available if a general arbitrary pure strain is to be produced by glide. A slip system is said to be independent of others if the pure strain that can be produced by its operation *cannot* be produced by combining suitably chosen amounts of slip on these other systems.

In a cubic crystal we have just seen that the glide systems $[0\bar{1}1](011)$ and $[\bar{1}10](110)$ are independent. The system $[\bar{1}01](101)$ is not independent of these two. This is easily seen by writing down together the pure strain components referred to the same cubic axes for amounts of slip γ_1 on $[0\bar{1}1](011)$, γ_2 on $[\bar{1}10](110)$ and γ_3 on $[\bar{1}01](101)$. These are:

$$\begin{pmatrix} 0 & 0 & 0 \\ 0 & -\gamma_1/2 & 0 \\ 0 & 0 & \gamma_1/2 \end{pmatrix}, \quad \begin{pmatrix} -\gamma_2/2 & 0 & 0 \\ 0 & \gamma_2/2 & 0 \\ 0 & 0 & 0 \end{pmatrix}, \quad \begin{pmatrix} -\gamma_3/2 & 0 & 0 \\ 0 & 0 & 0 \\ 0 & 0 & \gamma_3/2 \end{pmatrix} \tag{7.24}$$

Table 7.3 Independent slip systems in crystals

System	Class	Number of independent systems
$\langle 1\bar{1}0\rangle\{111\}$	$m\bar{3}m$	5
$\langle 1\bar{1}1\rangle\{110\}$	$m\bar{3}m$	5
$\langle 1\bar{1}0\rangle\{110\}$	$m\bar{3}m$	2
$\langle 001\rangle\{110\}$	$m\bar{3}m$	3
$\langle 1\bar{1}0\rangle\{001\}$	$m\bar{3}m$	3
$\langle 11\bar{2}0\rangle(0001)$	$6/mmm$	2
$\langle 11\bar{2}0\rangle\{1\bar{1}00\}$	$6/mmm$	2
$\langle 11\bar{2}0\rangle\{1\bar{1}01\}$	$6/mmm$	4
$\langle 11\bar{2}3\rangle\{11\bar{2}\bar{2}\}$	$6/mmm$	5
$\langle 10\bar{1}\rangle\{101\}$	$4/mmm$	4

Obviously, if $\gamma_1 = \gamma_2 = \gamma_3$, the pure strain produced by the last system can be duplicated by appropriate combinations of slip on the other two.

Since five independent systems produce a general pure strain (without volume change), it is clear that a crystal cannot possess more than five independent systems in this sense. The number of independent systems produced by the operation of all members of a given family is listed in Table 7.3 for various families of slip systems found in the more symmetric crystals.

When pencil glide occurs, the slip direction remains fixed but the slip plane can be any plane in the zone whose axis is the glide direction. It is then easily shown (see Problem 7.6) that shear in a given slip direction can provide two independent components of the pure strain tensor.

A straightforward procedure which establishes the number of independent slip systems for a given family of slip systems is to be systematic when examining the various strain tensors for each member of the family of slip systems. Thus, for example, if, once again, we consider crystals with the NaCl structure, the slip systems are $\langle 1\bar{1}0\rangle\{110\}$ at room temperature. There are six distinct glide systems (Section 7.2): each of the six $\{110\}$ planes contains one distinct $\langle 1\bar{1}0\rangle$ direction. It is convenient to label the six slip systems A–F for simplicity (Table 7.4).

Table 7.4 The six slip systems in crystals with the NaCl structure, labelled A – F

Label	Slip system
A	$[1\bar{1}0](110)$
B	$[110](1\bar{1}0)$
C	$[01\bar{1}](011)$
D	$[011](01\bar{1})$
E	$[10\bar{1}](101)$
F	$[101](10\bar{1})$

For each, we can write down the strain tensor produced by shearing an angle γ:

$$\varepsilon_A = \frac{\gamma}{2}\begin{pmatrix} 1 & 0 & 0 \\ 0 & -1 & 0 \\ 0 & 0 & 0 \end{pmatrix}; \quad \varepsilon_B = \frac{\gamma}{2}\begin{pmatrix} 1 & 0 & 0 \\ 0 & -1 & 0 \\ 0 & 0 & 0 \end{pmatrix}; \quad \varepsilon_C = \frac{\gamma}{2}\begin{pmatrix} 0 & 0 & 0 \\ 0 & 1 & 0 \\ 0 & 0 & -1 \end{pmatrix};$$

$$\varepsilon_D = \frac{\gamma}{2}\begin{pmatrix} 0 & 0 & 0 \\ 0 & 1 & 0 \\ 0 & 0 & -1 \end{pmatrix}; \quad \varepsilon_E = \frac{\gamma}{2}\begin{pmatrix} 1 & 0 & 0 \\ 0 & 0 & 0 \\ 0 & 0 & -1 \end{pmatrix}; \quad \varepsilon_F = \frac{\gamma}{2}\begin{pmatrix} 1 & 0 & 0 \\ 0 & 0 & 0 \\ 0 & 0 & -1 \end{pmatrix} \tag{7.25}$$

It is immediately apparent that:

$$\varepsilon_A = \varepsilon_B; \quad \varepsilon_C = \varepsilon_D; \quad \varepsilon_E = \varepsilon_F$$

so that there are at most three independent slip systems. However, an examination of ε_E shows that it can be produced by a linear combination of ε_A and ε_C:

$$\varepsilon_E = \varepsilon_A + \varepsilon_C$$

and so there are only two independent slip systems in crystals with the NaCl structure: for example, A and C. A proof using this procedure that there are five independent slip systems in c.c.p. and b.c.c metals is given in Appendix 5.

To decide formally for the general case whether the members of a given family, or of a set of families, of slip systems will together provide five independent systems, we can proceed as follows. Choose a set of orthogonal axes and write down, using Eq. (7.14), the components of the pure strain tensor produced by an arbitrary amount of glide on a given system. Call these components $\varepsilon_{11}, \varepsilon_{22}, \varepsilon_{33}, \varepsilon_{12}, \varepsilon_{13}, \varepsilon_{23}$. Since volume is conserved, $\varepsilon_{11} + \varepsilon_{22} + \varepsilon_{33} = 0$, and so we need only consider the five components $\varepsilon_{11}, \varepsilon_{22}, \varepsilon_{12}, \varepsilon_{13}$, ε_{23}, which we can write down as a 1×5 matrix; that is, as a single-row matrix with five columns.

Repeating this procedure for four other slip systems, referring the strain tensor to the same axes as before, we can then form the 5×5 determinant of the quantities $\varepsilon_{11}, \varepsilon_{22}, \varepsilon_{12}$, $\varepsilon_{13}, \varepsilon_{23}$ as rows for the five systems. If this determinant has a value other than zero, the five chosen slip systems are independent of one another, since a determinant will equal zero if any row can be expressed as a linear combination of other rows. If a crystal glides only on a family of slip systems that does not possess five independent members, there will be directions in which it cannot be extended or compressed; that is, there will be certain orientations of shear stress that cannot produce glide. These can be found easily from Eqs. (7.30) and (7.31) (Section 7.4).

If a random polycrystal (i.e. a body consisting of a large number of individual crystals (or grains) with the crystal axes in the various grains distributed randomly in space) is to deform by glide in each of the component grains without holes appearing at the grain boundaries, the average grain in the interior of the body must change its shape so as to conform with the changes of shape of the neighbouring grains, and with the change of shape of the body as a whole. Provided a grain possesses five independent glide systems, in the sense that we have defined, this grain can undergo a general arbitrary change of shape by glide. This condition is often stated as one that must be obeyed for all grains if the polycrystal is to be capable of large plastic changes of shape by glide, within the grains, without internal cavities forming. It is called the *von Mises condition* [35]. During deformation by slip of a polycrystal, the individual grains are constrained by their neighbours. Under these conditions, the *crystal lattice* often rotates in space.

The part played by lattice rotation can be appreciated by *supposing* that each grain of a polycrystal suffers the same small deformation e_{ij}, so that the deformation of the whole body is homogeneous.[1] We divide the small deformation e_{ij} into a pure strain and a rotation:

$$e_{ij} = \varepsilon_{ij} + \omega_{ij} \tag{7.26}$$

The pure strain ε_{ij} must be the same in each grain, and so must the rotation ω_{ij}. By operating five independent slip systems we can supply the same ε_{ij} to each grain. This will not in general produce the same rotation of each grain. The rotation produced directly by slip does not rotate the lattice, but due to the constraint of neighbouring grains a rigid-body rotation which rotates the lattice may always be added so that the total rotation, given by:

$$(\omega_{ij}) = (\omega_{ij})_{\text{slip}} + (\omega_{ij})_{\text{lattice}} \tag{7.27}$$

is the same for all grains. Each grain will therefore receive the lattice rotation that is needed to bring its total rotation to the common value.

This argument assumes that each grain may rotate freely. Whether this is so or not can only be decided by consideration of the constraints at the intercrystalline boundaries.

7.4 Large Strains of Single Crystals: The Choice of Glide System

In many experiments on glide a single crystal of the substance under investigation is pulled in tension or compressed along a given direction (Figure 7.12). It is easy to find the shear stress on the slip plane and resolved in the slip direction (called the *resolved shear stress*) by altering the axes of reference of the tensor representing the applied stress (see Problem 7.11). However, this can be very simply derived directly from Figure 7.12.

If a force $\mathbf{F}$ is applied to the crystal of cross-sectional area A_0, the tensile stress parallel to $\mathbf{F}$ is $\sigma = F/A_0$. The force $\mathbf{F}$ has a component $F \cos \lambda_0$ in the slip direction, where λ_0 is the angle between $\mathbf{F}$ and the slip direction; this force acts over an area $A_0/\cos \phi_0$, so that the resolved shear stress τ is:

$$\tau = \frac{F \cos \lambda_0}{A_0 / \cos \phi_0} = \sigma \cos \phi_0 \cos \lambda_0 \tag{7.28}$$

where ϕ_0 is the angle between $\mathbf{F}$ and the normal to the glide plane. The factor $\cos \phi_0 \cos \lambda_0$ is often known as the Schmid factor, from the work of Erich Schmid [36]. The stress normal to the slip plane is:

$$\sigma_{\text{n}} = \frac{F \cos \phi_0}{A_0 / \cos \phi_0} = \sigma \cos^2 \phi_0 \tag{7.29}$$

---

[1] In a real polycrystal, the deformation of each crystal grain may not be homogeneous; indeed, X-ray examination of individual grains in a polycrystal proves that usually it is not.

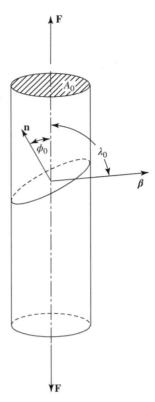

Figure 7.12 Deformation in tension of a single crystal. Glide is presumed to occur first in a direction β on a slip plane with normal **n**. β makes an angle of λ_0 with the force **F** while **n** makes an angle of ϕ_0 with **F**

It is worth emphasizing that in Figure 7.12 the directions of **F, n** and β are not necessarily coplanar, so that only in a special case is ϕ_0 equal to $(90° - \lambda_0)$.

The resolved shear stress on any glide system can be evaluated from Eq. (7.28). To a very good approximation, it is always found that when a crystal is subjected to an increasing uniaxial tensile or compressive load, as in Figure 7.12, slip always occurs first on that glide system on which the resolved shear stress is greatest – the Schmid criterion [36]. To a worse approximation, it is found that in a given pure crystal at a given temperature, glide starts when the resolved shear stress reaches a certain critical value. This last approximation, called the law of *critical resolved shear stress*, is moderately well obeyed in crystals of metals, where there is usually a high density of dislocations present (see Chapter 8), but in non-metals the resolved shear stress at which glide occurs is so dependent on the previous history of the crystal that such a rule is seldom of use.

The particular glide system with the highest resolved shear stress for a particular orientation of **F** with respect to the crystal axes may be found from Eq. (7.28). For common families of slip systems the result of examining all members to find which one is subject to the largest resolved shear stress is easily shown on a stereogram. A stereogram of a cubic crystal centred on 001 is shown in Figure 7.13. If a c.c.p. metal crystal is considered, the

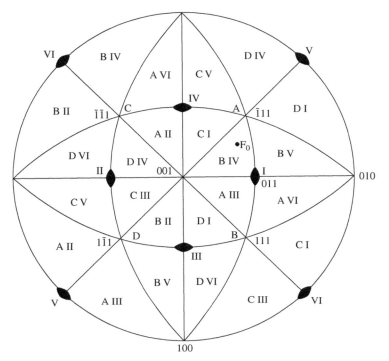

Figure 7.13   Standard stereogram of a c.c.p. metal crystal which glides on {111} in ⟨1$\bar{1}$0⟩ to show the particular slip system with the maximum resolved shear stress for any orientation of the tensile axis. (After Schmid and Boas [37], p. 88, who redrew a stereographic diagram of Taylor and Elam [38].)

point group is $m\bar{3}m$ and slip occurs on the ⟨1$\bar{1}$0⟩{111} system. There are 12 physically distinct glide systems.

If **F** is the direction of a uniaxial force applied to the crystal, the direction of **F** can be plotted on the stereogram, say as F_0 in Figure 7.13. The particular glide system with the largest resolved shear stress is indicated by the lettering of the unit triangle on the stereogram within which **F** falls. For instance, for the case shown in Figure 7.13, the triangle is lettered B IV, meaning that slip will occur first on the octahedral plane B, (111), in the direction IV, [$\bar{1}$01]. If **F** lies on a boundary between two unit triangles, then clearly two slip systems are equally stressed (see Problem 7.8). If **F** is exactly along ⟨110⟩, ⟨111⟩, or ⟨001⟩, there are four, six and eight slip systems respectively with the same largest resolved shear stress.

Similar diagrams for some other common slip systems are given in Figures 7.14 and 7.15, and are used in the same way as Figure 7.13. In b.c.c. crystals, slip can occur on planes other than {110} (Table 7.1), and so within a given stereographic triangle, the slip system at a particular position depends on the slip systems which can be activated and their relative critical resolved shear stresses. In the example shown in Figure 7.15, slip is presumed to occur in ⟨111⟩ directions on {1$\bar{1}$0} and {11$\bar{2}$} only, and for simplicity, it is assumed that the critical resolved shear stresses on these two sets of planes are equal. If the critically

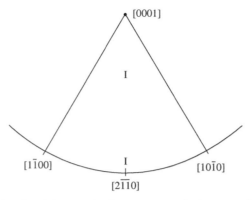

Figure 7.14   A unit triangle of the standard stereogram of a hexagonal crystal of point group 6/*mmm* which glides on (0001) in ⟨11$\bar{2}$0⟩, showing the slip direction of maximum resolved shear stress

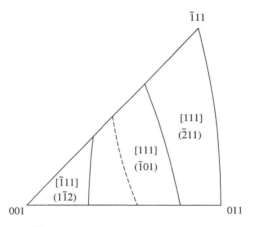

Figure 7.15   The 001–011–$\bar{1}$11 stereographic triangle for a b.c.c. metal crystal which glides on {1$\bar{1}$0} in ⟨111⟩ and on {11$\bar{2}$} in ⟨111⟩, showing the regions within this triangle where the resolved shear stresses on the various slip systems are greatest for a situation where the critically resolved shear stress is the same on the {1$\bar{1}$0} and {11$\bar{2}$} planes. The dashed line is the boundary between the two possible ⟨111⟩ {11$\bar{2}$} systems if slip did not occur on {1$\bar{1}$0}

resolved shear stress on the {11$\bar{2}$} planes were $<\sqrt{3}/2$ that on the {1$\bar{1}$0} planes, slip on {1$\bar{1}$0} would not occur, and the boundary between the two possible ⟨111⟩{11$\bar{2}$} systems shown would be the dashed line. Conversely, if the critically resolved shear stress on the {11$\bar{2}$} planes were $>2/\sqrt{3}$ that on the {1$\bar{1}$0} planes, slip only on the [111] ($\bar{1}$01) slip system would occur for all initial crystal orientations within this stereographic triangle [39, 40].

From Table 7.1, it is worth noting that, in crystals of the NaCl structure slipping on ⟨1$\bar{1}$0⟩{110}, at least two slip systems are always equally stressed. C.c.p. metal crystals have been very intensively studied. Even for crystals with the position of **F** near to the centre of the unit triangle, small amounts of glide on systems other than the most highly

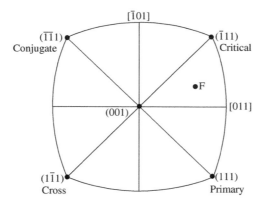

Figure 7.16 Nomenclature for the various {111} slip planes for a c.c.p. single crystal where the applied force **F** is initially in the stereographic triangle bounded by 001, 011 and $\bar{1}11$

stressed one are observed. Since the slip plane of a glide system is easily observed, these slip systems are usually identified by reference to the glide plane alone. The slip plane of the most highly stressed system is called the primary glide plane; the names given to this and the others are shown in Figure 7.16.

In cubic crystals slipping on {111} $\langle 1\bar{1}0 \rangle$, a simple rule has been given by Diehl [41] to identify the primary glide system when the direction of **F** lies in any of the unit triangles shown in Figure 7.13. This states: 'Reflect the $\langle 110 \rangle$ pole of the triangle in question in the opposite side of the triangle to find the glide direction, and reflect the {111} pole of the triangle in the opposite side to find the glide plane normal.' An equivalent rule to this is the OILS rule devised by Hutchings [42] as a non-graphical method for deducing the slip system with the highest Schmid factor. A mathematical proof of the validity of these two rules is given in Appendix 5.

7.5 Large Strains: The Change in the Orientation of the Lattice During Glide

Returning to Figure 7.12, if a small amount of glide, $d\gamma$, occurs on the slip system shown, the crystal will increase in length by dl in the direction of **F** where:

$$dl = l \cos\phi_0 \cos\lambda_0 d\gamma \qquad (7.30)$$

where $d\gamma$ is the increment of crystallographic glide strain and l the instantaneous length of the crystal.[2] If e is the (small) elongation parallel to **F** produced by $d\gamma$ then $e = dl/l$, and so it follows from Eq. (7.30) that:

$$e = \cos\phi_0 \cos\lambda_0 d\gamma \qquad (7.31)$$

---

[2] The identity of the geometrical factors in Eqs. (7.28) and (7.30) is worth noting in passing. This arises because, of course, if a given applied force produces no resolved shear stress on a given glide system, because $\cos\phi_0$ or $\cos\lambda_0$ is zero, then it does no work if that glide system operates. Certain orientations of an applied force **F** will give zero values for τ in Eq. (7.28)) or for dl in Eq. (7.30) if the crystal does not possess five independent slip systems.

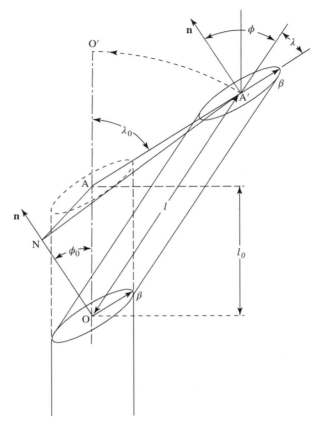

Figure 7.17   Change in orientation during a glide operation if there is no lateral constraint when a single crystal is pulled in tension

The relative translation of the parts of the crystal parallel to β implies that as the crystal gets longer its two ends move with respect to one another in a direction transverse to the applied force **F**. This effect is not important for very small strains, but when minerals are deformed large amounts in compression (usually under hydrostatic pressure), or metal crystals are extended by large amounts in tension, it must be taken into account. In compression, the ends of the crystal may be prevented from moving sideways by friction, and in tension, crystals are usually gripped at the ends.

Consider the tensile case, and suppose that there is no lateral constraint. The crystal will take up the form shown in Figure 7.17. If, however, due to the lateral constraints the point A′ is forced to lie above O at O′, then if there is glide solely on the system shown, the crystal lattice will be rotated with respect to the direction OO′. From Figure 7.17 the elongation of the crystal e equals $(l - l_0)/l_0$, so from the triangle OAA′:

$$\frac{l}{l_0} = 1 + e = \frac{\sin \lambda_0}{\sin \lambda} \qquad (7.32)$$

and from triangles NAO, NA'O:

$$\frac{l}{l_0} = \frac{\cos \phi_0}{\cos \phi} \tag{7.33}$$

where λ_0 and ϕ_0 are the initial values of λ and ϕ. The crystallographic glide strain γ is given by $AA'/(l_0 \cos \phi_0)$ and $AA' = l \cos \lambda - l_0 \cos \lambda_0$, so:

$$\gamma = \frac{l \cos \lambda - l_0 \cos \lambda_0}{l_0 \cos \phi_0} = \frac{\cos \lambda}{\cos \phi} - \frac{\cos \lambda_0}{\cos \phi_0} \tag{7.34}$$

At a particular instance during glide, the resolved shear stress τ [$=(F/A) \cos \phi \cos \lambda$], where A is the cross-sectional area of the ellipse whose centre is A', is:

$$\tau = \frac{F \cos \phi}{A} \left(1 - \frac{\sin^2 \lambda_0}{(1+e)^2} \right)^{1/2} \tag{7.35}$$

It can be seen from Figure 7.17 that the area of the glide plane remains constant during deformation (except for negligibly small changes due to the formation of slip steps). Therefore, the quantity $A/\cos \phi$ does not alter, and so Eq. (7.35) can be written as:

$$\tau = \frac{F \cos \phi_0}{A_0} \left[1 - \left(\frac{l_0}{l} \right)^2 \sin^2 \lambda_0 \right]^{1/2} \tag{7.36}$$

where A_0 is the original cross-sectional area of the specimen prior to tensile testing. As deformation proceeds, the value of λ decreases, and so for a given force $\mathbf{F}$ the resolved shear stress increases.

We can also derive Eqs. (7.32)–(7.34) directly from Eq. (7.7) and Figure 7.8. It is useful to do this because the vector formulae make the compression case easy to analyze. From Figure 7.8, a vector $\mathbf{r}$ is changed by glide of amount γ to the vector $\mathbf{r}'$, where:

$$\mathbf{r}' = \mathbf{r} + \gamma(\mathbf{r} \cdot \mathbf{n})\boldsymbol{\beta}$$

(see Eq. (7.7)). We take $\mathbf{r}$ parallel to the tensile axis when $\gamma = 0$; $\mathbf{r}'$ then represents the orientation of the tensile axis after a crystallographic glide strain γ:

$$\frac{l}{l_0} = (1+e) = \frac{|\mathbf{r}'|}{|\mathbf{r}|} \tag{7.37}$$

The quantity $\mathbf{n}$ is a unit vector normal to the glide plane, so:

$$\cos \phi = \frac{(\mathbf{r}' \cdot \mathbf{n})}{|\mathbf{r}'|} = [(\mathbf{r} \cdot \mathbf{n}) + \gamma(\mathbf{r} \cdot \mathbf{n})(\boldsymbol{\beta} \cdot \mathbf{n})] \frac{1}{|\mathbf{r}'|} = \frac{l_0 \cos \phi_0}{l} \tag{7.38}$$

since $\cos \phi_0 = (\mathbf{r} \cdot \mathbf{n})/|\mathbf{r}|$ and $\boldsymbol{\beta} \cdot \mathbf{n} = 0$. Therefore, we have Eq. (7.33). Also:

$$|\mathbf{r}' \times \boldsymbol{\beta}| = |\mathbf{r}'| \sin \lambda \tag{7.39}$$

since $\boldsymbol{\beta}$ is a unit vector. From Eq. (7.7):

$$\mathbf{r}' \times \boldsymbol{\beta} = \mathbf{r} \times \boldsymbol{\beta}, \qquad \text{since } \boldsymbol{\beta} \times \boldsymbol{\beta} = 0 \tag{7.40}$$

and so:

$$\frac{l}{l_0} = \frac{\sin \lambda_0}{\sin \lambda} \tag{7.41}$$

which is Eq. (7.32). Finally, if we form the dot product of both sides of Eq. (7.7) with $\boldsymbol{\beta}$, we obtain:

$$l \cos \lambda = l_0 \cos \lambda_0 + \gamma l_0 \cos \phi_0 \tag{7.42}$$

Therefore, using Eq. (7.33) we have:

$$\gamma = \frac{\cos \lambda}{\cos \phi} - \frac{\cos \lambda_0}{\cos \phi_0}$$

which is Eq. (7.34).

When a long, thin crystal is used in an experiment, such as that illustrated in Figure 7.12, the change in orientation of the tensile axis with respect to the crystal axes (which is easier to consider than what is the same thing, the change of the crystal axes with respect to the tensile direction) can be measured and plotted on a stereogram. Since the change of orientation is such that the tensile axis rotates towards the slip direction, this can be used to determine the direction of glide. An actual example for a c.c.p. metal crystal is given in Figure 7.18a [43]. When the orientation of the tensile axis in Figure 7.18a reaches the great circle joining the (001) and ($\bar{1}$11) poles, which is a mirror plane, the resolved shear stresses on two-slip systems are equal and slip might occur on both systems simultaneously. When two slip systems operate simultaneously, *double slip* or *duplex slip* is said to occur.

In many cases in c.c.p. cubic metals, a crystal continues to slip on the primary system after the orientation has reached the mirror plane. In this case the orientation of the tensile axis 'overshoots' into the neighbouring unit triangle, so that slip is not proceeding on the nominally most highly stressed system. An example is shown in Figure 7.18b. After

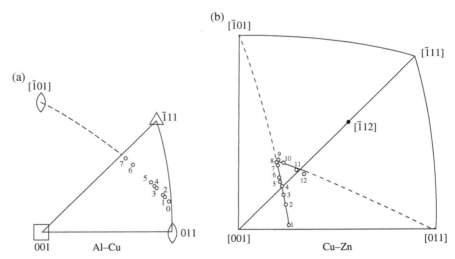

Figure 7.18 Changes in orientation of c.c.p. metal crystals during glide. The numbers indicate successive determinations of the crystal orientation. (Part (a) is based on a diagram in figure 6 of [43]; (b) is based on figure 16 of [44], redrawn as figure 69 of [45].)

a certain amount of this additional primary slip the conjugate system suddenly operates and further slip occurs on this system, sometimes followed by 'overshooting' on the conjugate system, as in Figure 7.18b. When the tensile axis reaches a ⟨112⟩ pole – [$\bar{1}$12] in the example shown in Figure 7.18b, the vector sum of the two slip directions [$\bar{1}$01] and [011] – no further orientation change should occur, provided each slip system contributes equally to the deformation.

There is also a change of orientation of the lattice of a crystal during a compression test. To deal with this case it is simplest to use vector formulae. We first use Eq. (7.7) to find how a given plane of normal **m** is altered by glide. Let $\mathbf{r}_1$, $\mathbf{r}_2$ be any two nonparallel unit vectors in the given plane. Write $\mathbf{m} = \mathbf{r}_1 \times \mathbf{r}_2$. Now transform each of $\mathbf{r}_1$ and $\mathbf{r}_2$ according to Eq. (7.7) to give $\mathbf{r}'_1$ and $\mathbf{r}'_2$. Then the new vector **m′** representing **m** after a crystallographic glide strain of amount γ is:

$$\mathbf{m}' = \mathbf{r}'_1 \times \mathbf{r}'_2 \tag{7.43}$$

Substituting for $\mathbf{r}'_1$ and $\mathbf{r}'_2$ from Eq. (7.7), we have:

$$\mathbf{m}' = [\mathbf{r}_1 \times \mathbf{r}_2] + \gamma(\mathbf{r}_1 \cdot \mathbf{n})\boldsymbol{\beta} \times \mathbf{r}_2 - \gamma(\mathbf{r}_2 \cdot \mathbf{n})\boldsymbol{\beta} \times \mathbf{r}_1 = \mathbf{m} + \gamma\boldsymbol{\beta} \times [(\mathbf{r}_1 \cdot \mathbf{n})\mathbf{r}_2 - (\mathbf{r}_2 \cdot \mathbf{n})\mathbf{r}_1] \tag{7.44}$$

Using the formula for the vector triple product (Appendix 1), this is seen to be:

$$\mathbf{m}' = \mathbf{m} + \gamma\boldsymbol{\beta} \times \mathbf{n} \times [\mathbf{r}_2 \times \mathbf{r}_1] \tag{7.45}$$

That is:

$$\mathbf{m}' = \mathbf{m} - \gamma\boldsymbol{\beta} \times [\mathbf{n} \times \mathbf{m}] \tag{7.46}$$

and again using the formula for the vector triple product:

$$\mathbf{m}' = \mathbf{m} - \gamma((\boldsymbol{\beta} \cdot \mathbf{m})\mathbf{n} - (\boldsymbol{\beta} \cdot \mathbf{n})\mathbf{m}) \tag{7.47}$$

or:

$$\mathbf{m}' = \mathbf{m} - \gamma(\boldsymbol{\beta} \cdot \mathbf{m})\mathbf{n} \tag{7.48}$$

since $(\boldsymbol{\beta} \cdot \mathbf{n}) = 0$. It is evident from Eqs. (7.46) and (7.48) that any plane moves to its final position in such a way that its normal always lies in the plane containing the initial position of the normal and **n**, the normal to the glide plane.

If we now consider the case of compression of a thin crystal between flat plates, as in Figure 7.19, we take the constraint to be that the plane initially normal to the stress axis must be maintained in this orientation with respect to the stress axis as the crystal deforms. To maintain positive values of γ during compression, we suppose slip to occur so that $(\boldsymbol{\beta} \cdot \mathbf{m})$ is negative (Figure 7.19). Eq. (7.48) then reads:

$$\mathbf{m}' = \mathbf{m} + \gamma(\boldsymbol{\beta} \cdot \mathbf{m})\mathbf{n} \tag{7.49}$$

From Eq. (7.49), the normal to the glide plane then approaches the normal to the compression plates by rotating about an axis which is parallel to the line of intersection of the glide plane and the compression plates (Figure 7.19b). The initial thickness of the crystal is h_0, which will be reduced during compression to a height h such that:

$$\frac{h_0}{h} = \frac{|\mathbf{m}'|}{|\mathbf{m}|} \tag{7.50}$$

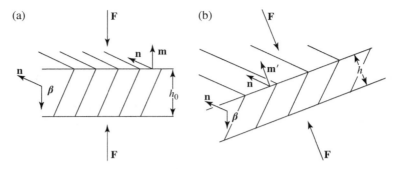

Figure 7.19 Compression of a single crystal between plates

because the volume of the crystal must be conserved. With the constraint that we have assumed, the change of orientation of the crystal lattice will be such that the normal to the slip plane rotates towards the axis of compression. If the initial and final angles, after a shear strain γ, between the direction of compression and the normal to the slip plane are ϕ_0, ϕ, and λ and λ_0 are the corresponding obtuse values of the angle between the compression axis and the slip direction, then we have, from Eq. (7.49):

$$\mathbf{m}' \cdot \boldsymbol{\beta} = \mathbf{m} \cdot \boldsymbol{\beta} \tag{7.51}$$

since:

$$\mathbf{n} \cdot \boldsymbol{\beta} = 0$$

Therefore:

$$\frac{h_0}{h} = \frac{\cos \lambda_0}{\cos \lambda} \tag{7.52}$$

and also, from Eq. (7.49):

$$\mathbf{m}' \times \mathbf{n} = \mathbf{m} \times \mathbf{n} \tag{7.53}$$

Hence:

$$\frac{h_0}{h} = \frac{\sin \phi_0}{\sin \phi} \tag{7.54}$$

Finding the modulus of each side of Eq. (7.49), we have:

$$|\mathbf{m}'|^2 = |\mathbf{m}|^2 + 2\gamma |\mathbf{m}|^2 \cos \phi_0 \cos \lambda_0 + \gamma^2 |\mathbf{m}|^2 \cos^2 \lambda_0 \tag{7.55}$$

and using Eq. (7.50):

$$\left(\frac{h_0}{h}\right)^2 = 1 + 2\gamma \cos \phi_0 \cos \lambda_0 + \gamma^2 \cos^2 \lambda_0 \tag{7.56}$$

Since λ_0 is obtuse, because we have specified that $(\boldsymbol{\beta} \cdot \mathbf{m})$ is negative, it is apparent from this equation that $h_0 < h$. Forming $\mathbf{m}' \cdot \mathbf{n} = \mathbf{m} \cdot \mathbf{n} + \gamma \, (\boldsymbol{\beta} \cdot \mathbf{m})$ and using Eqs. (7.50) and (7.52):

$$\gamma = \frac{\cos \phi}{\cos \lambda} - \frac{\cos \phi_0}{\cos \lambda_0} \tag{7.57}$$

The resolved shear stress τ is given by:

$$\tau = \frac{F}{A}\cos\phi\cos\lambda = \frac{F}{A}\left(\frac{h}{h_0}\right)\cos\lambda_0\left[1 - \left(\frac{h}{h_0}\right)^2\sin^2\phi_0\right]^{1/2} \tag{7.58}$$

Since $A_0 h_0 = Ah$, Eq. (7.58) may be written as:

$$\tau = \frac{F}{A}\left(\frac{h}{h_0}\right)^2\cos\lambda_0\left[1 - \left(\frac{h}{h_0}\right)^2\sin^2\phi_0\right]^{1/2} \tag{7.59}$$

In compression, the normal to the glide plane approaches the direction of compression, so that, for example, in Figure 7.13, $\mathbf{F}_0$ would approach the (111) pole until it meets the boundary between [001] and [011]; thereafter, systems B IV and A III would be equally stressed and the stable orientation of the normal to the glide plane would be the normal to (011), the plane midway between A and B lying in the same zone as A and B.

Formulae for the change in orientation of the crystal lattice with respect to the direction of an applied force have been given for *duplex* slip in both compression and tension by Bowen and Christian [46]. During double glide there is a change in orientation of the lattice of the crystal with respect to the direction of a tensile force until the applied force is symmetrically related to the two slip systems. Thus, as we have seen, in tension of c.c.p. metals, $\langle 112 \rangle$ directions represent stable orientations lying midway between two $\langle 110 \rangle$ directions.

Problems

7.1 α-ZnS has the point group $\bar{4}3m$.
 (a) Does it possess a centre of symmetry? Assume that the slip planes are $\{111\}$ and the slip directions $\langle 1\bar{1}0 \rangle$.
 (b) Is slip in the $[1\bar{1}0]$ direction crystallographically equivalent to that in $[\bar{1}10]$?

7.2 During the process of glide the displacements imposed upon the surfaces of a crystal are centrosymmetric. Use this notion and Figure 7.5 to deduce the rules (1)–(4) given in Section 7.2.

7.3 Enumerate the number of physically distinct glide systems for a cubic crystal (e.g. a c.c.p. metal) of point group $m\bar{3}m$ which slips on $\{111\}$ in $\langle 1\bar{1}0 \rangle$ directions. Along which directions should such a crystal be stressed in uniaxial tension if the resolved shear stress is to be the same on (a) eight systems, (b) six systems and (c) four systems? Show that for a given tensile axis $[uvw]$, the Schmid factor $\cos\phi\cos\lambda$ for the slip system which operates must satisfy the condition:

$$\frac{\sqrt{2}}{3\sqrt{3}} \leq \cos\phi\cos\lambda \leq 0.5$$

Hence or otherwise, determine the individual values of ϕ and λ when $\cos\phi\cos\lambda = 0.5$.

7.4 Establish the number of independent slip systems possessed by CsBr, which is cubic ($Pm\bar{3}m$) and slips on $\{110\}$ in $\langle 001 \rangle$. How many independent systems does such a crystal possess if pencil glide occurs with the same glide directions?

7.5 Consider an amount of glide γ to occur on the (011) plane in the $[01\bar{1}]$ direction of a cubic crystal. Take axes parallel to the cubic axes.
(a) Write down the components of the pure strain tensor.
(b) Write down the components of the pure strain tensor if glide of amount γ' occurs on (100) in the same direction, $[01\bar{1}]$, as before.
(c) How many components of the pure strain tensor can be altered independently of one another by glide on these two systems?
(d) If γ' was always equal to a constant fraction of γ so that $\gamma' = k\gamma$, what would be the answer to (c)?
(e) Use the above results to prove generally that if pencil glide occurs in a crystal then slip with a given slip direction can provide two independent components of the pure strain tensor.

7.6 Can you think of three slip directions which operating together would provide a contradiction of the statement 'When pencil glide occurs a crystal need only possess in general three non-coplanar slip directions in order for a general strain to be possible'?

7.7 Copper is cubic ($Fm\bar{3}m$) and slips on $\{111\}$ $\langle 1\bar{1}0 \rangle$. Sketch a stereogram of a crystal showing the components of the slip systems. Which slip system(s) are expected to operate if the tensile axis is (a) [123], (b) [113] and (c) [122]?

7.8 Rock salt is cubic ($Fm\bar{3}m$) and slips on $\{110\}$ $\langle 1\bar{1}0 \rangle$. Sketch a stereogram showing the components of the slip systems. Which slip system(s) are expected to operate if the tensile axis is (a) [123], (b) [213] and (c) [013]?

7.9 Magnesium oxide is cubic with the NaCl structure and slips on $\{110\}$ in $\langle 1\bar{1}0 \rangle$.
(a) List the directions along which a tensile or compressive force applied to the crystal cannot produce glide.
(b) Can the crystal be twisted plastically about [100]?
(c) List the same directions as in (a) for graphite. (The glide elements for graphite are listed in Table 7.1.)

7.10 The analysis of strain in Chapter 6 requires that the strain be infinitesimal. Explain why this restriction does not invalidate the analysis of Section 7.3 (which shows that five independent glide systems are necessary for a general deformation) because, when glide takes place, large strains occur in some volumes of the crystal.

7.11 A tetragonal crystal is subject to a tensile stress of magnitude σ along [100] and to a shear stress τ on (100) in [010].
(a) Write down the components of the stress tensor referred to the crystal axes.
(b) Find the shear stress on (110) in the $[1\bar{1}0]$ direction by altering the axes of reference of the stress tensor.
(c) Find, by the same procedure, the tensile stress normal to (110).

7.12 Prove Eq. (7.30).

7.13 A single crystal of zinc which slips on (0001) in $\langle 11\bar{2}0 \rangle$ directions is stressed in tension and is oriented so that the tensile axis makes an angle of 45° with [$11\bar{2}0$] and is at 60° to the normal to (0001).

(a) Along which direction will slip first occur?

(b) If slip starts when the tensile stress on the crystal is 2.1 MPa, calculate the critical resolved shear stress necessary to produce slip in zinc.

7.14 Copper slips on {111} in $\langle 1\bar{1}0 \rangle$. A cylindrical single crystal 10 cm long is stressed in tension. The tensile axis is [$\bar{1}23$].

(a) Which is the slip system with the highest resolved shear stress? The specimen is strained in tension until the tensile axis reaches a symmetry position where two slip systems have equal resolved shear stresses.

(b) Which are the two systems?

(c) What is the orientation of the tensile axis at this stage?

(d) What is the new length of the specimen?

7.15 Discuss the following statement. 'A general small homogeneous change of shape of a body can be described by a pure strain and a rotation. If the volume of the body is conserved the pure strain tensor has five independent components and the pure rotation tensor has three. Therefore, eight quantities must be specified to describe the change of shape of a single crystal within a polycrystalline body. It follows from this that eight slip systems must in general operate in all grains of a polycrystal which is subject to a general change of shape.'

Suggestions for Further Reading

Chin, G.Y., Thurston, R.N. and Nesbitt, E.A. (1966). Finite plastic deformation due to crystallographic slip. *Trans. Am. Inst. Min. Metall. Petrol. Engrs.* 236: 69–76.

Dieter, G.E. (1988). *Mechanical Metallurgy* (SI metric edition adapted by David Bacon). London: McGraw-Hill.

Groves, G.W. and Kelly, A. (1963). Independent slip systems in crystals. *Phil. Mag.* 8: 877–887.

McGregor Tegart, W.J. (1966). *Elements of Mechanical Metallurgy*. New York: Macmillan.

Schmid, E. and Boas, W. (1950). *Plasticity of Crystals*, London: F.A. Hughes and Co. Ltd.

References

[1] Reusch, E. (1867). Ueber eine besondere Gattung von Durchgängen im Steinsalz und Kalkspath. *Poggendorff's Annalen der Physik und Chemie* 132: 441–451.

[2] Ewing, J.A. and Rosenhain, W. (1899). Experiments in micro-metallurgy:— Effects of strain. Preliminary notice. *Proc. Roy. Soc. Lond.* A 65: 85–90.

[3] Ewing, J.A. and Rosenhain, W. (1900). The crystalline structure of metals. *Phil. Trans. Roy. Soc. Lond.* A 193: 353–375.

[4] Groves, G.W. and Kelly, A. (1963). Independent slip systems in crystals. *Phil. Mag.* 8: 877–887.

[5] Anderson, P.M., Hirth, J.P. and Lothe, J. (2017). *Theory of Dislocations*, 3rd Edition. New York: Cambridge University Press.

[6] Nabarro, F.R.N. (1987). *Theory of Crystal Dislocations*. New York: Dover Publications.

[7] Bacon, D.J. and Vitek, V. (2002). Atomic scale modelling of dislocations and related properties in the hexagonal-close-packed metals. *Metall. Mater. Trans A* 33: 721–733.

[8] Kelly, A. and Rowcliffe, D.J. (1966). Slip in titanium carbide. *Physica Status Solidi* 14: K29–K33.

[9] Eyre, B.L. and Bartlett, A.F. (1966). Determination of the Burgers vector of dislocations in deformed uranium carbide. *Phil. Mag.* 13: 641–643.

[10] Skrotzki, W., Tamm, R., Oertel, C.-G., Röseberg, J. and Brokmeier, H.-G. (2000). Microstructure and texture formation in extruded lead sulfide (galena). *J. Struct. Geol.* 22: 1621–1632.

[11] Rachinger, W.A. (1956). Glide in lead telluride. *Acta Metall.* 4: 647–649.

[12] Byron, J.F. (1968). The yield and flow of single crystals of uranium oxide. *J. Nucl. Mater.* 28: 110–114.

[13] Mitchell, T.E. (1979). Application of transmission electron microscopy to the study of deformation in ceramic oxides. *J. Am. Ceram. Soc.* 62: 254–267.

[14] Mitchell, T.E. (1999). Dislocations and mechanical properties of MgO–Al$_2$O$_3$ single crystals. *J. Am. Ceram. Soc.* 82: 3305–3316.

[15] Yang, K.-H., Ho, N.-J. and Lu, H.-Y. (1999). Deformation microstructure in (001) single crystal strontium titanate by Vickers indentation. *J. Am. Ceram. Soc.* 92: 2345–2353.

[16] Levade, C., Couderc, J.-J., Dudouit, I. and Garigue, J. (1986). The plastic behaviour of natural sphalerite crystals between 473 and 873 K. *Phil. Mag. A* 54: 259–279.

[17] Kedjar, B., Thilly, L., Demenet, J.-L. and Rabier, J. (2010). Plasticity of indium antimonide between −176 °C and 400 °C under hydrostatic pressure. Part II: Microscopic aspects of the deformation. *Acta Mater.* 58: 1426–1440.

[18] Androussi, Y., Vanderschaeve, G. and Lefebvre, A. (1989). Slip and twinning in high-stress-deformed GaAs and the influence of doping. *Phil. Mag. A* 59: 1189–1204.

[19] Amelinckx, S. and Delavignette, P. (1960). Electron optical study of basal dislocations in graphite. *J. Appl. Phys.* 31: 2126–2135.

[20] Yang, F. and Li, J.C.M. (2007). Deformation behavior of tin and some tin alloys. *J. Mater. Sci.: Mater. Electron.* 18: 191–210.

[21] Ashbee, K.H.G. and Smallman, R.E. (1963). The plastic deformation of titanium dioxide single crystals. *Proc. Roy. Soc. Lond. A* 274: 195–205.

[22] Suzuki, K., Ichihara, M. and Takeuchi, S. (1991). High-resolution electron microscopy of lattice defects in TiO$_2$ and SnO$_2$. *Phil. Mag. A* 63: 657–665.

[23] Georgieff, M. and Schmid, E. (1926). Über die Festigkeit und Plastizität von Wismutkristallen. *Z. Phys.* 36: 759–774.

[24] Slonaker, R.E., Smutz, M., Jensen, H. and Olson, E.H. (1965). Factors affecting the growth and the mechanical and physical properties of bismuth single crystals. *J. Less-Common Met.* 8: 327–338.

[25] Steegmuller, C. and Daniel, J.S. (1972). Slip in bismuth. *J. Less-Common Met.* 27: 81–85.

[26] Rider, J.G. and Heckscher, F. (1966). Slip in single crystals of mercury. *Phil. Mag.* 13: 687–692.

[27] Snow, J.D. and Heuer, A.H. (1973). Slip systems in Al$_2$O$_3$. *J. Am. Ceram. Soc.* 56: 153–157.

[28] Stokes, R.J., Johnston, T.L. and Li, C.H. (1961). Mechanical properties of tellurium single crystals. *Acta Metall.* 9: 415–427.

[29] Spooner, F.J. and Wilson, C.G. (1966). Slip and twinning in gallium. *J. Less-Common Met.* 10: 169–184.

[30] Daniel, J.S., Lesage, B. and Lacombe, P. (1971). The influence of temperature on slip and twinning in uranium. *Acta Metall.* 19: 163–173.

[31] Stokes, R.J. and Li, C.H. (1963). Dislocations and the strength of polycrystalline ceramics. In: *Materials Science Research, Volume 1: the proceedings of the research conference on structure and properties of engineering materials*, held March 12–13 1962 at North Carolina State College, (eds. H.H. Stadelmaier and W.W. Austin), pp. 133–158. New York: Plenum Press.

[32] Tyson, W.G. (1969). Slip modes of thallium. *Acta Metall.* 17: 863–868.

[33] Carnahan, T.G. and Scott, T.E. (1973). Deformation modes of hcp yttrium at 77, 298 and 497 K. *Metall. Trans.* 4: 27–32.

[34] Vitek, V. (1992). Structure of dislocation cores in metallic materials and its impact on their plastic behaviour. *Prog. Mater. Sci.* 36: 1–27.

[35] von Mises, R. (1928). Mechanik der plastischen Formänderung von Kristallen. *Zeitschrift für Angewandte Mathematik und Mechanik* 8: 161–185.

[36] Schmid, E. (1925). Neuere Untersuchungen an Metallkristallen. In: *Proc. 1st. International Congress for Applied Mechanics, Delft, 1924* (eds. C.B. Biezeno and J.M. Burgers), pp. 342–353. Delft: Technische Boekhandel en Drukkerij J. Waltman Jr.

[37] Schmid, E. and Boas, W. (1950). *Plasticity of Crystals*, London: F.A. Hughes and Co. Ltd. [Originally published as *Kristallplastizität mit besonderer Berücksichtigung der Metalle*, Julius Springer, Berlin in 1935.]

[38] Taylor, G.I. and Elam, C.F. (1925). The plastic extension and fracture of aluminium crystals. *Proc. Roy. Soc. Lond. A* 108: 28–51.

[39] Chin, G.Y. (1972). Competition among {110}, {112} and {123} <111> slip modes in bcc metals. *Metall. Trans.* 3: 2213–2216.

[40] Orlans-Joliet, B., Bacroix, B., Montheillet, F., Driver, J.H. and Jonas, J.J. (1988). Yield surfaces of b.c.c. crystals for slip on the {110}<111> and {112}<111> systems. *Acta Metall.* 36: 1365–1380.

[41] Diehl, J. personal communication to Seeger, A., acknowledged on p. 24 of Seeger, A. (1958) Kristallplastizität, *Handbuch der Physik*, Band VII.2. In *Kristallphysik II* (ed. S. Flügge), pp. 1–210. Berlin: Springer-Verlag.

[42] Hutchings, I.M. (1993). Quick non-graphical method for deducing slip systems in cubic close packed metals in tension or compression. *Mater. Sci. Tech.* 9: 929–930.

[43] Price, R.J. and Kelly, A. (1964). Deformation of age-hardened aluminium alloy crystals–I. Plastic flow. *Acta Metall.* 12: 159–169.

[44] von Göler, Freiherr and Sachs, G. (1929). Zugversuche an Kristallen aus Kupfer und α-Messing. *Z. Phys.* 55: 581–620.

[45] Seeger, A. (1958). Kristallplastizität, *Handbuch der Physik*, Band VII.2. In: *Kristallphysik II* (ed. S. Flügge), pp. 1–210. Berlin: Springer-Verlag.

[46] Bowen, D.K. and Christian, J.W. (1965). The calculation of shear stress and shear strain for double glide in tension and compression. *Phil. Mag.* 12: 369–378.

8

Dislocations

8.1 Introduction

In Chapter 7 the geometry of glide was studied on a macroscopic scale. The geometrical results on this scale are adequately described by the model of blocks of crystal sliding over one another rigidly. However, when glide occurs in a real crystal, all the atoms above the slip plane do not move simultaneously over those below. At any given time some of the atoms have moved into their new positions while others have not yet done so, and the displacement of the upper block of crystal relative to the lower varies from one region of the slip plane to another. Lines in the slip plane separating regions where slip has occurred from those where it has not are called *dislocations* [1–3].[1] This is the simplest way in which to introduce the idea of dislocations; we shall see later that they can be formed in other ways.

For ease of illustration and simplicity we will consider slip in a crystal which has the primitive cubic lattice and one atom associated with each lattice point, i.e. a simple cubic crystal. A block of such a crystal cut with faces parallel to {100} planes is shown in Figure 8.1. A (001) slip plane, PQRT, is shown, and slip in the [010] direction has occurred over only the region PQSS′. It is evident that the crystal lattice is distorted, especially close to the dislocation which lies along SS′. The distortion decreases as the amount of slip over PQSS′ is reduced. In Figure 8.1 the displacement over PQSS′ is equal to [010],[2] which is the smallest displacement that can be made without producing a fault in the crystal structure at the plane PQSS′. The displacement is a constant vector quantity over the whole region PQSS′ (except within a few atoms of the dislocation line SS′), and it is this quantity that characterizes the dislocation. As a dislocation moves through the crystal, the displacement

---

[1] The term 'dislocation' was first used in this context by Love [4, 5].
[2] A specific displacement is written in terms of its components along the crystal axes (Section 1.2). In cubic crystals the lattice parameter a is often included in the notation, although it is superfluous.

Crystallography and Crystal Defects, Third Edition. Anthony Kelly and Kevin M. Knowles.
© 2020 John Wiley & Sons Ltd. Published 2020 by John Wiley & Sons Ltd.
Companion website: www.wiley.com/go/kelly/crystallography3e

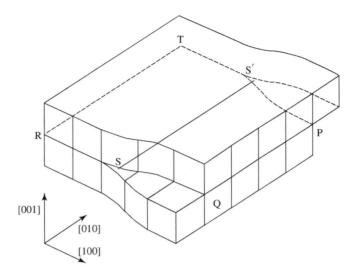

Figure 8.1 A screw dislocation in a primitive cubic lattice

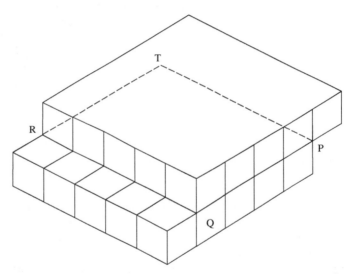

Figure 8.2 Primitive cubic lattice after the screw dislocation in Figure 8.1 has reached the surface at RT

associated with it does not change. This characteristic quantity is called the *Burgers vector* of the dislocation after Johannes (Jan) Martinus Burgers [6].[3]

As the dislocation SS′ in Figure 8.1 moves from right to left, the slipped region of the crystal increases. When the dislocation reaches the surface at RT, the whole crystal has been slipped by [010], as shown in Figure 8.2. A dislocation may turn into a new orientation;

---

[3] The term 'Burgers vector' was introduced into the scientific literature by Sir Charles Frank in 1949 [7].

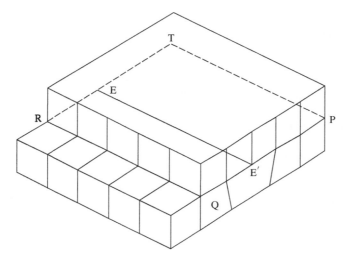

Figure 8.3   An edge dislocation in a primitive cubic lattice

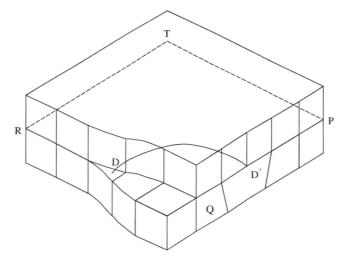

Figure 8.4   A mixed dislocation, DD′, in a primitive cubic lattice

this does not change its Burgers vector. The result of rotating SS′ in the slip plane through a right angle is shown in Figure 8.3. The dislocation which results, EE′, has the same [010] displacement vector associated with it, the crystal now having slipped by this amount over QREE′ and not having slipped over PTEE′. A length of dislocation which lies normal to its Burgers vector, such as EE′, is called an *edge dislocation*; one like SS′ which lies parallel to its Burgers vector is called a *screw dislocation*.[4] In general, a dislocation may lie at any angle to its Burgers vector or may be curved like the dislocation DD′ in Figure 8.4. At D′ the

---

[4] These two types of dislocations are also referred to as a Taylor dislocation and a Burgers dislocation, respectively, in early papers on dislocations, after Taylor [1] and Burgers [6] who first postulated their existence in crystals.

dislocation is in edge orientation and at D in screw orientation. Elsewhere the dislocation has both edge and screw components and is referred to as *mixed*.

The reason for the names 'edge' and 'screw' becomes clear when we look at the positions of the atoms around the dislocation. The positions close to the dislocation cannot be calculated exactly without a detailed knowledge of the laws of force between the atoms, although in some cases they can be observed at the surface of a crystal at atomic resolution with scanning tunnelling microscopes and atom probes, and within thin crystals using high-resolution transmission electron microscopy. For the present, it is enough to recognize that there is a small amount of arbitrariness in the location of the atoms shown in Figures 8.5 and 8.6. The arrangement of atoms in a plane normal to the edge dislocation EE' of Figure 8.3 is shown in Figure 8.5. There is an extra incomplete plane of atoms KE above the slip plane. The edge of the extra plane coincides with the dislocation and gives it its name. A similar view of a plane normal to a screw dislocation does not show the distortion of the lattice, because the displacements of the atoms are parallel to the dislocation in this case (Figure 8.6a). A view of the atom positions in the atom planes immediately above and below the slip plane is given by Figure 8.6b. The term 'screw' arises because the successive planes of atoms normal to the dislocation are converted by the presence of the dislocation into a single screw surface, or spiral ramp. This can be visualized from Figure 8.1 by imagining oneself to be walking around the dislocation on these planes. Beginning at atom 1 below the plane of the paper and continuing to atoms 2, 3, 4 and back to an atom equivalent to 1 below the plane of the paper, but displaced along SS', and so on, shows the spiral ramp nature of this surface.

Crystals normally contain a great number of dislocations, even before they are plastically deformed. The total length of dislocation line in 1 m^3 of a metallic crystal (the *dislocation density*) may typically be 10^{12} m, although a wide variation over orders of magnitude is possible, depending on the degree to which the metal has been either annealed or cold worked.

So far we have considered only dislocations occurring as a result of slip and lying in a slip plane. This is an unnecessary restriction, and dislocations can, in fact, be produced

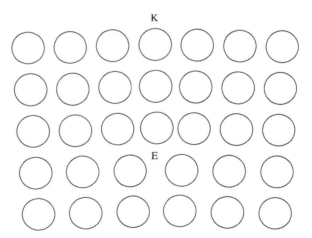

Figure 8.5   Atom positions around an edge dislocation in a simple cubic crystal (according to Eqs. (8.21a) and (8.21b))

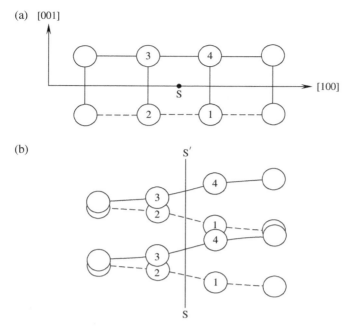

(a) [001]

[100]

(b)

Figure 8.6 Schematics of a screw dislocation in a simple cubic crystal (a) looking along the dislocation line SS′ in Figure 8.1 and (b) looking normal to SS′. In (b) SS′ is in the plane of the paper; atoms 1 and 2 are below the plane of the paper relative to SS′ and atoms 3 and 4 are above the plane of the paper. The convention used in (b) for dashed and dotted lines joining atoms has been shown in (a) to help relate these two diagrams to Figure 8.1

in ways other than by glide, as we shall see in Section 8.2. A general method is therefore needed for describing the Burgers vector – that characteristic discontinuity of displacement which defines a dislocation in a crystal.

The following method is often used. First, a closed circuit is made around the suspected dislocation, called a Burgers circuit [6]. This is a circuit made by jumping from one lattice point to a neighbouring lattice point until the starting point is reached again. The circuit must be made in 'good' material[5] – that is, in regions of the crystal in which, even though they may be strained, each jump is *recognizably* associated with a jump in a perfect crystal of the same structure. With this vital condition fulfilled, a corresponding set of jumps can then be traced in the perfect crystal; if it fails to close, the circuit in the real crystal does indeed surround a dislocation.

A little thought will show that if the circuit fails to close then it must do so by a lattice vector. The lattice vector that is needed to complete the circuit in the ideal perfect crystal is defined as the Burgers vector of the dislocation. A dislocation that can be dealt with by this procedure must have a Burgers vector equal to a translation vector of the Bravais lattice; such a dislocation is called a perfect dislocation.

---

[5] The term 'good' in this context was introduced by Sir Charles Frank following the usage suggested by John C. Slater at the request of William B. Shockley [8].

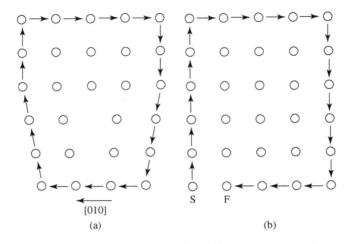

Figure 8.7 A Burgers circuit around an edge dislocation in a simple cubic crystal

This method of defining the Burgers vector is illustrated in Figure 8.7. This is a diagram of the edge dislocation at E in Figure 8.3, with the dislocation line EE' parallel to [100] running into the paper, with respect to the axes in Figure 8.1. The circuit of 17 successive jumps in the real crystal in (a) has a corresponding set of jumps in the perfect crystal in (b) which fails to close. The vector [010] is needed to complete the circuit in the ideal crystal; this is the Burgers vector.

Some sign convention is needed to define the *sense* of the Burgers vector. The FS/RH convention works as follows. First, a sense is arbitrarily attached to the dislocation line. This permits a choice of Burgers circuit with a right-hand screw relationship to the line (the RH of the notation). In Figure 8.7 the line is taken to run into the paper. The Burgers vector is then taken to run from the finish to the start of the path in the perfect crystal (the FS of the notation); for example, from F to S in Figure 8.7(b).[6] If the sense of the line is reversed, the Burgers vector is also reversed, so that some consistency in choosing a line sense is desirable.

As an example, consider a closed loop of dislocation in a primitive cubic lattice lying in a slip plane (Figure 8.8a). Suppose that within the loop the top half of the crystal has slipped by [010] with respect to the bottom half, as shown by the arrow. Then the positions of the atoms in the (100) plane (the plane α in Figure 8.8a) are as shown in Figure 8.8b. Since the edge dislocations at E_1 and E_2 would annihilate one another if they glided together (their extra planes would join), they are commonly referred to as edge dislocations of opposite sign. The Burgers vector at E_2 is in the opposite direction to that at E_1 if the direction assigned to the line at E_2 is the same as that assigned to the line at E_1, such as into the page. This would be perfectly fine if the dislocation lines at E_1 and E_2 were actually parallel lines.

If the whole loop is being considered, however, it is more logical to make the sense of the line continuous around the loop. Then, if the direction of the line at E_1 is into the page, the direction of the line at E_2 must be out of the page, and when the Burgers circuits round E_1 and E_2 are each taken in the RH sense, the same sense of Burgers vector is obtained

---

[6] The opposite convention, SF/RH, is used by some authors. For an account of where one convention has been used, and where the other, see de Wit [9].

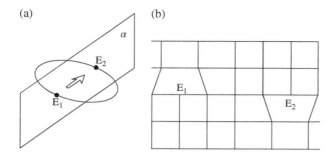

(a)   (b)

Figure 8.8   Schematic of a closed dislocation loop in a simple cubic crystal

at E_1 as at E_2. This second way of stating the Burgers vector emphasizes the fact that the displacement discontinuity, which the Burgers vector measures, remains constant along a given dislocation line. This follows because if we move the Burgers circuit along the dislocation line it must remain in good material.

The principle of the conservation of the Burgers vector is a general one, and from it, it follows that a dislocation line cannot end inside a crystal. It must end at a surface, form a closed loop or meet other dislocations. A point at which dislocation lines meet is called a *node* [8]. The conservation of the Burgers vector in this case can be stated in the following way. If the directions of all the dislocation lines are taken to run out from the node then the sum of the Burgers vectors of all the dislocations is zero. This is the dislocation analogue of Kirchhoff's law of the conservation of electric current in a network of conductors.

8.2   Dislocation Motion

We have seen that the motion of a dislocation on a slip plane changes the area over which slip has occurred. In this section the effect of dislocation motion in general will be discussed. The treatment here follows closely that given in [8].

Consider two neighbouring atoms, P and Q, lying at lattice points in a plane which is threaded by a dislocation DD′ (Figure 8.9a). Suppose the dislocation passes between P and Q so as to move inside a Burgers circuit, part of which is the step from P to Q. The corresponding circuit in the ideal crystal then acquires a closure failure equal to the Burgers vector, **b**. This must be attributed to a displacement of P relative to Q of **b** in the real crystal – it cannot be attributed to the relative displacement of the members of any other neighbouring pair of atoms in the Burgers circuit because it is possible to draw Burgers circuits through these that do not contain the dislocation either before or after its movement.

The determination of the sense of the relative displacement requires some care. Define a positive direction **QP** by the cross product $\mathbf{l} \times \mathbf{d}$ (Figure 8.9b); that is, **l**, **d** and **QP** are related as a right-handed set of axes x, y, z. Here **l** is the vector parallel to the dislocation line used in defining the Burgers vector **b** by the FS/RH convention; **d** is a vector in the direction of motion of the dislocation. With this definition, the displacement of the atom on the positive side, P, relative to that on the negative, Q, is +**b** when the dislocation passes between P and Q, as shown. We shall use this result to study the relative displacement of material on either side of the surface that a dislocation sweeps out during its motion.

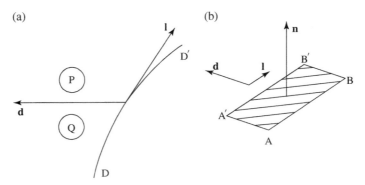

Figure 8.9 The motion of a dislocation DD′ in the direction represented by the vector **d** between two atoms, P and Q, lying at lattice points that are neighbours prior to the passage of the dislocation between them

Suppose a segment of a dislocation line AB moves to A′B′, sweeping out a plane surface whose unit normal is **n**. The sense of **n** is defined in accordance with the previous paragraph as the sense of **l** × **d** (Figure 8.9b). The atoms above ABB′A′ are displaced by **b** relative to those below.

If **b** · **n** is positive, the atoms above ABB′A′ move away from those below, by, for example, the production of lattice vacancies, thereby creating a void, whose volume is **b** · **n** per unit area swept out. Conversely, motion for which **b** · **n** < 0 requires the insertion of extra material of volume **b** · **n** per unit area. When **b** · **n** ≠ 0 the dislocation motion is normally accompanied by diffusion, which brings in atoms to fill the void when **b** · **n** > 0 and disperses the extra atoms when **b** · **n** < 0. Dislocation motion of this type is called 'climb'. The origin of this name can be seen from Figure 8.5. If the atoms on the edge of the extra plane are removed, the edge dislocation in Figure 8.5 will climb up out of its slip plane. Climb downwards could be accomplished by adding atoms to the extra plane.

Dislocation glide is motion that does not involve the addition or removal of material; that is, motion for which **b** · **n** = 0. The glide surface is defined accordingly as that surface which contains the dislocation line and whose normal is everywhere perpendicular to the Burgers vector. For a straight-edge dislocation, this defines a single plane, which is the slip plane. For a screw dislocation any plane passing through the dislocation fulfils the condition **b** · **n** = 0.

For a closed loop of dislocation a cylinder is defined as the glide surface. The surface of the cylinder is generated by lines parallel to the Burgers vector passing through all points on the dislocation (Figure 8.10a). The glide of the loop is not confined to the surface of its glide cylinder, because if it has any screw parts they can glide on any plane; however, the area of the loop in projection on a plane normal to the axis of the cylinder remains constant (Figure 8.10b). This area is zero for a loop on a slip plane (Figure 8.8a). When the area is not zero, the loop is called 'prismatic'. A special case is a pure-edge dislocation loop. This could be produced by prismatic punching [10], as shown in Figure 8.11, in which case its structure corresponds to a penny-shaped disc of extra atoms in the plane of the loop. With a Burgers vector of opposite sense, its structure corresponds to a disc of missing atoms. In the latter case, the loop could be formed by collecting together a disc of vacant lattice sites.

(a)

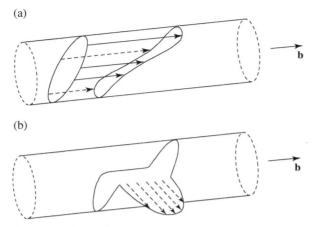

(b)

Figure 8.10   The dislocation loop shown in (a) can glide so that its area projected normal to the Burgers vector does not change. It moves in the surface of a cylinder. (b) This shows that a screw segment can leave the surface of this cylinder without changing the projected area

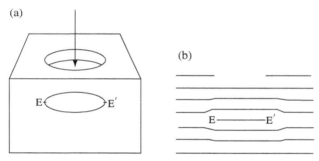

Figure 8.11   (a) Production of a prismatic dislocation loop by punching. (b) Section normal to the plane of the loop and passing through E and E′

When a dislocation glides, there is no problem of acquiring or disposing of extra atoms. This is in sharp distinction to climb, and it leads us to suppose that a dislocation will be able to glide much more quickly than it can climb. The ease of gliding may be quite different for different crystallographic surfaces, and may be so much easier for one crystallographic plane that slip is effectively confined to that plane.

8.3   The Force on a Dislocation

In Section 8.1 we found that when a dislocation moved on a slip plane, the area of the slip plane across which slip had occurred either increased or decreased, according to the direction of motion. As a result, work can be done by forces applied to the crystal. For example, suppose forces acting in the slip direction are applied to the top and bottom surfaces of the crystal in Figure 8.1, in the directions SS′ on the top surface and S′S on the bottom. When the dislocation, of unit length say, moves to the left and sweeps out an area 1.dx of slip plane, applied forces of magnitude σdx move through the slip distance b, if the applied forces are of strength σ per unit area of the top and bottom surfaces. The work done is then:

$$dw = \sigma b \, dx \tag{8.1}$$

We can use this result to define a force[7] acting on a unit length of the dislocation in the direction Ox:

$$F = -\frac{dE}{dx} \tag{8.2}$$

where E is the energy of the system, which comprises the crystal and the device applying the forces to it. If we examine the work done by the forces acting on the surfaces at the slip plane, which are displaced relative to one another when the dislocation moves a distance dx, we see that it has the same magnitude $\sigma b \, dx$ as the work done by the externally applied forces, but is opposite in sign. This work done at the slip plane is dissipated, like the work done at sliding surfaces in a friction problem. The dissipated energy is supplied by the device applying the forces to the crystal, which therefore loses energy:

$$-dE = dw = \sigma b \, dx \tag{8.3}$$

Substituting Eq. (8.2) in Eq. (8.3) gives:

$$F = \sigma b \tag{8.4}$$

This definition of the force per unit length acting on a dislocation can be extended to the general case of a dislocation moving in any direction in a crystal which is stressed internally or by external forces.

To do this we use the result of Section 8.2 that when a segment of dislocation line sweeps out an area of internal surface, the material on the positive side of that surface is displaced by **b** relative to the material on the negative side. The crystal is conceived to be cut at this surface. The force acting on one of the surfaces of the cut is calculated from the stress in the crystal, and the work done by this force on the crystal during the displacement is then calculated, the other surface being held fixed.

The dislocation will be driven in such a direction that this work is negative. In the absence of external forces the energy of the crystal will then decrease (its self-stress will be reduced), and the magnitude of the force on the dislocation is defined by the decrease in the energy of the crystal when the dislocation moves a unit distance. If the dislocation is being driven by externally applied forces, this energy loss is made up by the work done on the crystal by the applied forces; the energy of the crystal itself does not change. The decrease in energy of a larger system which includes the device applying the force can then be considered to define the force acting on the dislocation, and identical equations are obtained.

Using the sign convention of Section 8.2, the force **F** acting on a unit area of the surface bounding the negative side of the cut is given by (see Section 6.3):

$$F_i = \sigma_{ij} n_j \tag{8.5}$$

---

[7] Defining a 'force' exerted on a dislocation line is a way of describing the tendency of the configuration which constitutes the dislocation to move through the crystal. There is no physical force on a dislocation line in the sense in which there is a force on a rod when a string attached to it is pulled, for example.

or, in vector notation:

$$\mathbf{F} = \sigma \cdot \mathbf{n} \tag{8.6}$$

This force is exerted on the material on the negative side of the cut by the material on the positive side of the swept-out surface. The stress is σ and $\mathbf{n}$ is a unit vector normal to the swept-out surface. The sense of $\mathbf{n}$ is given by $\mathbf{l} \times \mathbf{d}$, where $\mathbf{l}$ is a unit vector parallel to the dislocation line and $\mathbf{d}$ is a unit vector in the direction of motion (Section 8.2 and Figure 8.9). Without loss of generality, we can take $\mathbf{d}$ perpendicular to $\mathbf{l}$, since any motion of the dislocation parallel to itself sweeps out no area. Therefore:

$$\mathbf{n} = \mathbf{l} \times \mathbf{d} \tag{8.7}$$

The movement of the dislocation displaces the surface bounding the negative side of the cut by $-\mathbf{b}$ if the opposite surface bounding the positive side of the cut remains at rest, so that the work done on the crystal is:

$$w = -(\sigma \cdot \mathbf{n}) \cdot \mathbf{b} \tag{8.8}$$

which, because σ is a symmetrical tensor, can be written in suffix notation (Appendix 1) as:

$$w = -\sigma_{ij} n_j b_i = -\sigma_{ji} b_i n_j$$

That is, in vector notation:

$$w = -(\sigma \cdot \mathbf{b}) \cdot \mathbf{n} \tag{8.9}$$

The decrease in energy, $-w$, is given by the negative of this, so:

$$-w = (\sigma \cdot \mathbf{b}) \cdot (\mathbf{l} \times \mathbf{d})$$

or:

$$-w = (\sigma \cdot \mathbf{b}) \times \mathbf{l} \cdot \mathbf{d} \tag{8.10}$$

using the theorems on the scalar triple product of vectors (Appendix 1).

We now have the work done by the crystal when a unit length of dislocation moves a unit distance in the form:

$$-w = \mathbf{F} \cdot \mathbf{d} \tag{8.11}$$

where $\mathbf{d}$ is a unit vector in the direction of motion. $\mathbf{F}$ is therefore the force per unit length acting on the dislocation, so:

$$\mathbf{F} = (\sigma \cdot \mathbf{b}) \times \mathbf{l} \tag{8.12}$$

This equation is often referred to as the Peach–Koehler equation [11]. In this equation, the expression $\sigma \cdot \mathbf{b}$ represents the force per unit length acting on a plane normal to the Burgers vector, of area b. Since $\mathbf{l}$ is a unit vector parallel to the dislocation, we see that the force is always normal to the dislocation line.

As an example of the application of this formula, consider the edge dislocation of Figure 8.12 in a region of general stress σ. The stress may be the result of forces applied to

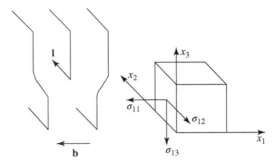

Figure 8.12 An edge dislocation in a region of general stress σ showing the stresses on the surface whose outward normal is **b**

the exterior of the crystal, or it may be a local self-stress.[8] Suppose that the dislocation runs along the x_2-axis and that its Burgers vector is in the direction $-x_1$. It is evident that the glide plane is the plane containing these two directions. The traction on the surface whose outward normal is **b** is the result of components due to σ_{11}, σ_{12} and σ_{13} (Figure 8.12). The component of $(\sigma \cdot \mathbf{b})$ due to σ_{12} is parallel to **l** and so makes no contribution to $(\sigma \cdot \mathbf{b}) \times \mathbf{l}$.

The component σ_{11} produces a traction in the direction of **b** and a component of force on the dislocation of magnitude $\sigma_{11}b$ acting downwards, perpendicular to the glide plane. If σ_{11} is negative, the force on the dislocation is directed upwards. This confirms the intuition that a compression along the x_1-axis will tend to squeeze the extra plane out of the crystal, and that a tensile stress will tend to draw it downwards. The component of the force normal to the glide plane is called the climb force.

The glide force acts in the glide plane, and in this case it is given by the shear component σ_{13}, which produces a force per unit length of magnitude $\sigma_{13}b$ directed along Ox_1. This is completely consistent with glide arising through volume-conserving shear processes (Chapter 7). Sometimes it is only important to know the glide force; that is, the component of force in the slip plane. As can be seen from Eq. (8.8), the glide force on any dislocation is simply b times the component of the traction on the slip plane that acts in the slip direction.

When an edge dislocation climbs under the influence of a stress, a void or extra material is created, and it is intuitively clear that the climb will be brought to a halt if the extra material is allowed to accumulate, or if the void is not filled in. This leads to the idea of a 'chemical' force on a dislocation, due, for instance, to the absence of atoms on some of the sites in the surrounding lattice. To understand this force we must anticipate some results in Chapter 10.

It is shown in Chapter 10 that a crystal in *equilibrium* contains a certain number of empty atomic sites, called vacancies. To define the chemical force in this instance we write down the equilibrium concentration of vacancies in the presence of an edge dislocation under stress. When the concentration has this equilibrium value the dislocation has no tendency to create or annihilate vacancies by climbing, and so we define a chemical force equal and opposite to the climb force exerted by the stress.

---

[8] A self-stress, sometimes called an internal stress, is a stress that exists within a crystal that is free from externally applied forces.

Suppose that when a vacancy is created by adding a lattice atom to the edge of the extra plane of an edge dislocation, a length αb of dislocation climbs by βb. Here α and β are constants of order unity which depend upon the structure of the crystal. Then when a vacancy is created, a stress that exerts a climb force F per unit length does work W[9] given by:

$$W = F\alpha\beta b^2 \tag{8.13}$$

Consequently, the equilibrium concentration of vacancies is increased from its value C_0 in the absence of the climbing force[10] to:

$$C = C_0 \exp\left(\frac{F\alpha\beta b^2}{kT}\right) \tag{8.14}$$

where k is Boltzmann's constant and T is absolute temperature. We define a chemical (or osmotic) force equal and opposite to F as being exerted by the concentration C of vacancies through the equation:

$$-F = -\frac{kT}{\alpha\beta b^2} \ln \frac{C}{C_0} \tag{8.15}$$

In the absence of stress, a concentration C which is not equal to its equilibrium value C_0 gives a chemical force that is not balanced by a force due to stress. The magnitude of this force is given by Eq. (8.15). The dislocation will then climb upwards in Figure 8.12 if $C/C_0 > 1$; that is, in the presence of a 'supersaturation' of vacancies.

It is interesting to compare the magnitude of the 'chemical stress' F/b with the magnitude of a typical mechanical stress. Setting $\alpha = \beta = 1$, $\mu b^3 \sim 5$ eV,[11] where μ is the shear modulus Eq. (6.82) and $kT \approx 0.025$ eV at room temperature, we obtain:

$$\frac{F}{b} \sim \frac{\mu}{200} \ln \frac{C}{C_0} \tag{8.16}$$

Many crystals cannot support an applied stress as large as $\mu/200$, so quite small supersaturations ($C/C_0 \sim 3$) produce a force larger than that arising from any applied stress. Larger supersaturations, such as might rise from quenching from close to the melting point, can cause small prismatic dislocation loops to be produced. These have been observed, for example, in aluminium [13, 14].

8.4   The Distortion in a Dislocated Crystal

It is important to know exactly how the structure of a crystal is distorted by a dislocation. This information is needed in tackling such problems as how dislocations interact with each other and with other lattice defects. An obvious way to approach the problem is to investigate an elastic continuum containing a dislocation. If a satisfactory solution is found for the dislocated continuum, the atoms in a crystal can then be assigned the displacements

---

[9] Here we are neglecting work that will be done by the hydrostatic component of the stress if the volume of the crystal changes when an atom is removed from an interior site. For the effect of this, see Weertman [12].

[10] See Section 10.1.

[11] The eV (electron volt) is a unit of energy equal to 1.602×10^{-19} J.

that are suffered by the corresponding points in the continuum. As a first approximation, isotropic elasticity can be assumed.

For a screw dislocation in an isotropic continuum, the problem is simple enough for an intuitive approach to give the correct solution. We expect the displacement at any point to be parallel to the direction of the line of the dislocation. In addition, because of the radial symmetry of the situation, a total displacement b must be spread out evenly over a circular path around the dislocation. This suggests a displacement u_z given by:

$$u_z = \frac{b\theta}{2\pi} \tag{8.17}$$

If a path at a radial distance r from the dislocation is imagined to be straightened out, as in Figure 8.13, the shear strain is seen to be:

$$\gamma_{z\theta} = \frac{b}{2\pi r} \tag{8.18}$$

The non-zero stress component is the corresponding shear stress, given by Hooke's law, Eq. (6.85), as:

$$\sigma_{z\theta} = \sigma_{\theta z} = \frac{\mu b}{2\pi r} \tag{8.19}$$

where μ is the shear modulus defined in Eq. (6.82).

It remains only to check that the stresses given by Eq. (8.19) are such that every part of the body containing the dislocation is in equilibrium. Because the stress $\sigma_{z\theta}$ does not change with z or θ, the opposite faces of all small elements, such as the one shown in Figure 8.14, centred at a distance r away from the z-axis, are acted on by equal and opposite forces, so that the element is in equilibrium. Furthermore, since σ_{zr}, $\sigma_{\theta r}$ and σ_{rr} are everywhere zero, the surface of a cylinder around the dislocation is free of force.

A difficulty with the shear strain given by Eq. (8.18) is that it becomes infinite as r approaches zero. Since Hooke's law applies only to small strains, it is usual to cut off the solution at a radius where the strains become large, say $r \sim 5b$. The material within a cylindrical surface of this radius is called the dislocation core. The core surface is in effect a

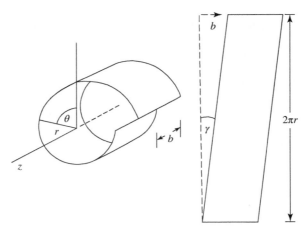

Figure 8.13 Strain due to a screw dislocation lying along the z-axis

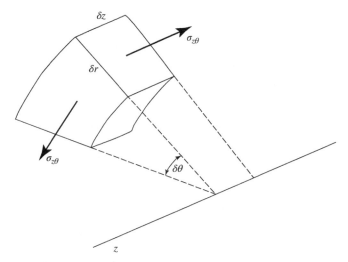

Figure 8.14 Stress due to a screw dislocation along the z-axis acting on a small element of volume $r\,dr\,d\theta\,dz$ at a distance r away from the z-axis

boundary of the elastic body. The forces that would be exerted from point to point on this boundary by the core material are not known, but it is known that no net force can be exerted in any direction, nor any couple. This condition is satisfied by Eq. (8.19), which gives a cylindrical surface that is entirely free of force. Equations (8.17) and (8.19) therefore give satisfactory displacements and stresses at distances greater than about $5b$ from a screw dislocation in an infinitely long isotropic cylinder. For a finite cylinder with free ends, the solution needs to be modified in order to remove the effect of the couple on the end surfaces that is produced by the stress $\sigma_{z\theta}$.

The problem of finding the displacements around other dislocations can be approached in the same way.[12] A displacement function is sought that gives the required discontinuity of displacement in a circuit around the dislocation and in addition satisfies the equations of equilibrium. Expressed in Cartesian coordinates, with the dislocation lying along the x_3-axis, the equations of equilibrium are the three equations:

$$\frac{\partial \sigma_{i1}}{\partial x_1} + \frac{\partial \sigma_{i2}}{\partial x_2} = 0 \quad (i = 1, 2, 3) \tag{8.20}$$

The stress σ_{ij} can be expressed in terms of the derivatives of the components of displacement u_1, u_2 and u_3 through Hooke's law (Section 6.5). Because the stress and strain given by any displacement having the required discontinuity become infinite at the dislocation, it is further necessary to ensure that a cylinder of material containing the dislocation is in equilibrium, or in other words, that the force exerted on the core surface vanishes.

The solution for an edge dislocation in an isotropic medium is more complicated than that for a screw. If the dislocation lies along the x_3-axis with its Burgers vector parallel to x_1,

---

[12] The remainder of this section is somewhat advanced and can be omitted at a first reading.

the displacements can be written in the forms [6]:

$$u_1 = \frac{b}{2\pi} \left[\tan^{-1} \frac{x_2}{x_1} + \left(\frac{\lambda + \mu}{\lambda + 2\mu} \right) \frac{x_1 x_2}{x_1^2 + x_2^2} \right] \tag{8.21a}$$

$$u_2 = \frac{b}{2\pi} \left[-\frac{\mu}{2(\lambda + 2\mu)} \ln(x_1^2 + x_2^2) + \left(\frac{\lambda + \mu}{\lambda + 2\mu} \right) \frac{x_2^2}{x_1^2 + x_2^2} \right] \tag{8.21b}$$

where λ is the Lamé constant (Section 6.5). The first term in Eq. (8.21a) has the same form as the solution for a screw dislocation, and provides the necessary cyclic discontinuity in u_1 equal to the Burgers vector b. The first term in u_2, the displacement normal to the slip plane, is needed in order that the three equations of equilibrium in Eq. (8.20) are satisfied. The second terms in Eqs. (8.21a) and b are needed to make the core force vanish. The pattern produced when these displacements are imposed on a square array of points in the $x_1 x_2$-plane is shown in Figure 8.5. The elastic constants used in Eqs. (8.21) to obtain Figure 8.5 were those of tungsten, which, to a good approximation, happens to be elastically isotropic.

A consideration of the relevant equations leading to Eqs. (8.21) shows that there is an element of choice in the constant of integration in the equations relating the displacements to the elastic strains. Results quoted elsewhere for these displacements [5, 15–18] are the same, apart from differences in the choice of the constant of integration.

The stresses which follow from the displacements given by Eqs. (8.21) are:

$$\sigma_{11} = -\frac{\mu b}{2\pi(1 - v)} \frac{x_2(3x_1^2 + x_2^2)}{(x_1^2 + x_2^2)^2}$$

$$\sigma_{22} = \frac{\mu b}{2\pi(1 - v)} \frac{x_2(x_1^2 - x_2^2)}{(x_1^2 + x_2^2)^2} \tag{8.22}$$

$$\sigma_{33} = v(\sigma_{11} + \sigma_{22})$$

$$\sigma_{12} = \frac{\mu b}{2\pi(1 - v)} \frac{x_1(x_1^2 - x_2^2)}{(x_1^2 + x_2^2)^2}$$

where v is Poisson's ratio.

The stresses and strains due to a mixed dislocation in an isotropic medium whose line makes an angle θ with its Burgers vector can be obtained by simply adding the stresses and strains due to a screw dislocation of the Burgers vector $b \cos \theta$ to those due to an edge dislocation of the Burgers vector $b \sin \theta$. This is because the terms in the stress fields of the screw and edge components are independent of one another.

Tungsten, together with aluminium, is an exceptional crystal in being, elastically, almost isotropic. For most crystals, obtaining accurate values for the positions of atoms outside the core of a dislocation requires the use of the true elastic constants of the crystal, rather than the averaged values, which describe the isotropic behaviour of a polycrystal of the material. This means that the correct form of Hooke's law must be used in converting the three equations of equilibrium in Eq. (8.20) into equations in the displacements. In the general case, this introduces all nine terms of the type $\partial^2 u_i / \partial x_\alpha \partial x_\beta$ into Eq. (8.20) ($i = 1, 2, 3; \alpha, \beta = 1, 2$).

The resultant mathematics is highly complex, requiring the solution of a sixth-order polynomial [18–20]. The cases that have been solved analytically are ones in which this number of terms is reduced because of symmetry. This requires not only that the crystal structure be of high symmetry, but also that the dislocation lies in some special direction in the crystal, such as along a twofold axis of symmetry. For example, full solutions have been published for an edge dislocation along $\langle 001 \rangle$ in a cubic crystal, a screw normal to a mirror plane and a dislocation lying in the basal plane of a hexagonal crystal [21, 22]. Yin et al. [23] have compared a number of approaches for numerical efficiency and accuracy to compute stresses, displacements, and forces on finite lengths of arbitrarily oriented dislocations in anisotropic media.

The simplest case is that of a screw dislocation lying normal to a mirror plane [21]. Then, if the x_3-axis is chosen to be parallel to the dislocation, the elastic constants $c'_{14}, c'_{15}, c'_{24}, c'_{25}, c'_{34}, c'_{35}, c'_{46}$ and c'_{56} vanish (see Table 6.2; primed elastic constants are constants referred to axes that are not chosen with a conventional relationship to the crystal axes) [18]. The displacement can be written in the form [18]:

$$u_3 = -\frac{b}{2\pi}\tan^{-1}\frac{(c'_{44}c'_{55} - c'^2_{45})^{1/2}x_2}{c'_{44}x_1 - c'_{45}x_2} \tag{8.23}$$

For a screw dislocation along $\langle 110 \rangle$ in a c.c.p. crystal, Eq. (8.23) reduces to:

$$u_3 = -\frac{b}{2\pi}\tan^{-1}\frac{A^{1/2}x_2}{x_1} \tag{8.24}$$

where the x_1- and x_2- axes are shown in Figure 8.15 and A is the anisotropy factor given by Eq. (6.87):

$$A = \frac{2c_{44}}{c_{11} - c_{12}}$$

The anisotropy factor destroys the radial symmetry of the displacement, which is otherwise of the same form as in an isotropic medium. When $A > 1$, as it is for c.c.p. metals, the shear

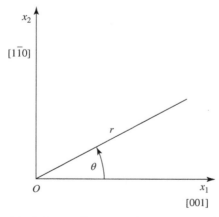

Figure 8.15   The relationship between the x_1- and x_2-axes, r and θ for a screw dislocation lying along [110] in a c.c.p. crystal

strain is greatest on the {110} plane; when $A < 1$, as it is for some alkali halides and some b.c.c. metals, the shear strain is greatest on the {100} plane. Because the shear modulus varies with orientation, a shear stress σ_{zr} is introduced:

$$\sigma_{zr} = \frac{c_{44}b(1 - A)\sin\theta\cos\theta}{2\pi A^{1/2}r(\cos^2\theta + A\sin^2\theta)} \tag{8.25}$$

in addition to the shear stress, $\sigma_{z\theta}$, which has the same form as in the isotropic case, namely [21]:

$$\sigma_{z\theta} = \frac{c_{44}b}{2\pi A^{1/2}r} \tag{8.26}$$

8.5 Atom Positions Close to a Dislocation

Although the solutions for dislocations in an elastic continuum can be used to assign displacements to the atoms around a dislocation in a crystal, these displacements are unlikely to be correct for atoms close to the dislocation, because the elastic continuum solutions may not be valid in this region of high strain. To determine the displacements in this core region of the dislocation, a detailed knowledge of the forces acting between atoms is needed. Atomistic simulations now enable these displacements to be determined using suitable interatomic potentials. Comprehensive reviews of progress in this area over the years can be found in the *Dislocations in Solids* series [24–26]. However, to convey the essential physical aspects of such displacements without recourse to such detailed simulations, our discussion here will be less rigorous.

As a starting point, consider a simple cubic lattice and take the displacements for a screw dislocation given by Eq. (8.17), treating it as an isotropic continuum:

$$u_z = \frac{b\theta}{2\pi}$$

If the position of the dislocation line is taken to be into the page along [010] at S in Figure 8.16, and θ is taken as the angle measured anticlockwise from the [100] direction, which can be considered in this diagram to be the trace of the (001) plane on which the screw

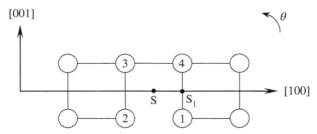

Figure 8.16 Schematic of a screw dislocation in a simple cubic crystal looking along the dislocation line going into the plane of the paper. The core of the dislocation is at S and θ is measured anticlockwise as shown relative to [100]. If the core of the dislocation moves to the position S_1, the 1–4 bonds would be highly strained and so the energy stored in the core of the dislocation would be higher at S_1 than at S. (See also Figures 8.1 and 8.6)

dislocation glides, the displacements of the atoms numbered 1–4, might be, for example:

$$\frac{b}{8}, \frac{3b}{8}, \frac{5b}{8}, \frac{7b}{8}$$

respectively. The displacement across the (001) slip plane is:

$$\Delta u = b - \frac{b\theta}{\pi} \tag{8.27}$$

No displacement has occurred across the slip plane to the far left of the dislocation ($\theta = \pi$), e.g. to the left of R in Figure 8.1, while the relative displacement is b to the far right ($\theta = 0$), e.g. to the right of Q in Figure 8.1. Recognizing that in the plane of the diagram atom 4 is at coordinates of x_1 along [100] and $b/2$ along [001], and so on for atoms 1–3, it is apparent that Eq. (8.27) can be written in the form:

$$\Delta u = b - \frac{b}{\pi}\tan^{-1}\frac{b}{2x_1} = b - \frac{b}{\pi}\left(\frac{\pi}{2} - \tan^{-1}\frac{2x_1}{b}\right)$$

or:

$$\Delta u = \frac{b}{2} + \frac{b}{\pi}\tan^{-1}\frac{2x_1}{b} \tag{8.28}$$

generating an equation for the displacement as a function of x_1, so that when x_1 is ∞, $\Delta u = b$, and when x_1 is $-\infty$, $\Delta u = 0$. This function is plotted in Figure 8.17. The distance on the slip plane over which the displacement lies between $\frac{1}{4}$ and $\frac{3}{4}$ of its extreme values is defined as the *width* of the dislocation on that plane. The width of the dislocation in Figure 8.17 is b, slightly more than the width of 0.75 b determined for the width of a screw dislocation core in aluminium in atomistic simulations [27].

Suppose now that we double the width of the dislocation by making Δu change with x_1 according to the equation:

$$\Delta u = \frac{b}{2} + \frac{b}{\pi}\tan^{-1}\frac{x_1}{b} \tag{8.29}$$

This is equivalent to giving the atoms the displacements that they would have, according to the symmetrical solution (8.17), if they were arranged on a tetragonal lattice with a spacing across the slip plane twice that within the plane. The displacements of the atoms 1–4 become approximately:

$$\frac{b}{6}, \frac{b}{3}, \frac{2b}{3}, \frac{5b}{6}$$

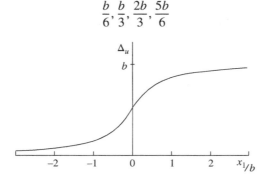

Figure 8.17   Plot of Eq. (8.28)

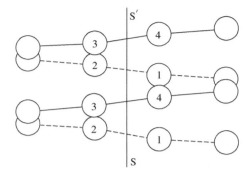

Figure 8.18   A screw dislocation of twice the width of the one shown in Figure 8.6

The widened dislocation is shown in Figure 8.18. The shear strain in bonds 1–4 and 2–3 is twice that in the bonds 1–2 and 3–4. A similar situation would be expected in a real crystal if the bonds between the slip planes were weaker than those between atoms within a slip plane.

A model which gives special consideration to the bonds across the slip plane is the Peierls dislocation [28]. In this model the crystal is supposed to be divided into two blocks at the slip plane, each of which is an elastic continuum. The atomic structure is taken into account only at the slip plane, where the two blocks join. A periodic force law is allowed to act between the rows of atoms immediately above the slip plane and those immediately below. For simplicity, a sine law is chosen:

$$\sigma = \frac{\mu b}{2\pi a} \sin \frac{2\pi \Delta u}{b} \tag{8.30}$$

The spacing between the slip planes is a, so that $\Delta u/a$ is the shear in the bond between two rows of atoms across the slip plane, while b is the atomic spacing within a slip plane. Equation (8.30) reduces to Hooke's law at small strains; that is, at small values of Δu. A dislocation is made by straining the elastic blocks and then joining them at the slip plane. The restoring force acting between sheared rows of atoms according to Eq. (8.30) tends to narrow the dislocation. The forces due to the elastic strains within the blocks tend to widen it. A balance is reached with a certain distribution of displacements. Either an edge or a screw dislocation can be made, although the model is more attractive for an edge dislocation, where there are large shear strains on a unique slip plane.

The displacements across the slip plane of a Peierls screw dislocation are identical to those predicted by the elastic continuum solution (8.28). That is:

$$\Delta u = \frac{b}{2} + \frac{b}{\pi} \tan^{-1} \frac{2x_1}{a} \tag{8.31}$$

The displacements for a Peierls edge dislocation differ from those for an elastic edge dislocation and give:

$$\Delta u = \frac{b}{2} + \frac{b}{\pi} \tan^{-1} \frac{2x_1(1 - v)}{a} \tag{8.32}$$

The Peierls edge is wider than the screw by the factor $1/(1 - v)$, where v is Poisson's ratio. As v increases towards its greatest possible value of 0.5, the resistance of the material to elastic shear decreases relative to its resistance to dilatation. An increase in the width of an edge dislocation as v increases therefore accords with the qualitative idea that if the atoms are hard, but shear over one another easily, as for metallic materials, then squeezing in the extra plane of atoms will spread the dislocation out over its slip plane. On the other hand, if there are strong directed bonds across the slip plane, their resistance to shear will lead to a narrow dislocation. Such directed bonds occur in covalent crystals like silicon.

An important property which is controlled by the atom positions in the dislocation core is the force needed to start a straight dislocation moving in an otherwise perfect lattice. As the dislocation moves, the displacements around it will change, and in general the energy stored in the core will change slightly. Consequently, a force is exerted on the dislocation, given by:

$$F_x = -\frac{dE}{dx} \tag{8.33}$$

where E is the energy due to the dislocation.

The maximum value of this force is called the Peierls force, because it was first investigated by Peierls using a model of an edge dislocation [28]. The method can be briefly illustrated with the help of Figure 8.16, taking the dislocation to be a Peierls screw. For any position of the core of the dislocation along the [100] direction in Figure 8.16, Eq. (8.31) can be used to assign displacements to the atoms. The energy stored in the bonds across the slip planes, 2–3, 1–4 and so on, can then be summed as a function of the position of the dislocation.

The energy sum E' turns out to be a minimum for position S in Figure 8.16 and a maximum at S_1 where the highly strained 1–4 bond makes a large contribution. At an intermediate position, dE'/dx_1 can be shown to have a maximum value of the form:

$$\frac{dE'}{dx_1} = F_p \approx \mu \exp\left(-\frac{\pi a}{b}\right) \tag{8.34}$$

This result has no quantitative significance because of the unrealistic nature of the model, and indeed it gives far too large a value for the force needed to move a dislocation in, say, a c.c.p. metal. The decrease in F_p with the parameter a, which on this model is equivalent to the width of the dislocation, is physically reasonable. As a dislocation becomes narrower, the change in displacement as it moves is distributed among fewer atoms. When the width is very small, a few core atoms have to make large jumps, for which there may be a substantial energy barrier, in order to move the dislocation by one atomic spacing.

8.6   The Interaction of Dislocations with One Another

When a crystal is plastically deformed, some of its dislocations are lost as they glide to the surfaces of the crystal. However, dislocations must also be replenished during plastic flow, because a strain of 10% typically increases their density by 10^{12}–10^{13} m^{-2} (dislocation density is defined in Section 8.1). As the density of dislocations increases, their interaction with one another becomes important. For example, the stress needed to deform the crystal plastically increases and becomes controlled by the forces that dislocations exert on one another, rather than by the stress needed to move an isolated dislocation.

A simple mechanism by which dislocations can multiply was suggested by Frank and Read [29]. Its operation requires only that a dislocation in a slip plane be pinned at two points. In Figure 8.19 the dislocation is pinned at the points x' and y' where it changes its plane. It is apparent from Figures 8.19a and b that this configuration could occur if a screw segment xy glided off its original slip plane to the position $x'y'$. This manoeuvre is called *cross-slip*. When the segment $x'y'$ changes back to a plane parallel to the original slip plane, it is easy to see that it can generate any number of expanding loops, using x' and y' as pivots. This variant of the Frank–Read mechanism is due to Koehler [30]. It allows slip to spread out sideways from a single active slip plane and so it can account for the commonly observed *slip band*, which is a set of very closely spaced parallel planes on which slip has occurred (see Figure 7.3).

When slip occurs in bands, or even when it occurs more diffusely, dislocations on nearby parallel slip planes must on occasion come close to one another. Then the stress due to the one dislocation will exert a force on the other dislocation, according to Eq. (8.12), where we substitute for σ the stress due to one dislocation at the position of the other. Two special cases, parallel screw dislocations and parallel edge dislocations, will be briefly reviewed, assuming isotropic elasticity.

Parallel screw dislocations a distance r apart experience a simple central force:

$$F = \frac{\mu b^2}{2\pi r} \tag{8.35}$$

The force per unit length, F, is repulsive when the screw dislocations are identical, so that their stress fields reinforce one another, and attractive when they are of opposite sign, so that their stress fields cancel as they come together. There is no position of equilibrium.

Parallel edge dislocations, on the other hand, have two positions of equilibrium in respect of the component of force which lies in the slip plane. Since an edge dislocation would

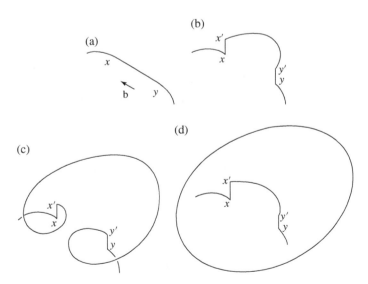

Figure 8.19 Dislocation multiplication

have to climb to move out of its slip plane, this is the only component of force that can cause an edge dislocation to move at low temperature. Its magnitude per unit length can be derived from Eq. (8.22). It is convenient to write it in terms of polar coordinates as follows:

$$F_g = \frac{\mu b^2}{2\pi(1-v)r} \cos\theta(\cos^2\theta - \sin^2\theta) \tag{8.36}$$

The angle θ is defined in Figure 8.20, in which Eq. (8.36) is illustrated for edge dislocations of *opposite* sign. There is an unstable equilibrium at $\theta = \pi/2$ and a stable one at $\theta = \pi/4, 3\pi/4$. The force per unit length, F_m, which would be needed to push the parallel edge dislocations past one another on slip planes a distance y apart is:

$$F_m = \frac{\mu b^2}{8\pi(1-v)y} \tag{8.37}$$

Since yield stresses are typically in the range 10^{-3}–10^{-4} μ, edge dislocations of opposite sign are likely to become stuck against one another if they lie on slip planes which are a few hundred Å or less apart. Close pairs of dislocations of opposite sign are in fact commonly observed in deformed crystals; they are called *dislocation dipoles*.

The forces acting between *like* edge dislocations are equal in magnitude, but opposite in sign to those acting between unlike edge dislocations. The stable equilibrium therefore occurs at $\theta = \pi/2$. Two like edge dislocations are less likely to trap one another during glide than two unlike ones, because the applied stress moves them both in the same direction. However, the equilibrium configuration at $\theta = \pi/2$ becomes particularly important when

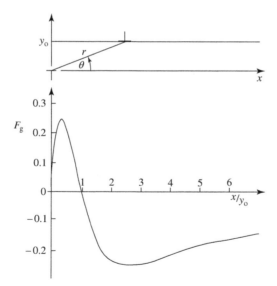

Figure 8.20 Force between parallel edge dislocations on slip planes a distance y_0 apart. The force is given in units of $\mu b^2/2\pi(1-v)y$

climb can occur. The component of force normal to the slip plane, the climb force, is given by:

$$F_c = \frac{\mu b^2}{2\pi(1 - v)r} \sin\theta(1 + 2\cos^2\theta) \tag{8.38}$$

This force is one of attraction between unlike edge dislocations and repulsion between like edge dislocations, for *all* θ. If the temperature is high enough to allow climb, unlike edge dislocations from different slip planes annihilate one another, while any excess edge dislocations of the same sign which are left over arrange themselves in walls perpendicular to the slip plane. This is known as *polygonization* and occurs during the process of recovery when metals are annealed.

The elastic interaction of two dislocations in an anisotropic medium can differ quite significantly from the corresponding interaction in an isotropic medium. In copper, for example, two parallel screw dislocations lying along $\langle 110 \rangle$ in the same $\{111\}$ slip plane exert forces on one another which are directed out of the slip plane, because of the shear stress σ_{rz} given by Eq. (8.25). Substitution of the appropriate constants into Eq. (8.25) shows that the component of force normal to the slip plane is no less than 60% of the component of attraction or repulsion in the slip plane, which would be the sole component if the crystal of copper were elastically isotropic.

Dislocations on intersecting slip planes also exert forces on one another due to their stress fields. For example, the stress field of a screw dislocation in an *isotropic* medium exerts a force on an orthogonal screw dislocation, given by:

$$F = \frac{\mu b_1 b_2}{2\pi} \frac{d}{r^2} \tag{8.39}$$

where F is the force per unit length at the point P on the orthogonal screw dislocation, defined in Figure 8.21. The direction of F is normal to both dislocations. Its sign can be checked by comparing the shear stresses due to each screw at the midpoint of the shortest line joining them (the line of length d in Figure 8.21). The screw dislocations attract if the two stresses cancel one another and repel if they reinforce one another.

Dislocations on intersecting slip planes interact in a second way, when they cut through one another. Because the material on one side of the surface swept out by a moving dislocation is displaced by the Burgers vector relative to the material on the other side, each dislocation acquires an extra length of line, equal in magnitude and direction to the Burgers vector of the other dislocation. If the extra length of line happens to lie on the plane on

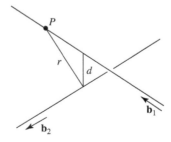

Figure 8.21 Orthogonal screw dislocations, a perpendicular distance d apart

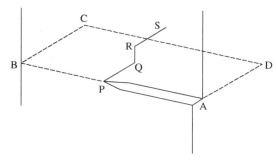

Figure 8.22 A jog QR on a dislocation PQRS where PQ and RS have screw dislocation character

which the remainder of the dislocation is already gliding, it is called a *kink*. A kink can be eliminated immediately by glide. When the extra length of line does not lie on this plane, it is called a *jog*.

When two orthogonal screw dislocations intersect, each acquires a jog which is a segment of edge dislocation. A jog of this type is shown in Figure 8.22, where the sign of the screw is indicated by means of the step that it produces on the surface of the crystal. If the jogged dislocation PQRS moves as a whole towards AD, the jog QR is forced to climb. Careful inspection of Figure 8.22 will show that a void is produced of volume QR times b times the distance moved. This void will take the form of a trail of vacant lattice sites. If PQRS moves in the opposite direction, towards BC, a trail of interstitial defects will be produced.

A third way in which two intersecting dislocations may interact with one another is that in certain circumstances they can come together and combine to form a new dislocation. This process is discussed in detail in the next chapter.

Problems

8.1 A screw dislocation of Burgers vector **b** passes straight through a crystal, meeting the front and back surfaces at 90°. Keeping this orientation, it moves so that its ends trace out circular paths and the dislocation line sweeps out a cylindrical surface. Describe the deformation produced by n revolutions of the dislocation.

8.2 Show how a pure edge dislocation loop, like that in Figure 8.11, can develop screw segments by gliding into a new shape. Hence, show that it is geometrically possible for an edge dislocation loop to become a source for glide on any plane containing its Burgers vector.

8.3 An edge dislocation in a crystal of NaCl lies along [001] and its Burgers vector is $\frac{1}{2}[110]$. If the dislocation is to be moved through a distance of 1 μm in the [100] direction, calculate the volume of material that must be added to or subtracted from the edge of its extra plane, per metre of dislocation.

8.4 An edge dislocation in a crystal of NaCl lies along [001] and its Burgers vector is $\frac{1}{2}[110]$. Calculate the glide force acting upon the dislocation under the action of a tensile stress σ when (a) when the tensile axis is [100] and (b) the tensile axis is [111].

8.5 A screw dislocation in a crystal of AgCl has the Burgers vector $\frac{1}{2}[110]$. If the dislocation could glide with equal ease on any plane, on which plane would it glide when under the action of a tensile stress (a) when the tensile axis is [100] and (b) when the tensile axis is [111]?

8.6 A screw dislocation lies along the axis of an isotropic cylinder of radius R. If the stress field were given by Eq. (8.19), there would be forces acting on the end faces of the cylinder. Obtain an expression for the couple which would then be acting on the cylinder. (In a free cylinder the stress field would not be exactly given by Eq. (8.19) but would contain an additional term, unimportant near the dislocation line, corresponding to a twisting of the cylinder, which neutralizes the couple implied by Eq. (8.19) alone.)

8.7 From either Eqs. (8.21) or Eqs. (8.22) and (6.87), obtain the strain at a point directly above or below an edge dislocation ($x_1 = 0$). Calculate the dilatations at points $5b$ above and $5b$ below an edge dislocation in aluminium.

8.8 Prepare a diagram of a plane which is normal to an edge dislocation in an isotropic medium, showing the lines in the plane on which one of the stress components σ_{11}, σ_{22} and σ_{12} vanishes (the dislocation lies along the x_3-axis with its Burgers vector parallel to Ox_1). Show the signs of the stresses in the various regions of the diagram.

8.9 Consider a screw dislocation in a c.c.p. crystal lying along [110]. Show (from Eq. (8.24)) that at equal distances from the dislocation, the shear strain on the $(1\bar{1}0)$ plane that passes through the dislocation is A times the shear strain on the (001) plane that passes through the dislocation, where $A = 2c_{44}/(c_{11} - c_{12})$.

8.10 Obtain an expression for the width on the (001) plane of a screw dislocation lying along [110] in a sodium chloride crystal. Assume that the elastic continuum solution for the displacements, Eq. (8.24), applies to all the atoms around the dislocation.

8.11 A cubic crystal is being stretched in a [100] direction. Its deformation is being produced by edge dislocations lying parallel to [001] and gliding on (110) and $(1\bar{1}0)$ planes. If the density of these dislocations is ρ, the magnitude of the Burgers vector of each one is b and their average velocity is $\bar{v}$, determine the rate of increase of tensile strain of the crystal ($= (1/l)(dl/dt)$, l = length of crystal).

8.12 Analyse the interaction between two edge dislocations whose lines are parallel to one another but whose Burgers vectors are orthogonal. Determine all positions of equilibrium with respect to glide forces and divide these into cases of stable, unstable and neutral equilibrium.

8.13 Find the total force experienced by an infinitely long screw dislocation with Burgers vector $\mathbf{b}_1$ due to the presence of an infinitely long orthogonal screw dislocation with Burgers vector $\mathbf{b}_2$.

8.14 In a crystal of sodium chloride, two orthogonal screw dislocations, with Burgers vectors $\frac{1}{2}[110]$ and $\frac{1}{2}[1\bar{1}0]$ respectively, are being driven towards one another by a tensile stress along [100]. Demonstrate that each dislocation acquires a jog which produces interstitial defects if the dislocations continue to move under the action of the applied stress.

8.15 What is the minimum number of dislocations each with Burgers vector equal to the smallest lattice translation vector that can meet at a node in the following crystals: (a) b.c.c. iron, (b) diamond, (c) zinc and (d) caesium chloride? Under what conditions will the node be glissile?

Suggestions for Further Reading

Anderson, P.M., Hirth, J.P. and Lothe, J. (2017). *Theory of Dislocations*, 3rd Edition. New York: Cambridge University Press.

Cottrell, A.H. (1953). *Dislocations and Plastic Flow in Crystals*. Oxford: Oxford University Press.

Hull, D. and Bacon, D.J. (2011). *Introduction to Dislocations*, 5th Edition. Oxford: Butterworth-Heinemann.

Nabarro, F.R.N. (1987). *Theory of Crystal Dislocations*. Mineola, NY: Dover Press.

Read, W.T. (1953). *Dislocations in Crystals*. New York: McGraw-Hill.

Steeds, J.W. (1973). *Introduction to Anisotropic Elastic Theory of Dislocations*. Oxford: Clarendon Press.

Weertman, J. and Weertman, J.R. (1992). *Elementary Dislocation Theory*. Oxford: Oxford University Press.

References

[1] Taylor, G.I. (1934). The mechanism of plastic deformation of crystals. Part I.—Theoretical. *Proc. R. Soc. London A* 145: 362–387.

[2] Orowan, E. (1934). Zur Kristallplastizitat. III. Über den Mechanismus des Gleitvorganges. *Z. Phys.* 89: 634–659.

[3] Polanyi, M. (1934). Über eine Art Gitterstörung, die einen Kristall platisch machen könnte. *Z. Phys.* 89: 660–664.

[4] Love, A.E.H. (1927). *The Mathematical Theory of Elasticity*. Cambridge: Cambridge University Press, pp. 221–228.

[5] Nabarro, F.R.N. (1952). The mathematical theory of stationary dislocations. *Adv. Phys.* 1: 269–394.

[6] Burgers, J.M. (1939). Some considerations on the fields of stress connected with dislocations in a regular crystal lattice. I. *Proc. Konin. Ned. Akad. Wet.* 42: 293–325.

[7] Frank, F.C. (1949). Sessile dislocations. *Proc. Phys. Soc. A* 62: 202–203.

[8] Frank, F.C. (1951). Crystal dislocations.—Elementary concepts and definitions. *Phil. Mag.* 42: 809–819.

[9] deWit, R. (1965). The direction of the force on a dislocation and the sign of the Burgers vector. *Acta Metall.* 13: 1210–1211.

[10] Seitz, F. (1950). Prismatic dislocations and prismatic punching in crystals. *Phys. Rev.* 79: 723–724.

[11] Peach, M. and Koehler, J.S. (1950). The forces exerted on dislocations and the stress fields produced by them. *Phys. Rev.* 80: 436–439.

[12] Weertman, J. (1965). The Peach–Koehler equation for the force on a dislocation modified for hydrostatic pressure. *Phil. Mag.* 11: 1217–1223.

[13] Hirsch, P.B., Silcox, J., Smallman, R.E. and Westmacott, K.H. (1958). Dislocation loops in quenched aluminium. *Phil. Mag.* 3: 897–908.

[14] Silcox, J. and Whelan, M.J. (1960). Direct observations of the annealing of prismatic dislocation loops and of climb of dislocations in quenched aluminium. *Phil. Mag.* 5: 1–23.

[15] Koehler, J.S. (1941). On the dislocation theory of plastic deformation. *Phys. Rev.* 60: 397–410.

[16] Read, W.T. (1953). *Dislocations in Crystals*. New York: McGraw-Hill.

[17] Kosevich, A.M. (1979). Crystal dislocations and the theory of elasticity. In: *Dislocations in Solids*, vol. 1 (ed. F.R.N. Nabarro). Amsterdam: North-Holland, pp. 33–141.

[18] Anderson, P.M., Hirth, J.P. and Lothe, J. (2017). *Theory of Dislocations*, 3rd Edition. New York: Cambridge University Press.

[19] Steeds, J.W. (1973). *Introduction to Anisotropic Elastic Theory of Dislocations*. Oxford: Clarendon Press.

[20] Steeds, J.W. and Willis, J.R. (1979). Dislocations in anisotropic media. In: *Dislocations in Solids*, vol. 1 (ed. F.R.N. Nabarro). Amsterdam: North-Holland, pp. 143–165.

[21] Eshelby, J.D., Read, W.T. and Shockley, W. (1953). Anisotropic elasticity with applications to dislocation theory. *Acta Metall.* 1: 251–259.

[22] Chou, Y.T. and Eshelby, J.D. (1962). The energy and line tension of a dislocation in a hexagonal crystal. *J. Mech. Phys. Solids* 10: 27–34.

[23] Yin, J., Barnett, D.M. and Cai, W. (2010). Efficient computation of forces on dislocation segments in anisotropic elasticity. *Modell. Simul. Mater. Sci. Eng.* 18, 045013-1–045013-14.

[24] Bullough, R. and Tewary, V.K. (1979). Lattice theories of dislocations. In: *Dislocations in Solids*, vol. 2 (ed. F.R.N. Nabarro). Amsterdam: North-Holland, pp. 1–65.

[25] Duesbery, M.S. (1989). The dislocation core and plasticity. In: *Dislocations in Solids*, vol. 8 (ed. F.R.N. Nabarro). Amsterdam: North-Holland, pp. 67–173.

[26] Cai, W., Bulatov, V.V., Chang, J., Li, J. and Yip, S. (2004). Dislocation core effects on mobility. In: *Dislocations in Solids*, vol. 12 (eds. F.R.N. Nabarro and J.P. Hirth). Amsterdam: Elsevier, pp. 1–80.

[27] Lu, G., Kioussis, N., Bulatov, V.V. and Kaxiras, E. (2001). Dislocation core properties of aluminium: a first-principles study. *Mater. Sci. Eng. A* 309–310: 142–147.

[28] Peierls, R. (1940). The size of a dislocation. *Proc. Phys. Soc.* 52: 34–37.

[29] Frank, F.C. and Read, W.T. (1950). Multiplication processes for slow moving dislocations. *Phys. Rev.* 79: 722–723.

[30] Koehler, J.S. (1952). The nature of work-hardening. *Phys. Rev.* 86: 52–59.

9

Dislocations in Crystals

9.1 The Strain Energy of a Dislocation

A considerable amount of strain energy is stored in the elastically distorted region around a dislocation line. The magnitude of the energy stored in an elastically strained region is always of the form $\frac{1}{2} \times$ (elastic modulus) $\times$ (strain)^2, per unit volume (see Eq. (6.60)). Since the strain at a given point is proportional to the magnitude of the Burgers vector of the dislocation $\mathbf{b}$, the total strain energy will be proportional to $\mathbf{b}^2$. To find the absolute magnitude of the energy stored, it is necessary to make a more detailed calculation. This can be done in two ways.

In the first method, the energy stored in a small element of volume is found and an integration is performed. For a screw dislocation the appropriate element is a thin cylindrical shell of radius r, thickness dr, because the shear strain γ within such a thin shell is constant (Eq. (8.18)):

$$\gamma = \frac{b}{2\pi r}$$

An elastically isotropic medium is assumed. The strain energy per unit length of shell is then:

$$dE = \left(\frac{1}{2}\mu\gamma^2\right) 2\pi r \ dr = \frac{\mu b^2}{4\pi r}dr \tag{9.1}$$

and so the energy per unit length of dislocation in the region from the core, radius r_0, out to a radius R is:

$$E = \int_{r_0}^{R} \frac{\mu b^2}{4\pi r}dr$$

or:

$$E = \frac{\mu b^2}{4\pi} \ln\left(\frac{R}{r_0}\right) \tag{9.2}$$

Crystallography and Crystal Defects, Third Edition. Anthony Kelly and Kevin M. Knowles.
© 2020 John Wiley & Sons Ltd. Published 2020 by John Wiley & Sons Ltd.
Companion website: www.wiley.com/go/kelly/crystallography3e

To be exact, Eq. (9.2) should be corrected to allow for the relaxation of stress at the free end surfaces of the cylinder. However, any uncertainty in the value of r_0 makes this correction unimportant.

The strain energy due to an edge dislocation is more easily calculated by a second method. Suppose that the dislocation is formed in initially unstrained material by making a cut from the surface of the material up to the site of the dislocation line, and then applying forces to the faces of the cut so as to give them a relative displacement of b. Then the work done by these forces equals the elastic energy stored around the completed dislocation. As the faces of the cut are displaced, the Burgers vector increases smoothly from zero to $\mathbf{b}$. The work done in increasing the Burgers vector of unit length of dislocation from β to $\beta + d\beta$ is:

$$dW = F(\beta)\, d\beta \tag{9.3}$$

where $F(\beta)$ is the force acting on one of the faces of the cut in the direction of β. This force is given by the stress field of the dislocation at the cut surface. For an edge dislocation which is made by cutting along the slip plane in a cylinder of radius R (Figure 9.1):

$$F = \int_{r_0}^{R} \sigma_{12}\, dx_1 \tag{9.4}$$

The fact that only the stress on a single plane is involved gives this method an advantage over volume integration when the stress field is complicated, as it is for the case of an edge dislocation in an isotropic medium. From the forms of Eqs. (8.22), with $x_2 = 0$ and the Burgers vector $= \beta$:

$$\sigma_{12} = \frac{\mu\beta}{2\pi(1-v)x_1} \tag{9.5}$$

and therefore:

$$F = \frac{\mu\beta}{2\pi(1-v)} \ln\left(\frac{R}{r_0}\right) \tag{9.6}$$

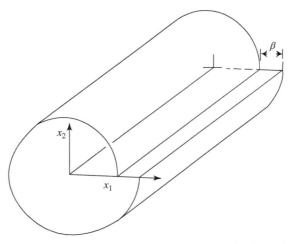

Figure 9.1 An edge dislocation formed by making a cut in a cylinder and then displacing the faces of the cut by a length β along its slip plane by applying a suitable force

The strain energy per unit length is then:

$$E = W = \int_0^b F \, d\beta = \frac{\mu b^2}{4\pi(1-v)} \ln\left(\frac{R}{r_0}\right) \tag{9.7}$$

In an isotropic medium, an edge dislocation has more elastic strain energy than a screw, in the ratio of $1:(1-v)$. For aluminium, $v = 0.34$ and so per unit length the edge dislocation has 1.5 times as much energy as the screw; for tungsten, $v = 0.17$ and the ratio is only 1.2.

The strain due to a mixed dislocation, with a Burgers vector at an angle θ to the dislocation line, is equal to the sum of the strains due to parallel edge and screw dislocations of Burgers vectors $b\sin\theta$ and $b\cos\theta$, respectively. Therefore, the energy of a mixed dislocation is given by:

$$E = \frac{Kb^2}{4\pi} \ln\left(\frac{R}{r_0}\right) \tag{9.8}$$

where:

$$K = \frac{\mu}{1-v}\sin^2\theta + \mu\cos^2\theta \tag{9.9}$$

To estimate the absolute magnitude of the strain energy per unit length of dislocation line, it is sufficient to consider Eq. (9.2):

$$E = \frac{\mu b^2}{4\pi} \ln\left(\frac{R}{r_0}\right)$$

The difficulty in applying this equation is that the strain energy increases without limit as R increases. In a real crystal, either the surfaces of the crystal, or more probably other dislocations, will limit R. Fortunately, because of its logarithmic dependence on R, the magnitude of E is not very sensitive to the particular value of R which is assumed.

For aluminium, which is approximately isotropic elastically, with $\mu = 28\,\mathrm{GPa}$, $b = 2.86 \times 10^{-10}$ m and an assumed $r_0 = b$ [1], the energy per unit length is $1.49 \times 10^{-9}\,\mathrm{J\,m^{-1}}$ when R is set at $1\,\mu\mathrm{m}$, and $2.75 \times 10^{-9}\,\mathrm{J\,m^{-1}}$ when R is set at $1\,\mathrm{mm}$. A value of $2 \times 10^{-9}\,\mathrm{J\,m^{-1}}$ corresponds to an energy per Å length of dislocation of $1.25\,\mathrm{eV}$. This energy is large enough to ensure that a crystal containing a single dislocation is never in thermodynamic equilibrium. (Even at high temperatures the entropy introduced by a dislocation subtracts only a relatively small amount from its energy, so that the free energy of a dislocated crystal of ordinary size is always greater than that of a perfect crystal.)

To obtain the total energy of a dislocation, the energy stored within the dislocation core must be added to the energy given by Eq. (9.8). The core energy can be roughly estimated in various ways, which agree in giving its order of magnitude as $\mu b^2/4\pi$ $\mathrm{J\,m^{-1}}$ [1, 2]. For aluminium, this amounts to $0.1\,\mathrm{eV}$ per Å, considerably less than the energy stored *outside* the core. In a real crystal, the importance of the core energy lies less in its magnitude than in the *variation* of this magnitude as the dislocation glides through the crystal. If the core energy increases sharply as the dislocation moves from its position of minimum energy, its motion is strongly resisted (Section 8.5).

It is always geometrically possible for two dislocations to combine to form a single dislocation, as shown in Figure 9.2. When this happens, the Burgers vector of the resultant dislocation, $\mathbf{b}_3$, is simply the vector sum of the Burgers vectors of the combining dislocations:

$$\mathbf{b}_3 = \mathbf{b}_1 + \mathbf{b}_2 \tag{9.10}$$

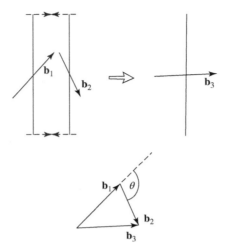

Figure 9.2 A dislocation reaction in which two dislocations combine to form a single dislocation

To determine whether such a reaction is energetically favourable, we must compare the energy of a crystal containing a dislocation of the Burgers vector $\mathbf{b}_3$ with that of a crystal containing two widely separated dislocations of the Burgers vectors $\mathbf{b}_1$ and $\mathbf{b}_2$. The former can be written as $K_3 b_3^2$ and the latter as $K_1 b_1^2 + K_2 b_2^2$.

Although it is evident from Eq. (9.8) that the constants K_1, K_2 and K_3 depend on the angles between the dislocation lines and their Burgers vectors, the difference between $K_3 b_3^2$ and $K_1 b_1^2 + K_2 b_2^2$ is usually so large that the sign of $K_3 b_3^2 - (K_1 b_1^2 + K_2 b_2^2)$ is not affected by setting $K_1 = K_2 = K_3$. Consequently, the sign of $b_3^2 - (b_1^2 + b_2^2)$ is a useful criterion for determining whether the reaction of two dislocations to form a third leads to a reduction in energy. Using this, a reduction will occur if:

$$b_3^2 < b_1^2 + b_2^2 \tag{9.11}$$

This rapid test for deciding whether two dislocations will react is known as Frank's rule [3]. It can also be expressed in terms of the angle θ between $\mathbf{b}_1$ and $\mathbf{b}_2$ (Figure 9.2). If $\theta < 90°$, the dislocations will not react; if $\theta > 90°$ they will attract and react with one another. The same rule can be applied in reverse to decide whether a dislocation with Burgers vector $\mathbf{b}_3$ is stable. If two lattice vectors summing to $\mathbf{b}_3$ and making an angle $\theta < 90°$ with each other are available, the dislocation will dissociate into two dislocations with these vectors as Burgers vectors. The rule is obviously most reliable when θ is far from $90°$.

It follows from Frank's rule that most dislocations other than those with the shortest lattice vector as the Burgers vector are unstable. For example, a dislocation with a Burgers vector $\frac{1}{2}\langle 211 \rangle$ in a c.c.p. crystal would decompose into two $\frac{1}{2}\langle 011 \rangle$ dislocations:

$$\tfrac{1}{2}[211] \rightarrow \tfrac{1}{2}[110] + \tfrac{1}{2}[101] \tag{9.12}$$

However, it sometimes happens that a dislocation whose Burgers vector is greater than the least lattice vector cannot decompose spontaneously. An example is the [100] dislocation in

Table 9.1 Definitely stable dislocations in some Bravais lattices

Lattice	Burgers vector	Number of equivalent vectors	Square of Burgers vector
Simple cubic	$\langle 100 \rangle$	6	a^2
B.c.c.	$\frac{1}{2}\langle 111 \rangle$	8	$3a^2/4$
	$\langle 100 \rangle$	6	a^2
F.c.c.	$\frac{1}{2}\langle 110 \rangle$	12	$a^2/2$
Hexagonal	$\frac{1}{3}\langle 11\bar{2}0 \rangle$	6	a^2
	$\langle 0001 \rangle$	2	c^2
Rhombohedral	$\langle 100 \rangle$	6	a^2
$\alpha < 90°$	$\langle 1\bar{1}0 \rangle$	6	$4a^2 \sin^2 \alpha/2$
	$\langle 1\bar{1}1 \rangle$	6	$a^2(1 + 4\sin^2 \alpha/2)$
Rhombohedral	$\langle 100 \rangle$	6	a^2
$\alpha > 90°$	$\langle 110 \rangle$	6	$4a^2 \cos^2 \alpha/2$
	$\langle 111 \rangle$	2	$9a^2\left(1 - \frac{4}{3}\sin^2\alpha/2\right)$
Simple tetragonal	$\langle 100 \rangle$	4	a^2
	$\langle 001 \rangle$	2	c^2
Body-centred tetragonal	$\frac{1}{2}\langle 111 \rangle$	8	$a^2/2 + c^2/4$
$c/a < \sqrt{2}$	$\langle 100 \rangle$	4	a^2
	$\langle 001 \rangle$	2	c^2
Body-centred tetragonal	$\frac{1}{2}\langle 111 \rangle$	8	$a^2/2 + c^2/4$
$c/a > \sqrt{2}$	$\langle 100 \rangle$	4	a^2

a b.c.c. crystal. Although the $\frac{1}{2}\langle 111 \rangle$ dislocation has a smaller energy, Frank's rule predicts that the decomposition:

$$[100] \rightarrow \tfrac{1}{2}[1\bar{1}\bar{1}] + \tfrac{1}{2}[111] \tag{9.13}$$

will not occur, and that no other geometrically possible decomposition is favourable energetically.

The Burgers vectors that are definitely stable in various lattices are listed in Table 9.1. Doubtful cases in which the vector is the sum of two orthogonal vectors ($\theta = 90°$) are omitted.

The fact that the dislocation with the smallest Burgers vector has the least elastic energy fits in well with the observation that the slip direction is almost always that of the shortest lattice translation vector in the crystal (Table 7.1). It is natural to enquire whether the observed slip planes are those on which a dislocation of minimum Burgers vector has the smallest elastic energy, when the elastic anisotropy of the crystal is taken into account. The elastic energy per unit length for an undissociated dislocation in an anisotropic medium is given by:

$$E = \frac{K'b^2}{4\pi} \ln\left(\frac{R}{r_0}\right) \tag{9.14}$$

where K' depends on the elastic constants. Foreman has computed the energy factor K' in a number of cases [4]. In copper, for example, an edge dislocation with $\frac{1}{2}\langle 110 \rangle$ Burgers vector has a lower elastic energy when it lies in a $\{110\}$ plane ($K' = 64.5$ GPa) than when it lies in a $\{111\}$ plane ($K' = 74.5$ GPa). Since the observed slip plane is $\{111\}$, it appears that the elastic energy alone does not determine which slip plane operates. In order to understand the choice of slip plane it is clearly necessary to take account of the atomic structure of the crystal (see also Section 7.1).

The fact that a dislocation line has a certain amount of energy *per unit length* associated with it causes it to try to shorten itself, or in other words, it gives it a line tension. It is not easy to define this line tension precisely because the energy is not concentrated at the line itself but is contained in its far-reaching strain field. Let us consider the tendency of a circular dislocation loop of radius r to shrink. To prevent it from shrinking, a radial force F_r per unit length must be applied, given by the condition that the energy change accompanying an infinitesimal contraction of the loop must vanish. Therefore:

$$F_r 2\pi r \, dr = E 2\pi \, dr \tag{9.15}$$

or:

$$F_r = \frac{E}{r}$$

where E is the energy per unit length of the dislocation loop. Since the strains due to the opposite sides of the loop cancel one another at distances that are large compared to r, we can determine E approximately by setting $R = r$ in Eq. (9.2):

$$E \approx \frac{\mu b^2}{4\pi} \ln\left(\frac{r}{r_0}\right) \tag{9.16}$$

If we consider a small segment of the loop, shown in Figure 9.3, subtending an angle $d\theta$ at the centre of the loop, we see that to keep this segment in equilibrium its ends must experience a tension T given by setting the sum of radial forces equal to zero:

$$2T \sin\frac{d\theta}{2} = F_r r \, d\theta \tag{9.17}$$

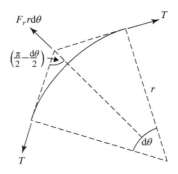

Figure 9.3 Forces on a small segment of a dislocation loop in equilibrium

Since dθ is very small, we therefore have from Eqs. (9.15) and (9.17):

$$T = E \tag{9.18}$$

We may extend this result with the aid of Eq. (9.16) to say that the line tension of a dislocation bowed to a radius of curvature ρ is approximately:

$$T \approx \frac{\mu b^2}{4\pi} \ln\left(\frac{\rho}{r_0}\right) \tag{9.19}$$

With $\rho \approx 500 r_0$, substitution into Eq. (9.19) gives a value of $T \approx 0.5\ \mu b^2$, which is often used as the value of the line tension of a dislocation in rough calculations.

A diagram which helps in the calculation of the stress needed to extrude a dislocation between obstacles a distance l apart is shown in Figure 9.4. The radial force on the dislocation is supplied by a shear stress τ on the slip plane, parallel to the Burgers vector. From Eq. (8.4):

$$F_r = \tau b \tag{9.20}$$

Using Eqs. (9.15) and (9.18), if the radius of curvature of the dislocation line is ρ:

$$\tau = \frac{T}{b\rho} \tag{9.21}$$

or, setting $T = 0.5\ \mu b^2$:

$$\tau = \frac{0.5\mu b}{\rho} \tag{9.22}$$

Since the minimum value of ρ is $0.5 l$, occurring when the dislocation has been bowed out to a semicircle, the stress required to force the dislocation between the obstacles is given by:

$$\tau \approx \frac{\mu b}{l} \tag{9.23}$$

[5]. It is apparent from this equation that this is also the shear stress needed to operate a Frank–Read source whose pinning points are a distance l apart [6]. In both these situations, an additional term can be added to the right-hand side of Eq. (9.23) to take account of the finite stress required to move the dislocation within its slip plane (e.g. Section 8.5) [6].

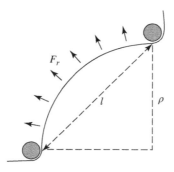

Figure 9.4   A dislocation line being extruded between obstacles a distance *l* apart

However, this additional term is often taken to be negligible or can be shown to be negligible for a particular physical situation.

An interesting improvement to Eq. (9.19) can easily be made by allowing for the fact that the strain energy of a dislocation varies with the angle θ between its line and its Burgers vector. (For an isotropic medium, this variation is given by Eq. (9.9).) As a consequence, if the equilibrium of an isolated segment of dislocation is considered, not only must a force E be applied to its ends to stop it shrinking, but also a couple Γ per unit length must be applied to it to stop it turning into an orientation of lower energy.

The value of Γ is found by considering an infinitesimal rotation $d\theta$ of the segment from its equilibrium position. Setting the energy change equal to zero:

$$\Gamma d\theta - dE = 0$$

or:

$$\Gamma = \frac{dE}{d\theta} \tag{9.24}$$

As the dislocation curves through the infinitesimal angle $d\theta$, the couple per unit length acting upon it changes by $(d^2E/d\theta^2)\, d\theta$. This results in unbalanced radial forces acting on the ends of a curved segment of length dS, as shown by Figure 9.5. The total radial force per unit length needed to keep the dislocation bowed out is now given by:

$$F_r\, dS = E\, d\theta + \frac{d^2E}{d\theta^2} d\theta$$

or:

$$F_r = \frac{E}{\rho} + \frac{1}{\rho}\frac{d^2E}{d\theta^2} \tag{9.25}$$

If the line tension T is *defined* by the equation:

$$F_r = \frac{T}{\rho} \tag{9.26}$$

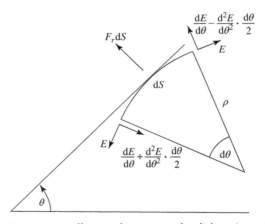

Figure 9.5 Forces acting on a small curved segment of a dislocation when account is taken of the variation of the strain energy E as a function of the angle θ between its line locally and its Burgers vector, **b**

then, from Eq. (9.25), T is given by:

$$T = E + \frac{d^2 E}{d\theta^2} \tag{9.27}$$

Substitution of Eq. (9.9) into Eq. (9.27) shows that a screw dislocation in an isotropic material is much stiffer than an edge, even though the edge has the greater elastic energy. Physically, this is because in order to bend a screw dislocation, it must be given some edge character, of relatively high energy. In aluminium, which as we have already noted is approximately isotropic elastically, $v = 0.34$, so that the screw dislocation is four times as stiff as the edge.

9.2 Stacking Faults and Partial Dislocations

So far, it has been assumed that the smallest Burgers vector that a crystal dislocation may possess is the shortest lattice vector. Although such a dislocation could reduce the elastic strain energy by dissociating into two dislocations with smaller Burgers vectors, this would create a fault in the crystal structure, which, in general, would have a high energy. In certain special cases, however, the fault has a rather small energy. When this is so, the dislocation does indeed dissociate into two dislocations, called *partial dislocations* or, briefly, partials, which leave a planar fault between them as they move apart. The mutual repulsion of the partials due to their stress fields decreases as the partials separate, but the attractive force exerted by the surface tension of the fault remains constant. An equilibrium separation is therefore reached at which the forces of repulsion and attraction balance.

It is now well established that $\langle 110 \rangle$ dislocations in c.c.p. metals dissociate into partials on the $\{111\}$ slip plane. The fault that lies between the partials is a fault in the stacking sequence of the $\{111\}$ planes, which does not change the relative positions of atoms that are nearest neighbours of one another. The production of this stacking fault can best be visualized with the aid of sheets of closely packed balls, stacked one on top of another (Section 3.2). By sliding an upper block of sheets through $a/6 \langle 211 \rangle$ over those underneath, the so-called intrinsic stacking fault is produced (Figure 9.6). For example, if the second ABC block in the sequence ABCABC is slid over the first, the sequence becomes ABCBCA ... The sheets at the fault are now in the sequence BCBC, which, if continued, produces a close-packed hexagonal structure. Such a fault occurs when the $\frac{1}{2} \langle 110 \rangle$ dislocation dissociates into $\frac{1}{6} \langle 211 \rangle$ dislocations, which are called Shockley partials or Heidenreich–Shockley partials [7]:

$$\tfrac{1}{2}[01\bar{1}] \rightarrow \tfrac{1}{6}[11\bar{2}] + \tfrac{1}{6}[\bar{1}2\bar{1}] \tag{9.28}$$

for dissociation of the $\frac{1}{2}[01\bar{1}]$ dislocation on the (111) plane. The pattern of atoms at the (111) slip plane around a $\frac{1}{2}[01\bar{1}]$ 60° dislocation that has dissociated into Shockley partials is shown in Figure 9.6. (For simplicity in drawing, the partials are shown to be very narrow; in a real crystal they would be wider.)

The separation of the Shockley partials is determined by the energy per unit area, γ, of the stacking fault between them. The fault draws the partials together with a force γ per

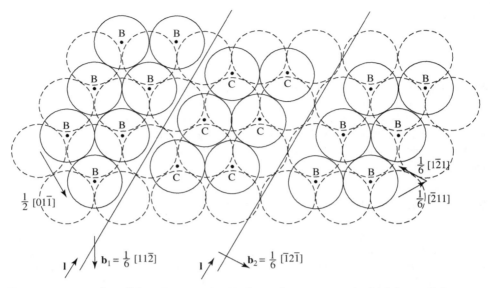

Figure 9.6 A perfect dislocation on a (111) plane of a c.c.p. metal which has split into two Shockley partials

unit length, so that the equilibrium separation r is given by:

$$\frac{G}{r} = \gamma \tag{9.29}$$

where the elastic repulsion factor G can be calculated. Assuming isotropic elasticity, the screw component of the first partial, $b_1 \cos \theta_1$, gives a shear stress on the slip plane:

$$\sigma_{23} = \frac{\mu b_1 \cos \theta_1}{2\pi r} \tag{9.30}$$

where Ox_2 is normal to the slip plane, Ox_3 is parallel to the dislocations and θ_1 is the angle between the partial dislocation line and its Burgers vector. From Eqs. (8.22), the edge component gives the stress:

$$\sigma_{12} = \frac{\mu b_1 \sin \theta_1}{2\pi(1-v)r} \tag{9.31}$$

The component of the traction on the slip plane in the direction of the Burgers vector of the second partial is given by[1]:

$$\tau = \sigma_{23} \cos \theta_2 + \sigma_{12} \sin \theta_2 \tag{9.32}$$

Since, by Section 8.3, $G = \tau b_2$, it follows that:

$$G = \frac{\mu b_1 b_2}{2\pi} \left(\cos \theta_1 \cos \theta_2 + \frac{\sin \theta_1 \sin \theta_2}{1-v} \right) \tag{9.33}$$

---

[1] Care must be taken to measure θ_1 and θ_2, the angles between the partial dislocation lines and their Burgers vectors, in the same sense.

The substitution of typical values for the constants in Eq. (9.33) shows that a relatively wide separation of the partials of the order of 100 Å occurs only when the stacking fault energy is as low as $10\,\mathrm{mJ\,m^{-2}}$. Most c.c.p. metals and alloys have a higher stacking fault energy than this, so the separation of partials will only be seen in the transmission electron microscope using weak beam imaging conditions [8].

The remaining sections of this chapter are devoted to the study of dislocations in specific crystal structures.

9.3   Dislocations in C.C.P. Metals

The $\langle 110 \rangle$ slip direction of c.c.p. metals is easy to understand, because $\frac{1}{2}\langle 110 \rangle$ is the shortest lattice vector. The $\{111\}$ slip plane is the plane on which a 'perfect' dislocation with a $\frac{1}{2}\langle 110 \rangle$ Burgers vector may split into two Shockley partial dislocations, as described in Section 9.2. The equilibrium separation of the two partials is determined by the energy of the stacking fault between them. The lower the stacking fault energy, the wider the split and the more rigorously should slip be confined to a particular $\{111\}$ plane. The stacking fault energy is therefore a property of particular physical importance because it determines the propensity of materials to twin and to cross-slip; unfortunately, it is also very difficult to measure accurately. Various values are listed in Table 9.2.

The production of a stacking fault during slip is effectively illustrated by a rigid ball model, such as that shown in Figure 9.6. In this model, the slip path that requires the least 'riding up' of the balls follows a zigzag route along the valleys in the $\{111\}$ surface, in two alternating $\langle 112 \rangle$ directions. The displacements along $\langle 112 \rangle$ alternately create and remove a stacking fault. From inspection of this model, one might wrongly conclude that only one type of stacking fault could occur on $\{111\}$. In particular, the stacking fault that would be obtained by interchanging the positions of the two partials in Figure 9.6 appears at first to be impossible, since it seems to involve A–A stacking, with a large ride-up of the balls when the leading dislocation advances. However, this conclusion can be modified if the necessary displacements are spread over two successive $\{111\}$ planes. For example, at the position of the dislocation on the left of Figure 9.6, starting with a perfect crystal, a displacement $\frac{1}{6}[\bar{2}11]$ of all the balls above the bottom $\{111\}$ layer to the right of this position, followed by a displacement $\frac{1}{6}[1\bar{2}1]$ of all the balls above the next $\{111\}$ layer, produces a net $\frac{1}{6}[\bar{1}\,\bar{1}2]$ displacement without violating nearest-neighbour relationships. Formally, the partial dislocation then consists of two Shockley partials, $\frac{1}{6}[2\bar{1}\,\bar{1}]$ and $\frac{1}{6}[\bar{1}2\bar{1}]$, on adjacent planes,

Table 9.2   The stacking fault energies of c.c.p. metals

Metal	Stacking fault energy (mJ m^{-2})	Ref.
Ag	16	[9]
Al	135	[10]
Au	32	[11]
Cu	41	[9, 12]
Ni	120–130	[13]

but, apart from at its core, the dislocation cannot be distinguished from a single $\frac{1}{6}[11\bar{2}]$ Shockley partial.

Therefore, by spreading the displacements over two planes, the $\frac{1}{6}[11\bar{2}]$ Shockley partial can be made the leading partial for glide from left to right in Figure 9.6. When an interchange of the order of the partials is accomplished in this way, they are separated by a 'double' stacking fault, which is called an extrinsic fault. The energy of this fault is expected to be slightly higher than that of the intrinsic (single) fault. Consequently, the form of splitting shown in Figure 9.6 is believed to be the usual one.

The distinction between the structures of intrinsic and extrinsic stacking faults can be neatly displayed by a notation due to Frank [14]. When a close-packed plane is stacked upon the plane beneath according to a pair in the sequence ABC ... , the stacking of that plane is depicted by a triangle Δ. When a plane is stacked so that its relation to the plane beneath it is a pair in the reverse sequence BAC ... , its stacking is then depicted by an inverted triangle ∇ (pronounced *nabla*). These symbols are derived from the stacking of balls – they represent the orientation of the triangles of balls onto which the balls in the next layer are stacked. The c.c.p. structure is then shown as $\Delta\Delta\Delta\Delta$ and the h.c.p. structure as $\Delta\nabla\Delta\nabla$ ... The intrinsic stacking fault is $\Delta\Delta\nabla\Delta\Delta$, while the extrinsic stacking fault is $\Delta\Delta\nabla\nabla\Delta\Delta$, emphasizing that the extrinsic fault is equivalent to two intrinsic faults on adjacent planes.

It is possible to produce stacking faults by means other than slip on the close-packed plane. An intrinsic stacking fault is produced by removing part of a {111} layer of atoms, as shown in Figure 9.7a, provided that the gap is then closed directly, without any shear parallel to the {111} plane. Physically, this may happen when excess vacancies collect together on a {111} plane. Squeezing in a piece of an extra {111} layer creates an extrinsic fault, as shown in Figure 9.7b. The faults shown in Figure 9.7 are both bounded by partial dislocations whose Burgers vectors are normal to the {111} plane and are of the magnitude of the spacing of these planes; that is, $\frac{1}{3}\langle 111\rangle$. Distinct from Shockley partials, these partials could glide only on surfaces that are normal to the usual {111} slip plane; if they did so they would leave a high-energy stacking fault behind them. They are called *Frank partials* or, because of their immobility, Frank sessiles.

Although a stacking fault must be bounded by a partial dislocation, or end at a surface, it is now clear that the Burgers vector of the partial is not uniquely determined by the type

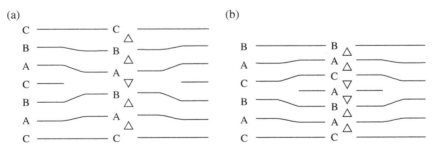

Figure 9.7 (a) Faulted vacancy loop in a c.c.p. metal, showing traces of the (111) planes in a plane normal to the loop. (b) Extra atom loop

of fault. In a given case, any lattice vector may be added to the Burgers vector of the partial without affecting the fault. For example, an intrinsic fault on (111) bounded by a $\frac{1}{6}[11\bar{2}]$ Shockley partial might equally well be bounded by two other Shockley partials, $\frac{1}{6}[\bar{2}11]$ or $\frac{1}{6}[1\bar{2}1]$. Hence:

$$\frac{1}{6}[\bar{2}11] = \frac{1}{6}[11\bar{2}] + \frac{1}{2}[\bar{1}01] \tag{9.34}$$

$$\frac{1}{6}[1\bar{2}1] = \frac{1}{6}[11\bar{2}] + \frac{1}{2}[0\bar{1}1] \tag{9.35}$$

or, by the Frank partial with the Burgers vector $\frac{1}{3}[\bar{1}\bar{1}\bar{1}]$:

$$\frac{1}{3}[\bar{1}\bar{1}\bar{1}] = \frac{1}{6}[11\bar{2}] + \frac{1}{2}[\bar{1}\bar{1}0] \tag{9.36}$$

Before going on to consider reactions between dislocations in c.c.p. metal crystals, it will be helpful to introduce a device known as Thompson's tetrahedron [15]. This gives a picture of various Burgers vectors in relation to the four {111} planes, which can be chosen to form the four faces of a tetrahedron. This tetrahedron and its relation to the crystal structure is shown in Figure 9.8a; the four faces of the tetrahedron, when they are opened out from the vertex D, are shown in Figure 9.8b. The lettering of the various points is conventional and follows Thompson's original paper.[2] An edge of the tetrahedron, such as CB, represents the Burgers vector of a perfect dislocation, $\frac{1}{2}\langle110\rangle$. Vectors given by Greek–Roman letter combinations such as δC represent Shockley partials, except when the Greek and Roman letters are in the same position in their respective alphabets, such as γC, which indicates a Frank partial.

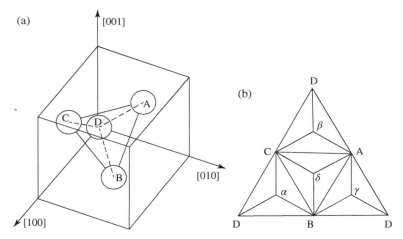

Figure 9.8 Thompson's tetrahedron. (After Thompson [15]). Source: Reproduced by permission of the Institute of Physics.

---

[2] An alternative convention would be to letter the tetrahedron as the mirror image of Figure 9.8a.

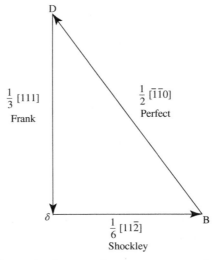

Figure 9.9 The vector relationship between the Burgers vectors of a Shockley partial, a perfect dislocation and a Frank partial in a c.c.p. metal

The reaction between a Frank partial, a Shockley partial and a perfect dislocation given by Eq. (9.36) becomes, in Thompson's notation, the reaction:

$$\delta D = \delta B + BD$$

The relationship between these three Burgers vectors is shown in Figure 9.9. This reaction has some interesting applications. The stacking fault within the 'vacancy loop' of Figure 9.7a can be removed if the Frank dislocation decomposes according to Eq. (9.36) and if then its Shockley component shrinks to nothing by gliding in the plane of the loop. This removes the stacking fault at the expense of an increase in elastic energy due to the increase in the Burgers vector of the loop from $\frac{1}{3}[\bar{1}\,\bar{1}\,\bar{1}]$ to $\frac{1}{2}[\bar{1}\,\bar{1}0]$. In comparing the energies of faulted and unfaulted loops, allowance must be made for the shortening of the unfaulted loop when it glides into a $(\bar{1}\,\bar{1}0)$ orientation normal to its Burgers vector. The unfaulted loop is favoured when the stacking fault energy is high and also when the loop is large, because the energy gained by destroying the fault is proportional to the *area* of the unfaulted region, whereas the extra elastic energy is approximately proportional to its *perimeter*.

Another application is to the possible dissociation of a perfect dislocation into a Shockley partial and a Frank partial:

$$BD \rightarrow B\delta + \delta D$$

For example:

$$\frac{1}{2}[\bar{1}\,\bar{1}0] \rightarrow \frac{1}{6}[\bar{1}\,\bar{1}2] + \frac{1}{3}[\bar{1}\,\bar{1}\,\bar{1}] \tag{9.37}$$

Since the Burgers vectors of these two partials are orthogonal, and since a stacking fault would have to be created between them, this dissociation does not ordinarily occur. It may occur in special cases, such as when the Shockley components from the two members of a perfect dislocation dipole can annihilate one another, leaving behind a Frank dipole. This reaction is simply the reverse of the unfaulting of a prismatic loop described above.

In order to define the *sense* of the Burgers vector of a dislocation, when it is written in Thompson's notation it is necessary to introduce a convention that will assign a *sense* to the slip displacement signified by the symbol δC, for instance. The following convention can be used.

The sense of the slip displacement signified by δC is defined as follows. The observer looks at the slip plane ABC from outside the tetrahedron; that is, with the apex D pointing away from him or her (Figure 9.8a). Then the atoms above the slip plane ABC closest to the observer are displaced by δC. Referring to the crystal structure, it is clear that the displacements δA, δB and δC produce an intrinsic stacking fault, whereas Aδ, Bδ and Cδ produce an extrinsic fault. The dissociation of the perfect dislocation CB into Shockley partials on ABC can be written as:

$$CB \rightarrow C\delta + \delta B \tag{9.38}$$

The Burgers vectors of the partials in the Thompson notation are then as shown in Figure 9.10, according to the FS/RH convention. (The line vectors have been chosen to run down the page, so that the observer looks in the negative line direction, in order to match the figure with the logical form of Eq. (9.38).) Reversing the order of the partials would produce an extrinsic fault between them.

The order of the Shockley partials is an important factor in dislocation reactions. The only reaction between perfect dislocations that definitely leads to a reduction in energy is of the type:

$$CB + BA \rightarrow CA \tag{9.39}$$

If the dislocations of Burgers vectors BA and CB meet on their common slip plane, they will react over a certain length, forming two nodes (Figure 9.11). It is apparent from a

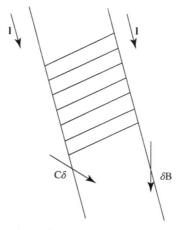

Figure 9.10   The dissociation of a perfect dislocation CB into two partial dislocations Cδ and δB separated by a stacking fault which is shown partially shaded

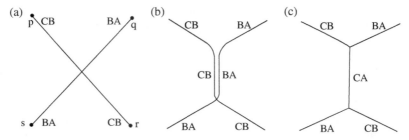

Figure 9.11   Reaction of two dislocations on a common slip plane. The dislocations are assumed to be pinned at the points p, q, r and s

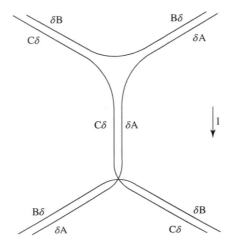

Figure 9.12   Same reaction as Figure 9.11 but showing splitting into partials

consideration of Figure 9.12 that the nodes will have different structures, because of the sequence of the Shockley partials (it is assumed that only intrinsic faults occur). Alternate expanded and contracted nodes are seen in stainless steel, for example [16].

The same reaction (9.39) can also occur when the dislocations are on different slip planes. For example, suppose that dislocations CB and BA lie on the planes opposite vertices A and C respectively, and let their lines lie parallel to the intersection of these planes, BD, as they glide towards one another. This event is likely to occur during the later stages of the tensile deformation of a crystal, because it corresponds to the intersection of primary and conjugate slip (Section 7.4). If the two dislocations on the {111} planes are dissociated into Shockley partials, as shown in Figure 9.13, and if they come together in such a way that just before they meet their stacking faults make an acute angle ($\cos^{-1}(1/3) = 70.53°$) with one another, the leading partials αB and $B\gamma$ will react to form a new partial along BD:

$$\alpha B + B\gamma \rightarrow \alpha\gamma$$

$$\tfrac{1}{6}[12\bar{1}] + \tfrac{1}{6}[\bar{2}\bar{1}1] \rightarrow \tfrac{1}{6}[\bar{1}10]$$

(9.40)

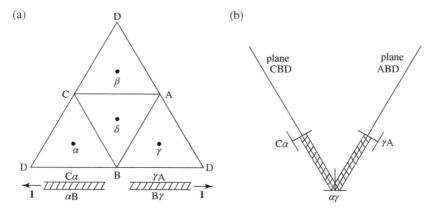

Figure 9.13 Lomer–Cottrell lock

The edge dislocation $\alpha\gamma$ is called a stair-rod, a name suggested to Thompson by Nabarro (quoted by Thompson in a footnote in [15]), from the way in which it joins two stacking faults at a bend rather like a stair rod holding down a stair carpet. The whole configuration is called either a *Lomer–Cottrell lock* or a *Cottrell–Lomer lock* [17, 18]; here, we shall adopt the former terminology. Since it cannot glide, it is an important obstacle to other dislocations.

The stair-rod dislocation of the type $\alpha\gamma$ can also be produced by the dissociation of a Frank partial:

$$\alpha A \rightarrow \alpha\gamma + \gamma A$$
$$\tfrac{1}{3}[\bar{1}11] \rightarrow \tfrac{1}{6}[\bar{1}10] + \tfrac{1}{6}[\bar{1}12] \tag{9.41}$$

This reaction makes it possible to understand the small tetrahedra of stacking faults that can be produced experimentally by quenching gold from near its melting point to room temperature, and then ageing at $100\,^{\circ}\mathrm{C}$ [19].[3] Imagine that the excess vacancies which are trapped by the quench first form a Frank loop similar to that shown in Figure 9.7a. Let the loop be a triangle, with sides along the $\langle 110 \rangle$ directions AB, BC, CA. Each side can then dissociate on a different $\{111\}$ plane, as shown in Figure 9.14. The Shockley dislocations produced by the dissociations attract one another and react at the intersections of the $\{111\}$ planes to form stair-rod dislocations. The final product is a tetrahedron whose faces are intrinsic stacking faults and whose edges are stair-rod dislocations. Each edge of the tetrahedron is exactly like the corner of a Lomer–Cottrell lock. The formation of the tetrahedron decreases the elastic energy at the expense of additional stacking fault energy.

The $\tfrac{1}{6}\langle 110 \rangle$ dislocation is one of four possible stable stair rods. Other dislocations at an acute-angle bend can be derived by adding lattice vectors to $\tfrac{1}{6}[110]$. The only resultant which is stable, according to Frank's rule, is $\tfrac{1}{3}[1\bar{1}0]$, formed by adding $\tfrac{1}{2}[1\bar{1}0]$. One of the interesting occurrences that would produce an *obtuse-angle* stacking fault bend is the

---

[3] Ageing means holding at a constant temperature for a period of time, usually in order to allow some change in the structure of the material to occur.

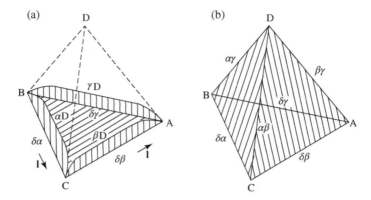

Figure 9.14 Formation of a stacking fault tetrahedron from a Frank vacancy loop

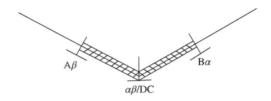

Figure 9.15 Hirth lock

reaction of the leading partials of the dislocations BD on the plane opposite the vertex A and AC on the plane opposite the vertex B:

$$BD \rightarrow B\alpha + \alpha D$$
$$AC \rightarrow A\beta + \beta C \tag{9.42}$$

$$\tfrac{1}{6}[\bar{2}\,\bar{1}\,\bar{1}] + \tfrac{1}{6}[2\bar{1}1] \rightarrow \tfrac{1}{6}[0\bar{1}0] \tag{9.43}$$

The resulting configuration, shown in Figure 9.15, is called the *Hirth lock* [20, 21], even though, as Nabarro [22] notes in his book on p. 220, it first appeared in a tabulation given by Friedel [23]. In Thompson's notation, Eq. (9.43) may be written as:

$$\alpha D + \beta C \rightarrow \alpha\beta/DC \tag{9.44}$$

The $\tfrac{1}{3}\langle 010 \rangle$ vector denoted by $\alpha\beta/DC$ is in the direction of the line joining the point bisecting $\alpha\beta$ to the point bisecting DC and has twice the magnitude of this line.

Some of the properties of dislocations in c.c.p. metals are summarized in Table 9.3.

9.4 Dislocations in the Rock Salt Structure

The dislocations that have been identified in crystals of the rock salt (NaCl) structure have a $\tfrac{1}{2}\langle 110 \rangle$ Burgers vector, as would be expected (Table 9.1). The slip plane is usually $\{110\}$, but not always (Table 7.1).

Table 9.3 Dislocations in c.c.p. metals

Name of dislocation	Burgers vector **b**		Relative b^2
	Indices	Thompson notation	
Perfect	$\frac{1}{2}\langle 110 \rangle$	AB	1
Shockley	$\frac{1}{6}\langle 211 \rangle$	$A\delta$	$\frac{1}{3}$
Frank	$\frac{1}{3}\langle 110 \rangle$	$A\alpha$	$\frac{2}{3}$
Thompson stair-rod	$\frac{1}{6}\langle 110 \rangle$	$\alpha\delta$	$\frac{1}{9}$
Acute stair-rod	$\frac{1}{3}\langle 110 \rangle$	–	$\frac{4}{9}$
Obtuse stair-rod	$\frac{1}{3}\langle 100 \rangle$	$\delta\alpha/CB$	$\frac{2}{9}$
Obtuse stair-rod	$\frac{1}{6}\langle 310 \rangle$	–	$\frac{5}{9}$

When slip on $\{110\}$ is examined, it is impossible to find a stacking fault that looks as though it might have a low energy and could serve as the halfway stage of an indirect $\frac{1}{2}\langle 110 \rangle$ displacement. Therefore, the $\frac{1}{2}\langle 110 \rangle$ dislocation is not expected to split into partials on $\{1\bar{1}0\}$. The fact that only perfect dislocations need be considered greatly simplifies the analysis of dislocation reactions. The relationship between the $\langle 110 \rangle$ Burgers vectors and the $\{1\bar{1}0\}$ slip planes is shown in Figure 9.16, which may be compared with Thompson's tetrahedron. A consideration of this figure confirms that, as expected from the crystallography, each Burgers vector lies in only one slip plane (e.g. AC lies in OAC only) and each slip plane contains only one Burgers vector (OAC contains only AC). This one-to-one correspondence contrasts with the slip systems of the c.c.p. metals, and still further simplifies the study of dislocation interactions. In fact, there is only one dislocation reaction to consider – that between dislocations on planes at 60° to one another:

$$\left.\begin{array}{c} AB + BC \rightarrow AC \\ \frac{1}{2}[\bar{1}01] + \frac{1}{2}[01\bar{1}] \rightarrow \frac{1}{2}[\bar{1}10] \end{array}\right\} \tag{9.45}$$

If the reacting dislocations AB and BC are confined to their slip planes, (101) and (011) respectively, the resultant AC is laid down along the line of intersection of the slip planes, OB, $[11\bar{1}]$. Hence, the glide plane of the resultant AC is (112), and for this reason it is very immobile at low temperatures. In LiF at room temperature, the dislocations on one plane will block slip on another which intersects it at 60° so efficiently that, after 1% shear strain in single slip, a resolved shear stress of about 15 times that which was needed to activate the first system must be put upon the intersecting system in order to bring it into operation [24].

Although less complex in their interactions than dislocations in c.c.p. metals, dislocations in crystals with the rock salt structure do have a property of special interest – they can carry an electric charge. The edge dislocation on $\{110\}$ has an 'extra plane' consisting of a sheet of Na^+ ions and a sheet of Cl^- ions (Figure 9.17). This sheet of NaCl 'molecules' has been isolated in Figure 9.18. Its edge defines the dislocation line, which has a number of jogs in it. At A, an extra anion defines two jogs. The section BC has a net charge of $+e$; therefore the

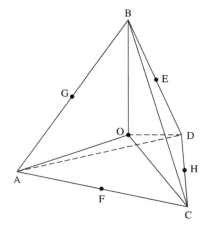

Figure 9.16 Relation between slip planes and directions in {110} ⟨1$\bar{1}$0⟩ or {110} ⟨1$\bar{1}$1⟩ slip. The point O lies at the centre of the tetrahedron ABCD whose edges are parallel to ⟨110⟩. The planes OAB, OAC, and so on are {110} and the directions OA, OB, OC, OD are ⟨111⟩

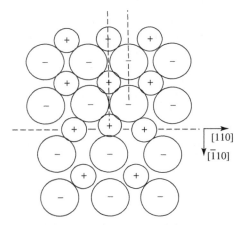

Figure 9.17 Edge dislocation in NaCl. The slip plane and the two sheets of ions constituting the extra plane are shown dotted

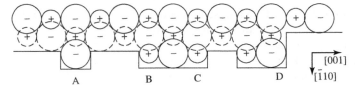

Figure 9.18 Edge dislocation in NaCl viewed in a different section to that shown in Figure 9.17. Here, the dislocation line runs left to right and has a number of jogs at A, B, C and D

jogs at B and C are each assigned a charge of $+ e/2$. The 'whole' jog at D is uncharged. This is the type of jog which is produced by intersection with an orthogonal screw dislocation; charged jogs of types B and C may be called half-jogs or, because they are produced when a single ion jumps on to or off the edge of the extra plane, diffusional jogs. It is apparent that the line as a whole can be charged, up to a maximum of $e/2$ per atom length.

Physically, a line may become charged by acting as a source or sink for vacancies (Section 10.1). For example, because the energy needed to make a single cation vacancy is less than that needed to make an anion vacancy in NaCl, either a surface or an edge dislocation raised to a high temperature will generate excess cation vacancies which can diffuse into the crystal until their emission is throttled by the electric field that is built up as a consequence of the resultant space charge [25, 26]. Relative to the adjacent crystal, the dislocation or surface acquires a positive charge, which is balanced by the nearby vacant cation sites in the crystal. On the other hand, a dislocation in an impure crystal of NaCl can acquire a negative charge in order to compensate, partly, for the extra positive charge carried by divalent impurity cations, such as Ca^{2+}, by acting as a sink at which excess cation vacancies caused by the impurity cations can condense [25, 26]. At low temperatures the effect of relatively low p.p.m. levels of impurities will predominate over the equilibrium properties that the dislocation would have in a pure crystal.

9.5 Dislocations in Hexagonal Metals

By analogy with c.c.p. metals, the slip plane of hexagonal metals should be the closest-packed basal plane (0001), the Burgers vector of a slip dislocation should be the **a** vector of the hexagonal lattice, $\frac{1}{3}[11\bar{2}0]$, and, furthermore, this dislocation may be expected to split into two $\frac{1}{3}\langle 10\bar{1}0 \rangle$ partials, the analogues of Shockley partials. However, the analogy is certainly incomplete, because other slip planes, and another slip direction, sometimes operate (Table 7.2). With this reservation in mind, we will first examine dislocations on the basal plane, produced by slip and by the aggregation of point defects.

Since the three $\langle 11\bar{2}0 \rangle$ slip vectors are coplanar, reaction between slip dislocations can only form a network of dislocations in the basal plane. Written in the notation of Figure 9.19, the reaction is:

$$\left. \begin{array}{c} AB + BC \rightarrow AC \\ \frac{1}{3}[\bar{1}2\bar{1}0] + \frac{1}{3}[\bar{1}\,\bar{1}20] \rightarrow \frac{1}{3}[\bar{2}110] \end{array} \right\} \tag{9.46}$$

The splitting of a perfect dislocation can also be shown on Figure 9.19:

$$\left. \begin{array}{c} AB \rightarrow A\sigma + \sigma B \\ \frac{1}{3}[\bar{1}2\bar{1}0] \rightarrow \frac{1}{3}[\bar{1}100] + \frac{1}{3}[01\bar{1}0] \end{array} \right\} \tag{9.47}$$

Between the partials $A\sigma$ and σB, which are analogous to Shockley partials, there is a small layer of c.c.p. stacking. The sequence ABABABA becomes ABACBCB, or in Frank's notation, $\triangle \triangledown \triangle \triangledown \triangle \triangledown$ is changed to $\triangle \triangledown \triangledown \triangledown \triangle \triangledown$. The order of the partials on successive basal planes must alternate; that is, if it is $A\sigma - \sigma B$ on one plane it must be $\sigma B - A\sigma$ on the next.

(a)   (b)

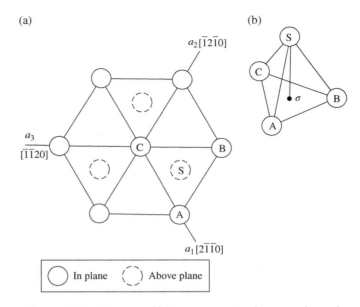

Figure 9.19 Atoms and lattice vectors in a hexagonal metal

This is in contrast to the splitting of perfect dislocations in a c.c.p. metal, where the order of the Shockley partials must always be the same if the same type of stacking fault is to be produced.

In metals of small c/a ratio, the $\frac{1}{3}\langle\overline{2}110\rangle$ dislocation slips quite easily on a prism plane, $\{10\overline{1}0\}$, indicating that it is not widely split on the basal plane. Zirconium and titanium, for example, slip much more easily on prism planes than on the basal plane. It is interesting to note that the reaction $AB + BC \rightarrow AC$ between prism plane dislocations cannot produce a strong barrier, because the product, formed along the intersection of two of the prism planes, is always oriented so that it can glide on the third.

The experimentally determined stacking fault energy of Co determined by weak beam transmission electron microscopy is relatively low, $27 \pm 4\,\mathrm{mJ\,m^{-2}}$ at room temperature, with dissociations observed of 3.5 ± 0.5 nm for perfect screw dislocations and 9.5 ± 0.5 nm for perfect edge dislocations, respectively [27]. (Cobalt transforms to a c.c.p. structure at 420 °C – see Section 12.3.) Stacking fault energies in other h.c.p. metals are generally regarded as being much higher (see for example [27–29]), of the order of $200\,\mathrm{mJ\,m^{-2}}$, consistent with $\frac{1}{3}\langle\overline{2}110\rangle$ dislocations not being split to any significant extent on the basal plane.

As with c.c.p. metals, stacking faults can arise as a result of point defect condensation. Such a fault is different from the fault produced by the splitting of a $\frac{1}{3}\langle11\overline{2}0\rangle$ dislocation. When a disc of vacancies collects on a basal plane, a prismatic dislocation loop forms at its perimeter. If the disc is one vacancy thick, and if A–A stacking is to be avoided, the loop may be faulted in one of two ways [30]. Either a single basal layer or the whole crystal on one side of the loop must be sheared over by a vector of the $A\sigma$ type. These alternatives are shown in Figure 9.20. In case (a), the single layer can be shifted from an A into a C position by passing $A\sigma$ partials of opposite sign on either side of it. The Burgers vector

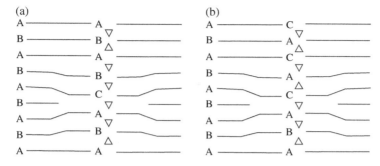

Figure 9.20 Two possible structures for a vacancy loop in a hexagonal metal

of the prismatic dislocation remains $\frac{1}{2}[0001]$, σS in Figure 9.19b, at the cost of a stacking fault which is triple, in terms of the number of next-nearest-neighbour stacking violations [30]. In case (b), a single $A\sigma$ partial is passed, so that the loop encloses a single stacking fault, but its Burgers vector increases to AS, as follows:

$$\left.\begin{array}{c} A\sigma + \sigma S = AS \\ \frac{1}{3}[\bar{1}100] + \frac{1}{2}[0001] \rightarrow \frac{1}{6}[\bar{2}203] \end{array}\right\} \tag{9.48}$$

Similar reasoning applies to faulted loops produced by discs of extra atoms [30]. There is clear experimental evidence for these single stacking fault vacancy loops in zinc and magnesium [30–33]; such stacking faults have a smaller energy than the double fault produced by the splitting of a perfect slip dislocation.

A more surprising slip system is the $\{11\bar{2}2\}\langle11\bar{2}3\rangle$, called second-order pyramidal glide. This occurs in a number of h.c.p. metals, such as Zn, Cd, Mg and Be [34]. The remarkable feature is the large size of the slip vector, $\frac{1}{3}\langle11\bar{2}3\rangle$; that is, $\mathbf{c}+\mathbf{a}$. Even more puzzling is the fact that this system operates, second only to basal slip, in zinc and cadmium, which have the largest c/a ratios. A slip plane and its slip vectors are shown in the hexagonal cell in Figure 9.21. Each $\{11\bar{2}2\}$ slip plane contains one $\frac{1}{3}\langle11\bar{2}3\rangle$ slip vector and each slip vector lies in one slip plane.

The relationship between the six slip planes and six Burgers vectors, by means of a surface analogous to that used to illustrate the rock salt slip systems (Figure 9.16), is shown in Figure 9.22. The slip vectors are A′B, A′C, B′C, B′A, C′A, C′B, and their antiparallel counterparts; the six slip planes are represented by A′BσA′ for the slip vector A′B, and so on.

In a review of deformation modes in h.c.p. metals, Yoo et al. [35] noted that little is known about the dislocation cores of the $\mathbf{c}+\mathbf{a}$ dislocations apart from results obtained through atomistic simulation: there is a distinct lack of detailed experimental observations on dislocation interactions and their role in pyramidal slip. More recent work has confirmed this trend – while there have been improvements in the choice of atom potential to model $\mathbf{c}+\mathbf{a}$ dislocations in h.c.p. metals [36], experimental observations of the pyramidal glide system in h.c.p. metals still remain few in number. In addition, as Hull and Bacon note [37], there seems to be a close relationship between the occurrence of $\{11\bar{2}2\}\langle11\bar{2}3\rangle$ slip and twinning (Chapter 11) in h.c.p. metals, which serves to complicate the situation.

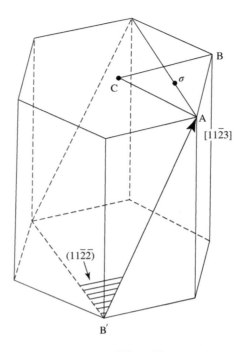

Figure 9.21 A $\{11\overline{2}\overline{2}\}\langle11\overline{2}3\rangle$ slip system

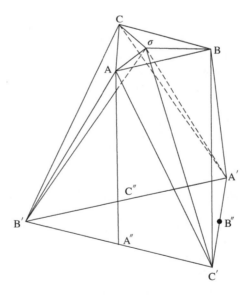

Figure 9.22 Relation between slip planes and directions in $\{11\overline{2}\overline{2}\}\langle11\overline{2}3\rangle$ slip

Possible dislocation reactions between $\mathbf{c} + \mathbf{a}$ dislocations on $\{11\bar{2}\bar{2}\}$ planes in this system can be dealt with easily by decomposing each $\frac{1}{3}\langle 11\bar{2}3 \rangle$ vector into its $\mathbf{c}$ and $\mathbf{a}$ components. The $\mathbf{c}$ components either add or cancel; provided $c > a$, there is an increase in energy when they add and a decrease when they cancel, whatever the $\mathbf{a}$ components may be. There are three reactions, then, corresponding to the three possible combinations of the $\mathbf{a}$ components when the $\mathbf{c}$ components cancel. The reaction is least favourable energetically when the $\mathbf{a}$ components are alike:

$$\left. \begin{array}{c} \mathrm{C'A + AB' \to C'B'} \\ \frac{1}{3}[\bar{1}\bar{1}23] + \frac{1}{3}[\bar{1}\bar{1}2\bar{3}] \to \frac{2}{3}[\bar{1}\bar{1}20] \end{array} \right\} \tag{9.49}$$

The product $\mathrm{C'B'}$ is formed along the line $\mathrm{A}\sigma$ in the basal plane common to the two slip planes $\mathrm{C'A\sigma C'}$ and $\mathrm{B'A\sigma B'}$ and can immediately split into two perfect $\frac{1}{3}[\bar{1}\bar{1}20]$ dislocations. The most favourable reaction is:

$$\left. \begin{array}{c} \mathrm{C'A + BC' \to BA} \\ \frac{1}{3}[\bar{1}\bar{1}23] + \frac{1}{3}[\bar{2}\bar{1}\bar{1}3] \to \frac{1}{3}[1\bar{2}10] \end{array} \right\} \tag{9.50}$$

A consideration of Figure 9.22 shows that the product BA is formed along the line $\mathrm{C'}\sigma$ common to $\mathrm{C'A\sigma C'}$ and $\mathrm{C'B\sigma C'}$, so that it is likely to be immobile and an obstacle to other dislocations. The result of the third reaction is less easy to see from Figure 9.22. The reaction is:

$$\left. \begin{array}{c} \mathrm{C'A + BA' \to C'C''} \\ \frac{1}{3}[\bar{1}\bar{1}23] + \frac{1}{3}[\bar{2}11\bar{3}] \to \frac{1}{3}[\bar{3}030] \end{array} \right\} \tag{9.51}$$

The product $\mathrm{C'C''}$ should split into two $\mathbf{a}$ dislocations, $\mathrm{AC + BC}$, but since its line does not lie in the basal plane, these dislocations are also likely to be immobile.

It is also conceivable in principle that the $\frac{1}{3}\langle 11\bar{2}3 \rangle$ dislocation might itself split into partials of the type AS (Figure 9.19):

$$\frac{1}{3}[11\bar{2}3] \to \frac{1}{6}[02\bar{2}3] + \frac{1}{6}[20\bar{2}3] \tag{9.52}$$

However, there is no reason to assign a low energy to the fault, which would have to form on $\{11\bar{2}\bar{2}\}$. Furthermore, $\frac{1}{3}[11\bar{2}3]$ screw dislocations in zinc and cadmium are observed to cross-slip frequently, which is evidence against such splitting.

There are two reactions between a basal-plane dislocation and a second-order pyramidal dislocation. The first is simply a rearrangement of (9.50):

$$\left. \begin{array}{c} \mathrm{C'A + AB \to C'B} \\ \frac{1}{3}[\bar{1}\bar{1}23] + \frac{1}{3}[\bar{1}2\bar{1}0] \to \frac{1}{3}[\bar{2}113] \end{array} \right\} \tag{9.53}$$

The product lies along $\mathrm{A}\sigma$, out of its pyramidal slip plane $\mathrm{C'B\sigma C'}$. Therefore it can act as a barrier. The second reaction is simply the annihilation of the $\mathbf{a}$ component of the pyramidal:

$$\left. \begin{array}{c} \mathrm{C'A + A''C' \to A''A} \\ \frac{1}{3}[\bar{1}\bar{1}23] + \frac{1}{3}[11\bar{2}0] \to [0001] \end{array} \right\} \tag{9.54}$$

The extra plane of the product A″A consists of two basal layers of atoms. These may be thought of as tending to wedge open the basal plane, which is also a cleavage plane. In this respect the reaction is like that which produces the $\langle 100 \rangle$ dislocation in b.c.c. metals (Section 9.6).

9.6 Dislocations in B.C.C. Crystals

It is now well established that a slip dislocation in a b.c.c. metal has the expected Burgers vector $\frac{1}{2}\langle 111 \rangle$. The plane on which this dislocation prefers to glide is far less certain (Table 7.1). We shall first assume that slip occurs somewhat more easily on $\{110\}$ than on other planes in the $\langle 1\bar{1}1 \rangle$ zone.

The relationship between the $\frac{1}{2}\langle 111 \rangle$ Burgers vectors and the $\{110\}$ slip planes can be shown on the same diagram that was used for rock salt structures (Figure 9.16). It is evident from Figure 9.16 that any two different Burgers vectors meet either at 70.53° or at 109.47°. In the latter case the dislocations will react:

$$\left. \begin{array}{c} \text{OA} + \text{OC} \rightarrow \text{EF} \\ \frac{1}{2}[1\bar{1}1] + \frac{1}{2}[11\bar{1}] \rightarrow [100] \end{array} \right\} \tag{9.55}$$

The character of the product dislocation depends on the slip planes of the reactants, because these determine the direction in which its line lies. If the reacting dislocations are on the same slip plane then the glide plane of the product will of course be this plane as well. If they are on planes such as AOB and AOC, which intersect at 60° along a $\langle 111 \rangle$, then the product, formed along this line, will still be able to glide in a $\{110\}$ plane. However, if the reacting dislocations are on orthogonal planes, such as AOD and COB, their product will lie along a $\langle 001 \rangle$, such as GH in Figure 9.16. It is then an edge dislocation with a $\{100\}$ glide plane. The extra plane of this dislocation is parallel to the usual cleavage plane of a b.c.c. crystal, and it has been suggested that the crystal may be cracked open for a short distance beneath the extra plane [38].

The fact that slip is not very rigorously confined to any one plane is strong evidence that the $\frac{1}{2}\langle 111 \rangle$ dislocation is not split widely into partials, and that therefore stacking faults have a high energy. Nevertheless, it is possible to find plausible-looking stacking faults in a rigid-ball model of the structure. A ball model of the $\{110\}$ planes is shown in Figure 9.23. The arrows starting at P point out the natural path in which to slide the balls of the upper layer over the lower. This corresponds to the dissociation:

$$\frac{1}{2}[111] \rightarrow \frac{1}{8}[011] + \frac{1}{4}[211] + \frac{1}{8}[011] \tag{9.56}$$

The fault which is formed on both sides of the $\frac{1}{4}[211]$ dislocation occurs quite naturally in a rigid-ball structure, because it leads to closer packing (for this reason it may be important in the martensitic transformation of a b.c.c. to a h.c.p. structure – see Section 12.10). Thus, the $\frac{1}{8}[011]$ which produces the fault drops the ball P from a saddle position into a valley. However, nearest-neighbour relationships are violated and the fault must ordinarily have a rather high energy, since it has never been detected in b.c.c. metals.

Computer calculations using atomistic simulations have confirmed the high energy of faults in the b.c.c. crystal structure. For example, the early work of Vitek [39] using a

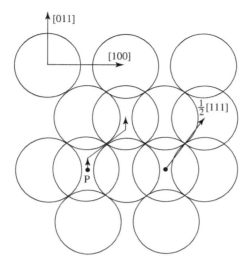

Figure 9.23 A ball model of the {110} planes in a b.c.c. metal

simple central force potential and limited computer power showed that there are no stable intrinsic faults on either the {110} or the {112} planes in b.c.c. crystals. As Ventelon and Willaime [40] have noted, atomistic simulations of screw dislocations in b.c.c. metals do not show splitting into well-defined partials and stacking faults, but instead reveal core configurations that are able to spread into the three {110} planes containing the ⟨111⟩ zone axis of the screw dislocation. The power of such calculations highlights the limitations of further hard-sphere models showing splitting of $\frac{1}{2}\langle111\rangle$ dislocations into partials separated by stacking faults on {110} and {211} planes, such as those considered in the first edition of this book [41]: such models are purely geometric in nature and therefore cannot predict the energies of the postulated stacking faults. By comparison, atomistic simulations are becoming increasingly sophisticated in terms of the choice of potential, even when they apply only to rigid dislocations at absolute zero; as Duesbery and Vitek argue [42], it is expected that results of such calculations will carry through, at least qualitatively, to conditions at finite temperature.

9.7 Dislocations in Some Covalent Solids

In crystals of diamond, Ge and Si, the atoms are held together by strong, directional covalent bonds (Figure 3.12). Dislocations in these crystals are immobile at low temperatures. This is because the Peierls stress is high; that is to say, the energy of the dislocation varies sharply with its position in the crystal. Extensive slip occurs only at elevated temperatures – above about 60% of the absolute melting temperature in Ge and Si. The slip plane is then {111} and the slip direction ⟨1$\bar{1}$0⟩.

It is possible to imagine the way in which glide might occur. The dislocation will lie, as much as it can, in a position of low energy, probably along some prominent crystallographic direction. It advances when, aided by thermal activation, a small length of line surmounts

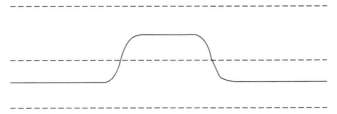

Figure 9.24 Double kink in a dislocation line. The dotted lines represent positions of high energy for the dislocation

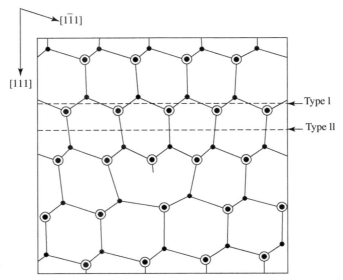

Figure 9.25 A 60° dislocation in sphalerite. The structure of a 60° dislocation in the diamond structure is identical, except that all the atoms are then the same

its energy barrier and spills over into the next energy trough (Figure 9.24). The connecting segment which is stranded on the energy hill is called a kink, distinguished from a jog by the fact that it lies wholly in the slip plane (Section 8.6). The kinks can glide apart relatively easily and, as they do so, the dislocation is advanced by the trough separation.

The dislocation lying on a $\{111\}$ plane at 60° to its Burgers vector, $\frac{1}{2}\langle 1\bar{1}0\rangle$, happens to be easy to draw in projection on to a $\{10\bar{1}\}$ plane normal to the dislocation; it is shown in this way in Figure 9.25. In this figure, which serves equally well for the sphalerite (zinc blende) structure and for the diamond crystal structure, the small black dots represent a column of one atom species (e.g. Zn) and the encircled dots a second atom species (e.g. S); these two species are identical in the diamond crystal structure. Here, the dislocation line is along the vector common to (111) and $(1\bar{1}1)$ – that is, $[10\bar{1}]$ – while the Burgers vector of the dislocation is $\frac{1}{2}[01\bar{1}]$.

There are two possible $\{111\}$ slip planes, type I and type II. The 60° dislocation in Figure 9.25 is shown with its extra plane ending at a type II plane, where it breaks the lower number of bonds. The one bond per atom which is broken is described as a dangling bond. Intuitively, it might seem that slip should occur at the widely spaced type II planes, by the glide of dislocations like that of Figure 9.25. However, this is before the possibility of splitting into partials is examined.

There can be stacking faults on the {111} plane that do not violate tetrahedral bonding and are exactly analogous to intrinsic and extrinsic faults in c.c.p. metals. The stacking sequence of the perfect crystal can be written as $a\,ab\,\beta c\,\gamma a\,\alpha \dots$ (see Sections 3.4 and 3.5). In this sequence, the diatom $a - \alpha$ can be treated as a unit, denoted by A. Type I planes separate sheets of diatoms packed in the c.c.p. sequence ABCA ... An intrinsic fault is formed by passing a Shockley partial, $\frac{1}{6}\langle 211\rangle$, over a type I plane; passing two different Shockley partials over successive type I planes produces the extrinsic fault. The $\frac{1}{2}\langle 110\rangle$ dislocation can split into partials on a type I plane:

$$\tfrac{1}{2}[01\bar{1}] = \tfrac{1}{6}[11\bar{2}] + \tfrac{1}{6}[\bar{1}2\bar{1}] \tag{9.57}$$

so that one dislocation is a 30° partial and the other a 90° partial. Experimental evidence shows that moving dislocations are indeed split into partials in this way [43], with the 90° partials being more mobile than the 30° partials [44]. This set of partials is known as the 'glide set' of dislocations [2]. Partials dissociating on the alternative type II planes are termed the 'shuffle set' [2]; these have a more complicated geometry and are believed to be less mobile [2], although they may play an important role in low-temperature deformation [45].

A rich variety of stacking faults is found in Si films grown epitaxially from the vapour phase [46]. Faults originating at irregularities in the substrate grow along with the film, sometimes joined together as pieces of polyhedra. A schematic of a piece of an octahedral stacking fault in a film, which is parallel to {100}, is shown in Figure 9.26. The faces of the half-octahedron are alternate intrinsic and extrinsic faults, joined by a $\frac{1}{6}\langle 110\rangle$ stair-rod dislocation. The configuration at a corner can be derived from the Lomer–Cottrell lock (Figure 9.13) by gliding one of the Shockley partials of the lock to the other side of the stair-rod dislocation.

An alternative tetrahedrally bonded structure is that of wurtzite, in which ZnS units are stacked in the hexagonal sequence ABA ... Stacking faults in sphalerite consist of thin

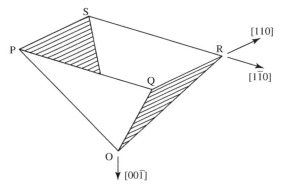

Figure 9.26 Stacking fault in a silicon film. Point O lies at the bottom surface of the film and points P, Q, R and S at the top surface. OP, OQ, OR and OS are stair-rod dislocations

layers of the wurtzite structure. Similarly, stacking faults in wurtzite can occur which are thin layers of the sphalerite structure and are exactly analogous to stacking faults in a hexagonal metal. Dislocations in AlN, which has this structure, are widely split, the fault energy being only about $4 \, \text{mJ m}^{-2}$ [47].

Neither wurtzite nor sphalerite possesses a centre of symmetry, with the interesting consequence that the core structures of dislocations of opposite sign can be distinguished from one another in these structures. Suppose that an undissociated 60° dislocation in, say, GaAs, which has the sphalerite structure, always lies on a type II plane (Figure 9.25). Its extra plane then ends either on a Ga atom or on an As atom, according to whether it comes in from the top or the bottom of the crystal. Dislocations of opposite sign can therefore be labelled Ga and As dislocations, and because of their different core structures they may have quite different Peierls stresses. In this context, it is of interest to note that the set of (111) planes on which dislocations reside and move has been a topic of controversy since the 1960s in crystals with the sphalerite and diamond structures [45].

Tetrahedrally bounded crystals contain a three-dimensional network of strong bonds, but in many crystals the atoms are bonded together strongly only within layers which are themselves connected by relatively weak forces. Graphite is such a crystal. The dislocations and stacking faults on the basal plane of graphite are exactly analogous to those on the basal plane of a hexagonal metal. Figure 9.19b is applicable; its relationship to the structure of graphite is shown in Figure 9.27. As a consequence of the weak interlayer bonding, the energies of the three basic types of stacking fault described in Section 9.5 are very small. One of these faults is produced by the dissociation of a perfect dislocation, as follows:

$$\left. \begin{array}{c} \text{AC} \rightarrow \sigma\text{C} + \text{A}\sigma \\ \tfrac{1}{3}[\bar{2}110] \rightarrow \tfrac{1}{3}[\bar{1}010] + \tfrac{1}{3}[\bar{1}100] \end{array} \right\} \tag{9.58}$$

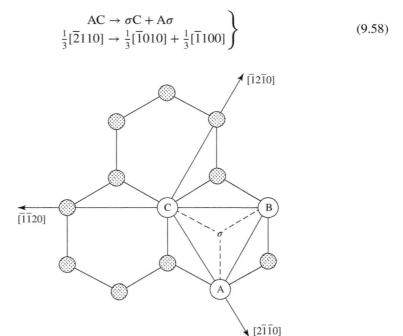

Figure 9.27   Atoms and lattice vectors in the basal plane of graphite

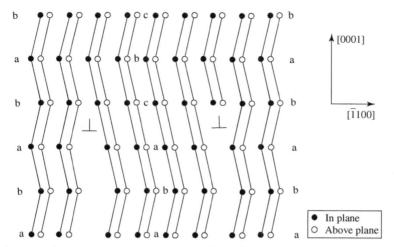

Figure 9.28   A schematic side view of a split 60° dislocation in graphite. At the fault between the partials a layer of rhombohedral stacking has been produced

By direct measurement of the separation of the $\frac{1}{3}\langle 10\bar{1}0\rangle$ partials, which are 1500 Å apart in the edge orientation and 800 Å apart in the screw orientation [48], the energy of the fault has been found to be only 0.5 mJ m^{-2}. A schematic side view of a split 60° dislocation can be seen in Figure 9.28, showing the layer of rhombohedral stacking at the fault.

Although slip dislocations are strictly confined to the basal planes in graphite, their reactions with one another are a little more complex than might at first be expected. The complexities are a result of the wide separation of the partials and of the fact that, because of this wide separation, the individual partials on nearby planes can, in effect, react with one another. As an example, consider a reaction between the dislocations AB and AC, which would simply repel one another if they were not split. Suppose that they lie on adjacent planes, so that if AB splits in the order Aσ + σB then AC must split in the order of σC + Aσ. If the split dislocations are caused to approach, AB from the left and AC from the right, the partials σB and σC will come over one another and react:

$$\sigma B + \sigma C \rightarrow A\sigma \qquad\qquad (9.59)$$

The result is a stable double ribbon, shown in Figure 9.29 [49].

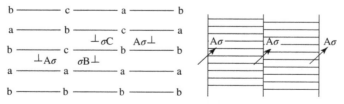

Figure 9.29   Reaction between partial dislocations on adjacent planes in graphite to form a stable double ribbon

9.8 Dislocations in Low Symmetry Crystal Structures

In comparison with the level of research interest in dislocations in metallic, ceramic, and semiconducting materials with relatively high symmetry, such as those in the crystal structures discussed in Sections 9.3–9.7, dislocations in low symmetry crystal structures are much more of a niche interest. For example, in a recent review of dislocations in molecular crystals, Olson et al. [50] suggest that these remain *terra incognita*, principally because of their structural architecture: these materials consist of flexible building units of low symmetry (e.g. Section 3.10), in which the intermolecular attractions between molecules are anisotropic. Of particular interest in such materials is the mechanism of screw dislocation generation because this is a key enabler in their crystal growth. It has been known since the early 1950s that the addition of new material to surface steps caused by the screw components of dislocations intersecting surfaces produces growth spirals or hillocks in materials with simple crystal structures and also in those with more complex crystal structures, such as molecular crystals (e.g. [51]). With the advent of atomic force microscopy, the growth of such hillocks can now be seen in-situ and the full complexity of their growth characteristics in molecular crystals can be understood. Thus, for example, apparently large steps ~6 nm in height on (0001) surfaces in *L*-cystine crystals can be seen to be composed of more elementary layers ~1 nm in height, caused by the presence of a 6_1 screw axis in the crystal [52].

The study of dislocations in molecular crystals such as hexahydro-1,3,5-trinitro-1,3,5-triazinane (also known as cyclotrimethylene trinitramine), $C_3H_6N_6O_6$, which crystallizes in an orthorhombic unit cell with eight molecules per unit cell is relevant to the use of this material as an explosive. Large-scale molecular dynamics modelling of the response of this material to conditions of shock loading shows how this involves both the creation of dislocation loops homogeneously and the subsequent motion of these loops [53].

Research into dislocations in other materials such as pure metal crystal structures with low symmetry and mineral crystal structures with low symmetry is also a highly specialized subject area. A definitive account of the state of knowledge of dislocations in minerals is to be found in [54]. While many minerals have high symmetry crystal structures (e.g. Sections 3.5 and 3.6), there are also many with rhombohedral, orthorhombic, monoclinic, and triclinic crystal structures. In addition, until very recently, the often complex chemistry of minerals has made it difficult for dislocation cores to be modelled in such materials, but there are now computational methods available to tackle such problems [55].

By comparison with minerals, relatively little progress has been made in the understanding of dislocations in pure metal and metal alloy crystal structures with low symmetry since the 1970s. For Bi, Ga, Hg and U, the references given in Table 7.1 for their glide elements are still very useful source references, in addition to which there is some work in the later 1970s and early 1980s examining various aspects of dislocations in these materials, such as their elastic energy, stress fields and dislocation interactions [56–60]. Examples of dislocation analyses in orthorhombic and monoclinic metal alloys are shown in work by Kumai and Sato et al. [61, 62]. In Al_3Ni, dislocations have large 1 nm $\langle 110 \rangle$ Burgers vectors and slip on four $\{110\}$ glide planes of this orthorhombic phase, while in the complex monoclinic alloy $Al_{13}Fe_4$ with 102 atoms per unit cell, the [010] Burgers vector of dislocations seen are also large, 0.8 nm, simply because of the size of the unit cell.

Research into quasicrystals triggered by the 1984 paper by Shechtman et al. discussed in Section 4.6 has shown how dislocations seen in these materials are always associated with other defects termed phasons, as discussed comprehensively in a recent review [63]. It is noteworthy that an understanding of the behaviour of dislocations in quasicrystals helps when studying dislocations in complex metallic alloys, such as the quasicrystalline approximant phases (Section 4.6).

9.9 Dislocations in Other Crystal Structures

The detailed analysis of the geometry of dislocations, which has been carried out in Sections 9.1–9.7 for a number of specific crystal structures, could be repeated for many other crystal structures. It is hoped that the principles employed have now been demonstrated sufficiently that the reader is able to deal with any case that may be encountered in metals, ceramics, semiconductors, molecular crystals, mesomorphic phases, quasicrystals, and superlattice structures.

Problems

9.1 In a b.c.c. metal, two dislocations of Burgers vector $\frac{1}{2}[111]$ and $\frac{1}{2}[1\bar{1}\bar{1}]$ respectively have reacted over a certain length (as in Figure 9.11). If the reacting dislocations are pinned at points p, q, r and s, which lie at the corners of a square of side L, make an estimate of the length of the product dislocation (neglect the variation of the energy of a dislocation with the orientation of its line).

9.2 A dislocation is being extruded between small precipitate particles which lie in a row, each particle being a distance l from its neighbours. Determine the force acting on each particle as a function of the shear stress σ applied on the slip plane of the dislocation parallel to the Burgers vector **b** and estimate the maximum force that can be exerted on each particle.

9.3 Using Frank's rule, make a study of reactions between stable perfect dislocations in a body-centred tetragonal crystal. Consider the effect of the value of the c/a ratio on the reactions.

9.4 Compare the number of atoms that are improperly surrounded by nearest neighbours (i.e. as an atom in the h.c.p. structure instead of as in the c.c.p. structure) at intrinsic and in extrinsic faults in c.c.p. metals. What structure is formed if both the Shockley partials in a Lomer–Cottrell lock are forced to glide to the opposite side of the stair-rod dislocation?

9.5 Using a dislocation corresponding to one particular slip plane and slip direction as one of the reacting dislocations forming a Lomer–Cottrell lock, how many different locks can be formed? For the slip system (111) $[\bar{1}01]$, give the Burgers vector of each partial dislocation in all of the possible Lomer–Cottrell locks.

9.6 If the concentration of vacancies in aluminium at its melting point is 9.4×10^{-4} and if after quenching the aluminium from its melting point the vacancies condense into discs on the close-packed plane and form Frank loops, determine the density of dislocations introduced (a) when the loops are 50 Å in radius and (b) when the loops are 500 Å in radius.

9.7 Assuming isotropic elasticity, calculate the separation of the Shockley partials of a screw dislocation in Ag, Al and Au. Use the stacking fault energies given in Table 9.2. For Ag take $\mu = 28$ GPa, $v = 0.38$; for Al take $\mu = 26$ GPa, $v = 0.34$; for Au take $\mu = 28$ GPa, $v = 0.42$.

9.8 For the first four lattices listed in Table 9.1, find the Burgers vectors of any dislocations whose stability, according to Frank's rule, is doubtful; that is, dislocations that might dissociate into two perfect dislocations whose Burgers vectors are at 90° to one another.

9.9 Lead sulfide, PbS, has the same crystal structure as NaCl, but the usual slip system is $\{100\}\langle 011\rangle$ instead of $\{011\}\langle 0\bar{1}1\rangle$.
   (a) Compare dislocation reactions that may occur as a result of slip in PbS with those occurring in NaCl.
   (b) Sketch the structure of the edge of the extra plane of an edge dislocation in PbS.

9.10 Describe the dislocation loops that might be produced in zinc by the aggregation of interstitial atoms into discs of atoms parallel to the basal plane. Also describe the dislocation loops that might be produced in graphite by the aggregation of vacant sites into discs parallel to the basal plane.

9.11 Project the structure of silicon on to a $(10\bar{1})$ plane and show, in this projection, the structure of an intrinsic stacking fault and the structure of an extrinsic stacking fault in silicon.

9.12 Show how, in graphite, two dislocation ribbons (i.e. dislocations that are split into widely separated partials) that are on next-nearest-neighbour basal planes can interact to form a threefold ribbon.

9.13 An alloy of 50 at% Cu–50 at% Zn contains dislocations of Burgers vector $\frac{1}{2}\langle 111\rangle$ and $\langle 100\rangle$. The alloy is then ordered to produce the *B*2 superlattice (Section 3.8). Describe the effect of gliding the $\frac{1}{2}\langle 111\rangle$ dislocation on $\{1\bar{1}0\}$ and the $\langle 100\rangle$ dislocation on $\{010\}$ after ordering has occurred. From your result, how would you expect two identical $\frac{1}{2}\langle 111\rangle$ dislocations to interact in the ordered alloy?

Suggestions for Further Reading

See the suggestions for Chapter 8 and the following:

Nabarro, F.R.N., Duesbery, M.S., Hirth, J.P. and Kubin, L. (1979–2010). *Dislocations in Solids*, vol. 1–16. Amsterdam: Elsevier.
Suzuki, T., Takeuchi, S. and Yoshinaga, H. (1991). *Dislocation Dynamics and Plasticity*. Berlin: Springer (translated from the Japanese edition published by Shokabo in 1985).

References

[1] Lu, G., Kioussis, N., Bulatov, V.V. and Kaxiras, E. (2001). Dislocation core properties of aluminum: a first-principles study. *Mater. Sci. Eng. A* 309–310: 142–147.
[2] Anderson, P.M., Hirth, J.P. and Lothe, J. (2017). *Theory of Dislocations*, 3rd Edition. New York: Cambridge University Press.

[3] Frank, F.C. (1949). Written answer to a remark by E.N. da C. Andrade on the paper by N.F. Mott (1949) Mechanical properties of metals, *Physica* 15: 119–128: *Physica* 15: 131–133.

[4] Foreman, A.J.E. (1955). Dislocation energies in anisotropic crystals. *Acta Metall.* 3: 322–330.

[5] Orowan, E. (1948). Discussion on internal stresses. In: *Symposium on Internal Stresses in Metals and Alloys*, pp. 451–453. London: Institute of Metals.

[6] Orowan, E. (1954). Dislocations and mechanical properties. In: *Dislocations in Metals* (ed. M. Cohen), pp. 69–195. New York: American Institute of Mining and Metallurgical Engineers.

[7] Heidenreich, R.D. and Shockley, W. (1948). Study of slip in aluminum crystals by electron microscope and electron diffraction methods. In: *Bristol Conference on the Strength of Solids, The Physical Society, London*, pp. 57–75.

[8] Williams, D.B. and Carter, C.B. (1996). *Transmission Electron Microscopy: A Textbook for Materials Science*. New York: Plenum Press.

[9] Cockayne, D.J.H., Jenkins, M.L. and Ray, I.L.F. (1971). The measurement of stacking-fault energies of pure face-centred cubic metals. *Phil. Mag.* 24: 1383–1392.

[10] Dobson, P.S., Goodhew, P.J. and Smallman, R.E. (1967). Climb kinetics of dislocation loops in aluminium. *Phil. Mag.* 16: 9–22.

[11] Jenkins, M.L. (1972). Measurement of the stacking-fault energy of gold using the weak-beam technique of electron microscopy. *Phil. Mag.* 26: 747–751.

[12] Stobbs, W.M. and Sworn, C.H. (1971). The weak beam technique as applied to the determination of the stacking-fault energy of copper. *Phil. Mag.* 24: 1365–1381.

[13] Carter, C.B. and Holmes, S.M. (1977). The stacking-fault energy of nickel. *Phil. Mag.* 35: 1161–1172.

[14] Frank, F.C. (1951). The growth of carborundum: dislocations and polytypism. *Phil. Mag.* 42: 1014–1021.

[15] Thompson, N. (1953). Dislocation nodes in face-centred cubic lattices. *Proc. Phys. Soc. B* 66: 481–492.

[16] Whelan, M.J. (1959). Dislocation interactions in face-centred cubic metals, with particular reference to stainless steel. *Proc. Roy. Soc. Lond. A* 249: 114–137.

[17] Lomer, W.M. (1951). A dislocation reaction in the face-centred cubic lattice. *Phil. Mag.* 42: 1327–1331.

[18] Cottrell, A.H. (1952). The formation of immobile dislocations during slip. *Phil. Mag.* 43: 645–647.

[19] Silcox, J. and Hirsch, P.B. (1959). Direct observations of defects in quenched gold. *Phil. Mag.* 4: 72–89.

[20] Hirth, J.P. (1961). On dislocation interactions in the fcc lattice. *J. Appl. Phys.* 32: 700–706.

[21] Bilde-Sørensen, J.B., Leffers, T. and Barlow, P. (1977). Dissociated structure of dislocation loops with Burgers vector $a<100>$ in electron-irradiated Cu–Ni. *Phil. Mag.* 36: 585–595.

[22] Nabarro, F.R.N. (1987). *Theory of Crystal Dislocations*. New York: Dover Publications.

[23] Friedel, J. (1955). On the linear work hardening rate of face-centred cubic single crystals. *Phil. Mag.* 46: 1169–1186.

[24] Alden, T.H. (1963). Extreme latent hardening in compressed lithium fluoride crystals. *Acta Metall.* 11: 1103–1105.

[25] Eshelby, J.D., Newey, C.W.A., Pratt, P.L. and Lidiard, A.B. (1958). Charged dislocations and the strength of ionic crystals. *Phil. Mag.* 3: 75–89.

[26] Whitworth, R.W. (1975). Charged dislocations in ionic crystals. *Adv. Phys.* 24: 203–304.

[27] Korner, A. and Karnthaler, H.P. (1983). Weak-beam study of glide dislocations in h.c.p. cobalt. *Phil. Mag. A* 48: 469–477.

[28] Fleischer, R.L. (1986). Stacking fault energies of hcp metals. *Scr. Metall.* 22: 223–224.

[29] Wen, L., Chen, P., Tong, Z.-F., Tang, B.-Y., Peng, L.-M. and Ding, W.-J. (2009). A systematic investigation of stacking faults in magnesium via first-principles calculation. *Eur. Phys. J. B* 72: 397–403.

[30] Berghezan, A., Fourdeux, A. and Amelinckx, S. (1961). Transmission electron microscopy studies of dislocations and stacking faults in a hexagonal metal: zinc. *Acta Metall.* 9: 464–490.

[31] Harris, J.E. and Masters, B.C. (1966). The annealing of faulted loops in magnesium and zinc. *Proc. Roy. Soc. Lond. A* 292: 240–263.

[32] Dobson, P.S. and Smallman, R.E. (1966). The climb of dislocation loops in zinc. *Proc. Roy. Soc. Lond. A* 293: 423–431.

[33] Hales, R., Smallman, R.E. and Dobson, P.S. (1968). The formation and growth of vacancy loops in magnesium. *Proc. Roy. Soc. Lond. A* 307: 71–81.

[34] Yoo, M.H. (1981). Slip, twinning, and fracture in hexagonal close-packed metals. *Metall. Trans. A.* 12: 409–418.

[35] Yoo, M.H., Morris, J.R., Ho, K.M. and Agnew, S.R. (2002). Nonbasal deformation modes of hcp metals and alloys: role of dislocation source and mobility. *Metall. Trans. A.* 33: 813–822.

[36] Nogaret, T., Curtin, W.A., Yasi, J.A., Hector, L.G. and Trinkle, D.R. (2010). Atomistic study of edge and screw $<c + a>$ dislocations in magnesium. *Acta Mater.* 58: 4332–4343.

[37] Hull, D. and Bacon, D.J. (2011). *Introduction to Dislocations*, 5th Edition. Oxford: Butterworth-Heinemann.

[38] Cottrell, A.H. (1958). Theory of brittle fracture in steel and similar metals. *Trans. Am. Inst. Min. Metall. Petrol. Engrs.* 212: 192–203.

[39] Vítek, V. (1968). Intrinsic stacking faults in body-centred cubic crystals. *Phil. Mag.* 18: 773–786.

[40] Ventelon, L. and Willaime, F. (2010). Generalized stacking-faults and screw-dislocation core-structure in bcc iron: a comparison between *ab initio* calculations and empirical potentials. *Phil. Mag.* 90: 1063–1074; corrigendum, 4071.

[41] Kelly, A. and Groves, G.W. (1970). *Crystallography and Crystal Defects*. London: Longman Group Limited.

[42] Duesbery, M.S. and Vitek, V. (1998). Plastic anisotropy in b.c.c. transition metals. *Acta Mater.* 46: 1481–1492.

[43] Olsen, A. and Spence, J.C.H. (1981). Distinguishing dissociated glide and shuffle set dislocations by high resolution electron microscopy. *Phil. Mag. A* 43: 945–965.

[44] Kolar, H.R., Spence, J.C.H. and Alexander, H. (1996). Observation of moving dislocation kinks and unpinning. *Phys. Rev. Lett.* 77: 4031–4034.

[45] Cai, W., Bulatov, V.V., Chang, J., Li, J. and Yip, S. (2004). Dislocation core effects on mobility. In: *Dislocations in Solids, Volume 12* (eds. F.R.N. Nabarro and J.P. Hirth), pp. 1–80. Amsterdam: Elsevier.

[46] Booker, G.R. (1964). Crystallographic imperfections in silicon. *Disc. Faraday Soc.* 38: 298–304.

[47] Delavignette, P., Kirkpatrick, H.B. and Amelinckx, S. (1961). Dislocations and stacking faults in aluminum nitride. *J. Appl. Phys.* 32: 1098–1100.

[48] Baker, C., Chou, Y.T. and Kelly, A. (1961). The stacking-fault energy of graphite. *Phil. Mag.* 6: 1305–1308.

[49] Delavignette, P. and Amelinckx, S. (1962). Dislocation patterns in graphite. *J. Nucl. Mater.* 5: 17–66.

[50] Olson, I.A., Shtukenberg, A.G., Kahr, B. and Ward, M.D. (2018). Dislocations in molecular crystals. *Rep. Prog. Phys.* 81: 096501-1–096501-37.

[51] Forty, A.J. (1954). Direct observations of dislocations in crystals. *Adv. Phys.* 3: 1–25.

[52] Shtukenberg, A.G., Zhu, Z.-N., An, Z.-H., Bhandari, M., Song, P.-C., Kahr, B. and Ward, M.D. (2013). Illusory spirals and loops in crystal growth. *Proc. Natl. Acad. Sci. U.S.A.* 110: 17195–17198.

[53] Cawkwell, M.J., Ramos, K.J., Hooks, D.E. and Sewell, T.D. (2010). Homogeneous dislocation nucleation in cyclotrimethylene trinitramine under shock loading. *J. Appl. Phys.* 107, 063512-1–063512-11.

[54] Barber, D.J., Wenk, H.-J., Hirth, G. and Kohlstedt, D.L. (2010). Dislocations in minerals. In: *Dislocations in Solids*, vol. 16 (eds. J.P. Hirth and L. Kubin), pp. 171–232. Amsterdam: Elsevier.

[55] Walker, A.M., Carrez, P. and Cordier, P. (2010). Atomic-scale models of dislocation cores in minerals: progress and prospects. *Mineral. Mag.* 74: 381–413.

[56] Motohashi, Y. and Otake, S. (1975). Elastic energy, stress field and interaction of dislocations in bismuth crystal. *Jpn. J. Appl. Phys.* 14: 435–444.

[57] Motohashi, Y. and Ohtake, S. (1978). Elastic energy, stability and interaction of slip dislocations and the choice of slip systems in rhombohedral crystals, Bi, Sb, As and Hg. *Physica Status Solidi (a).* 50: 335–347.

[58] Otake, S., Namazue, H. and Matsuno, N. (1980). Critical resolved shear stresses of two slip systems in bismuth single crystals. *Jpn. J. Appl. Phys.* 19: 433–437.

[59] Durbec, N., Pichaud, B. and Minari, F. (1981). Glide systems in gallium: justification by dislocation specific energies. *J. Less-Common Met.* 82: 373–377.

[60] Burle-Durbec, N., Pichaud, B. and Minari, F. (1984). Glide systems in α-Ga: further results deduced from transmission electron microscopy observations. *J. Less-Common Met.* 98: 79–84.

[61] Yamashita, K., Fujimoto, I., Murakumo, T., Kumai, S. and Sato, A. (2000). Observation of glide dislocations in DO_{20} ordered Al_3Ni. *Phil. Mag. A* 80: 219–235.

[62] Miyazaki, S., Kumai, S. and Sato, A. (2010). Plastic deformation of a monoclinic $Al_{13}Fe_4$ at 300 $\sim$ 873 K. *Phil. Mag.* 90: 2597–2618.

[63] Feuerbacher, M. (2012). Dislocations in icosahedral quasicrystals. *Chem. Soc. Rev.* 41: 6745–6759.

10

Point Defects

10.1 Introduction

Disturbances in a crystal which, apart from elastic strains associated with them, extend for no more than a few interatomic distances in any direction are called point defects. Although impurity atoms in solution can be classified as point defects, we shall be mainly concerned with structural point defects, of which there are two elementary types. The first is the vacancy, which consists of an atomic site from which the atom is missing. The second consists of a small region of crystal which contains an extra atom. This is called an interstitial defect, from the idea that the extra atom is squeezed into an interstice between the others.

A point defect differs from a dislocation and from a two-dimensional defect such as a crystal boundary in two important respects. The first is that, even with modern transmission electron microscopes, scanning tunneling microscopes and atom probe field-ion microscopes, point defects are relatively difficult to observe directly, so that in the main they can be detected and studied only through their effect upon some physical property of the crystal. The second distinction is that point defects may be present in appreciable concentrations when the crystal is in *thermodynamic equilibrium*. While dislocations and interfaces always raise the free energy of a crystal, adding a certain number of point defects to an otherwise perfect crystal reduces its free energy to a minimum value. This is because of a gain of entropy which arises from the many possible sets of places in the crystal where the point defects can be put. This configurational entropy is given by:

$$S = k \ln W \tag{10.1}$$

where k is Boltzmann's constant (1.38×10^{-23} J K^{-1} or 8.62×10^{-5} eV K^{-1}) and W is the number of different arrangements of point defects. When n defects are distributed

Crystallography and Crystal Defects, Third Edition. Anthony Kelly and Kevin M. Knowles.
© 2020 John Wiley & Sons Ltd. Published 2020 by John Wiley & Sons Ltd.
Companion website: www.wiley.com/go/kelly/crystallography3e

amongst N sites, the number of arrangements is $^NP_n = N(N-1)\ldots(N-n+1) = N!/(N-n)!$ if each defect is distinct. In fact, all the defects are identical, and the n labels needed to distinguish them from one another can be distributed in $n!$ ways. Therefore:

$$W = {}^NC_n = \frac{N!}{(N-n)!n!} \tag{10.2}$$

Manipulating Eq. (10.2) into a form suitable for the application of Stirling's approximation ($\ln x! \approx x \ln x - x$, for large x) gives:

$$\ln W = N \ln N - (N-n) \ln (N-n) - n \ln n \tag{10.3}$$

so that in Eq. (10.1):

$$S = k[N \ln N - (N-n) \ln (N-n) - n \ln n] \tag{10.4}$$

If the presence of one defect increases the internal energy of the crystal by E_f, the change in free energy, ΔF, of a crystal containing n identical defects, at a temperature of T K, is:

$$\Delta F = nE_f - T(S + nS') \tag{10.5}$$

Here it is assumed that n/N is so small that the defects do not interact with one another. The extra entropy term nS' in Eq. (10.5) represents the fact that each defect may add a certain entropy S' to the crystal because of its effect on the vibration of the atoms in its neighbourhood. The number of defects producing the minimum free energy, n_e, is found by setting $d\Delta F/dn = 0$, giving:

$$\frac{n_e}{N - n_e} = \exp\left(\frac{S'}{k}\right) \exp\left(-\frac{E_f}{kT}\right) \tag{10.6}$$

If the defect in question is a vacancy then $N - n_e$ is simply the number of atoms in the crystal. Since in fact $n_e \ll N$, so that terms $O(n_e^2)$ or higher can be neglected, Eq. (10.6) can be written as:

$$\frac{n_e}{N} = \exp\left(\frac{S'}{k}\right) \exp\left(-\frac{E_f}{kT}\right) \tag{10.7}$$

In c.c.p. metals, S'/k is of the order of unity, so that the atomic concentration of vacancies, n_e/N, is essentially given by $\exp(-E_f/kT)$. For example, in the c.c.p. metals Cu, Ag and Au, the energy of formation of a vacancy, E_f, is approximately 1 eV, leading to vacancy concentrations of the order of 10^{-4} at temperatures close to the melting point. By comparison, the term in S'/k only contributes a factor of the order of three to Eq. (10.7).

The vacancies can be generated at surfaces, grain boundaries or dislocation lines. An atom jumping onto a surface or a grain boundary leaves a vacant site into which deeper atoms can jump, driving the vacancy into the interior (Figure 10.1). Similarly, an atom can leave a site vacant by jumping onto the edge of the extra plane of an edge dislocation. If an interior atom were to jump into an interstitial position, both a vacancy and an interstitial would be created at the same time. Such a defect, called a *Frenkel defect*, does not occur in appreciable concentration in equilibrated c.c.p. metals because of the high internal energy increase produced by an interstitial. This is not necessarily true for other crystals;

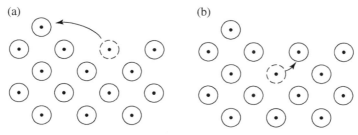

Figure 10.1 Crystal surface acting as a vacancy source. In (a) an atom jumps onto a surface, leaving a site into which a deeper atom can jump, as happens in (b), driving the vacancy into the interior

for example, crystals of AgCl and AgBr contain an atomic concentration of Ag in interstitial positions of the order of 10^{-3} and 10^{-2} respectively, at their melting temperatures.

Since the energy of formation of a point defect determines its concentration in a crystal which is in equilibrium, it would clearly be helpful to be able to calculate this quantity. Unfortunately, this is a very difficult task, although it is easy to make a very simple estimate. For example, the energy needed to form a vacancy can be estimated very roughly as follows. Assume that the cohesive energy of a crystal can be represented fairly as the sum of the energies of interaction of nearby pairs of atoms. Starting with any one atom, we sum the energies of the bonds made with each of its neighbours. This summation is repeated for a second atom, and so on. The final total sum gives twice the cohesive energy of the crystal, since the energy of each bond is counted twice. If each atom is identically attached to bonds of total energy $2E_c$, the cohesive energy per atom is E_c. To form a vacancy, we remove an atom from the interior of the crystal and place it on the surface. The first step breaks bonds of energy $2E_c$, while the second reforms half the bonds (on average), leaving E_c as the energy to form the vacancy. This estimate is very inaccurate, because it makes no allowance for the possibility of a change in the energies of the bonds between the atoms near the vacant site. In fact, when an atom is removed, the surrounding nuclei will be displaced and the electrons will be redistributed in such a way as to release energy. Consequently, a relaxation energy E_{relax} must be subtracted from E_c:

$$E_f = E_c - E_{relax} \qquad (10.8)$$

Since the value of E_{relax} is of the same order of magnitude as E_c and is very difficult to calculate, the formation energy E_f is correspondingly difficult to obtain accurately. Fortunately, however, a number of careful calculations of the formation energy of a vacancy have been made since the pioneering work of Huntington and Seitz in 1942 (e.g. [1–8]). In agreement with experimental results (e.g. [9–11]), calculations cluster in the range 0.4–1 eV. Johnson [6] states that most experimental values for E_f for b.c.c. and c.c.p. transition elements and noble metals lie in the range 8.5–10 kT_m, where k is Boltzmann's constant and T_m is the melting temperature in Kelvin of the metal. Calculations (e.g. [5, 7]) set a much higher value of 3–4 eV on the formation energy of an interstitial defect in c.c.p. metals, in agreement with the lack of evidence for the existence of interstitials in c.c.p. metals which have been brought to equilibrium at high temperatures.

Point defects with a high energy of formation can occur in crystals that are *not* in equilibrium. Both vacancies and interstitials are produced by any radiation that can knock atoms out of their normal sites, since the displaced atom leaves a vacancy behind and itself becomes an interstitial defect. For this reason, the study of point defects is particularly relevant to materials used in the nuclear industry. Similarly, plastic deformation can be expected to produce both vacancies and interstitials through the action of dislocations upon one another (Section 8.6).

The ability of a crystal to retain a point defect produced in one of these ways depends upon the mobility of the defect in the crystal, since a highly mobile defect that is not in equilibrium will quickly be lost. Point defects can destroy themselves at a surface or a grain boundary, or at a dislocation, or by combining with other point defects. The mobility of a defect present in equilibrium is also an important property, because atoms can be transported through a crystal by the movement of equilibrium point defects. For example, self-diffusion in c.c.p. metals occurs by the movement of vacancies.

The mobility of a defect is governed by the increase in the free energy of the crystal as the defect passes through the position of maximum energy on the way from one rest position to the next. The frequency with which the defect jumps to a new position is given by:

$$v = v_0 \exp\left(\frac{S_M}{k}\right) \exp\left(-\frac{E_M}{kT}\right) \tag{10.9}$$

where v_0 is the 'attempt frequency', or the frequency of vibration of the defect in the appropriate direction. The free energy increase has been divided into an entropy increase, S_M, and an increase in internal energy, E_M. The number of jumps that can be made in 1 second is shown in Table 10.1 for various values of the migration energy E_M, assuming that $v_0 = 10^{13}$ s^{-1} and $\exp(S_M/k) = 1$; these are both reasonable values for a vacancy in a c.c.p. metal.

The energy of migration of a defect is even more difficult to calculate than its formation energy. When the defect is responsible for self-diffusion, as a vacancy is in c.c.p. metals, the migration energy can be obtained from measurements of rates of self-diffusion. The self-diffusion coefficient in an equilibrated c.c.p. metal is proportional to the equilibrium concentration of vacancies and to their mobility, so that the activation energy for self-diffusion, E_{SD}, is given by:

$$E_{SD} = E_f + E_M \tag{10.10}$$

From Eq. (10.10), using a calculated value of E_f, the migration energy of a vacancy in copper is deduced to be about 1 eV, similar to that of E_f. Indeed, for c.c.p. metals, Johnson [12] quotes a ratio of $E_M/(E_f + E_M)$ of 0.4–0.5 derived from lattice defect calculations. As

Table 10.1 Estimated number of jumps made per second by a point defect having migration energy E_M for an attempt frequency, v_0, of 10^{13} s^{-1}

E_M(eV)	v at 77 K	v at 300 K	v at 773 K	v at 1273 K
0.1	2.9×10^6	2.1×10^{11}	2.2×10^{12}	4.0×10^{12}
0.5	$\ll 1$	4.0×10^4	5.5×10^9	1.0×10^{11}
1.0	$\ll 1$	$\ll 1$	3.0×10^6	1.1×10^9
2.0	$\ll 1$	$\ll 1$	1	1.2×10^5

Note: 1 eV = 96.5 kJ mol^{-1}.

a consequence, it is evident from Table 10.1 that the mobility of vacancies in copper is insignificant at room temperature.

By contrast, an interstitial copper atom in copper has a lower migration energy, typically about 0.1 eV [4, 6, 7, 13]. However, there is an added complication: such interstitials tend to have split configurations, such as the ⟨100⟩ split configuration or ⟨100⟩ dumb-bell configuration [4, 6, 7], in which two atoms share the same host site in the crystal, and where the centre of mass of the two atoms is at the crystal host site [6, 7]. Thus, for example, instead of having a single atom at $\left(\frac{1}{2},\frac{1}{2},0\right)$, there might instead be two atoms at $\left(\frac{1}{2},\frac{1}{2},\frac{3}{10}\right)$, and $\left(\frac{1}{2},\frac{1}{2},-\frac{3}{10}\right)$. The low migration energy is associated with a movement of the dumb-bell to a neighbouring host atom site, with a 90° rotation of the axis of the dumb-bell [4, 7]. In principle, therefore, an interstitial atom in copper is too mobile to be retained at room temperature, since its low migration energy means it can cover macroscopic distances in 1 second. Similar considerations apply to self-interstitials in b.c.c. metals, as we will discuss in more detail in Section 10.4, when structural configurations of point defects are further considered.

In practice, of course, although they are relatively mobile, isolated interstitials formed in a metal at room temperature as a result of particle irradiation will interact with other defects in a material. Such interactions will trap isolated interstitials, preventing them from escaping to the free surface, and give rise to aggregates of point defects. These will be considered in Sections 10.3 and 10.4.

10.2 Point Defects in Ionic Crystals

In Section 10.1, examples of point defects were drawn mostly from c.c.p. metals. Defects in alkali halides have also been studied intensively. They have additional complexities; for example, both anion and cation defects may occur. Suppose in a pure alkali halide that the formation energy E^+ of a cation vacancy is less than that of an anion vacancy, E^-, as is the case in sodium chloride. In coming to equilibrium at high temperatures, the crystal will initially produce more cation than anion vacancies at surfaces or dislocation lines. This will set up an electric field at the surfaces and dislocation lines opposing the issue of further cation vacancies and aiding the formation of anion vacancies. The region over which this electric field is significant is known as a space charge region.

At equilibrium far away from any space charge region, the crystal will contain equal numbers n of both types. An equation analogous to Eq. (10.7) can be derived:

$$\frac{n}{N} = \exp\left(\frac{S^+ + S^-}{2k}\right)\exp\left(-\frac{(E^+ + E^-)}{2kT}\right) \tag{10.11}$$

where N is the number of lattice points (equal to the number of ion pairs). A defect consisting of an anion vacancy and a cation vacancy (separated from one another) is called a Schottky defect. In sodium chloride, a concentration of Schottky defects of 10^{-4} has been found near the melting point, with a temperature dependence indicating that $(E^+ + E^-) = 2.7\,\text{eV}$ (see Table 10.3, below).

Schottky defects in sodium chloride are formed because of the entropy which they add to the crystal. Ionic crystals frequently contain point defects for a different reason. For example, crystals of ferrous oxide, which has the same crystal structure as sodium chloride, always contain a rather high concentration of cation vacancies. This is a result of the fact that some of the iron is always present in the ferric state, Fe^{3+}. The extra oxygen associated with the ferric iron is accommodated in the normal oxygen ion sublattice, leaving some cation sites unoccupied. In this instance the presence of the point defect is a consequence of a departure from the stoichiometric composition FeO.[1] If the composition is $Fe_{1-x}O$, the concentration of cation vacancies is x, and since x may be as high as 0.1, the concentration of vacancies is orders of magnitude greater than the equilibrium concentration of Schottky defects would be in a stoichiometric crystal. The presence of an impurity can introduce point defects into ionic crystals in a similar way. For example, sodium chloride containing a small number of divalent impurity cations such as Ca^{2+} contains an equal number of vacant cation sites, which compensate for the extra positive charge introduced by the Ca^{2+} ions. Such considerations are important in the crystal chemistry of electrolytes for batteries and fuel cells, enabling efficient conduction of the mobile ion in the electrolyte between electrodes of opposite polarity.

Although pure alkali halides normally have a precisely stoichiometric composition, it is possible to force an alkali halide crystal to take up an excess of cations by heating it in the vapour of the alkali metal. The extra cations are incorporated in the normal cation sublattice, leaving a number of anion sites vacant. The electrons which maintain the electrical neutrality of the crystal are associated with the anion vacancies. By exciting these electrons, electromagnetic radiation in the frequency range of the visible spectrum is absorbed as it passes through the crystal, and the crystal becomes coloured; sodium chloride becomes a yellowish-brown, for example. The anion vacancy with its trapped electron is called an F-centre (after the German world for colour, *Farbe*). Exposure of an alkali halide to a damaging radiation also produces F-centres, but various other defects are produced at the same time, so that the colour of an irradiated crystal is not the same as that of a crystal heated in the vapour of the alkali metal.

10.3 Point Defect Aggregates

An interstitial atom and a vacant site of the same atom obviously reduce the internal energy of the crystal when they come together, since by doing so they annihilate one another and restore the perfect crystal. Even defects that are not complementary to one another can often lower the internal energy of the crystal by joining together. For example, if an atom next to a vacancy in a c.c.p. metal is removed so as to form a second vacancy, the number of bonds that must be broken is one less than the number that would be broken in forming the second vacancy in an isolated position.

Therefore, less energy is needed to form a divacancy than is needed to form two isolated vacancies. The difference in energy is equal to the decrease in internal energy that would result from joining two vacancies together. This energy is defined as the binding energy of the vacancies. The binding energy of a pair of vacancies in aluminium, for example, is believed to be about 0.17 eV (see Table 10.4).

---

[1] Point defects due to nonstoichiometry can also occur in metallic systems. For example, when Al atoms are added to the superlattice Ni–Al, which has the CsCl structure, they extend the Al sublattice, leaving Ni sites vacant [14].

In general, the energy of formation of a divacancy may be written as $(2E_F - E_b)$, where E_F is the formation energy of a single vacancy and E_b is the binding energy. The equilibrium concentration of divacancies, n_2, is given by:

$$n_2 \propto \exp\left(-\frac{(2E_f - E_b)}{kT}\right) \tag{10.12}$$

and so, using Eq. (10.7), the equilibrium numbers of vacancies and divacancies, n_1 and n_2 respectively, in a crystal having N atomic sites are related by the equation:

$$\frac{n_1^2}{n_2 N} = q \exp\left(-\frac{E_b}{kT}\right) \tag{10.13}$$

The number q depends on the entropies of formation of the defects and on the number of possible divacancy positions in a crystal of N atomic sites. From Eqs. (10.13) and (10.7):

$$\frac{n_1}{n_2} = q \exp\left(-\frac{S'}{k}\right) \exp\left(\frac{E_f - E_b}{kT}\right) \tag{10.14}$$

Since $E_f > E_b$, the ratio of single vacancies to divacancies increases as the temperature is lowered. Even at the melting point of a c.c.p. metal, probably less than 20% of its vacancies are associated.

The situation in a crystal that has come to equilibrium is described by Eq. (10.14). When a crystal is cooled, it may be possible to maintain a 'local' equilibrium between vacancies and divacancies, while the total number of vacant sites, n_T, is unable to decrease, because of a scarcity of nearby sinks, such as free surfaces, grain boundaries and dislocations. In this case, Eq. (10.13) still applies, together with the constraint:

$$n_1 + 2n_2 = n_T \tag{10.15}$$

Dividing Eq. (10.15) by Eq. (10.13) gives:

$$\frac{n_2}{n_1} + 2\left(\frac{n_2}{n_1}\right)^2 = \frac{n_T}{Nq} \exp\left(\frac{E_b}{kT}\right) \tag{10.16}$$

A consideration of this equation shows that, under these circumstances, the ratio of single vacancies to divacancies *decreases* as the temperature is lowered. Physically, this is because a fixed number of vacant sites greater than the equilibrium value have a higher energy and higher entropy when in the form of single vacancies than when in the form of divacancies.

In general, both interstitials and vacancies that are trapped at low temperatures in excess of their equilibrium number will condense into aggregates of two or more defects. Ultimately, an aggregate may grow into a form that can be observed in the transmission electron microscope, such as a prismatic dislocation loop, or, in the case of vacancies in some c.c.p. metals, a stacking fault tetrahedron (Figure 9.14). In the intermediate stages of aggregation when the cluster contains only a few defects, it is impossible to observe the cluster directly and it is difficult to decide theoretically which of several alternative configurations should actually be adopted. Whether or not a given configuration occurs may depend on how easily it is nucleated.

10.4   Point Defect Configurations

The positions of the atoms around a point defect can only be observed directly under the special conditions that prevail in microscopes such as the atom probe field-ion microscope or in the scanning tunneling microscope. In both cases the atoms in a surface can be imaged, and it has been possible to identify vacant sites. In general, however, it is necessary to turn to indirect experiments or to theoretical treatments.

In the theoretical treatment of point defect configurations there is an important difference between vacancies and interstitials. The basic configuration of a vacancy is known, whereas that of an interstitial defect is uncertain. In the case of a vacancy, all that needs to be determined is the displacement of the atoms surrounding the unoccupied site. In the case of an interstitial defect, the extra atom might be located at any one of a number of *non-equivalent* positions in the crystal. Several different configurations must therefore be compared in order to discover which has the lowest energy.

These problems have been attacked by treating the energy of a crystal containing the defect as the sum of the energies of interaction of atoms. This is a two-body interaction problem in its most simple form. A function is chosen to describe the dependence of the energy φ of two atoms on their separation r. One example is the Morse function [15]:

$$\phi = D\{\exp[-2\alpha(r - r_0)] - 2\exp[-\alpha(r - r_0)]\} \tag{10.17}$$

In Eq. (10.17), D is the dissociation energy of the pair of atoms and r_0 their equilibrium separation. The constant α can be derived from the elastic compressibility of the crystal. Other workers have used the Born–Meyer repulsive potential [16]:

$$\phi = A \exp\left[-\frac{B(r - r_0)}{r_0}\right] \tag{10.18}$$

together with a pressure applied to the boundaries of the crystal in order to hold the atoms together. Minimum-energy configurations have then been determined by summing the energies of pairs of atoms for a large number of trial configurations. The various authors of these calculations up to the early 1960s, and prior to the advent of computers becoming available for large-scale calculations, pointed out that they were based on an inexact treatment of the energy of the crystal, but that more exact treatments were difficult. Later calculations used more sophisticated empirical potentials and pseudopotentials for the two-body interaction potentials [6, 7, 12]. The increasing power of computers has now enabled a variety of computationally intensive approaches to be used, such as an *ab initio* non-local pseudopotential or embedded atom potentials [17, 18]. A recent review has compared state-of-the-art methods in this area using examples from a wide range of materials [19].

Examples of calculated values of the displacements of the atoms around a vacancy are shown in Table 10.2, taken from the work of Girifalco and Weizer [20]. The atoms in the innermost shell move inwards, whereas those in the next shell move away from the vacancy. It should be remembered that this reversal may well be merely a consequence of the model used, and that it might not occur in a real crystal. The relaxations in b.c.c. metals are considerably larger than those in c.c.p. metals. A relatively small relaxation is to be expected in a 'full' metal like copper, in which the ion cores of the atoms are close to one another,

Table 10.2 Relaxation displacements around a vacancy expressed as a percentage of the normal distance from the vacant site [20]

Crystal		Shell	
Metal	Structure	First	Second
Pb	c.c.p.	1.42	−0.43
Ni	c.c.p.	2.14	−0.39
Cu	c.c.p.	2.24	−0.40
Ca	c.c.p.	2.73	−0.41
Fe	b.c.c.	6.07	−2.12
Ba	b.c.c.	7.85	−2.70
Na	b.c.c.	10.80	−3.14

A positive sign indicates a displacement towards the vacancy.

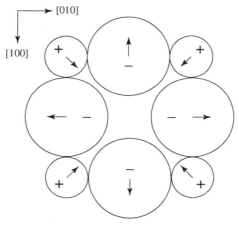

Figure 10.2 {100} plane of NaCl, showing the sense of the displacements of the ions surrounding a Na$^+$ ion vacancy

since the inward collapse of the shell of atoms next to the vacancy is very quickly arrested by the repulsive force that the atoms within that shell exert upon one another.

The form of the displacements around a vacancy in a sodium chloride crystal can be deduced by considering the electric charges involved. A positive ion vacancy, being a region deficient in positive charge, draws the surrounding cations inwards and repels the anions, which are its nearest neighbours (Figure 10.2). According to Mott and Littleton [21], the anions next to a sodium ion vacancy in sodium chloride are displaced outwards by 0.07d, where d is the interionic distance (equal to half the lattice parameter). In addition to being displaced as a whole, the ions themselves are deformed and polarized by the electric field of the vacancy.

Calculations of the energies of interstitial configurations in metals have yielded some surprising results, as has already been mentioned in Section 10.1. Intuitively, it might be expected that an extra atom forced into a c.c.p. metal would occupy the centre of a unit cube, where it would be surrounded by atoms lying at the corners of an octahedron. Examination of a ball model of the structure shows that this octahedral interstice is the largest

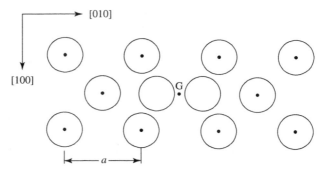

Figure 10.3 Hypothetical split interstitial in a c.c.p. metal. The plane of the diagram is (001). Source: Reproduced by permission of the American Physical Society [4].

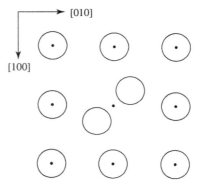

Figure 10.4 Hypothetical split interstitial in a b.c.c. metal. The plane of the diagram is (001). Source: Reproduced by permission of the American Physical Society [4].

hole available (see Section 3.2). However, a less symmetrical configuration is reported to have a lower computed energy [4, 6, 7, 17]. In this configuration, shown in Figure 10.3, two atoms are displaced equally from a normal lattice site, each by about $0.3a$ along $a\langle 100\rangle$, where a is the lattice parameter. The energy difference which favours this 'split' interstitial is small, but a number of studies using different atom potentials have shown that this configuration is definitely preferred over other possible configurations. Similarly, in a b.c.c. metal, an obvious site for an extra atom is the octahedral interstice at the midpoint of an edge or face of the unit cube, but the configuration that is computed to have the least energy for α-Fe is the $\langle 110\rangle$ split configuration, shown in Figure 10.4 [22].

An interstitial defect that is similar to the split configurations described was originally postulated for b.c.c. metals by Paneth in 1950 [23]. An extra atom is supposed to be located between two atoms in a close-packed $\langle 111\rangle$ row, where it produces extensive relaxation outwards along the row. Paneth coined the word 'crowdion' for this defect. Thus, for example, six atoms might be squeezed into a length normally occupied by five, as in Figure 10.5. The existence of crowdions has never been established experimentally, even though recent density-functional calculations suggest that they have the lowest formation energy in the

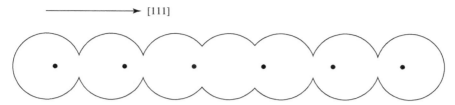

Figure 10.5 Crowdion in an alkali metal, as postulated by Paneth [23]. The line of atoms is ⟨111⟩ and the dots mark lattice sites

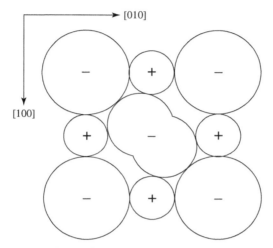

Figure 10.6 H-centre in a KCl crystal. The plane of the diagram is (001). (No attempt is made to depict relaxation displacements.)

six b.c.c. metals V, Nb, Ta, Cr, Mo and W [24]. Variants of the Paneth crowdion model can also be found in postulated configurations for interstitial configurations in both c.c.p. and b.c.c. metals considered by Johnson and Dederichs et al. [6, 7].

A defect resembling the crowdion occurs in alkali halides that have been damaged by ionizing radiation at very low temperatures. In potassium chloride, this defect, a chlorine ion interstitial, called an H-centre, consists of a chlorine *atom* lodged in a close packed ⟨110⟩ row of Cl^- ions so that a $(halogen_2)^-$molecule-ion is formed [25] (Figure 10.6). In this instance, the electronic properties of the defect provide strong experimental evidence for the configuration.

When a point defect is in the course of moving from one rest position to a neighbouring one, its atomic configuration changes. The accompanying increase in internal energy determines the frequency of movement, through Eq. (10.9). While these changes are difficult to calculate with precision without extensive computations, some interesting qualitative observations can be made relatively easily. For example, when a b.c.c. metal vacancy moves in α-Fe, two maximum-energy configurations must be overcome before it achieves its new rest position. This can be seen from Figure 10.7, which shows that the atom that changes place with the vacancy must squeeze successively through the centres of two triangles of atoms. Between these two triangles, the atom has a little more room, so that a local energy

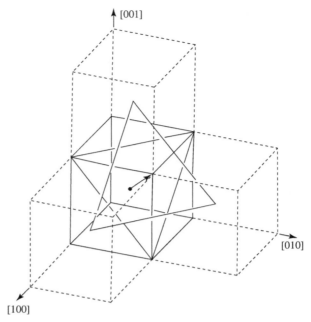

Figure 10.7 Path followed by an atom jumping into a vacant nearest-neighbour site in a b.c.c. crystal. Any displacements of atoms near the vacancy from their normal sites have been neglected in this drawing

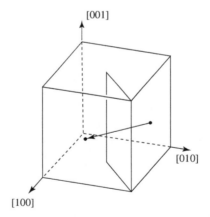

Figure 10.8 Path followed by an atom jumping into a vacant nearest-neighbour site in a c.c.p. crystal

minimum occurs at the midpoint of its jump. Computer simulations confirm this qualitative expectation [22]. When a c.c.p. metal vacancy moves, the replacing atom need only pass through a single rectangle of atoms (Figure 10.8). By examining a model of close-packed spheres, which is appropriate for a 'full' metal like copper, it can be seen that this rectangle is forced to expand to allow the atom to pass through it. The configuration of greatest energy is expected to occur when the rectangle is fully expanded, at the midpoint of the jump.

An interstitial defect that consists simply of an extra atom or ion lying in an interstice between normally situated atoms or ions can move in two basically distinct ways. The extra atom or ion may either jump into a neighbouring, equivalent, interstice or it may replace a normally situated atom or ion which itself moves into an interstice. The replacement mechanism is called an interstitialcy jump. It is believed that interstitial Ag^+ ions move through AgBr and AgCl crystals by this mechanism [26].

The mobility of an unsymmetrical defect can be expected to be anisotropic. A crowdion, for example (Figure 10.5), would move very easily along its own line by means of a set of quite small atom movements. The mobility of simpler defects will be anisotropic because of crystallographic anisotropy. For example, a vacancy in the graphite structure (Figure 3.14) must move more easily within the sheet of strongly bonded carbon atoms forming the basal plane of the crystal structure than from one sheet to another. Calculations bear out this prediction [27]. It must be noted that the predicted formation energy of vacancies in graphite is very high (7.3 eV), so a very high temperature of the order of 3000 °C is required to enable a vacancy concentration of $\sim 10^{-10}$ to be obtained at equilibrium [27].

10.5 Experiments on Point Defects in Equilibrium

There are various ways of putting point defects into a crystal. The most selective method is to heat the crystal to a temperature at which its point defects can come to equilibrium. Under such circumstances, there may be only one type of defect with a small enough energy of formation to be present in large numbers. Point defects can be produced at low temperatures by irradiating or plastically deforming a crystal. These two methods have the drawback that they introduce several different types of defect whose effects upon the crystal and upon one another are difficult to disentangle. This is especially the case when the defects are being studied through some property of the crystal that does not discriminate between one type of defect and another; experiments that do have some power of discrimination are very valuable.

The least ambiguous experiments are those performed at high temperatures upon a crystal containing an equilibrium number of point defects. Simmons and Balluffi [28–31] performed experiments of this type by measuring the lengths and lattice parameters of rods of c.c.p. metals as a function of their temperatures. The principle of these experiments is as follows. Suppose that vacancies are formed as the temperature is raised. Each time a vacancy is created by means of an atom jumping out on to a surface, grain boundary or dislocation, one atomic site is added to the metal. The volume of the metal would then increase by the volume occupied by one atom in a perfect crystal, were it not for the fact that the atoms around the vacancy move slightly from their normal lattice sites. This relaxation is propagated to the surfaces of the metal as an elastic strain, which causes the surfaces of the metal to move inwards. The total volume change produced by adding n vacancies to N atoms is then:

$$\frac{\Delta V}{V} = \frac{n}{N} + \left(\frac{\Delta V}{V} \right)_e \tag{10.19}$$

where $(\Delta V/V)_e$ is the volume change due to elastic strain. Now it happens that the change in lattice parameter measured by X-rays, which results from the elastic strain caused by evenly distributed point defects, is simply that which would be given by the homogeneous dilatation $(\Delta V/V)_e$; that is, the change in lattice parameter $\Delta a/a$ is given by:

$$\frac{\Delta a}{a} = \frac{1}{3}\left(\frac{\Delta V}{V}\right)_e \tag{10.20}$$

The small fractional change in length of the rod, $\Delta L/L$, is given by:

$$\frac{\Delta L}{L} = \frac{1}{3}\left(\frac{\Delta V}{V}\right)$$

From Eqs. (10.19 and 10.20):

$$\frac{\Delta L}{L} = \frac{n}{3N} + \frac{\Delta a}{a}$$

That is:

$$\frac{n}{N} = 3\left(\frac{\Delta L}{L} - \frac{\Delta a}{a}\right) \tag{10.21}$$

The changes in length and lattice parameter diverge as the temperature is raised and vacancies are generated in a rod of Al, as can be seen from Figure 10.9. Although very precise

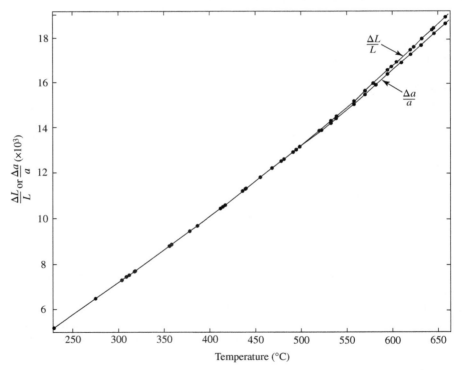

Figure 10.9 Effects of changes in length and lattice parameter with temperature in aluminium [28]. Source: Reproduced by permission of the American Physical Society.

Table 10.3 Fraction of sites vacant at the melting point, n/N, in equilibrium

Material	n/N at melting point	Method	Ref.
Al	9.4×10^{-4}	Length–lattice parameter	[28]
Ag	1.7×10^{-4}	Length–lattice parameter	[29]
Au	7.2×10^{-4}	Length–lattice parameter	[30]
Cu	1.9×10^{-4}	Length–lattice parameter	[31]
Na	7.5×10^{-4}	Length–lattice parameter	[32]
Pb	1.7×10^{-4}	Length–lattice parameter	[33]
KCl	1.6×10^{-4}	Electrical conductivity	[34]
NaCl	2.8×10^{-4}	Electrical conductivity	[34]

measurements are needed, the experiments are successful in giving a direct measure of the equilibrium concentration of vacancies in a number of c.c.p. metals; examples of some values are listed in Table 10.3.

It is evident that Eq. (10.21) can be used to describe the effect of any defect whose production adds or removes atomic sites. If ΔN sites are added:

$$\frac{\Delta N}{N} = 3 \left(\frac{\Delta L}{L} - \frac{\Delta a}{a} \right) \tag{10.22}$$

The creation of a single interstitial atom at a surface, grain boundary or dislocation destroys one atomic site, so that the measured value of $(\Delta L/L - \Delta a/a)$ would be negative if interstitial atoms were the dominant point defects.

Other studies of point defects in equilibrium are based on measurements of electrical conductivity. This method can be applied more easily to ionic crystals than to metals, since all of the conductivity of a high-band-gap ionic crystal is likely to be due to the mobility of its charged point defects, whereas only a small part of the high-temperature resistivity of a metal is due to the scattering of electrons by point defects. For example, the conductivity of NaCl at high temperatures is due to Schottky defects. If an electric field of $100 \, \text{V m}^{-1}$ drives the sodium and chlorine ion vacancies at speeds of μ^+ and μ^- m s^{-1} respectively, the conductivity, σ, due to a concentration n of Schottky defects per m^3 is:

$$\sigma = ne(\mu^+ + \mu^-) \tag{10.23}$$

where e is the electronic charge. Simply from the relative sizes of the anions and cations in all alkali halides except the caesium salts, it is reasonable to infer that the cation vacancy is more mobile than the anion vacancy. In CsCl, calculations suggest that while the chloride anion is more mobile than the caesium cation for single-vacancy migration, cation migration is actually favoured overall because of a low migration energy for a vacancy-pair mechanism of migration [35, 36]. The immobility of the chloride ion in NaCl has been confirmed by measuring the amounts of material transported to the anode and cathode when a current is passed through a crystal of NaCl. The fraction of the current carried by sodium ions moving to the cathode (the transport number of Na$^+$) is found to be close to unity [37]. Under such circumstances, Eq. (10.23) can be approximated by:

$$\sigma = ne\mu^+ \tag{10.24}$$

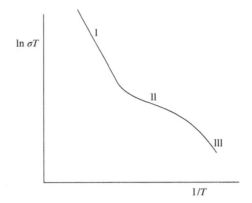

Figure 10.10 Schematic conductivity plot for a NaCl crystal containing a small concentration of a divalent cation (for example, a cation fraction of 10^{-4})

Provided that the mobility μ^+ is known, the conductivity σ measures the concentration of Schottky defects directly. The mobility μ^+ can be derived from the conductivity of crystals containing a known concentration n_i of a divalent cation impurity, such as Ca^{2+}. Each divalent cation introduces a sodium ion vacancy, and the total of these can be made to exceed by far the equilibrium number of vacancies in a pure crystal. The conductivity is then given by:

$$\sigma = n_i e \mu^+ \tag{10.25}$$

In order to apply Eq. (10.25), the temperature should be sufficiently low that the equilibrium concentration of Schottky defects is small compared to n_i, but sufficiently high that few of the sodium ion vacancies are bound to divalent cations, which attract them because of their opposite charge.

A consideration of how Eqs. (10.24) and (10.25) vary as a function of temperature shows that a plot of $\ln \sigma T$ against $1/T$ enables an effective activation energy for the dominant charge-carrying mechanism to be determined from the slope of the plot [34]. Such a schematic for a slightly impure crystal is shown in Figure 10.10. Even though this is a plot of $\ln \sigma T$ against $1/T$, rather than $\ln \sigma$ against $1/T$, such a plot is commonly called a conductivity plot.

In region I, at high temperatures where Schottky defects are abundant, the conductivity is almost identical to that of a pure crystal. Here, the slope of the line gives an activation energy E_I composed of both the Schottky defect formation energy, $(E^+ + E^-)/2$, and the energy barrier to the motion of a cation vacancy, E_M^+, which enters through the mobility μ^+:

$$E_I = E_M^+ + \tfrac{1}{2}(E^+ + E^-) \tag{10.26}$$

In region II, the number of Schottky defects is small in comparison with n_i, and the slope in this region gives E_M^+ alone. The increase in the gradient of the $\ln \sigma T$ against $1/T$ graph as the temperature is lowered into region III is due to the association of cation vacancies and impurity ions, and possibly to further clustering of defects [34].

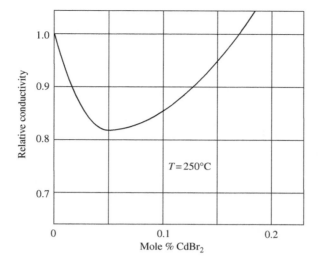

Figure 10.11   Effect of CdBr$_2$ additions on the electrical conductivity of AgBr [38]

A striking example of a discriminatory conductivity experiment at high concentrations of a second cation is provided by measurements made by Teltow at 250 °C on AgBr doped with CdBr$_2$ [38]. As Cd^{2+} ions are added, the conductivity at first decreases, but then passes through a minimum as the Cd^{2+} content is increased (Figure 10.11). This curious behaviour can be explained by supposing that the current in a pure AgBr crystal is carried mainly by interstitial Ag$^+$ ions. The Ag$^+$ ion vacancies which are introduced by the divalent Cd^{2+} ions destroy a number of Ag$^+$ interstitials, but, at sufficiently high concentration, carry enough current themselves to make up for this loss.

10.6   Experiments on Quenched Metals

While measurements made near the melting point provide the most direct evidence for the existence of point defects in equilibrium, many data have been obtained from quenched metals. By cooling rapidly enough, point defects can be 'frozen in' at a low temperature, where they can be studied more conveniently. A practical difficulty is that the population of defects will change during the initial stages of a quench if the quench is too slow, while a quench that is too fast will produce thermal gradients severe enough to cause plastic deformation or fracture. These conflicting conditions have tended to restrict quenching experiments to metals, which are generally less susceptible to thermal shock than non-metals.

The property of a quenched metal that is most frequently studied is its electrical resistivity. This can be measured quickly and accurately, but it has the drawback of being non-discriminatory, since all lattice defects increase the resistivity of metals.

Typically, a fine wire of the metal is quenched into brine, or cold helium, where it cools at a rate of 10^4 or 10^5 °C s^{-1}. An advantage of working at liquid helium temperature is that the resistivity of the perfect crystal is then so low that most of the measured resistivity is caused by defects, whereas at room temperature quenched-in defects add only a few per cent to the resistivity. An activation energy can be derived from the resistivities of a metal quenched from various temperatures, by fitting them to the equation:

$$\Delta\rho = A \exp\left(-\frac{E}{kT}\right) \tag{10.27}$$

where $\Delta\rho$ is the extra resistivity measured after quenching from a temperature T. By assuming the presence of one type of defect, which increases the resistivity in direct proportion to its concentration, it is possible to identify the activation energy E in Eq. (10.27) with the formation energy of the defect.

Measurements of this type have been made chiefly in c.c.p. metals, and in these the vacancy is almost certainly the predominant defect. Successful quenching experiments on b.c.c. metals are more difficult, partly because of the ease with which b.c.c. metals pick up impurities. It is also commonly believed that the ratio of the formation energy to the motion energy of a vacancy is relatively large for b.c.c. metals, which would make it relatively difficult to retain a substantial concentration in a quench. However, some results have been obtained for b.c.c. metals, which are included in Table 10.4. A more comprehensive tabulation of data for both self-diffusion and impurity diffusion can be found in the book by Neumann and Tuijn [39], together with a summary of the various techniques other than quenching experiments now available to measure the extent of self-diffusion and impurity diffusion.

Most of the energies of motion listed in Table 10.4 were obtained from studies of the rate at which the resistivity returns to its normal value when the quenched metal is reheated. As soon as the temperature allows the point defects to migrate, they will reduce their concentration towards the equilibrium value by combining with other point defects or destroying

Table 10.4 Energies of formation and migration of vacancy defects in metals, E_f and E_M, respectively, and activation energies for self-diffusion, E_{SD}

Material	Defect	E_f (eV)	Ref.	E_M (eV)	Ref.	E_{SD} (eV)	Ref.
Ag	V_1	1.1	[40]	0.83	[40]	1.91	[41]
	V_2	$E_b = 0.38^a$	[40]	0.57	[40]		
Al	V_1	0.76	[42]	0.65	[43]	1.4	[44]
	V_2	$E_b = 0.17$	[45]	0.5	[45]		
Au	V_1	0.98	[46]	0.82	[46]	1.81	[47]
	V_2	$E_b < 0.15$	[48]	0.7	[48]		
Cu	V_1	1.14	[49]	1.08	[50]	2.04	[51]
	V_2	0.19	[52]	0.71	[49]		
Ni	V_1	1.4	[53]	1.5	[54]	2.90	[55]
Pt	V_1	1.51	[56]	1.38	[56]	2.89	[57]
Na	V_1	0.42	[32]	0.03	[58]	0.45	[59]
Mo	V_1	3.24	[60]	1.3	[61]	~4.5	[61]
W	V_1	3.51	[62]	1.78	[63]	5.5 ± 0.5	[64]
Mg	V_1	0.58	[65]	0.52	[66]	1.39^b	[67]
Sn	V_1	0.51	[68]	0.68	[68]	1.09 ∥ **a**;	[69]
						1.11 ∥ **c**	[69]

[a]A reanalysis of the data by Seeger [70] has led him to suggest that this cannot be substantially larger than 0.19 eV.
[b]This discrepancy between measured E_{SD} and measured $(E_f + E_M)$ is discussed in [65].
V_1 = single vacancy, V_2 = divacancy, E_b = binding energy of divacancy.

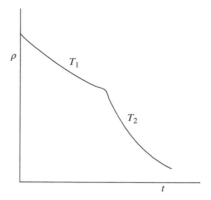

Figure 10.12 Annealing out of the quenched-in resistivity ρ of a metal as a function of time, t, with a sudden increase in temperature from T_1 to T_2

themselves at sinks such as surfaces or dislocations. The recovery of the resistivity can be followed in various ways. In an isochronal anneal, the specimen is heated at a constant rate, or for a short, fixed, time, at a succession of increasing temperatures; in an isothermal anneal, the recovery is followed at a constant temperature. A third method is to step up the temperature suddenly after a period of isothermal recovery (Figure 10.12). The ratio of the recovery rates at the intersection of the curves for the two temperatures, T_1 and T_2, can be used to define an activation energy through the equation:

$$\frac{(d\rho/dt)_1}{(d\rho/dt)_2} = \exp\left[-\frac{E_M}{k}\left(\frac{1}{T_1} - \frac{1}{T_2}\right)\right] \tag{10.28}$$

Equation (10.28) can easily be derived when recovery occurs by a *single* thermally activated mechanism, such as the diffusion of vacancies to sinks, since in this case:

$$\frac{dn}{dt} = f(n)\exp\left(-\frac{E_M}{kT}\right)$$

and:

$$\frac{d\rho}{dt} \propto \frac{dn}{dt} \tag{10.29}$$

The function $f(n)$ depends on the type and number of sinks. The advantage of finding the activation energy, E_M, by suddenly changing the temperature, rather than by comparing complete curves at different temperatures, is the greater probability that the function $f(n)$ will remain the same.

Unfortunately, there are grounds for doubting that recovery occurs, in general, by a single mechanism. Unless their concentration is below some value, which is difficult to assign, isolated vacancies will combine to form divacancies or larger clusters during recovery. The process of recovery might then be a complex one, consisting of the combination of vacancies, divacancies and perhaps larger clusters with one another, as well as the removal of some defects at sinks. When this happens, it is difficult to interpret an experimentally determined activation energy. A touchstone in the interpretation of results for c.c.p. metals has been the principle that the motion energy, E_M, of an isolated vacancy obeys Eq. (10.10):

$$E_{SD} = E_f + E_M$$

This equation is based on the belief that self-diffusion in a c.c.p. metal occurs by the motion of its equilibrium vacancies. Discrepancies with Eq. (10.10) have led to the interpretation of certain results in terms of the motion of divacancies, giving the values of the energy of binding of the divacancies and of the motion of divacancies shown in Table 10.4.

10.7 Radiation Damage

In the ease with which lattice defects are produced by radiation, metals can be distinguished rather sharply from non-metals. Only particles that can directly knock atoms out of their normal sites will produce lattice defects in metals. Radiation which merely excites electrons does not produce lattice defects unless the excited electrons themselves can displace atoms, as is the case with electrons excited by γ-irradiation. By contrast, many non-metals are severely damaged by an ionizing radiation such as X-rays. This is because of the greater permanence of an electronic disturbance in a non-metal [71].

Since a displaced atom leaves a vacant site behind it, irradiation normally introduces equal numbers of vacancies and interstitials. An interstitial may be produced at some depth by means of *replacement collisions*, in which an atom struck at the surface replaces a deeper neighbour, and so on, until finally an atom lacking the energy to replace another becomes lodged in an interstice [71]. This process occurs most effectively along a close-packed row of atoms in metals [72].

Fast neutrons produce damage evenly throughout specimens of ordinary size, because they travel several centimetres between collisions. When a head-on collision does occur, the struck atom is displaced with such violence that it travels for a considerable distance, perhaps as much as a thousand atom spacings, and in doing so displaces hundreds of other atoms. The majority of displacements occur at the end of the track of this 'primary knock-on', where they produce a very disturbed region called a *spike*. As a result of replacement collisions, interstitials are likely to be concentrated in the outer regions of the spike, with vacancies clustered in the centre.

A more uniform distribution of vacancy–interstitial pairs can be produced by irradiating a metal foil with electrons that have only enough energy to displace one atom. The maximum energy that an electron of energy E can transfer to an atom of mass M is given by:

$$E_{max} = \frac{2E(E + 2mc^2)}{Mc^2} \tag{10.30}$$

where m is the rest mass of the electron and c is the velocity of light [73]. The energy that must be imparted to an atom in order to displace it lies between 20 and 40 eV for most metals. Comparison with the energy given by Eq. (10.30) shows that each electron accelerated through a potential difference of 1 MeV can typically displace one or two atoms. As a consequence of electron irradiation in a transmission electron microscope, radiation damage can occur in specimens being studied: in aluminium for example, the threshold voltage for radiation damage is about 170 kV, while for copper it is about 400 kV [73, 74].

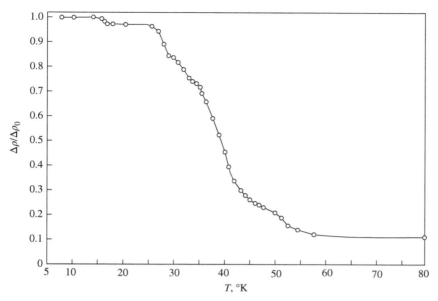

Figure 10.13 Isochronal recovery of electron-irradiated copper containing an atomic concentration of Frenkel defects of about 10^{-6} [75]

Despite the relatively simple nature of the damage produced by 1 MeV electrons at low temperatures, its removal by annealing is a complex process. An outstanding feature is that a great deal of recovery occurs at a remarkably low temperature. The isochronal recovery of the electrical resistivity of electron-irradiated copper in the range 10–80 K is shown in Figure 10.13 [75]. This is called stage I recovery and can be seen to be composed of five substages. The small extra resistivity left at the end of stage I anneals out in three further stages at successively higher temperatures. Recovery in neutron-irradiated copper follows a broadly similar pattern, with the addition of a fifth stage. It is impossible to deal concisely with the problem of assigning mechanisms to the various stages, but it is presumed that some part of stage I represents the self-annihilation of close vacancy–interstitial pairs [75]. Stage V probably involves the removal of prismatic dislocation loops produced by the clustering of like defects [76]. Such loops are commonly observed when heavily irradiated materials of all kinds are examined in the electron microscope [73, 74, 76, 77].

10.8 Anelasticity and Point Defect Symmetry

The configuration of atoms about a point defect can be such that its orientation with respect to the crystal axes must be specified. For example, the interstitial defect, centred on the point G with coordinates $(\frac{1}{2}, \frac{1}{2}, 0)$ in Figure 10.3, can lie along any of the three $\langle 100 \rangle$ directions. In Figure 10.3 it lies along [010]. These three possible orientations are crystallographically equivalent. Ordinarily, the free energies of all crystallographically equivalent configurations are the same, but when a stress is applied to the crystal, their energies may become different. The defects will then tend to reorient themselves so as to increase the proportion of

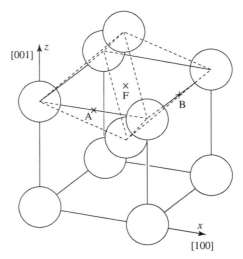

Figure 10.14 Interstitial sites occupied by C or N atoms in ferrite, α-Fe. The dotted lines show the squashed octahedron of Fe atoms surrounding the site F

those with less energy. Whether a stress can distinguish between different orientations of a defect or not depends on the symmetry of the configuration of the defect. Consequently, an experiment that detects a reorientation of defects under stress has the power to discriminate between different configurations of the defect.

The best-known example of a defect reorientation induced by stress is the redistribution of interstitial carbon atoms (or nitrogen atoms) present as impurity in ferrite, α-iron (Section 3.2.4). Each carbon atom lies at the centre of a squashed octahedron of Fe atoms, as at the site marked F in Figure 10.14. The symmetry of the atomic arrangement around F is such that the point group symmetry at F is 4/mmm; that is, the site F has tetragonal symmetry. At the site marked F, the fourfold rotation axis is parallel to the z-axis. Sites A and B are equivalent in position to F in the unit cell and also possess point group symmetry 4/mmm, but their fourfold axes are parallel to the x- and y-axes, respectively. This is very easily seen by drawing the orientations of the squashed octahedra of iron atoms forming the neighbours of sites A and B, respectively. At each of these sites, the two iron atoms that are closest to the carbon atoms lie along the fourfold axis, and so the carbon atom is expected to force them apart.

As a consequence, an interstitial at the F site will have a smaller energy than one at either the A or the B site when a *tensile* stress is applied along the z-axis. Therefore, it is energetically favourable for some of the carbon atoms in A and B sites to jump into F sites, causing the tensile strain in the z direction to increase. The rate at which this occurs will be determined by the frequency of jumping of the carbon atom. When the stress is removed, the distribution of carbon atoms will tend to become random again, because the F sites are no longer preferred, and so the strain will return to zero. If a tensile stress were applied along $\langle 111 \rangle$ instead of $\langle 001 \rangle$, the three sites A, B and F, being symmetrically disposed about this tensile axis, would remain equal in energy, and there would be no extra strain.

A strain which is time-dependent, but reversible, is called *anelastic*. When an anelastic strain ε_a due to a small constant stress σ is produced by a single simple process, such as that described above, it is found that ε_a increases with time according to the equation:

$$\frac{d\varepsilon_a}{dt} = \frac{1}{\tau}(\varepsilon_a^\infty - \varepsilon_a) \tag{10.31}$$

This equation shows that the strain approaches the value ε_a^∞ asymptotically. The quantity ε_a^∞ is proportional to the applied stress, so we can write:

$$\varepsilon_a^\infty = \sigma\delta s \tag{10.32}$$

The constant of proportionality, δs, is called the *relaxation of the compliance* and the quantity τ, with the dimensions of time, is called the relaxation time at constant stress. When the anelastic strain is produced by the motion of single atoms, the relaxation time is proportional to the time in which an atom makes one jump. Integrating Eq. (10.31) and adding to ε_a the instantaneous elastic strain at $t = 0$, σs, where s is the elastic compliance, gives:

$$\text{Total strain} = \sigma s + \sigma\delta s \left[1 - \exp\left(-\frac{t}{\tau}\right)\right] \tag{10.33}$$

where s is usually at least an order of magnitude larger than δs. The behaviour is shown schematically in Figure 10.15.

Anelastic strain in a crystal is often detected experimentally through the viscous drag that it exerts upon any mechanical vibration of the crystal, causing such a vibration to die out. This effect is called damping or *internal friction*. The magnitude of the damping experienced by the vibration is a maximum at a certain frequency. When the frequency is very high, the anelastic strain never has time to come into play and there is no damping. On the other hand, if the period of vibration (the reciprocal of the frequency) is very large compared to the relaxation time τ defined in Eq. (10.31) or (10.33), the anelastic strain has ample time to maintain its full value, given by Eq. (10.32), and is therefore indistinguishable from the elastic strain so far as that particular vibration is concerned. Again, there is no damping. It is a straightforward exercise to show that the maximum damping occurs when the period of the vibration is comparable to the relaxation time or, more precisely, when it is equal to $2\pi\tau$.

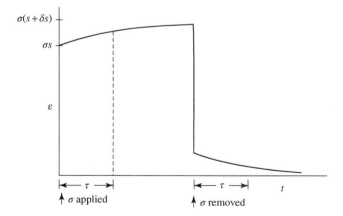

Figure 10.15   Effect of relaxation on the strain caused by a constant stress σ.

The effect of an applied stress on the number of point defects in various different orientations depends upon the strains that the defects produce in the crystal. If the macroscopic strain due to a dispersion of defects of one orientation is such that the forces applied to the crystal do work when this strain occurs, this orientation will be favoured in comparison to orientations producing strains that do not allow the applied forces to do work. If two different orientations happen to produce identical strains (referred to fixed axes, of course), they cannot be distinguished from one another by an applied stress of any kind. The strain that a defect can produce is not entirely arbitrary, because it must be consistent with the *symmetry of the defect*.

This is defined as follows. Examine an infinite crystal containing a single defect for the presence of point group symmetry operations at the centre of the defect. The centre of the defect, if it is not obvious, may be determined as being that point within the defect volume at which the number of symmetry operations is a maximum. Space group tables such as those in the *International Tables for Crystallography* [78] are of great help here, since they list the point group symmetries at different positions in the unit cell.

The group of symmetry elements that is then found defines the point symmetry of the defective crystal, which for brevity can be called the *symmetry of the defect*. Clearly, it is impossible for the homogeneous, macroscopic strain due to a uniform dispersion of a defect of a given orientation to be less symmetrical than the defect itself, because the symmetry of the defect is limited by the displacements of the atoms around it, which in turn determine the strain. However, the symmetry of the strain may be greater than that of the defect. A homogeneous strain is inherently centrosymmetric (Section 6.2) and so, for example, the strain produced by a dispersion of cation vacancy–anion vacancy pairs in an ionic crystal must be centrosymmetric, even though the defect itself has no centre of symmetry. The limitations on the type of strain that can be produced by a defect of any given symmetry can be found by consulting Table 5.2, reading 'defect system' in place of 'crystal system'.

For example, from Table 5.2 we see that the only type of strain quadric consistent with cubic symmetry is a sphere (i.e. pure dilatation). Therefore a defect of cubic symmetry can never be a source of anelasticity under homogeneous stress, because its strain quadric is the same for all orientations. To illustrate this, a defect of cubic symmetry is shown in Figure 10.16 – a hypothetical tetravacancy in a c.c.p. metal. Its symmetry is $\bar{4}3m$ and it

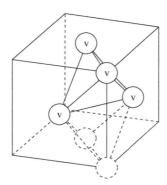

Figure 10.16   Hypothetical tetravacancy in a c.c.p. metal. The dotted lines show the alternative orientation of the tetrahedron of vacancies

has two distinct orientations (one is shown by solid lines and the other by broken lines in Figure 10.16). The relative numbers of these two orientations cannot be changed by an applied stress of any kind. An example of a defect of lower symmetry is the carbon atom in α-iron, whose symmetry is that of the site of the type F (Figure 10.14), provided that the displacements of the surrounding iron atoms conform to the site symmetry. In this case the defect point group is the tetragonal one, $4/mmm$. The strain quadric which is consistent with a tetragonal defect is a quadric of revolution about the fourfold axis of the defect (Table 5.2). The three orientations of the interstitial carbon (F, A and B in Figure 10.14) are distinguished by their differently oriented fourfold axes and so a general applied stress will distinguish between them.

Whether a *particular* stress will distinguish between defect orientations, each of which produces a different strain, depends on whether it does a different amount of work on the crystal as the strain is introduced in each case. Sometimes it is possible to pick out by inspection a special stress which cannot produce a change in defect orientation. An example, already mentioned, is the absence of an anelastic response to a tensile stress in the $\langle 111 \rangle$ direction of a crystal of iron containing interstitial carbon atoms. Nowick and Heller [79] have tabulated complete 'selection rules' for deciding whether or not there can be an anelastic response to a particular tensile stress in a crystal containing a defect of given symmetry.

Problems

10.1 If the activation energy for self-diffusion in copper is 2.04 eV and the activation energy for the migration of a vacancy is 1.08 eV (Table 10.4), determine the ratio of the concentration of vacancies present in equilibrium at 1000 °C to the concentration present in equilibrium at 500 °C.

10.2 Experimental determinations of the concentration of vacancies in copper gave the values of 8×10^{-5} at 980 °C and 16×10^{-5} at 1060 °C. From these values, compute the energy of formation and entropy of formation of a vacancy in copper.

10.3 Compare the concentration of positive ion vacancies in a NaCl crystal due to the presence of a mole fraction of 10^{-4} of $CaCl_2$ impurity with the concentration present in equilibrium in a pure NaCl crystal at 400 °C. The formation energy of a Schottky defect $E^+ + E^- = 2.12$ eV and the concentration of Schottky defects at the melting point, 800 °C, is 2.8×10^{-4}.

10.4 A crystal of ferrous oxide, $Fe_{1-x}O$, which is isostructural with NaCl, is found to have a lattice parameter $a = 4.30$ Å and a density of 5.72 Mg m^{-3}. Estimate x, stating clearly the assumptions you make about the structure of the crystal.

10.5 Simmons and Balluffi [29] observed that when a rod of silver was heated to its melting point, 960 °C, the fractional increase in the length of the rod exceeded the fractional increase in its lattice parameter by 5.6×10^{-5}. Assuming that the only defects present are isolated vacancies and taking the entropy of formation of a vacancy to be $1.5\,k$, calculate the formation energy of a vacancy E_f.

If in fact there were divacancies present as well as vacancies, and if there were one-tenth as many divacancies as vacancies, what error would this introduce into the value of E_f calculated on the assumption that only isolated vacancies occur?

10.6 Assume that a divacancy in a c.c.p. metal consists of two vacant sites which are nearest neighbours of one another. Determine the number of possible divacancy positions in one unit cell of the c.c.p. lattice. Hence show that, in equilibrium, the number of divacancies per lattice site is given by:

$$\frac{n_2}{N} = 6 \exp\left(-\frac{(2E_f - E_b)}{kT}\right)$$

Ignore any entropy which a divacancy may add to a crystal by virtue of its effect on the vibrations of the atoms in its neighbourhood.

10.7 A crystal of gold is cooled from its melting point to a temperature of 500 °C. Calculate the ratio of the number of vacancies to the number of divacancies in the crystal at 500 °C, first on the assumption that complete equilibrium is maintained and, second, on the alternative assumption that the total number of vacant sites remains constant but that the vacancies and divacancies achieve an equilibrium amongst themselves. Take the formation energy of a vacancy to be 0.98 eV, the binding energy of a divacancy to be 0.3 eV and the ratio of the number of vacancies to the number of divacancies at the melting point, 1063 °C, to be 20.

10.8 Consider an atom in a close-packed hexagonal metal jumping into a vacant nearest-neighbour site. How many essentially different types of jump are there? In each case, sketch the ring of neighbouring atoms that must be squeezed through on the way to the vacant site and compare this with the rectangle of atoms that must be squeezed through when an atom jumps into a nearest-neighbour vacancy in a c.c.p. metal (Figure 10.8).

10.9 What is the most likely position for an interstitial atom in a crystal of silicon or germanium? Sketch and describe the shape of the ring of atoms which the interstitial atom must pass through in moving from a site of this type to a nearest-neighbouring site of the same type.

10.10 A rod of gold is heated to its melting point, at which the concentration of vacancies is 7.2×10^{-4} per atomic site. The rod is then quenched so that all the vacancies generated at the melting point are trapped in the metal. Upon annealing at a slightly higher constant temperature, the rod is observed to shrink as the vacancy concentration is reduced towards its very low equilibrium value. If the total contraction is 1.1×10^{-4}, what change in lattice parameter should be observed during the annealing process?

10.11 What is the point group symmetry of the following defects:
(a) A vacancy pair consisting of an anion vacancy and a nearest-neighbour cation vacancy in NaCl?
(b) A vacancy pair consisting of vacant nearest-neighbour sites in a b.c.c. metal?
In each case determine by inspection the direction of a tensile stress that will not produce any anelastic strain due to the presence of the defect.

10.12 What is the multiplicity of the general form for the point group of (a) a b.c.c. crystal and (b) an interstitial atom of carbon in iron? What is the relationship between the number of orientations in which a point defect exists and the multiplicities of the general form for the point groups of the crystal and of the defect?

10.13 When a crystal of thoria, ThO_2 (which has the fluorite structure), is doped with calcium, the Ca^{2+} ions occupy Th^{4+} sites, the lack of positive charge being compensated for by O^{2-} ion vacancies. The Ca^{2+} ion and O^{2-} ion vacancy are believed to be associated as a pair on nearest-neighbour sites. Determine:
   (a) The number of different orientations of the pair.
   (b) The orientations that can be distinguished by an electric field applied along [100].
   (c) The orientations that can be distinguished by a tensile stress applied along [110].

Suggestions for Further Reading

Agulló-López, F., Catlow, C.R.A. and Townsend, P.D. (1988). *Point Defects in Materials*. London: Academic Press.

Cotterill, R.M.J., Doyama, M., Jackson, J.J. and Meshii, M. (1965). *Lattice Defects in Quenched Metals*. London: Academic Press.

Flynn, B.P. (1972). *Point Defects and Diffusion*. Oxford: Clarendon Press.

Gittus, J. (1978). *Irradiation Effects in Crystalline Solids*. Barking, Essex: Applied Science.

Neumann, G. and Tuijn, C. (2009). *Self-Diffusion and Impurity Diffusion in Pure Metals. Handbook of Experimental Data, Pergamon Materials Science Series,* Volume 14, Amsterdam: Elsevier.

Was, G.S. (2017). *Fundamentals of Radiation Materials Science: Metals and Alloys*. New York: Springer-Verlag.

References

[1] Huntington, H.B. and Seitz, F. (1942). Mechanism for self-diffusion in metallic copper. *Phys. Rev.* 61: 315–325.

[2] Huntington, H.B. (1942). Self-consistent treatment of the vacancy mechanism for metallic diffusion. *Phys. Rev.* 61: 325–338.

[3] Fumi, F.G. (1955). Vacancies in monovalent metals. *Phil. Mag.* 46: 1007–1020.

[4] Johnson, R.A. and Brown, E. (1962). Point defects in copper. *Phys. Rev.* 127: 446–454.

[5] March, N.H. (1973). Displaced charge and formation energies of point defects in metals. *J. Phys. F: Met. Phys.* 3: 233–247.

[6] Johnson, R.A. (1973). Empirical potentials and their use in the calculation of energies of point defects in metals. *J. Phys. F: Met. Phys.* 3: 295–321.

[7] Dederichs, P.H., Lehmann, C., Schober, H.R., Scholz, A. and Zeller, R. (1978). Lattice theory of point defects. *J. Nucl. Mater.* 69–70: 176–199.

[8] Cottrell, A.H. (2007). A commentary on Fumi's paper on vacancies in monovalent metals. *Phil. Mag. Lett.* 87: 1–4.

[9] McKee, B.T.A., Triftshäuser, W. and Stewart, A.T. (1972). Vacancy-formation energies in metals from positron annihilation. *Phys. Rev. Lett.* 28: 358–360.

[10] McGervey, J.D. and Triftshäuser, W. (1973). Vacancy-formation energies in copper and silver from positron annihilation. *Phys. Lett. A* 44: 53–54.

[11] Seeger, A. (1973). Investigation of point defects in equilibrium concentrations with particular reference to positron annihilation techniques. *J. Phys. F: Met. Phys.* 3: 248–294.

[12] Johnson, R.A. (1969). Sensitivity of lattice defect calculations for fcc metals to empirically derived interatomic potentials. *Radiation Effects* 2: 1–9.

[13] Schilling, W. (1978). Self-interstitial atoms in metals. *J. Nucl. Mater.* 69–70: 465–489.

[14] Hao, Y.L., Song, Y., Yang, R., Cui, Y.Y., Li, D. and Niinomi, M. (2003). Concentration of point defects in binary NiAl. *Phil. Mag. Lett.* 83: 375–384.

[15] Morse, P.M. (1929). Diatomic molecules according to the wave mechanics. II. Vibrational levels. *Phys. Rev.* 34: 57–64.

[16] Born, M. and Mayer, J.E. (1932). Zur Gittertheorie der Ionenkristalle. *Z. Phys.* 75: 1–18.

[17] De Vita, A. and Gillan, M.J. (1991). The *ab initio* calculation of defect energetics in aluminium. *J. Phys. Cond. Matt.* 3: 6255–6237.

[18] Tajima, N., Takai, O., Kogure, Y. and Doyama, M. (1999). Computer simulation of point defects in fcc metals using EAM potentials. *Comput. Mater. Sci.* 14: 152–158.

[19] Bredow, T., Dronskowski, R., Ebert, H. and Jug, K. (2009). Theory and computer simulation of perfect and defective solids. *Prog. Solid State Chem.* 37: 70–80.

[20] Girifalco, L.A. and Weizer, V.G. (1960). Vacancy relaxation in cubic crystals. *J. Phys. Chem. Solids* 12: 260–264.

[21] Mott, N.F. and Littleton, M.J. (1938). Conduction in polar crystals. I. Electrolytic conduction in solid salts. *Trans. Farad. Soc.* 34: 485–499.

[22] Johnson, R.A. (1964). Interstitials and vacancies in α iron. *Phys. Rev.* 134: A1329–A1336.

[23] Paneth, H.R. (1950). The mechanism of self-diffusion in alkali metals. *Phys. Rev.* 80: 708–711.

[24] Derlet, P.M., Nguyen-Manh, D. and Dudarev, S.L. (2007). Multiscale modeling of crowdion and vacancy defects in body-centered-cubic transition metals. *Phys. Rev. B* 76: 054107.

[25] Känzig, W. and Woodruff, T.O. (1958). The electronic structure of an H-center. *J. Phys. Chem. Solids* 9: 70–92.

[26] Weber, M.D. and Friauf, R.J. (1969). Interstitialcy motion in the silver halides. *J. Phys. Chem. Solids* 30: 407–419.

[27] Xu, C.H., Fu, C.L. and Pedraza, D.F. (1993). Simulations of point-defect properties in graphite by a tight-binding-force model. *Phys. Rev. B* 48: 13273–13279.

[28] Simmons, R.O. and Balluffi, R.W. (1960). Measurements of equilibrium vacancy concentrations in aluminum. *Phys. Rev.* 117: 52–61.

[29] Simmons, R.O. and Balluffi, R.W. (1960). Measurement of the equilibrium concentration of lattice vacancies in silver near the melting point. *Phys. Rev.* 119: 600–605.

[30] Simmons, R.O. and Balluffi, R.W. (1962). Measurement of equilibrium concentrations of lattice vacancies in gold. *Phys. Rev.* 125: 862–872.

[31] Simmons, R.O. and Balluffi, R.W. (1963). Measurement of equilibrium concentrations of vacancies in copper. *Phys. Rev.* 129: 1533–1544.

[32] Feder, R. and Charbnau, H.P. (1966). Equilibrium defect concentration in crystalline sodium. *Phys. Rev.* 149: 464–471.

[33] Feder, R. and Nowick, A.S. (1967). Equilibrium vacancy concentration in pure Pb and dilute Pb–Tl and Pb–In alloys. *Phil. Mag.* 15: 805–812.

[34] Dreyfus, R.W. and Nowick, A.S. (1962). Energy and entropy of formation and motion of vacancies in NaCl and KCl crystals. *J. Appl. Phys.* 33 (Suppl. to issue no. 1): 473–477.

[35] Ramdas, S., Shukla, A.K. and Rao, C.N.R. (1973). Mechanism of ion movement in alkali halides. *Phys. Rev. B* 8: 2975–2981.

[36] Murthy, C.S.N. and Murti, Y.V.G.S. (1971). Schottky defect energies in CsCl structure type crystals. *J. Phys. C* 4: 1108–1116.

[37] Tubandt, C. (1932). Leitfähigkeit und Überführungszahlen in festen Elektrolyten. In: *Handbuch der Experimentalphysik, Vol. 12, Part 1: Elektrochemie* (eds. W. Wien, F. Harms and K. Fajans), pp. 383–469. Leipzig: Akademische Verlagsgesellschaft m. b. h.

[38] Teltow, J. (1949). Zur Ionenleitung und Fehlordnung von Silberbromid mit Zusätzen zweiwertiger Kationen. I Leitfähigkeitsmessungen und Zustandsdiagramme. *Ann. Phys.* 5: 63–70.

[39] Neumann, G. and Tuijn, C. (2009). *Self-Diffusion and Impurity Diffusion in Pure Metals. Handbook of Experimental Data, Pergamon Materials Science Series*, Volume 14, Amsterdam: Elsevier.

[40] Doyama, M. and Koehler, J.S. (1962). Quenching and annealing of lattice vacancies in pure silver. *Phys. Rev.* 127: 21–31.

[41] Tomizuka, C.T. and Sonder, E. (1956). Self-diffusion in silver. *Phys. Rev.* 103: 1182–1184.

[42] Panseri, C. and Federighi, T. (1958). Isochronal annealing of vacancies in aluminium. *Phil. Mag.* 3: 1223–1240.

[43] DeSorbo, W. and Turnbull, D. (1959). Kinetics of vacancy motion in high-purity aluminum. *Phys. Rev.* 115: 560–563.

[44] Spokas, J.J. and Slichter, C.P. (1959). Nuclear relaxation in aluminum. *Phys. Rev.* 113: 1462–1472.

[45] Doyama, M. and Koehler, J.S. (1964). Quenching and annealing of zone-refined aluminum. *Phys. Rev.* 134: A522–A529.

[46] Bauerle, J.E. and Koehler, J.S. (1957). Quenched-in lattice defects in gold. *Phys. Rev.* 107: 1493–1498.

[47] Makin, S.M., Rowe, A.H. and LeClaire, A.D. (1957). Self-diffusion in gold. *Proc. Phys. Soc. London* 70: 545–552.

[48] Kino, T. and Koehler, J.S. (1967). Vacancies and divacancies in quenched gold. *Phys. Rev.* 162: 632–648.

[49] Wright, P. and Evans, J.H. (1966). Formation and migration energies of vacancies in copper. *Phil. Mag.* 13: 521–531.

[50] Ramsteiner, F., Lampert, G., Seeger, A. and Schüle, W. (1965). Die Erholung des elektrischen Widerstandes in abgeschrecktem Kupfer. *Physica Status Solidi* 8: 863–879.

[51] Kuper, A., Letaw, H., Slifkin, L., Sonder, E. and Tomizuka, C.T. (1955). Self-diffusion in copper. *Phys. Rev.* 98: 1870.

[52] Seeger, A., Gerold, V., Chik, K.P. and Rühle, M. (1963). Migration energy and binding energy of di-vacancies in copper. *Phys. Lett.* 5: 107–109.

[53] Nakamura, Y. (1961). Effect of quenched-in vacancies on the weak magnetization of nickel. *J. Phys. Soc. Jpn.* 16: 2167–2171.

[54] Schumacher, D., Schüle, W. and Seeger, A. (1962). Untersuchung atomarer Fehlstellen in verformtem und abgeschrecktem Nickel. *Z. Naturforsch. A* 17: 228–235.

[55] Hoffman, R.E., Pikus, F.W. and Ward, R.A. (1956). Self-diffusion in solid nickel. *Trans. Am. Inst. Min. Metall. Petrol. Engrs.* 206: 483–486.

[56] Jackson, J.J. (1965). Point defects in quenched platinum. In: *Lattice Defects in Quenched Metals* (eds. R.M.J. Cotterill, M. Doyama, J.J. Jackson and M. Meshii), pp. 467–479. New York: Academic Press.

[57] Cattaneo, F., Germagnoli, E. and Grasso, F. (1962). Self-diffusion in platinum. *Phil. Mag.* 7: 1373–1383.

[58] Hesketh, R.V. (1971). A comment on diffusion in sodium. *Phil. Mag.* 24: 1233–1237.

[59] Nachtrieb, N.H., Catalano, E. and Weil, J.A. (1952). Self-diffusion in solid sodium. I. *J. Chem. Phys.* 20: 1185–1188.

[60] Suezawa, M. and Kimura, H. (1973). Quenched-in vacancies in molybdenum. *Phil. Mag.* 28: 901–914.

[61] Balluffi, R.W. (1978). Vacancy defect mobilities and binding energies obtained from annealing studies. *J. Nucl. Mater.* 69–70: 240–263.

[62] Mundy, J.N. (1982). Electrical resistivity–temperature scale of tungsten. *Phil. Mag. A* 46: 345–349.

[63] Rasch, K.-D., Schultz, H. and Siegel, R.W. (1978). Electrical resistivity–temperature scale and vacancy parameters of tungsten. *Phil. Mag. A* 37: 567–569.

[64] Mundy, J.N., Ockers, S.T. and Smedskjaer, L.C. (1987). Vacancy migration enthalpy in tungsten at high temperatures. *Phil. Mag. A* 56: 851–860.

[65] Janot, C., Malléjac, D. and George, B. (1970). Vacancy-formation energy and entropy in magnesium single crystals. *Phys. Rev. B* 2: 3088–3098.

[66] Beevers, C.J. (1963). Electrical resistivity observations on quenched and cold-worked magnesium. *Acta Metall.* 11: 1029–1034.

[67] Shewmon, P.G. and Rhines, F.N. (1954). Rate of self-diffusion in polycrystalline magnesium. *Trans. Am. Inst. Min. Metall. Petrol. Engrs.* 200: 1021–1025.

[68] DeSorbo, W. (1960). Quenched-in defects in tin and the superconducting transition temperature. *J. Phys. Chem. Solids* 15: 7–12.

[69] Coston, C. and Nachtrieb, N.H. (1964). Self-diffusion in tin at high pressure. *J. Phys. Chem.* 68: 2219–2229.

[70] Seeger, A. (1964). The interpretation of quenching experiments and the binding energy of divacancies in silver. *Phys. Lett.* 9: 311–313.

[71] Kinchin, G.H. and Pease, R.S. (1955). The displacement of atoms in solids by radiation. *Rep. Prog. Phys.* 18: 1–51.

[72] Schilling, W. (1978). The physics of radiation damage in metals. *Hyperfine Interact.* 4: 636–644.

[73] Makin, M.J. (1968). Electron displacement damage in copper and aluminium in a high voltage electron microscope. *Phil. Mag.* 18: 637–653.

[74] Wolfenden, A. (1972). Electron radiation damage near the threshold energy in aluminum. *Radiation Effects* 14: 225–229.

[75] Corbett, J.W., Smith, R.B. and Walker, R.M. (1959). Recovery of electron-irradiated copper. I. Close pair recovery. *Phys. Rev.* 114: 1452–1459.

[76] Jenkins, M.L. and Kirk, M.A. (2000). *Characterization of Radiation Damage by Transmission Electron Microscopy*. Bristol: IoP Publishing.

[77] Zinkle, S.J., Horsewell, A., Singh, B.N. and Sommer, W.F. (1994). Defect microstructure in copper alloys irradiated with 750 MeV protons. *J. Nucl. Mater.* 212–215: 132–138.

[78] Aroyo, M.I. (ed.) (2016). *International Tables for Crystallography*, 6th, Revised Edition, Vol. A: *Space-Group Symmetry*, published for the International Union of Crystallography. Chichester: Wiley International.

[79] Nowick, A.S. and Heller, W.R. (1965). Dielectric and anelastic relaxation of crystals containing point defects. *Adv. Phys.* 14: 101–166.

11

Twinning

11.1 Introduction

When a crystal is composed of parts that are oriented with respect to one another according to some symmetry rule, the crystal is said to be twinned. The most frequently occurring symmetry rule of twinning, but not the only one, is that the crystal structure of one of the parts is the mirror image of the crystal structure of the other part, in a certain crystallographic plane called the *twinning plane*. Often, the plane of contact between the two parts, called the *composition plane*, coincides with the twinning plane.

Twinned crystals are often produced during growth, from the vapour, liquid, or solid. Alternatively, a single crystal may be made to become twinned by mechanically deforming it. Twinning can also arise as a result of a phase transformation from one crystal structure to another within a material: martensite phases produced as a result of martensitic transformations often have twinning as their lattice invariant shear. Martensitic transformations are considered in more detail in Chapter 12. It is also usual for ferroelectric phases to exhibit twinning as a result of phase transitions on cooling from a higher-temperature phase, which may, or may not, be ferroelectric. Twinned crystals of minerals are often found in nature; in such cases, it is not always clear whether the twins were produced by growth or by deformation. In this chapter we shall be dealing mainly with deformation twins, but it is important to realize that this represents only a fraction of the general phenomenon of twinning. It is relevant that the formal crystallographic description of deformation twinning that will be developed in this chapter can be applied to all twinning operations irrespective of how they are produced physically.

A simple example of twinning which can be produced in a variety of ways is provided by c.c.p. metals. In this case the symmetry rule connecting the differently oriented parts is that one part is the *mirror image* of the other in a {111} plane. Because the composition

Crystallography and Crystal Defects, Third Edition. Anthony Kelly and Kevin M. Knowles.
© 2020 John Wiley & Sons Ltd. Published 2020 by John Wiley & Sons Ltd.
Companion website: www.wiley.com/go/kelly/crystallography3e

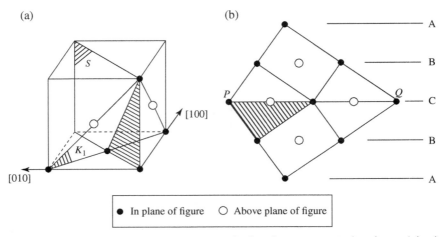

Figure 11.1 Structure of a twin in a c.c.p. metal. The plane *S* in (a) is the plane of the figure in (b), with the corresponding triangles in (a) and (b) shaded. The plane K_1 is the composition plane, PQ in (b)

plane is usually itself a {111} plane, the structures can be shown as being mirror images of one another in this plane, as in Figure 11.1, in which the composition plane is (111). It can be seen that nearest-neighbour relations are preserved at the boundary, but that an error in the stacking of the (111) planes occurs, so that the stacking sequence ABCABC is turned into ABCBAC; for example, ABCABCBACBAC ... As a result, there is a thin layer of h.c.p. material at the twin boundary – the sequence BCB. This stacking fault will have associated with it a characteristic energy, just as in stacking faults generated between partial dislocations in c.c.p. metals (Chapter 9).

It is natural to suppose that such an error might occur during the growth of the crystal. The experimental evidence is that the formation of growth twins is quite sensitive to the conditions of growth. Twins in c.c.p. crystals grown from the melt are not common, but gold films produced by vapour deposition contain a multitude of twins, as do layers of electrodeposited copper, and in both cases, the faster the rate of deposition, the higher the density of twins. Twinned grains are often seen in copper, α-brass and stainless steel polycrystals that have been cold-worked and then heated so as to produce recrystallization.[1] In a recrystallized material, a straight-sided lamella which is in a twin orientation to the rest of a grain is called an *annealing twin*; it can be regarded as a twin produced during crystal growth from the solid state. It is interesting that growth twins in aluminium are rare. This is because the energy of a stacking fault, and therefore of a twin boundary, in aluminium is relatively high (Table 9.2).

Examination of Figure 11.2 shows that the twinned crystal could have been produced by homogeneously shearing part of a single crystal. This immediately suggests that twin formation may be a mechanism of plastic deformation. In fact, this mechanism is not very easy to observe in c.c.p. metals, because of the relative ease with which they glide. However, Cu, Ag and Au crystals all twin prolifically during the later stages of tensile tests at low

---

[1] When a plastically deformed metal is heated to a temperature at which self-diffusion can occur, new strain-free grains nucleate among the deformed grains and grow so as to consume them. This process is called recrystallization.

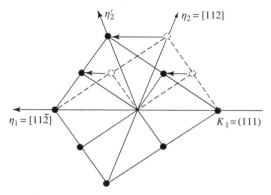

Figure 11.2 Formation of a twin in a c.c.p. metal by shear. The dotted lines indicate the structure above the composition plane, K_1, before the shear occurs

temperatures, and Cu has been found to twin when shock-loaded at ordinary temperatures. The minimum atom movements needed to accomplish twinning by shear in a c.c.p. crystal are shown in Figure 11.2. A displacement of $\frac{1}{6}[11\bar{2}]$ applied to the upper part of an originally single crystal produces a stacking fault. The same displacement applied at successively higher (111) layers produces a twin. This could be achieved by passing a partial dislocation with the Burgers vector $\frac{1}{6}[11\bar{2}]$ over each (111) plane above the composition plane. The geometry of deformation twinning will be described in more general terms in the next section.

11.2 Description of Deformation Twinning

Deformation by glide preserves the crystal structure in the same orientation (Chapter 7). Deformation by twinning reproduces the crystal structure, but in a specific new orientation. Macroscopically, it is evident that the deformed regions that have taken up a twin orientation with respect to the original crystal are lamellae that have undergone a homogeneous simple shear. Before proceeding further, it will be useful to enlarge upon the different levels at which deformation twinning can be described.

First, a description can be given of the change of shape of the region which becomes a twin lamella. This purely macroscopic description gives the elements of a simple shear. The interface between the twin and the matrix is the plane that is neither distorted nor rotated by the shear. Since we are dealing with simple shear, this plane will be both the composition plane, K_1, and the twin plane. The direction and magnitude of the shear can, ideally, be obtained from the displacements due to the lamella at two surfaces of the crystal, as shown in Figure 11.3.

Next, the relationship between the orientations of the lattice in the lamella and the lattice in the matrix can be described. Then, it can be decided whether the homogeneous simple shear found from the shape change adequately describes the lattice reorientation when the shear is applied to the points of the Bravais lattice. Often, in specific cases, it is found to do so. The effectiveness of the simple shear in describing the lattice reorientation has led

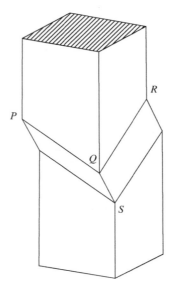

Figure 11.3 Displacements produced by a twin lamella. The traces *PQ* and *QR* define the unrotated plane of the shear (K_1). The magnitude and direction of the shear can be determined from *SQ*

to a general description of deformation twins in terms of a simple shear of the Bravais lattice, and this is the treatment we are about to pursue, partly in order to introduce the nomenclature that has grown up around it.

The ultimate level of description is to describe the atom movements that occur during twinning. In contrast to the description of shape change and orientation change, description at this level is not amenable to geometrical considerations alone and requires atomistic modelling to provide insight into the atom movements. Further consideration of this level of description will be postponed until actual examples are discussed, but it may be emphasized in advance that a simple shear applied to atom positions, as distinct from lattice points, is *not* always capable of producing all the atom movements that are needed to form a twin.

Confining ourselves, for the present, to the problem of shearing the Bravais lattice, we will ask the following question: can a homogeneous simple shear be applied to a Bravais lattice so as to change its orientation, but not its structure?

The geometry of a simple shear is illustrated in Figure 11.4. All points of the lattice on the upper side of the plane K_1 are displaced in the shear direction η_1 by an amount u_1 proportional to their distance above K_1. Thus:

$$u_1 = sx_2 \tag{11.1}$$

where s is the magnitude of the simple shear. The plane containing η_1 and the normal to K_1 is called the plane of shear, S. It can be seen that a vector parallel to η_2 in S will be the same length after the shear has been applied if the angle α which it makes with the normal to K_1 is given by:

$$s = 2\tan\alpha \tag{11.2}$$

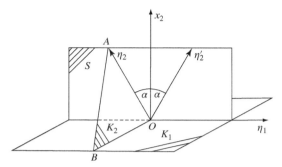

Figure 11.4 The elements of deformation twinning. O is an origin, K_1 is the composition plane (twin plane), η_1 is the shear direction in K_1 and K_2 is a plane which is rotated but undistorted as a result of the twinning operation. S is the plane of shear perpendicular to K_1 which intersects K_1 in η_1. η_2 is a rotated, but undistorted direction common to S and K_2. 2α is the angle between η_2 and the direction η_2' to which η_2 is rotated by the simple shear. x_2 is parallel to the normal to K_1 and bisects the angle 2α. OB defines the direction in K_1 perpendicular to η_1

The angle α is simply the angle between η_2 and the normal to K_1. With the exception of vectors parallel to OB, which are not rotated by the simple shear, all vectors in the plane containing η_2 and OB are unchanged in length, although rotated. This plane, AOB in Figure 11.4, is conventionally labelled K_2 and is called the second undistorted plane. K_1 is neither rotated nor distorted; it is both the composition plane and the twin plane.

So far it has not been specified whether the elements K_1, K_2, η_1 and η_2 are rational or not; that is, whether or not they pass through sets of points of the Bravais lattice. In this respect, the condition that the *lattice is to be reproduced* can be fulfilled in two ways.

In the first, K_1 is rational. Therefore, we can pick two lattice vectors $\mathbf{l}_1$ and $\mathbf{l}_2$ in K_1 (Figure 11.5a). These vectors are not affected by the shear. If then η_2 is rational, there is a third lattice vector $\mathbf{l}_3$ parallel to η_2 that is unchanged in length by being sheared to $\mathbf{l}_3'$. Remembering that $-\mathbf{l}_3$ is also a lattice vector (lattices are centrosymmetric), it can be seen from Figure 11.5a that the new cell defined by the basis vectors $\mathbf{l}_1$, $\mathbf{l}_2$ and $\mathbf{l}_3'$ is a reflection in the twin plane K_1 of the cell defined by the basis vectors $\mathbf{l}_1$, $\mathbf{l}_2$ and $-\mathbf{l}_3$ in the unsheared lattice. If this cell is a possible primitive unit cell of the lattice, the whole lattice is reconstructed by the shear. If it is not a primitive unit cell then only a superlattice made up of a fraction of the lattice points is necessarily reconstructed. Twins whose shear elements K_1 and η_2 are rational, while K_2 and η_1 are irrational, are called type I twins or reflection twins. In the nomenclature used in mineralogy, such twins are termed normal twins [1, 2].

Further consideration of the geometry of type I twinning in Figure 11.5a shows that this twinning process is equivalent to a rotation of $180°$ about the normal to K_1. Under these circumstances, the lattice vectors $\mathbf{l}_1$, $\mathbf{l}_2$ and $\mathbf{l}_3$ parallel to the sides of the cell outlined in solid lines in Figure 11.5a are rotated into vectors $-\mathbf{l}_1$, $-\mathbf{l}_2$ and $\mathbf{l}_3'$ parallel to the sides of the cell outlined in dashed lines in Figure 11.5a.

A second possibility is shown in Figure 11.5b. In this case, η_1 and K_2 are rational, but K_1 and η_2 are not. Two lattice vectors $\mathbf{l}_1$ and $\mathbf{l}_2$ can be chosen in K_2 together with a third lattice vector, $\mathbf{l}_3$, in the rational direction η_1. These vectors define the basis vectors of a cell prior to the shearing process. After shear has taken place, this cell is defined by the basis vectors

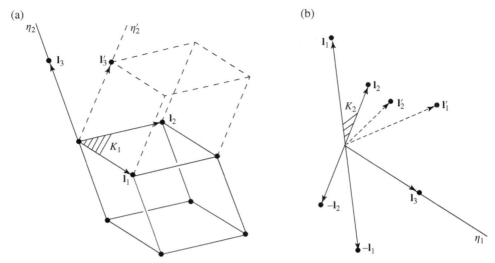

Figure 11.5 (a) Type I twin. (b) Type II twin. In (a) the lattice vector l_3 parallel to η_2 is sheared to l_3'. In (b) the lattice vector l_1 is sheared to l_1', and the lattice vector l_2 is sheared to l_2'. The vectors l_1' and l_2' in (b) are related to $-l_1$ and $-l_2$ respectively by a rotation of 180° about η_1

l_1', l_2' and l_3. Its orientation relationship to the cell defined by the basis vectors $-l_1$, $-l_2$ and l_3 in the unsheared lattice is a rotation of 180° about η_1. This is called a type II twin or a rotation twin. In the nomenclature used in mineralogy, such twins are termed parallel twins [1, 2].

As Crocker notes in his analysis of the possible twinning modes in orthorhombic uranium, α-U, degenerative type I and type II twinning modes occur in which one of the two irrational elements becomes rational [3]. Such situations arise if either the plane of shear S is rational or the normal to S is a rational direction. Thus, for example, for type I twins, η_1 must be a rational direction if S is rational, whereas K_2 is rational if the normal to S can be expressed as a rational direction.

More commonly, all four elements K_1, K_2, η_1 and η_2 are rational and the two types merge. Then the twin may be called compound (or degenerate), which is the usual type in the more symmetrical crystal structures. In a cubic lattice, *only* compound twins are possible. To prove this, consider a type I twin for which K_1 and η_2 are rational. In the cubic lattice, the normal to a rational plane is itself rational, so that if K_1 is rational, the plane of shear S is also rational, because it contains two rational directions: η_2 and the normal to K_1. Therefore, η_1, being the intersection of two rational planes K_1 and S, is also rational. Finally, K_2 is rational because it contains two rational directions, η_2 and the normal to S. A similar argument can be framed for cubic crystals starting with a type II twin for which K_2 and η_1 are rational.

α-U shows, among other twin modes, a type I twin mode with $K_1 = (112)$, $\eta_2 = [312]$ and a type II twin mode with $K_2 = (112)$, $\eta_1 = [312]$ [4]. In these two examples it happens that the elements of the two twins are related by interchange of K_1 and K_2, and of η_1 and η_2. Two twins related in this way are said to be *reciprocal* or *conjugate* to one another. The crystallography of two such twinning modes is distinct, but they have the same shear magnitude since the angle between K_1 and K_2 is the same for each mode.

A set of lattice points that forms a plane will still form a plane after the twinning shear has been applied to the lattice, but except in certain special cases, the pattern of lattice points in the plane will be changed. The plane therefore becomes a lattice plane of a different kind, so that its Miller indices change. It is useful to be able to write down the new Miller indices in the general case. The relevant transformation equations were first derived by Mügge [5, 6]. Special forms of these equations were derived at the same time for deformation twinning in calcite by Liebisch [7]. Alternative derivations and discussion of these equations have been given by, amongst others, Bell [8] and Pabst [9].

If the twin mode is a type I mode, the indices of K_1 are taken to be (HKL) and those of η_2 are taken to be $[UVW]$. If the twin mode is type II then the indices of K_2 are (HKL) and those of η_1 are $[UVW]$. Then, if (hkl) are the Miller indices of a plane before its transformation by twinning and $(h'k'l')$ its indices after the twinning operation:

$$h' = 2H\,(hU + kV + lW) - h(HU + KV + LW)$$

$$k' = 2K\,(hU + kV + lW) - k(HU + KV + LW) \qquad (11.3)$$

$$l' = 2L\,(hU + kV + lW) - l(HU + KV + LW)$$

The values of h', k' and l' given by Eq. (11.3) may contain a common factor, which must be removed to convert them to Miller indices. The indices $(h'k'l')$ are of course referred to the usual crystal axes *in the twin*[2]; these axes are in general *not* the directions into which the original crystal axes are transformed by the twinning shear.

Equations for the transformation of the indices of a direction $[uvw]$ into $[u'v'w']$ are obtained by simply interchanging $[UVW]$ and (HKL) and substituting $[uvw]$ for (hkl) and $[u'v'w']$ for $(h'k'l')$ in Eq. (11.3). Hence:

$$u' = 2U\,(Hu + Kv + Lw) - u\,(HU + KV + LW)$$

$$v' = 2V\,(Hu + Kv + Lw) - v\,(HU + KV + LW) \qquad (11.4)$$

$$w' = 2W\,(Hu + Kv + Lw) - w\,(HU + KV + LW)$$

These transformation equations may be of value in working out how a defect is changed when the lattice is twinned, assuming that the atoms defining the defect move only according to the twinning shear.

A convenient way of writing down transformation equations for specific cases is to use a transformation matrix (see Appendix 4). For example, in the case of a twin in a c.c.p. crystal with $K_1 = (111)$ and $\eta_2 = [112]$, Eq. (11.3) become:

$$h' = -2h + 2k + 4l$$

$$k' = 2h - 2k + 4l$$

$$l' = 2h + 2k$$

---

[2] In the case of a type I twin, these axes are the reflection in K_1 of the axes of the crystal before shear has occurred (often referred to simply as the 'crystal axes in the matrix'); in the case of a type II twin they are obtained from the crystal axes in the matrix by a rotation of 180° about η_1.

which, after removing the common factor of 2, can be written in the form:

$$\begin{pmatrix} h' \\ k' \\ l' \end{pmatrix} = \mathbf{Q} \begin{pmatrix} h \\ k \\ l \end{pmatrix} \tag{11.5}$$

where the transformation matrix $\mathbf{Q}$ is:

$$\mathbf{Q} = \begin{pmatrix} -1 & 1 & 2 \\ 1 & -1 & 2 \\ 1 & 1 & 0 \end{pmatrix}$$

The special planes and directions that do not change the form of their indices, when referred to the usual crystal axes in the twin, can easily be found. For example, in a type I twin, the plane K_1 obviously stays the same, and so does any plane in the zone of η_2, because it contains two lattice vectors that do not change, namely η_2 and the vector parallel to its own intersection with K_1. The three equations in Eq. (11.3) confirm this result.

11.3 Examples of Twin Structures

11.3.1 C.C.P. Metals

The structure of a twin in a c.c.p. metal has already been described (Figure 11.1). From studies of deformation twin lamellae in Cu and in Ag–Au alloys, it has been shown that a shear of magnitude 0.707 in the $[11\bar{2}]$ direction describes both the macroscopic shear and the lattice reorientation of a twin on a (111) plane [10, 11]. This agreement is not entirely trivial, because the same lattice reorientation could be produced by a shear of double the magnitude, that is, 1.414, in the reverse sense. For the shear of 0.707, the twinning elements are:

$$K_1 = (111), \quad \eta_1 = [11\bar{2}], \quad K_2 = (11\bar{1}), \quad \eta_2 = [112]$$

These elements are shown in Figure 11.2. Since the structure of a c.c.p. metal consists of an atom located at each of the lattice points, the atom movements are most compactly described by the same shear. This amounts to displacing each (111) layer in the twin by $\frac{1}{6}[11\bar{2}]$ over the layer underneath. In this example, the same simple shear describes the macroscopic shape change, the lattice reorientation and the most plausible atom movements needed to accomplish the twin.

11.3.2 B.C.C. Metals

Twinning is a relatively important mode of deformation in many b.c.c. transition metals such as Fe, V and Nb. Twinning is favoured relative to slip by a low temperature and a high strain rate. Purity appears to be important – for instance, 200 ppm of C has been found to suppress twinning in Nb at 77 K [12]. Some solid solution alloys, such as Mo with 35 at% Re, twin much more freely than the pure metal [13]. Only one type of twin is commonly found, the elements of which are:

$$K_1 = (112), \quad \eta_1 = [\bar{1}\bar{1}1], \quad K_2 = (11\bar{2}), \quad \eta_2 = [111]$$

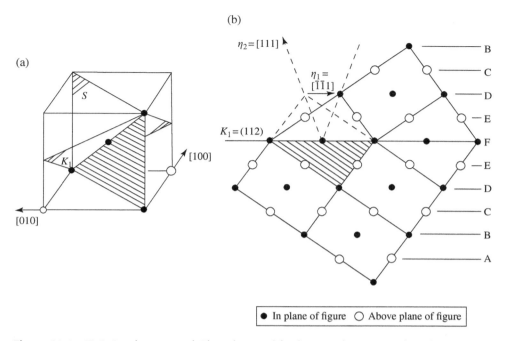

Figure 11.6 Twin in a b.c.c. metal. The scheme of the figure is the same as that of Figure 11.1

The magnitude of the twinning shear is 0.707, the same magnitude as in {111} twinning in c.c.p. metals. This twin is depicted in Figure 11.6. Just as for c.c.p. metals, it has been confirmed that the macroscopic shape change agrees with that predicted from the twinning elements [14]. The atom movements described by this shear are quite plausible. They correspond to a shift of $\frac{1}{6}[\bar{1}11]$ on successive (112) planes. In forming the b.c.c. structure from horizontal (112) layers, six layers are stacked before the seventh falls vertically on top of the first; the stacking sequence can be written as … ABCDEFAB …

The passage of a single partial dislocation with the $\frac{1}{6}[\bar{1}\bar{1}1]$ Burgers vector produces a stacking fault of the form ABCDEFEFAB … Little is known about the energy of such a fault, except that it is not so low as to permit extended dislocations to be readily observed in the transmission electron microscope. The passage of $\frac{1}{6}[\bar{1}\bar{1}1]$ partials on successive planes would produce the twin sequence ABCDEFEDCB … While such purely geometric considerations are very useful for establishing how twins might be produced, we have already noted in Section 9.6 that it is difficult for perfect $\frac{1}{2}<111>$ dislocations to dissociate into well-defined partial dislocations and stacking faults. Therefore, if the production of twins in b.c.c. metals involves $\frac{1}{6}<111>$ partials, a mechanism for their formation has to be invoked in any theory of twin formation in b.c.c. metals. Theories of how twins in b.c.c. metals form will be discussed further in Section 11.5.

11.3.3 Sphalerite (Zinc Blende)

We now turn to structures in which there is more than one atom per lattice point. Twinning in sphalerite, cubic ZnS, is an interesting example. The lattice is face-centred cubic

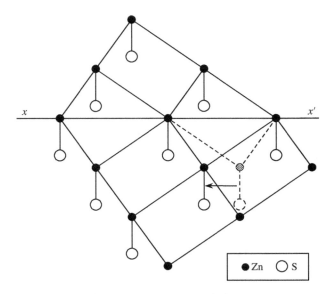

Figure 11.7 Twin in sphalerite

and there are two atoms associated with each lattice point, typified by S at $(0, 0, 0)$ and Zn at $\left(\frac{1}{4}, \frac{1}{4}, \frac{1}{4}\right)$ (Figure 3.18). The Zn and S atoms lie on interleaved face-centred cubic sublattices. Twins in this structure are closely related to twins in c.c.p. metals. A plausible structure at a twin–matrix interface is shown in Figure 11.7, which should be compared with Figure 11.1. The twin boundary in Figure 11.7 consists of Zn atoms rather than S atoms. The twinning shear is identical to that shown in Figure 11.1, namely $K_1 = (111)$, $\eta_1 = [\bar{2}11]$, $K_2 = (\bar{1}11)$, $\eta_2 = [211]$. The movement of the Zn atoms could be described by this shear, with the plane xx' as origin. However, the same shear would not place the S atoms correctly.

The simplest way of producing the necessary atom movements is to apply the twinning shear to *layers* of ZnS. Each layer consists of a sheet of Zn atoms lying on top of a sheet of S atoms, the sheets being parallel to (111). In Figure 11.7 the Zn atoms are shown connected to the S atoms directly beneath them in the same layer. It can be seen that the structure, as distinct from the lattice, is *not* mirrored in a K_1 plane.

A mirrored structure would change the $[111]$ direction to $[\bar{1}\,\bar{1}\,\bar{1}]$ in the twin – two distinct directions in this noncentrosymmetric structure (Chapter 3). It has been confirmed that this does not happen and that the structures of twin and matrix are related by a rotation of $180°$ about the normal to K_1, as shown [15].

The structure at the twin–matrix interface can be described as a fault in the stacking of the ZnS layers described above. The stacking sequence of these layers in sphalerite is the same as the stacking sequence of close-packed planes in a c.c.p. metal; that is, ... ABCA ... At the twin boundary the stacking sequence is ... ABCBA ..., which contains three layers in the sequence of the wurtzite structure, ... BCB ..., as a consequence of the rotation of $180°$ about the normal to K_1.

11.3.4   Calcite

Deformation twinning in calcite crystals has long been recognized and is quite easy to produce. For example, large pieces of a cleaved calcite rhombohedron can be forced into twin orientation by pressing a knife blade into one of their edges [1, 8, 16–19]. Referred to the rhombohedral cleavage pseudo unit cell (see Figure 3.26), the elements of the twinning shears are:

$$K_1 = (101), \quad K_2 = (010) \quad \eta_1 = [010], \quad \eta_2 = [101]$$

and the magnitude of the shear is 0.694. Referred to four-index hexagonal indices using the obverse setting for the rhombohedral unit cell shown in Figure A4.2, these twin elements can be seen to be:

$$K_1 = (1\bar{1}02), \quad K_2 = (\bar{1}101) \quad \eta_1 = [\bar{1}101], \quad \eta_2 = [1\bar{1}02]$$

The structure of the twin is shown in Figure 11.8. The twinning shear moves the Ca^{2+} ions and the centres of the CO_3^{2-} ions to their correct positions in the twin, but leaves the plane of the triangular CO_3^{2-} group at the wrong angle. To complete the reconstruction of the crystal in the twin, each CO_3^{2-} group must be rotated through 52.5° about an axis through its centre and normal to the plane of shear.

11.3.5   Hexagonal Metals

Twinning in hexagonal metals is of particular technical importance, because the limited nature of the common slip modes in these metals makes twinning a necessary component of any plastic deformation. Many types of twinning are exhibited; to some extent the type can be related to the *c/a* ratio of the metal (see Table 11.1). Broadly speaking, the lower the

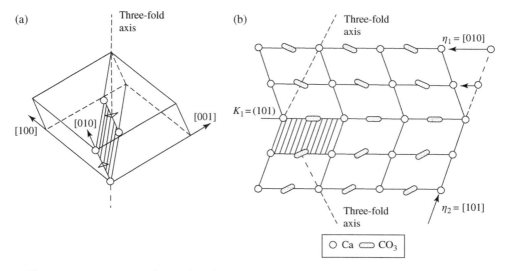

Figure 11.8   Twin in calcite. The scheme of the figure is the same as that of Figure 11.1

c/a ratio, the greater the variety of modes exhibited. All hexagonal metals twin by the mode:

$$K_1 = (10\bar{1}2), \quad \eta_1 = [\bar{1}011], \quad K_2 = (10\bar{1}\bar{2}), \quad \eta_2 = [10\bar{1}1]$$

The magnitude and sense of the shear, s, for this mode varies with the *c/a* ratio through the formula [20]:

$$s = \left(\frac{c^2}{a^2} - 3\right)\frac{a}{\sqrt{3}c}$$

It is always small, ranging from 0.171 for Cd (*c/a* = 1.886) to −0.199 for Be (*c/a* = 1.568).

$(10\bar{1}2)$ twins in Zr and in Zn are shown in Figures 11.9 and 11.10, respectively. The sense of the shear in Zn (*c/a* = 1.857) is opposite to the sense of the shear in Zr, for which *c/a* = 1.593. In both cases, it is evident that the twinning shear is not capable of describing all the atom movements. In these cases there is no obvious, simple way in which the atom movements can be described. Arrows have been drawn connecting the original position of an atom with that atom in the twin which is closest to it after it has been displaced by the twinning shear. There is no evidence whatever that the atoms actually follow these paths. In Figure 11.9, only the atoms at *P*, *Q* and *R* are carried directly to their new positions by the twinning shear. A movement that must be added to the twinning shear to describe the displacement of an atom to a position in the twinned crystal is usually termed a *shuffle* or *reshuffle*. When assumed atom movements are formally split up into a twinning shear followed by shuffles, the shuffles required will depend on the position of that K_1 plane

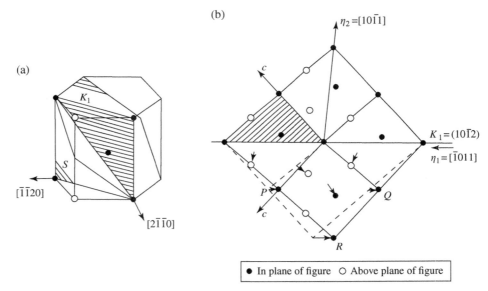

Figure 11.9 The $(10\bar{1}2)$ twin in zirconium. The scheme of the figure is the same as that of Figure 11.1

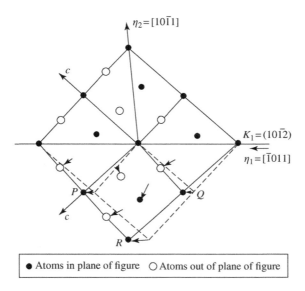

Figure 11.10   The (10$\bar{1}$2) twin in zinc

chosen as the origin of the shear. This aspect of twinning in hexagonal metals is still far from being fully understood [44].

11.3.6   Graphite

Twinning in graphite offers a beautiful illustration of the distinction that may exist between, on the one hand, the twinning shear, which describes the macroscopic shape change and lattice reorientation, and, on the other, a physically satisfying description of the atom movements. The elements of the twin are:

$$K_1 = (11\bar{2}1), \quad \eta_1 = [\bar{1}\bar{1}26], \quad K_2 = (0001), \quad \eta_2 = [11\bar{2}0]$$

and the magnitude of the shear is 0.367. The twinning shear reconstructs the Bravais lattice, but not the crystal structure, as shown in Figure 11.11 [30]. Atom shuffles would be required if a description of the atom movements were based on the twinning shear. Furthermore, any description of a mode of generation based on dislocations moving across (11$\bar{2}$1) planes would not be satisfactory, because it would require a disruption of the very strong bonds within the basal plane.

These difficulties are elegantly resolved by a description in terms of shear on the {0001} basal planes. A shear displacement of alternate basal planes by $\frac{1}{3}$[10$\bar{1}$0] and $\frac{1}{3}$[01$\bar{1}$0] fully describes the atom movements if accompanied by an appropriate rigid body rotation. Partial dislocations with these Burgers vectors are known to exist on the basal planes of graphite. A wall of such alternating partial dislocations at the matrix–twin interface constitutes a tilt boundary, which provides the required rotation (see Section 13.2). The lateral growth of a

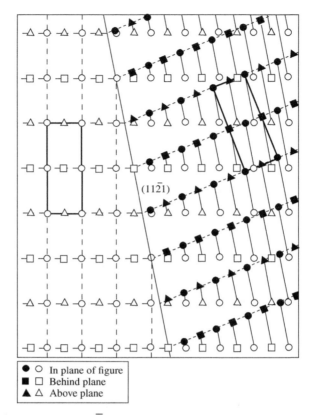

Figure 11.11 Plan view of the $(1\overline{1}00)$ plane in graphite, showing the structure produced by applying a twinning shear to the atoms. The outlined cells show that a new structure is produced and that shuffles must therefore be added to produce the actual structure within the twin. Source: Reproduced from [30] by permission of the Royal Society

twin lamella can occur by the glide of the wall of partials into the matrix. This convincingly accounts for experimental observations of twin movements in graphite; the model is shown in Figure 11.12 [30].

11.4 Twinning Elements

The twinning elements of various crystals are presented in Table 11.1. Only some of the twinning elements need to be experimentally determined, such as K_1, η_1 and the magnitude of the shear, s, or K_1 and η_2. The remaining elements then follow from the geometry of a simple shear (Figure 11.4). Most twins observed experimentally in systems other than cubic are either compound twins, or type I twins, in which K_1 and η_2 are rational. Type II twinning is more rare. The table includes examples of martensitic transformation twins (Chapter 12), as well as a number of examples of the more rare type II twins. Rational approximations to indices of irrational planes and directions are indicated in Table 11.1 by

Table 11.1 The twinning elements of various crystals

Material	Unit cell	K_1	η_1	K_2	η_2	s	Type of twin	Ref.
Fe, V, Nb, W, Mo, Cr	b.c.c.	112	$\bar{1}\bar{1}1$	$11\bar{2}$	111	0.707	C	[14]
Cu, Ag, Au	c.c.p.	111	$11\bar{2}$	$11\bar{1}$	112	0.707	C	[10, 11]
Cd	h.c.p	$10\bar{1}2$	$\bar{1}011$	$10\bar{1}\bar{2}$	$10\bar{1}\bar{1}$	0.171	C	[20]
Zn	h.c.p.	$10\bar{1}2$	$\bar{1}011$	$10\bar{1}\bar{2}$	$10\bar{1}\bar{1}$	0.140	C	[20]
Co	h.c.p.	$10\bar{1}2$	$\bar{1}011$	$10\bar{1}\bar{2}$	$10\bar{1}\bar{1}$	−0.128	C	[21]
		$11\bar{2}1$	$\bar{1}\bar{1}26$	0001	$11\bar{2}0$	0.614	C	[22]
Mg	h.c.p.	$10\bar{1}2$	$\bar{1}011$	$10\bar{1}\bar{2}$	$10\bar{1}\bar{1}$	−0.129	C	[20]
		$10\bar{1}1$	$30\bar{3}2$	$10\bar{1}3$	$30\bar{3}2$	0.136	C	[23]
		$\bar{1}013$	$\bar{1}012$	$10\bar{1}1$	$\bar{1}012$	0.136	C	[23]
Re	h.c.p.	$11\bar{2}1$	$11\bar{2}6$	0001	$11\bar{2}0$	0.619	C	[24]
Zr	h.c.p.	$10\bar{1}2$	$\bar{1}011$	$10\bar{1}\bar{2}$	$10\bar{1}\bar{1}$	−0.169	C	[25]
		$11\bar{2}1$	$\bar{1}\bar{1}26$	0001	$11\bar{2}0$	0.628	C	[26]
		$11\bar{2}2$	$\bar{1}\bar{1}23$	$11\bar{2}4$	$\bar{2}\bar{2}43$	0.225	C	[25]
Ti	h.c.p.	$10\bar{1}2$	$\bar{1}011$	$10\bar{1}\bar{2}$	$10\bar{1}\bar{1}$	−0.174	C	[27, 28]
		$11\bar{2}1$	$\bar{1}\bar{1}26$	0001	$11\bar{2}0$	0.630	C	[27, 28]
		$11\bar{2}2$	$\bar{1}\bar{1}23$	$11\bar{2}4$	$\bar{2}\bar{2}43$	0.219	C	[27, 28]
		$11\bar{2}4$	$\bar{2}\bar{2}43$	$11\bar{2}\bar{2}$	$\bar{1}\bar{1}23$	0.219	C	[28, 29]
Be	h.c.p.	$10\bar{1}2$	$\bar{1}011$	$10\bar{1}\bar{2}$	$10\bar{1}\bar{1}$	−0.199	C	[20]
Graphite	Hexagonal	$11\bar{2}1$	$\bar{1}\bar{1}26$	0001	$11\bar{2}0$	0.367	C	[30]
Calcite, $CaCO_3$	Hexagonal cleavage cell	$\bar{1}102$	$\bar{1}\bar{1}01$	0001	$\bar{1}102$	0.694	C	[7–9]
	Rhombohedral cleavage cell	101	010	010	101	0.694	C	
Sapphire, Al_2O_3	Rhombohedral; (Hexagonal cell)	$10\bar{1}2$	$\bar{1}011$	$\bar{1}014$	$20\bar{2}1$	0.202	C	[31]
	Morphological hexagonal cell	$10\bar{1}1$	$\bar{1}012$	$\bar{1}012$	$10\bar{1}1$	0.202	C	[31]
Bi	Rhombohedral face-centred cell	101	010	010	101	0.119	C	[20, 32]
	Primitive rhombohedral cell	112	$11\bar{1}$	110	001	0.119	C	

Material	Cell					Shear	Type of twin	Ref.
Hg	Rhombohedral face-centred cell	'$\bar{1}35$'	$\bar{1}21$	$\bar{1}11$	'$0\bar{1}1$'	0.633	II	[33]
	Rhombohedral primitive cell	'122'	$2\bar{1}0$	100	'$01\bar{1}$'	0.633	II	[33]
	Hexagonal unit cell	'$\bar{1}431$'	$3\bar{1}21$	$10\bar{1}1$	'$\bar{1}2\bar{1}0$'	0.633	II	[33]
β-Sn	Tetragonal I	301	$\bar{1}03$	$\bar{1}01$	101	0.119	C	[20]
In	Tetragonal F	101	$10\bar{1}$	$\bar{1}01$	101	0.150	C	[20]
α-U	Orthorhombic C	130	$3\bar{1}0$	$1\bar{1}0$	110	0.299	C	[4]
		'172'	312	112	'372'	0.228	II	[4]
		112	512	'172'	312	0.228	I	[4]
		'$17\bar{6}$'			'$12\bar{3}$'	0.214	II	[3, 34]
Cu–Al–Ni	Orthorhombic P	101	$\bar{1}01$	$\bar{1}01$	101	0.074	C	[35]
		121	'$\bar{5}43$'	'$\bar{2}3\bar{1}$'	111	0.261	I	[35]
		'$\bar{2}3\bar{1}$'	111	121	'$\bar{5}43$'	0.261	II	[36]
Ni–Ti	Monoclinic P	$11\bar{1}$	'112'	'124'	$21\bar{1}$	0.310	I	[37, 38]
		'$57\bar{7}$'	011	011	'$85\bar{5}$'	0.280	II	[38]
		001	100	100	001	0.238	C	[38–40]
		100	001	001	100	0.238	C	[39, 40]
Devitrite, $Na_2Ca_3Si_6O_{16}$	Triclinic P	'010'	100	$20\bar{1}$	'$1\,12\,2$'	0.327	II	[41]
Albite, $NaAlSi_3O_8$	Triclinic C^a	010	'$\bar{1}0\,10$'	'$\bar{3}05$'	010	0.148	I	[1, 2, 42]
		'$\bar{3}05$'	010	010	'$\bar{1}0\,10$'	0.148	II	[43]

Note: In the 'Type of twin' column in this table, 'C' is a compound twin, 'I' is a type I twin and 'II' is a type II twin. Rational approximants to irrational planes and directions for Type I and Type II twinning elements are denoted by quotation marks.

[a] Although albite is triclinic, it is conventionally described in terms of a non-primitive triclinic C unit cell with the space group $C\bar{1}$ to denote a centrosymmetric C-face-centred unit cell because of the similarity of this unit cell to other phases such as the potassium feldspar, sanidine, $KAlSi_3O_8$, which has a monoclinic C Bravais lattice. The description of the twin plane parameters are with respect to this C-face-centred unit cell.

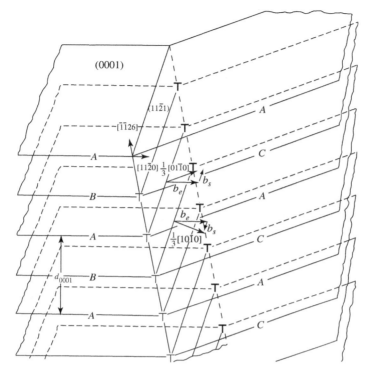

Figure 11.12 The formation of a twin in graphite by a partial dislocation on each basal plane. Source: Reproduced from [30] by permission of the Royal Society

quotation marks, such as for the type II deformation twinning mode in crystalline mercury reported by Guyoncourt and Crocker [33].

Some of the twinning modes tabulated are reciprocal to one another, so that the K_1 and η_2 of one mode become the K_2 and η_1 of the reciprocal mode, and vice versa. Note has already been made in Section 11.2 of the example of this reciprocity in α–uranium shown in Table 11.1. The $\{10\bar{1}1\}$ and $\{10\bar{1}3\}$ twinning modes in magnesium are another example of this reciprocity, as are the $\{11\bar{2}2\}$ and $\{11\bar{2}4\}$ twinning modes in titanium and the type I and type II twinning modes found in Cu–Al–Ni martensite. Further tabulations of twinning elements for different twinning modes in crystals can be found in the References at the end of this chapter. It should be noted that for crystals with rhombohedral unit cells, the description of the twinning elements is a function of the unit cell chosen; there is scope for confusion if the wrong unit cell is assumed. Albite, $NaAlSi_3O_8$, is one end member of the family of plagioclase feldspars which are the most common rock-forming mineral series [1, 2]. In plagioclases, the type I twin specified in Table 11.1 is referred to as an albite twin, while the type II twin (Problem 11.8) is referred to as a pericline twin.

From Table 11.1 it can be seen that crystals of the same structure tend to twin in the same mode. The reason for the choice of mode in a given case is less well understood than might be supposed. Consideration of Section 11.2 shows that, in principle, the number of possibilities from which to choose is enormous. To define a twinning shear that completely reconstructs the lattice, any three lattice vectors that form a primitive unit cell can be chosen; from these, two must be chosen to define K_1 (for type I twins) or K_2 (for type II twins).

A common-sense criterion of selection is to make the magnitude of the shear as small as possible, so as to reduce the size of the atom movements needed to form the twin. At first sight, this criterion might not appear to be appropriate, since the shears in Table 11.1 cover a wide range, and some are quite large. However, when the possibilities for a given structure are carefully examined, it is found that a minimum shear criterion is in fact quite effective [45]. The criterion favours low-index twinning planes for type I or compound twins.

The reason for this can be seen from Figure 11.13. Here, using the nomenclature in Figure 11.5, a vector $\mathbf{l}_3$ parallel to η_2 has sheared to $\mathbf{l}_3'$ as a result of the twinning operation. The higher the index of the plane K_1, the smaller the separation between K_1 planes. Since the vector $\mathbf{l}_3$, parallel to η_2, cannot be less than the smallest vector of the lattice, the angle that it makes with K_1 must decrease as the spacing of the K_1 planes decreases. This implies that the shear must increase. More precisely, if the vector $\mathbf{l}_3$ defines a lattice point in the nth K_1 plane above O and d is the separation of the K_1 planes:

$$\mathbf{l}_3^2 = (nd)^2 + \left(\tfrac{1}{2}snd\right)^2 \tag{11.6}$$

where s is the magnitude of the shear. If $\mathbf{E}$ is the smallest lattice vector, the inequality

$$\mathbf{l}_3^2 \geq \mathbf{E}^2 \tag{11.7}$$

has to be satisfied. Hence,

$$\frac{s^2 + 4}{\mathbf{E}^2} \geq \frac{4}{n^2 d^2}$$

As an illustration of the use of this inequality, consider twinning in the b.c.c. lattice. Any twin in a cubic lattice must be compound; therefore η_2 is rational and so Figure 11.13 applies. Here, $\mathbf{E}^2 = 3a^2/4$, where a is the lattice parameter. Searching for twins with $s \leq 1$ gives:

$$\frac{3a^2}{n^2 d^2} \leq 5 \tag{11.8}$$

In order to secure the reconstruction of the whole lattice, or a reasonable fraction of the lattice, n should be small. Furthermore, we can reject solutions where the planes under

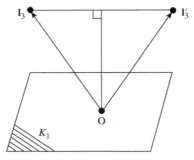

Figure 11.13 Shear and the geometry of deformation twinning: a vector $\mathbf{l}_3$ parallel to η_2 has sheared to $\mathbf{l}_3'$ as a result of the twinning operation. K_1 is the composition plane (twin plane)

consideration are mirror planes of the lattice (where, formally, $s = 0$). This immediately rules out {100} and {110} planes as possible twin planes for b.c.c. crystals. We can also clarify what we mean by d for planes in b.c.c. crystals: d is the separation for which there is the possibility of Bragg reflection. Thus, for example, for the spacings of planes parallel to (111), the required d is:

$$d_{222} = \frac{a}{\sqrt{12}} = \frac{a}{2\sqrt{3}} \tag{11.9}$$

whereas for planes parallel to (112), d is:

$$d_{112} = \frac{a}{\sqrt{6}} \tag{11.10}$$

For $n = 1$, there is no solution to Eq. (11.8) for b.c.c. crystals. For $n = 2$, there is only one solution of Eq. (11.8): the one for twinning on {112} planes. The (112) plane can therefore be selected as K_1. The direction of the shear on K_1 may be chosen by projecting the crystal structure onto K_1 and selecting the smallest translation needed to set up a mirror-image relationship in a K_1 plane. This leads to $\eta_1 = [\bar{1}\bar{1}1]$ and the twin depicted in Figure 11.6.

In structures where there is more than one atom per lattice point, such as the hexagonal metals, the twinning shear is not capable of describing all the atom movements that are needed to construct the twin. The choice of mode may still be rationalized in some cases by considering the effects of a shear applied homogeneously to atoms or *small groups* of atoms. A shear is sought that is small, and which demands only a small, plausible shuffling of the atoms to complete the twin structure, after it has been applied. This shear is then identified with the twinning shear. Some shuffles are inherently more likely than others; for example, the rotation of CO_3 radicals in calcite is obviously a more probable physical occurrence than a shuffle involving the disruption of a CO_3 radical.

Optical observations deduced from habit plane traces of very unusual twinning modes, such as {30$\bar{3}$4} bands in Mg [46, 47], may perhaps be explained by the operation of more than one simple twinning mode in the same region of crystal [44]. There is little direct evidence for this interpretation, although in many metals doubly twinned regions are observed where one twin lamella is intersected by another. Electron back-scattered diffraction (EBSD) measurements now make the analysis of such complex twinning behaviour seen within plastically deformed materials feasible at a sub-μm resolution [48].

The relative importance of twinning as a mode of deformation can sometimes be changed by adding an alloying element. Mo–Re alloys have already been mentioned (Section 11.3.2); another example is that of Be in Fe. Alloys containing 15–30 at% Be, quenched to preserve a solid solution, deform almost entirely by twinning, even at slow strain rates [49, 50]. The explanation for this phenomenon has similarities with the explanation for pseudoelasticity in materials which exhibit martensitic transformations (Chapter 12) [44]. It was previously believed that at the composition Fe_3Be an ordered $D0_3$ structure (Table 3.5) can be formed [49], as described in previous editions of this book. However, as Christian and Mahajan have noted [44], work by Morris and Cohen has been able to show that instead the microstructure at this composition consists of a two-phase mixture of an α b.c.c. Fe-rich solid solution and an ordered $B2$ (FeBe) crystal structure able to exist in various stages of decomposition depending on the precise details of the heat treatment [50].

11.5 The Morphology of Deformation Twinning

When a crystal has deformed by twinning, the twinned regions are usually in the form of plates parallel to K_1. Sometimes the plate is very thin and the twin is a lamella whose faces are accurately parallel to K_1. This is common in non-metals such as calcite or graphite, where lamellae only a few micrometres thick may run right through the crystal. Iron also twins in this fashion, producing what are called Neumann bands [51–54]. Measuring the orientations of the intersections of one lamella with two crystal surfaces immediately gives the twin plane K_1, which is accordingly the easiest of the twinning elements to determine.

The occurrence of perfect lamellae seems to be associated with the difficulty of deforming the matrix by slip. In metals that deform readily by slip, relatively short slabs of twin with a somewhat irregular interface are common. Twins in calcite assume a shorter, thicker and less regular form at higher temperatures when slip can occur. If a rigid twin is embedded in a perfectly rigid matrix, and if it is everywhere firmly bonded to that matrix, then the only possible interface is the undistorted, unrotated plane K_1. The twin would be a lamella, parallel to K_1, extending right through the matrix. (Possible twin intersections are neglected for the moment.)

For a twin of any other form, the matrix has to accommodate itself to the shape change of the twinned region. If the accommodation required is small enough, it may be obtained by elastic strain. Under this condition, a finite lamella must taper to an edge at its sides, and be lens-shaped. The elastic strain field can be represented by an appropriate array of dislocations, such as that shown in Figure 11.14 for a type I twin or a compound twin. The Burgers vector b and the vertical spacing of the dislocations d are together defined by the twinning shear s:

$$s = \frac{b}{d} \tag{11.11}$$

For example, an appropriate model for a c.c.p. twin consists of a twinning partial of the Burgers vector $\frac{1}{6}[11\bar{2}]$ on every (111) plane. The interface contains no dislocations when it is parallel to K_1, and it is then called *coherent*. The boundary becomes *incoherent* when it deviates from K_1 and its slope is defined by the spacing of dislocations in the K_1 plane. The thickness of the lamella at any point, h, is given by:

$$h = nd \tag{11.12}$$

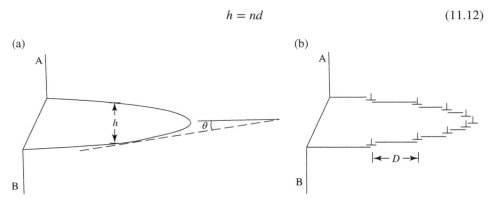

Figure 11.14 (a) Twin lamella intersecting a surface AB. (b) Dislocation model of the same lamella

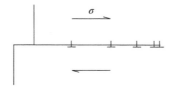

Figure 11.15   Dislocation model of a thin twin lamella

where n is the number of dislocations between the point in question and the tip of the lamella. The inclination θ of the interface to the twin plane is therefore given by:

$$\tan \theta = \frac{d}{D} \tag{11.13}$$

where D is the spacing of the dislocations in one interface (Figure 11.14b).

If the lamella is thin and tapered, a pile-up of dislocations on a single plane will represent the stress field adequately at distances large compared to the thickness of the lamella. This model is shown in Figure 11.15. The shear stress due to a pile-up of n screw dislocations at sufficiently large distances from the head of the pile-up is given by:

$$\sigma = \frac{\mu n b}{2\pi r} \tag{11.14}$$

From Eqs. (11.11) and (11.12), this can be written in the form:

$$\sigma = \frac{\mu h s}{2\pi r}$$

The product hs determines the magnitude of the accommodation stresses and strains in this case, and also in the more general case of mixed dislocations where both tensile and shear strains are produced. At the tip of a twin that has been blocked by some obstacle, the tensile strain may be large enough to nucleate a crack. Cracks of this type have been seen in b.c.c. metals, for which s is large ($s = 0.707$), as discussed by Christian and Mahajan in their comprehensive review of deformation twinning [44].

If there are no lattice friction forces opposing the motion of the dislocations in Figure 11.15, they will run back to the surface when the applied stress is removed. Behaviour of this sort has been observed in calcite, where small twins nucleated by indentation disappear when the load is removed. This phenomenon is called elastic twinning. More generally, twinning is not reversible and twins remain in a crystal after it has been unloaded. The reason is often that accommodation has occurred by slip, relieving the stresses at the edge of the twin or, in terms of the model of Figure 11.15, neutralizing the net Burgers vector of the pile-up. Under these conditions, blunt twin plates with quite irregular interfaces are possible.

An interesting feature of accommodation by slip in b.c.c. metals is that the slip is sometimes concentrated in a band stretching far ahead of the lamella. A model for this process, due to Sleeswyk [55], is shown in Figure 11.16. Every third twinning dislocation in the incoherent boundary dissociates into a perfect slip dislocation and a complementary partial. The perfect dislocations, called emissary dislocations, can glide ahead of the twin,

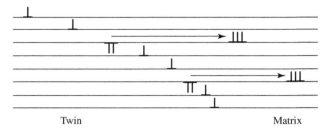

Twin   Matrix

Figure 11.16 Emissary dislocations. The dislocations shown by a single line are $\frac{1}{6}[\bar{1}\bar{1}1]$ twinning dislocations, the triple lines represent $\frac{1}{2}[\bar{1}\bar{1}1]$ perfect dislocations and the double lines represent $\frac{1}{3}[11\bar{1}]$ partials

carrying the twinning shear macroscopically, but not twinning the lattice. The long-range stress field of the twin boundary is thereby completely relieved, as can be seen from the fact that the Burgers vectors of its dislocations sum to zero.

Accommodation by slip is usually required at the intersection of two twin lamellae. In rare instances, the shears can be matched by twinning alone. In b.c.c. metals this is possible, because for a single shear direction of the type [111] there is a choice of three {211} K_1 planes. Zigzag lamellae are sometimes seen, composed alternately of the twins $K_1 = (2\bar{1}1)$ and $K_1 = (\bar{1}21)$ with a common $\eta_1 = [\bar{1}\bar{1}1]$, for example [56].

Detailed dislocation mechanisms for the nucleation and growth of a type I twin or compound twin lamella have been mainly based on the concept of a pole mechanism [57], illustrated schematically in Figure 11.17. A dislocation PQ has a Burgers vector such that it creates a twin by gliding over successive planes parallel to the twin plane K_1. This effect is produced by the intersecting dislocation PP', whose Burgers vector has a component normal to the twin planes that is equal to their spacing. The twin planes are therefore turned into a spiral ramp on which the twinning dislocation PQ glides. The dislocation PP' about which PQ spirals is called the *pole*. In a c.c.p. crystal, PP' may be an AC dislocation on the plane ACD, while PQ is a Shockley partial αC, gliding on the plane BCD (see Section 9.3). The Shockley partial may be produced by the dissociation of AC on plane BCD:

$$AC \rightarrow A\alpha + \alpha C$$

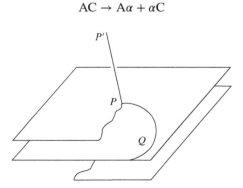

Figure 11.17 Pole mechanism for the growth of a twin

The Frank partial Aα cannot glide in the twin plane BCD and so only the Shockley partial αC spirals around the pole. In their review of deformation twinning, Christian and Mahajan [44] go further into the possible mechanisms invoked in the literature for nucleation and growth of type I and compound deformation twins.

In comparison with the knowledge base from experimental studies and atomic-scale simulations on the nucleation and growth of type I and compound deformation twins, the nucleation and growth of type II deformation twins has received far less attention in the literature because of the complexity of their crystallography. However, a recent topological model has been proposed by Pond and Hirth [58] to account for the formation and growth of type II twins and tested successfully against experimental observations of type II twinning in α–U [3, 4, 34]. Pond and Hirth note that in type I twins, nucleation and growth occurs by the generation and subsequent motion of disconnection defects (or twinning dislocations) along the low-energy K_1 twin interfaces. These disconnection defects have a combined dislocation and step nature evident in the schematic of Figure 11.14. If these disconnection defects are sufficiently mobile, a type I twin will form. If, however, the defect mobility is limited, the defects accumulate into a tilt wall and, following accommodational relaxation of the defect distortion fields, form a type II twin with the twin plane being K_2, i.e. the conjugate undistorted plane of a type I twinning operation.

Some care is needed in specifying the overall deformation of a crystal due to the formation of twins within it. The largest strain that can be obtained from one specific twin is of course the twinning shear itself, which is obtained when the entire crystal has become twinned. However, an unlimited amount of strain can often be obtained by multiple twinning. For example, if part of a c.c.p. twin lamella with $K_1 = (111)$ and $\eta_1 = [11\bar{2}]$ is retwinned on the same twin plane (111), with $\eta_1 = [\bar{1}2\bar{1}]$, the orientation of the doubly twinned region becomes the same as that of the original matrix, since the twinning shears add up to the slip displacement $\frac{1}{2}[011]$. In principle, a very large strain could be built up by the repeated application of twinning shears.

In a tensile test, the formation of a lamella of a given twin will contribute to the overall elongation of the specimen, provided that the tensile axis lies within the quadrant defined by η_1 and the normal to K_1 (Figure 11.4), or in other words, provided that the shear stress on the twin plane K_1, resolved in the direction of the twinning shear η_1, is positive. It should be noted that this condition is not the same as the condition for the tensile axis to be lengthened by the conversion of the entire crystal into its twin [59], because within the twinned region, any vector originally lying in the obtuse sector between K_2 and K_1 is lengthened, while only vectors lying in the acute sector of angle $(\pi/2 - \alpha)$ are shortened (Figure 11.4). When the tensile axis lies in the sector of angle between K_2 and the plane normal to η_1, the formation of a twin *lamella* is inhibited, even though the tensile force would do work on the crystal if it could be *completely* twinned.

11.6 Friedel's Classification of (Growth) Twinning

In X-ray diffraction work on nominally single crystals, occasions arise where the crystal under examination is actually twinned, most usually as a consequence of crystal growth rather than as a consequence of deformation twinning. In the X-ray diffraction community where it is routine to examine crystals grown in the laboratory, the nomenclature used for twinning is that introduced by Georges Friedel in his 1926 textbook *Leçons de Cristallographie* [60]. Friedel identified four ways in which twinning could arise:

- twinning by *merohedry*
- twinning by *reticular merohedry*
- twinning by *pseudomerohedry*
- twinning by *reticular pseudomerohedry*

Since this description of twinning prevails within the X-ray diffraction community, each of these four twinning operations will be described in the following sub-sections to show how they relate to the formal description of deformation twinning described in Sections 11.2–11.5.

11.6.1 Twinning by Merohedry

Merohedry, a work derived from Greek (meros, μέρος, meaning part, and ἕδρα meaning side) implies that each component crystal has fewer faces than it would have if the crystal structure had the highest point group symmetry possible for the Bravais lattice on which it is based [61]. The twin operation belongs to the point group of the lattice, but not the point group of the crystal.

Two examples demonstrate this principle. The first example, pyrite, FeS_2, has a point group $m\bar{3}$ and a primitive cubic lattice, which has $m\bar{3}m$ point symmetry. Therefore, {110} planes are mirror planes of the lattice, but not of the crystal structure. Such planes are candidate twin planes, and in fact both penetration and contact twins are found in pyrite with this twinning law. In the former, the interface between the two twinned regions can be highly irregular (e.g. [60], p. 429); in the latter, the interfaces are the {110} mirror planes of the lattice.

The crystal structure has four Fe atoms per unit cell located at $(0,0,0)$, $\left(\frac{1}{2},\frac{1}{2},0\right)$, $\left(0,\frac{1}{2},\frac{1}{2}\right)$ and $\left(\frac{1}{2},0,\frac{1}{2}\right)$ and eight S atoms per unit cell located at $\pm (x,x,x,)$, $\pm \left(-x+\frac{1}{2},-x,x+\frac{1}{2}\right)$, $\pm \left(-x,x+\frac{1}{2},-x+\frac{1}{2}\right)$ and $\pm \left(x+\frac{1}{2},-x+\frac{1}{2},-x\right)$, with $x = 0.384$ [62]. A diagram of 2×2 unit cells of the crystal structure projected on (001) is shown in Figure 11.18a and a hard sphere model of a contact twin is shown in Figure 11.18b after a similar diagram from Cahn [61]. A consideration of the (110) contact twin in Figure 11.18b shows that while the Fe–S distances in the bulk crystals are retained at the twin interface, the S–S distances of 2.177 Å in the bulk crystals either side of the twin interface reduce to 1.778 Å at the twin interface in this model of the (110) twin boundary. More recent work by Donnay et al. [63] has shown that on an atomic scale the twin boundary surface of the (110) contact twin is highly irregular. After an analysis of a number of possible hard sphere models of this twinning operation, they favoured a model first suggested by Strunz and Tennyson [64] in which there is a glide component parallel to (110) of $(\mathbf{a}_1 + \mathbf{a}_2)/2$, between the two twins. Such a glide component is completely compatible with this particular twinning operation and enables a twin interface to be produced in which each crystal either side of the interface contributes half of the sulfur doublets whose centres lie in the interface.

The second example, zinc oxide, ZnO, has a point group $6mm$ and a primitive hexagonal lattice which has $6/mmm$ point symmetry. ZnO is isostructural with wurtzite (Section 3.5.4).

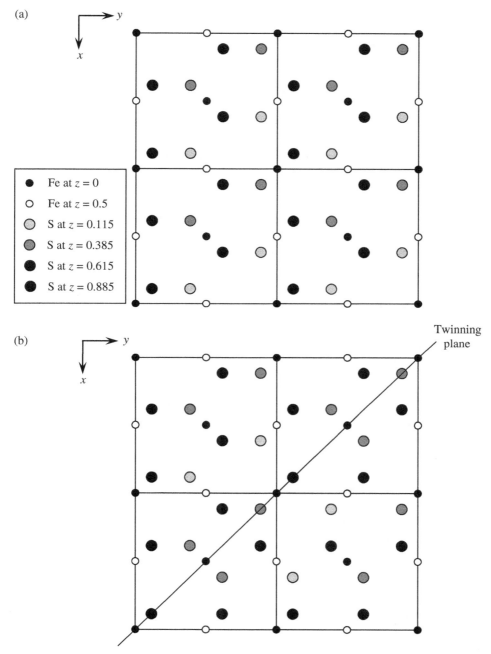

(a)

● Fe at $z = 0$
○ Fe at $z = 0.5$
◔ S at $z = 0.115$
◓ S at $z = 0.385$
◕ S at $z = 0.615$
◔ S at $z = 0.885$

(b)

Twinning plane

Figure 11.18 (a) Projection of 2×2 unit cells of pyrite projected onto (001) with a key for the heights of atoms in both diagrams and (b) a hard sphere model of a (110) contact twin in pyrite projected onto (001) after Cahn [61]

Hence, (0001) in ZnO can be a twin plane [65]. In the literature, this particular twinning operation is termed *inversion* twinning or *racemic* twinning.

A particular feature of merohedral twinning important to X-ray single crystal workers is that the direct lattices and the reciprocal lattices of the different domains coincide with one another exactly, so that the twinned crystal appears to be a single crystal until the end of any single crystal X-ray diffraction procedure aimed at determining the crystal structure [66].

It also follows that twinning by merohedry cannot be achieved by a shear process, i.e. twinning by merohedry cannot occur in deformation twinning. Therefore, unless special circumstances prevail, merohedral twins are twins which can *only* arise during crystal growth. In this context, the evolution of Dauphiné twinning during compression tests on polycrystalline aggregates of quartz is a good example where special circumstances prevail [67]: while there is no significant macroscopic shape change as a consequence of this evolution, Dauphiné twinning, produced by a 180° rotation about [0001], is activated in compression tests in quartz crystals to bring a direction of greater compliance parallel to the compression axis [68]. Any shape change arising as a consequence of this process in a single crystal or a polycrystalline aggregate is due to minute changes of elastic strain caused by this change in compliance [61].

11.6.2 Twinning by Reticular Merohedry

Here, a multiple lattice exists which is continuous in both lattices. In the terminology to be introduced and described in Chapter 13, twinning by reticular merohedry is twinning where a coincidence site lattice (CSL) common to the two components occurs. In such a description, the fraction of common lattice sites for two interpenetrating lattices related to one another by such a rotation is termed Σ. One restriction, however, in this interpretation in terms of CSLs is that the CSL under consideration has to be amenable to a description in terms of a 180° rotation about a particular axis [uvw].

A good example of twinning by reticular merohedry is twinning on {111} planes in c.c.p. metals, such as annealing twins in α-brass. A second example is deformation twinning on {112} planes in b.c.c. metals such as Fe, V and Nb. For c.c.p. metals, twinning on (111) produces a stacking sequence of the form … BCABCBACBAC … in the vicinity of the boundary (Section 11.1), from which it is evident that the atoms on the C planes will be common to the original c.c.p. material and its twin. For b.c.c. metals, twinning on (112) produces a stacking sequence of the form … ABCDEFEDCBA … in the vicinity of the boundary (Figure 11.6); here atom sites on the C and F planes will be common to the original b.c.c. material and its twin. Therefore, both of these examples have $\Sigma = 3$ CSLs.

This type of twinning mainly occurs in cubic crystals, but will also arise in other crystal systems under special circumstances, e.g. (120) reflection twinning in a tetragonal crystal for which $\Sigma = 5$ [69]. Obverse/reverse twinning in rhombohedral crystals is a further example of twinning by reticular merohedry. Here, the two possible settings of a rhombohedral crystal with respect to hexagonal axes, the reverse setting and the obverse setting, are related to one another by a rotation of 60°, or equivalently, 180°, about [0001] along which the two settings have a common threefold axis. These are also $\Sigma = 3$ CSLs. Examples in rhombohedral crystals of such twins are obverse/reverse growth twins in calcite, iron borate, and α-Al$_2$O$_3$ (corundum) [69].

11.6.3 Twinning by Pseudomerohedry

In twinning by pseudo-merohedry, the twinning is such that the symmetry possessed by the assembly of twinned crystals appears to have a symmetry higher than the symmetry of the individual crystals.

Thus, for example, a monoclinic crystal might have an angle β close to 90°. If this angle were exactly 90°, then the array of lattice points would have a point group symmetry of *mmm*, i.e. that of an orthorhombic lattice. Hence, a rotation of 180° about either the **a** axis or the **c** axis of this monoclinic crystal may produce two domain variants twinned with respect to one another. Since the assembly of twinned domains will appear to a good approximation to have the higher orthorhombic symmetry, rather than monoclinic symmetry, this type of twinning is known as *twinning by pseudomerohedry* to differentiate it from twinning by (strict) merohedry.

Likewise, if in an orthorhombic crystal the lattice parameters a and b are very similar, then, to a very good approximation, there is a fourfold axis parallel to [001]. Hence, the twin law in this particular case for twinning by pseudo-merohedry would be 90° about [001].

In triclinic crystals where the b axis is almost normal to the a and c axes, twinning by pseudo-merohedry arises by (i) reflection in (010) and (ii) rotation around [010]. In the former case this is known as *albite* twinning in feldspars; in the latter it is known as *pericline* twinning. Of course, for such twinning to be able to occur physically, the ionic interactions across the twin boundaries must be sufficiently energetically similar to those which would have formed in a single crystal, i.e. the twin boundaries must be of suitably low energy [70].

11.6.4 Twinning By Reticular Pseudomerohedry

In essence, this is twinning arising in crystal growth processes which is not describable in terms of any of the other three categories. It can be considered to be related to twinning by (strict) reticular merohedry in the same way as twinning by pseudomerohedry relates to twinning by (strict) merohedry. Thus, for example, while twinning in devitrite in Table 11.1 can be described within the framework of deformation twinning, it is actually an example of growth twinning by reticular pseudomerohedry. A second example of twinning by reticular pseudo-merohedry is provided by the thortveitite-related M phase, $(Mn_xZn_{1-x}V_2O_7)$ $(0.75 < x < 0.913)$ in which there is approximate overlap of one out of two X-ray reflections because of twinning within crystals chosen for single-crystal X-ray work [71].

11.7 Atomistic Modelling of Twin Boundaries

It is evident that for twinning to occur, either through shear or as a consequence of crystal growth, the twinning mode under consideration must have a low surface energy, so that the energy penalty of introducing the twin boundary is reasonable. Calculations of energies of twin boundaries have progressed from early, relatively simple, geometrical models, some of which have been considered in this Chapter to help rationalize why one twinning mode might be preferred over another, to sophisticated molecular dynamic simulations, with and without the deliberate introduction of defects.

To illustrate the principles of how molecular dynamic simulations can be applied, we will consider here twinning in devitrite [41, 70]. Devitrite, $Na_2Ca_3Si_6O_{16}$, is a centrosymmetric triclinic silicate phase with 54 atoms per unit cell. This phase nucleates heterogeneously on the surfaces of suitably heat treated soda-lime-silica float glass, growing as spherulites consisting of thin needles, within which a single Type II twinning mode has been observed [41]. Analysis of this twinning mode has shown that it has a rotation axis, η_1, of [100] and that, with respect to the triclinic unit cell of devitrite, the indices of the twin plane, K_1, are $(0, -2b\cos\gamma/a, -1-(2c\cos\beta/a))$, i.e. a plane which does not have rational indices. However, for devitrite, this formally irrational twin plane is actually the plane $(0\ 1\ \overline{0.0004})$, $0.02°$ from (010), using the unit cell parameters for devitrite in the form $a = 7.2291$ Å, $b = 10.1728$ Å, $c = 10.6727$ Å, $\alpha = 95.669°$, $\beta = 109.792°$, $\gamma = 99.156°$ [72]. Therefore, qualitatively, it might be thought that this because K_1 has a low interfacial energy.

Detailed calculations using molecular dynamic simulations comparing this twinning mode with other possible Type I twinning modes in devitrite confirm this qualitative expectation. Such simulations are time consuming, and so a balance has to be struck between the number of atoms in the simulations and the time taken for the simulations to show stable twin structures, i.e. ones where further relaxations of atomic positions do not decrease the energy of the system and where the system is stable to small perturbations. The potential used to model short range forces in this work was a Buckingham pair potential frequently used to model ionic materials [70].

For geometries where it is evident that a stable configuration cannot be obtained, and where instead running the simulations produced highly unstable, highly energetic, and highly unlikely atomic configurations, with significant atomic movements, simulations can be terminated earlier than for simulations where a stable configuration is achieved.

For each initial twin boundary configuration prior to the subsequent molecular dynamics simulation runs for devitrite, the construction of the twin models was broken down into three separate procedures:

(i) Firstly, the origin of the unit cell of devitrite was moved so that the position of the twin plane under consideration was a surface of the newly redefined unit cell of devitrite.

(ii) This newly defined unit cell was then replicated using translational symmetry to construct a grouping of $4 \times 2 \times 2$ unit cells representing the matrix and containing 864 atoms in total (16 unit cells).

(iii) Finally, atom positions in the twin were obtained from those in the matrix using the appropriate geometric transformation of either a rotation of $180°$ around a [100] position vector for Type II twinning, or reflection in the twin plane for possible alternative Type I twins.

In addition, periodic boundary conditions were imposed at the boundaries of the grouping of ions used to model the various twin boundary configurations considered.

Two examples of stable configurations from molecular dynamic simulations are shown in Figure 11.19 for the observed growth mode. Examination of the crystal structure of devitrite shows that the coordination octahedra around the calcium ions in the crystal structure share common edges to form one-dimensional double bands running parallel to [100] in devitrite. The way in which the silica tetrahedra in the crystal structure are configured in double bands with respect to these coordination octahedra around the calcium ions helps to explain why devitrite occurs as needles along the [100] direction. It also explains why twinning by a rotation of $180°$ about the [100] growth direction is a favourable twinning mode during

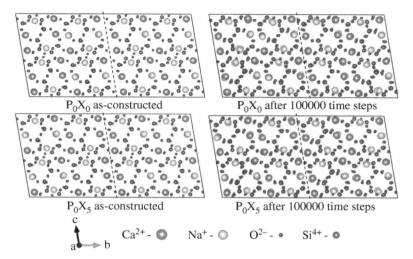

Figure 11.19 Structures for Type II twinning with stable twin modes for devitrite before and after molecular dynamics simulation. The (010) twin plane is shown by the dashed line in the two as-constructed models and the two models after 10^5 time steps. Periodic boundary conditions ensure that the two external boundaries of the atom grouping parallel to this twin plane are also (010) twin boundaries. P_0X_0 and P_0X_5 refer to two possible starting geometries

growth. In essence, it is found that the resulting (010) twin plane does indeed have a low interfacial energy – the simple qualitative interpretation turns out to be correct.

Problems

11.1 Determine the transformation matrix relating the indices of vectors in the matrix and twin in a deformation twinning operation in a c.c.p. crystal where $K_1 = (111)$ and $\eta_2 = [112]$. Confirm that $\langle 112 \rangle$ vectors lying in K_1 do not change the form of their indices as a result of this deformation twinning operation.

11.2 One of the ways of describing the orientation relationship of the twin and the matrix in a type I twin is as a tilt about an axis in the twinning plane K_1 and normal to the shear direction η_1. Determine the magnitude of the angle of tilt for the case of a twin in a c.c.p. metal.

11.3 Draw a plan of the K_1 plane in a c.c.p. metal showing two layers of atoms with the positions of the atoms in the upper layer before and after they are sheared to form the first layer of a twin.

11.4 Repeat Problem 11.3 for the case of a twin in a b.c.c. metal.

11.5 A twin in a b.c.c. metal has $K_1 = (112)$ and $\eta_1 = [\bar{1}\bar{1}1]$. Determine the Miller indices of the planes into which the following lattice planes are transformed by the twinning shear: (001), (010), (100). Hence write down the matrix for the transformation of planes and determine the plane into which the (101) plane transforms.

11.6 Show that if crystal axes are chosen non-conventionally, with Ox and Oy in the K_1 plane of a type I twin and Oz parallel to η_2, the transformation of any plane (hkl) is given by $h' = -h$, $k' = -k$, $l' = l$.

11.7 Using the geometry of simple shear shown in Figure 11.4, show that if $K_1 = (h_1 k_1 l_1)$ and $\eta_2 = [p_2 q_2 r_2]$, the indices of $K_2 = (h_2 k_2 l_2)$ are given by:

$$h_2 = h_1 - N\frac{[100] \cdot [p_2 q_2 r_2]}{|[p_2 q_2 r_2]|^2}$$

$$k_2 = k_1 - N\frac{[010] \cdot [p_2 q_2 r_2]}{|[p_2 q_2 r_2]|^2}$$

$$l_2 = l_1 - N\frac{[001] \cdot [p_2 q_2 r_2]}{|[p_2 q_2 r_2]|^2}$$

where $N = h_1 p_2 + k_1 q_2 + l_1 r_2$. Determine a corresponding set of equations to find the indices of $\eta_1 = [p_1 q_1 r_1]$, given K_1 and η_2.

11.8 Show that in pericline twinning in plagioclase (type II twinning), the composition plane K_1 has intersections with the (110) and ($1\bar{1}0$) planes which form the sides of a rhombus. (For this reason, K_1 in pericline twinning is known as the 'rhombic section'.) Show further that the angle σ that the trace of K_1 makes on (010) with the vector [100] is given by:

$$\cot \sigma = \cos \alpha^* \tan \gamma = \cot \beta - \frac{\cos \alpha}{\cos \gamma \sin \beta}$$

11.9 Which lattice rows remain rows of the same type after the lattice has suffered the twinning shear of a type I twin? By inspection or otherwise, determine the lattice directions into which the directions [001], [010] and [100] transform as a result of the shear of a twin in a b.c.c. metal for which $K_1 = (112)$ and $\eta_2 = [111]$. Hence write down the matrix for the transformation of directions and determine the direction into which [$1\bar{1}1$] transforms.

11.10 Twin lamellae parallel to {111} planes have been observed in germanium. Sketch the probable structure of this twin.

11.11 Calculate the angle between (0001) and ($10\bar{1}2$) in h.c.p. metals as a function of c/a. Hence on the supposition that h.c.p. metals can deform only by basal slip or ($10\bar{1}2$) twinning (not true of course), divide the metals into those that could be elongated, but not compressed, parallel to the c-axis, and those that could be compressed, but not elongated, in this direction. What are the corresponding results for ($11\bar{2}1$) and ($11\bar{2}2$) twinning?

11.12 Is the observed twinning mode in c.c.p. metals the mode with the smallest shear (the lattice being completely reconstructed by the shear)?

11.13 The copper–gold alloy CuAu containing equal numbers of copper and gold atoms has a superlattice structure in which (001) planes are composed alternately of copper and gold. Show that the passage of a dislocation of Burgers vectors $\frac{1}{2}[\bar{1}01]$ (the indices refer to the c.c.p. lattice of the disordered structure) over successive (101) planes produces a twinned superlattice with (101) as the twin plane.

11.14 The surface of a certain crystal of zinc is a basal plane, (0001). If this surface has been intersected by $\{01\bar{1}2\}$ twins (Table 11.1), at what angle will the surface within a twin lamella be tilted with respect to the surface of the surrounding matrix? If the measured angle of tilt turned out to be less than the expected value, what explanation would you suggest?

11.15 A rod-shaped single crystal of iron has the pole of the (112) plane (which is also the composition plane of the twin) at an angle of 30° to the axis of the rod, which lies in the plane defined by the (112) pole and $[\bar{1}\,\bar{1}1]$. The crystal is deformed in a tensile test by twinning on the (112) plane over one-third of its length. What is the tensile elongation if the ends of the rod are free to move laterally (that is, if the tensile axis does not rotate in the untwinned part of the crystal)?

11.16 Zinc crystals twin by shear on the plane $(10\bar{1}2)$ in the direction $[10\bar{1}1]$. Draw the arrangement of the atoms in the $(1\bar{2}10)$ plane for several unit cells. This plane is normal to $(10\bar{1}2)$ and contains the shear direction of the twin. Draw in the trace of one of the $(10\bar{1}2)$ planes. Suppose the part of your drawing on one side of the trace of the $(10\bar{1}2)$ plane represents the matrix crystal. Move the atoms situated precisely at the lattice points on the other side of the trace to the lattice points of the twin formed by reflection $(10\bar{1}2)$. Find graphically the magnitude of the shear involved and confirm that it is equal to:

$$\left(\frac{c^2}{a^2} - 3 \right) \frac{a}{\sqrt{3c}}$$

Note carefully from your drawing that the same shear applied to atoms situated originally at points $\left(\frac{2}{3}, \frac{1}{3}, \frac{1}{2} \right)$ will not reproduce the crystal structure exactly but that additional displacements of these atoms are required.

Suggestions for Further Reading

Bevis, M. and Crocker, A.G. (1968). Twinning shears in lattices. *Proc. Roy. Soc. Lond.* A 304: 123–134.

Bevis, M. and Crocker, A.G. (1969). Twinning modes in lattices. *Proc. Roy. Soc. Lond.* A 313: 509–529.

Cahn, R.W. (1954). Twinned crystals. *Adv. Phys.* 3: 363–445.

Hardouin Duparc, O.B.M. (2017). A review of some elements for the history of mechanical twinning centred on its German origins until Otto Mügge's K_1 and K_2 invariant plane notation. *J. Mater. Sci.* 52: 4182–4196.

Nabarro, F.R.N. (1992). *Dislocations in Solids*, vol. 9. *Dislocations and Disclinations*, Amsterdam: North Holland.

Reed-Hill, R.E., Hirth, J.P. and Rogers, H.C. (1965). Deformation twinning. In: *Metallurgical Society Conferences*, vol. 25. New York: Gordon and Breach.

Yoo, M.H. and Wuttig, M. (1994). *Twinning in Advanced Materials*. Warrendale, PA: TMS.

References

[1] Battey, M.H. and Pring, A. (1997). *Mineralogy for Students*, 3rd Edition. London: Longman.

[2] Deer, W.A., Howie, R.A. and Zussman, J. (2001). *Rock-Forming Minerals, Volume 4A, Framework Silicates: Feldspars*, 2nd Edition. The Geological Society London.

[3] Crocker, A.G. (1965). The crystallography of deformation twinning in alpha-uranium. *J. Nucl. Mater.* 16: 306–326.

[4] Cahn, R.W. (1953). Plastic deformation of alpha-uranium; twinning and slip. *Acta Metall.* 1: 49–70.

[5] Mügge, O. (1889). Ueber durch Druck entstandene Zwillinge von Titanit nach den Kanten [110] und [1$\bar{1}$0]. *Neues Jahrb. Mineral. Geol. Palaeontol.*, Band II: 98–115.

[6] Mügge, O. (1889). Ueber homogene Deformationen (einfache Schiebungen) an dem triklinen Doppelsalz BaCdCl$_4$.4aq. *Neues Jahrb. Mineral. Geol. Palaeontol.*, Beilage-Band VI: 274–304.

[7] Liebisch, T. (1889). Ueber eine besondere Art von homogenen Deformationen. *Neues Jahrb. Mineral. Geol. Palaeontol.*, Beilage-Band VI: 105–120.

[8] Bell, J.F. (1941). Morphology of mechanical twinning in crystals. *Am. Mineral.* 26: 247–261.

[9] Pabst, A. (1955). Transformation of indices in twin gliding. *Bull. Geol. Soc. Am.* 66: 897–912.

[10] Blewitt, T.H., Coltman, R.R. and Redman, J.K. (1957). Low-temperature deformation of copper single crystals. *J. Appl. Phys.* 28: 651–660.

[11] Suzuki, H. and Barrett, C.S. (1958). Deformation twinning in silver–gold alloys. *Acta Metall.* 6: 156–165.

[12] McHargue, C.J. and McCoy, H.E. (1963). Effect of interstitial elements on twinning in columbium. *Trans. Am. Inst. Min. Metall. Petrol. Engrs.* 227: 1170–1174.

[13] Jaffee, R.I., Sims, C.T. and Harwood, J.J. (1959). The effect of rhenium on the fabricability and ductility of molybdenum and tungsten. In: *Plansee Proceedings 1958* (ed. F. Benesovsky), pp. 380–411. New York: Pergamon Press.

[14] Paxton, H.W. (1953). Experimental verification of the twin system in alpha-iron. *Acta Metall.* 1: 141–143.

[15] Buerger, M.J. (1928). The plastic deformation of ore minerals, part 2 (concluded). *Am. Mineral.* 13: 35–51.

[16] Baumhauer, H. (1879). Ueber künstliche Kalkspath-Zwillinge nach – ½R. *Z. Krystall.* 3: 588–591.

[17] Johnsen, A. (1914). Schiebungen und Translationen in Kristallen. *Jahrb. der Radioaktivität und Elektronik* 11: 226–262.

[18] Klassen-Neklyudova, M.V. (1964). *Mechanical Twinning of Crystals* (trans. J.E.S. Bradley). New York: Consultants Bureau.

[19] Hardouin Duparc, O.B.M. (2017). A review of some elements for the history of mechanical twinning centred on its German origins until Otto Mügge's K_1 and K_2 invariant plane notation. *J. Mater. Sci.* 52: 4182–4196.

[20] Hall, E.O. (1953). Some observations on the crystallography of deformation twins. *Acta Crystallogr.* 6: 570–571.

[21] Hall, E.O. (1957). Twinning in cobalt. *Acta Metall.* 5: 110.

[22] Davis, K.G. and Teghtsoonian, E. (1962). Deformation twins in cobalt. *Acta Metall.* 10: 1189–1191.

[23] Reed-Hill, R.E. (1960). A study of the {10$\bar{1}$1} and {10$\bar{1}$3} twins modes in magnesium. *Trans. Am. Inst. Min. Metall. Petrol. Engrs.* 218: 554–558.

[24] Jeffery, R.A. and Smith, E. (1966). Deformation twinning in rhenium single crystals. *Phil. Mag.* 13: 1163–1168.

[25] Rapperport, E.J. and Hartley, C.S. (1960). Deformation modes of zirconium at 77°, 575° and 1075°K. *Trans. Am. Inst. Min. Metall. Petrol. Engrs.* 218: 869–877.

[26] Reed-Hill, R.E., Slippy, W.A. and Buteau, L.J. (1963). Determination of alpha zirconium {11$\bar{2}$1} twinning elements using grain boundary rotations. *Trans. Am. Inst. Min. Metall. Petrol. Engrs.* 227: 976–979.

[27] Rosi, F.D., Dube, C.A. and Alexander, B.H. (1953). Mechanism of plastic flow in titanium – determination of slip and twinning elements. *Trans. Am. Inst. Min. Metall. Petrol. Engrs.* 197: 257–265.

[28] Lainé, S.J. and Knowles, K.M. (2015). {$11\bar{2}4$} deformation twinning in commercial purity titanium at room temperature. *Phil. Mag.* 95: 2153–2166.

[29] Rosi, F.D., Perkins, F.C. and Seigle, L.L. (1956). Mechanism of plastic flow in titanium at low and high temperatures. *Trans. Am. Inst. Min. Metall. Petrol. Engrs.* 206: 115–122.

[30] Freise, E.J. and Kelly, A. (1961). Twinning in graphite. *Proc. Roy. Soc. Lond.* A 264: 269–276.

[31] Heuer, A.H. (1966). Deformation twinning in corundum. *Phil. Mag.* 13: 379–393.

[32] Gough, H.J. and Cox, H.L. (1932). The behaviour of single crystals of bismuth subjected to alternating torsional stresses. *J. Inst. Met.* 48: 227–247.

[33] Guyoncourt, D.M.M. and Crocker, A.G. (1968). The deformation twinning mode of crystalline mercury. *Acta Metall.* 16: 523–534.

[34] Lloyd, L.T. and Chiswik, H.H. (1955). Deformation mechanisms of alpha-uranium single crystals. *Trans. Am. Inst. Min. Metall. Petrol. Engrs.* 203: 1206–1214.

[35] Otani, N., Funatsu, Y., Ichinose, S., Miyazaki, S. and Otsuka, K. (1983). Orientation dependence of the deformation modes in a γ'_1 martensite single crystal in Cu–Al–Ni alloy. *Scr. Metall.* 17: 745–750.

[36] Ichinose, S., Funatsu, Y. and Otsuka, K. (1985). Type II deformation twinning in γ'_1 martensite in a Cu–Al–Ni alloy. *Acta Metall.* 33: 1613–1620.

[37] Otsuka, K., Sawamura, T. and Shimizu, K. (1971). Crystal structure and internal defects of equiatomic TiNi martensite. *Physica Status Solidi (a)* 5: 457–470.

[38] Knowles, K.M. and Smith, D.A. (1981). The crystallography of the martensitic transformation in equiatomic nickel–titanium. *Acta Metall.* 29: 101–110.

[39] Onda, T., Bando, Y., Ohba, T. and Shimizu, K. (1992). Electron microscopy study of twins in martensite in a Ti–50.0 at%Ni alloy. *Trans. JIM* 33: 354–359.

[40] Otsuka, K. and Ren, X. (2005). Physical metallurgy of Ti–Ni-based shape memory alloys. *Prog. Mater. Sci.* 50: 511–678.

[41] Knowles, K.M. and Ramsey, C.N.F. (2012). Type II twinning in devitrite, $Na_2Ca_3Si_6O_{16}$. *Phil. Mag. Lett.* 92: 38–48.

[42] Smith, J.V. and Brown, W.L. (1988). *Feldspar Minerals*, vol. 2, 1. Heidelberg: Springer-Verlag.

[43] Smith, J.V. (1958). The effect of composition and structural state on the rhombic section and pericline twins of plagioclase felspars. *Mineral. Mag.* 31: 914–928.

[44] Christian, J.W. and Mahajan, S. (1995). Deformation twinning. *Prog. Mater. Sci.* 39: 1–157.

[45] Jaswon, M.A. and Dove, D.B. (1960). The crystallography of deformation twinning. *Acta Crystallogr.* 13: 232–240.

[46] Couling, S.L. and Roberts, C.S. (1956). New twinning systems in magnesium. *Acta Crystallogr.* 9: 972–973.

[47] Reed-Hill, R.E. and Robertson, W.D. (1957). Additional modes of deformation twinning in magnesium. *Acta Metall.* 5: 717–727.

[48] Xu, S., Gong, M., Jiang, Y., Schuman, C., Lecomte, J.-S. and Wang, J. (2018). Secondary twin variant selection in four types of double twins in titanium. *Acta Mater.* 152: 58–76.

[49] Bolling, G.F. and Richman, R.H. (1965). Continual mechanical twinning, Part I: Formal description. *Acta Metall.* 13: 709–722.

[50] Green, M.L. and Cohen, M. (1979). Pseudotwinning and pseudoelasticity in β Fe–Be alloys. *Acta Mater.* 27: 1523–1538.

[51] Neumann, J.G. (1850). Ueber die krystallinische Structur des Meteoreisens von Braunau. *Naturwissenschaftliche Abhandlungen* 3 (Part II): 45–56.

[52] Smith, S.W.J., Dee, A.A. and Young, J. (1928). The mode of formation of Neumann bands. Part I. – The mechanism of twinning in the body-centred cubic lattice. *Proc. Roy. Soc. Lond.* A 121: 477–486.

[53] Smith, S.W.J., Dee, A.A. and Young, J. (1928). The mode of formation of Neumann bands. Part II. – The evidence that the bands are twins. *Proc. Roy. Soc. Lond.* A 121: 486–500.

[54] Smith, S.W.J., Dee, A.A. and Young, J. (1928). The mode of formation of Neumann bands. Part III. – The movement from which the twinning results. *Proc. Roy. Soc. Lond. A* 121: 501–514.

[55] Sleeswyk, A.W. (1962). Emissary dislocations: theory and experiments on the propagation of deformation twins in α-iron. *Acta Metall.* 10: 705–725.

[56] Sleeswyk, A.W. and Helle, J.N. (1961). Zigzag configurations of twins in α-iron. *Acta Metall.* 9: 344–351.

[57] Cottrell, A.H. and Bilby, B.A. (1951). A mechanism for the growth of deformation twins in crystals. *Phil. Mag.* 42: 573–581.

[58] Pond, R.C. and Hirth, J.P. (2018). Topological model of type II deformation twinning. *Acta Mater.* 151: 229–242.

[59] Frank, F.C. and Thompson, N. (1955). On deformation by twinning. *Acta Metall.* 3: 30–33.

[60] Friedel, G. (1926). *Leçons de Cristallographie*. Nancy, Paris, Strasbourg: Berger-Levrault.

[61] Cahn, R.W. (1954). Twinned crystals. *Adv. Phys.* 3: 363–445.

[62] Brostigen, G. and Kjekshus, A. (1969). Redetermined crystal structure of FeS_2 (pyrite). *Acta Chem. Scand.* 23: 2186–2188.

[63] Donnay, G., Donnay, J.D.H. and Iijima, S. (1977). A high-resolution electron micrograph of the twin boundary in pyrite. *Acta Crystallogr. A* 33: 622–626.

[64] Strunz, H. and Tennyson, C. (1965). Strukturelle Deutung der Pyrit- und Markasitzwillinge. *Neues Jahrb. Mineral. Monatshefte* 1965: 247–248.

[65] Kim, J.C. and Goo, E. (1990). Inversion twin boundaries in zinc oxide. *J. Am. Ceram. Soc.* 73: 877–884.

[66] Klapper, H. and Hahn, T. (2010). The application of *eigensymmetries* of face forms to anomalous scattering and twinning by merohedry in X-ray diffraction. *Acta Crystallogr. A* 66: 327–346.

[67] Minor, A., Rybacki, E., Sintubin, M., Vogel, S. and Wenk, H.-R. (2018). Tracking mechanical Dauphiné twin evolution with applied stress in axial compression experiments on a low-grade metamorphic quartzite. *J. Struct. Geol.* 112: 81–94.

[68] Tullis, J. (1970). Quartz: preferred orientation in rocks produced by Dauphiné twinning. *Science* 168: 1342–1344.

[69] Klapper, H. and Hahn, T. (2012). The application of *eigensymmetries* of face forms to X-ray diffraction intensities of crystals twinned by 'reticular merohedry'. *Acta Crystallogr. A* 68: 82–109.

[70] Li, B. and Knowles, K.M. (2013). Molecular dynamics simulation of albite twinning and pericline twinning in low albite. *Model. Simul. Mater. Sci. Eng.* 21, 055012 (18pp).

[71] Knowles, K.M., Sil, A., Stöger, B. and Weil, M. (2018). Crystal structure of the thortveitite-related *M* phase, $(Mn_xZn_{1-x})_2V_2O_7$ $(0.75 < x < 0.913))$: a combined synchrotron powder and single-crystal X-ray study. *Acta Crystallogr. C* 74: 1079–1087.

[72] Kahlenberg, V., Girtler, D., Arroyabe, E., Kaindl, R. and Többens, D.M. (2010). Devitrite $(Na_2Ca_3Si_6O_{16})$ – structural, spectroscopic and computational investigations on a crystalline impurity phase in industrial soda-lime glasses. *Mineral. Petrol.* 100: 1–9.

12

Martensitic Transformations

12.1 Introduction

When a steel is quenched from a temperature of, say, 950 °C, at which it is austenite, a c.c.p. phase, it transforms abruptly to a very hard body-centred tetragonal (b.c.t.) phase called martensite, named after Adolf Martens (1850–1914), a German metallurgist. Because of its enormous technical importance in connection with the tempering of martensite, this phase transformation in steels has been studied thoroughly over the years. More recently, the technological interest in shape memory materials has led to an interest in a wide variety of non-ferrous alloys in which martensitic transformations occur. Shape-memory materials rely on martensitic transformations for the ability of components made from them to 'remember' their physical shape by simply heating them after they have been deformed permanently in the martensitic state [1].

The two most striking characteristics of a martensitic transformation are that, first, the transformation takes place very rapidly, implying that long-range diffusion plays no part in the transformation, and, second, the shape of a transforming region alters [2]. On quenching from the austenite phase field, the transformation in steels of austenite to martensite begins at a certain temperature M_s, which decreases as the carbon content of the steel increases. When the steel is quenched to a temperature below M_s, a certain fraction of it transforms almost instantaneously to martensite. There is then little or no increase in the amount of martensite formed until the temperature is lowered further. The change of shape of a region that transforms can be detected by optical microscopy or atomic force microscopy as a distortion of the surface of the steel. The shape change is very significant, because it implies that the iron atoms have moved in a regular, systematic way in order to build the new structure.

Crystallography and Crystal Defects, Third Edition. Anthony Kelly and Kevin M. Knowles.
© 2020 John Wiley & Sons Ltd. Published 2020 by John Wiley & Sons Ltd.
Companion website: www.wiley.com/go/kelly/crystallography3e

Some phase transformations possess one of the above characteristics but not the other. For example, when a Cu–24 at% Ga alloy is cooled moderately quickly, it transforms rapidly from a b.c.c. phase to a hexagonal phase of the same composition [3]. A sharp interface moves rapidly through the b.c.c. phase, leaving transformed material in its wake. Clearly, no long-range diffusion is involved. However, no change of shape occurs, which suggests that the new structure is not built by systematic atom movements, but rather by atoms moving across a disordered interface in an irregular fashion. This conclusion is reinforced by the fact that a single grain of the new phase can extend across grain boundaries of the old, showing that a single orientation of the new phase can be produced from randomly different orientations of the old. Such transformations are usually referred to as massive transformations, and are distinguished both from martensitic transformations and from transformations that require diffusion.

Transformations that do require diffusion, which disqualifies them from the title of martensitic, sometimes exhibit a shape change. An example is provided by a Cu–4 wt% Be alloy. The high-temperature c.c.p. phase in this alloy can be preserved at room temperature by quenching. When the alloy is aged, a Be-rich precipitate forms. As a result of the shape change accompanying this precipitation, optically visible relief effects are produced on the surfaces of polished specimens [4], even though the phase transformation is clearly diffusion-controlled.

We shall define transformations that are both distortive and diffusionless as martensitic. In this chapter, we shall focus mainly on the crystallographic aspects of such transformations while referring as well to recent developments in the theory of martensitic transformations and to the mechanism of the shape memory effect. Discussion of further aspects of the materials science of martensitic transformations can be found in the Suggestions for Further Reading at the end of this chapter.

12.2 General Crystallographic Features

A crystal that has partly undergone a martensitic transformation often looks somewhat like a crystal containing twins. Plates of the new phase intersect the surface in lines that lie in a few well-defined directions. The surface within the plate is tilted with respect to the rest of the surface, just as it would be in a twin lamella produced as a consequence of deformation twinning. Even the clicking noise characteristic of twinning may be heard during a martensitic transformation. These observations suggest that the crystallography of martensitic transformations should be approached in the same way as that of twinning. Thus, the question could be asked: can a homogeneous simple shear transform the lattice of the old phase into the lattice of the new phase?

In most cases, including the martensitic transformation in steels, it is found that a more complex strain than a simple shear is needed to describe the lattice change. The crystallographic theory of martensitic transformations is concerned with finding a plausible strain that will carry some or all of the lattice points of the old phase into positions where they form part or all of the new lattice. This strain must be consistent with the observable features of the transformation, which will now be defined more carefully.

Assume that a crystal has partly transformed, and that the transformed regions are in the form of plates, as is often the case. The most easily measured parameter with an optical microscope is the orientation of the plane of the faces of the plate, which is normally

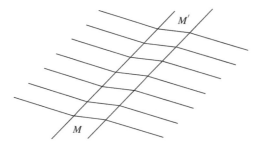

Figure 12.1 Scratched surface intersected by a martensite plate *MM'*

expressed as a plane in the matrix (austenite) lattice. This is called the *habit plane*, and it is found in the same way as the composition plane of a twin (Figure 11.3). The relationship between the orientation of the matrix lattice and the lattice in the plate can be found by acquiring X-ray reflections from both. The most difficult quantity to determine is the macroscopic strain undergone by the material in the plate. The displacement of fiduciary lines can be used to evaluate this strain. A corner of the specimen (as in Figure 11.3), scratches on the surface, or slip lines that existed in the matrix before the transformation can all be used as fiduciary lines.

A schematic of the deflection of scratches in a typical case is shown in Figure 12.1. The straightness of the scratches within the plate seen with an optical microscope demonstrates that the strain is homogeneous, at least on the macroscopic (optical) scale; that is, on the scale of micrometres. The fact that the scratches in the matrix remain straight as they approach the plate shows that the matrix around the plate has not been deformed inhomogeneously. This in turn implies that the plate has not been rotated into its present position, because that would severely deform the matrix. The continuity of the scratches across the interface implies that the interface, or habit plane, is not distorted. If it were, the scratches would no longer match up at the interface.

A plane that is left unrotated and undistorted is called an *invariant plane*. A strain that leaves a certain plane unrotated and undistorted can be described by the following equation for the displacement **u** of a point of position vector **r**:

$$\mathbf{u} = \mathbf{d}(\mathbf{r}\cdot\mathbf{h})$$

where **h** is the unit normal to the invariant plane and **d** is a constant vector in the direction of the displacement. The quantity $\mathbf{r}\cdot\mathbf{h}$ is the perpendicular distance from the point of position vector **r** onto the invariant plane passing through the origin. When **d** is parallel to the invariant plane, so that $\mathbf{d}\cdot\mathbf{h}$ is zero, the strain is a simple shear. This special case is shown in Figure 7.7 in the context of glide in crystals. In general, an invariant plane strain can be resolved into a simple shear given by the component of **d** normal to **h**, coupled with an extension or compression normal to the invariant plane, given by the component of **d** parallel to **h**. To determine **d** in magnitude and direction experimentally, *two* non-parallel fiduciary lines are required.

Although few precise determinations of the macroscopic strain in a martensite plate have been made, and although martensite does not always form well-defined, regular plates, it is now generally assumed that the characteristic strain in bulk specimens is at least approximately an invariant plane strain. The belief in an invariant plane strain is undoubtedly reinforced by the consideration that, when this type of strain occurs in a very thin plate that is embedded in a matrix, the matrix does not itself have to undergo any strain in order to make way for the strain suffered by the plate, or, in other words, it has no difficulty in *accommodating* the strain, except at the edges of the plate.

With the understanding that the macroscopic strain is one in which the habit plane is invariant, it is reasonable to ask whether any such strain, applied *homogeneously* to the points of the matrix (austenite) lattice, would convert this lattice into the martensite lattice? The answer is usually that it would not. The inevitable conclusion is that the lattice strain is *not* the same as the macroscopic strain. Therefore, the strain can be homogeneous only on a macroscopic scale; on a fine scale it must be inhomogeneous. In some cases the inhomogeneity can be directly observed, as when the martensite plate can be seen to be *internally twinned*. Such cases are amenable to observation by transmission electron microscopy (TEM). In other cases the inhomogeneity is less readily detected, even by TEM.

In cases where the type of inhomogeneity is not known, the theoretical approach has been first to postulate a plausible *homogeneous* strain that *will* transform the matrix lattice into the martensite lattice, and then to postulate an *added* deformation in the form of slip or twinning, which preserves the martensite lattice, but alters the macroscopic strain and brings it to the invariant plane strain that is observed. This treatment is illustrated, in two dimensions, by the diagrams in Figure 12.2. The test of the theory is that the final macroscopic strain must be identical with the observed strain, and that the orientation relationship between the martensite and matrix lattices must be predicted correctly.

The limitations of such a theory are obvious. It predicts only the final atom positions, given the initial ones, and is therefore phenomenological, in the sense that it does not attempt to predict the actual paths of the atoms. In principle, this prediction can be tested directly: either the atoms do end up in the specified positions, or they do not. In practice, it is only possible to test the theory indirectly, typically through the habit plane and orientation relationship. This leaves open the possibility that some other strain, based on different postulates, would also be compatible with the observed features of the transformation. There is no absolute guarantee of uniqueness.

The general considerations of this section will now be illustrated by a more detailed discussion of a number of specific transformations. Although the transformations in steels, Ni–Ti-based alloys and Cu–Zn–Al-based alloys are by far the most important cases technologically, the complexity of the crystallography of these particular transformations makes it easier to convey the essence of the crystallographic theory of martensitic transformations using other, more simple examples. This consideration has governed the choice of the transformations to be discussed in the next three sections; however, the section on transformations in steels can be read on its own without difficulty.

12.3 Transformation in Cobalt

Cobalt is c.c.p. at high temperature; on cooling, it transforms at 420 °C to a h.c.p. structure. The crystallography of this transformation is particularly simple, since it merely produces a change in the stacking sequence of close-packed planes from ABCAB ... to ABABA ... This change can be achieved by passing a partial dislocation of Burgers vector $\frac{1}{6}\langle 211 \rangle$

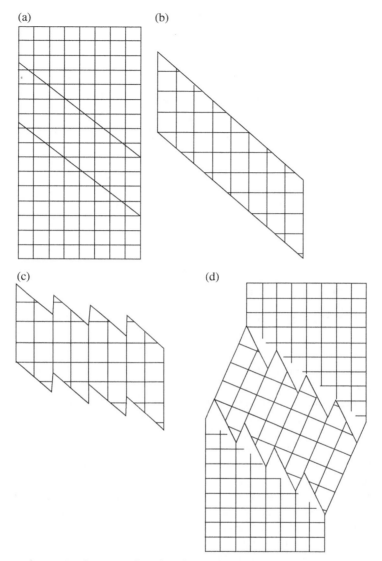

Figure 12.2 The square lattice within the plate outlined in (a) is strained into a rectangular lattice in (b), lengthening the upper and lower edges of the plate. In (c), slip within the plate has shortened its edges so that macroscopically they match the untransformed matrix, as shown in (d)

across *every other* {111} plane [5]. If each $\frac{1}{6}\langle 211 \rangle$ displacement is in the same $\langle 211 \rangle$ direction, there accumulates a macroscopic simple shear of magnitude g given by:

$$g = \frac{\left| \frac{1}{6}\langle 211 \rangle \right|}{2d_{(111)}} \tag{12.1}$$

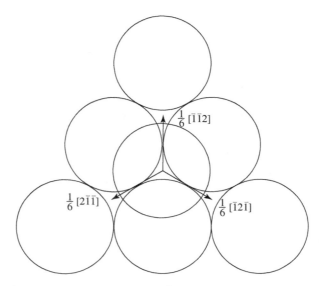

Figure 12.3   The three possible $\frac{1}{6}\langle 211 \rangle$ vectors in a (111) plane

where $d_{(111)}$ is the spacing between adjacent (111) planes. Therefore:

$$g = \sqrt{2}/4 \approx 0.354 \tag{12.2}$$

The simple shear on the (111) plane leaves this plane unrotated and undistorted, and so the (111) is expected to be the habit plane.

It is interesting to see how this result might be described more formally. It is obvious that no homogeneous strain, applied to the atoms, can turn a c.c.p. structure into a h.c.p structure, because the atoms lie at the points of a Bravais lattice in the c.c.p. structure, whereas only half of them do so in the h.c.p. structure. Any homogeneous strain of the c.c.p. structure would generate a new lattice with one atom at each lattice point. However, a homogeneous simple shear of magnitude $\sqrt{2}/4$ in the $\langle 211 \rangle$ direction on the $\{1\bar{1}\bar{1}\}$ plane does generate the correct hexagonal lattice from *half* the points of the c.c.p. lattice. Additional atom movements, producing no change of shape, are needed to complete the structure change.

Experimental study of the transformation has been hindered by the difficulty of observing partially transformed cobalt crystals. Striations parallel to the traces of {111} are observed after transformation, suggesting that {111} is the habit plane, as predicted. By alloying with Ni, the transformation temperature may be reduced so that some cubic phase is retained at room temperature. On such specimens, the close-packed planes and directions of the two phases have been found to be parallel to one another, with high precision (better than $\pm \frac{1}{2}°$) [6]. However, the shape change is quite irregular, and well-defined uniform tilts of macroscopic areas of surface are seldom observed. This can be accounted for by assuming that the direction of the $\langle 211 \rangle$ displacement varies between the three possibilities which exist for any given {111} plane, as shown in Figure 12.3. If the three possibilities were employed equally overall, the macroscopic shear would disappear entirely.

It is observed that a very small decrease in volume accompanies the transformation of c.c.p. cobalt to h.c.p. cobalt. The fractional volume change is -3.6×10^{-3}. It follows that the glide of $\frac{1}{6}\langle 211 \rangle$ partial dislocations does not by itself quite complete the transformation, because it does not provide this change of volume. Further, more detailed, considerations of the martensitic phase transformation in cobalt can be found in a paper by Tolédano et al. [7]. A similar c.c.p. → h.c.p. transformation occurs in Fe–Mn–Si alloys [8] and in Fe–Cr–Ni stainless steels [9], where the transformation is between austenite, γ, and ε-martensite.

12.4 Transformation in Zirconium

Zirconium, titanium and hafnium all undergo a transformation from a high-temperature b.c.c. phase, β, to a h.c.p. structure, α. Lithium has a similar transformation at about 77 K, as does sodium at 35 K [10], although the crystallographic details are more complex in these latter two metals, because of the possibility that transformation to more than one martensitic phase can take place.

The first step in considering these transformations is to compare the b.c.c. and h.c.p. lattices, in order to pick out a small strain that will convert the one into the other. A b.c.c. lattice is shown in Figure 12.4, with a distorted hexagonal cell picked out. The horizontal $(011)_\beta$ plane forms the $(0001)_\alpha$ basal plane of the distorted hexagonal cell, while the close-packed $\langle 1\bar{1}1 \rangle_\beta$ directions in this plane correspond to the close-packed $\langle 11\bar{2}0 \rangle_\alpha$ directions. This accounts for only four of the six $\langle 11\bar{2}0 \rangle_\alpha$ directions in the hexagonal lattice. The other two $\langle 11\bar{2}0 \rangle_\alpha$ directions within the basal plane are each derived from an edge of the b.c.c. cell; that is, from $[100]_\beta$ and $[\bar{1}00]_\beta$ in Figure 12.4.

In the particular case of zirconium, the dimensions of the hexagonal cell can be brought to their correct values by contracting the b.c.c. lattice by 10% along $[100]_\beta$, which becomes $[2\bar{1}\bar{1}0]_\alpha$, expanding by 10% along $[01\bar{1}]_\beta$, which becomes $[01\bar{1}0]_\alpha$, and expanding by 2% along $[011]_\beta$, which becomes $[0001]_\alpha$. This strain we will call the pure lattice strain, S_{ij}. Referred to unit vectors parallel to the orthogonal directions $[100]_\beta$, $[01\bar{1}]_\beta$ and $[011]_\beta$ as x-, y- and z-axes respectively, S_{ij} is found to be the tensor:

$$
S_{ij} = \begin{pmatrix} -0.10 & 0 & 0 \\ 0 & 0.10 & 0 \\ 0 & 0 & 0.02 \end{pmatrix} \tag{12.3}
$$

The only justification for choosing the particular correspondence between the b.c.c. and h.c.p. lattices shown in Figure 12.4 as a basis for a theory of the transformation is that the magnitudes of the strains in **S** given by Eq. (12.3) are either zero or quite small. Any number of other correspondences could be postulated because any cell in the b.c.c. lattice containing two lattice points can be chosen and strained until the points at its corners form one of the possible unit cells of the hexagonal lattice. (Note that, because the b.c.c. structure has one atom per lattice point while the h.c.p. structure has two atoms per lattice point, half of its atoms have to be reshuffled after the homogeneous lattice strain has been applied, in order to complete the structure change.)

In practice, all but a few correspondences are entirely implausible, because of the large strain that they demand, and for the b.c.c. → h.c.p. martensitic transformation it is possible to pick out the most plausible correspondence by inspection. A lattice correspondence can

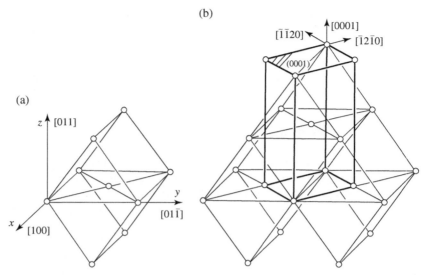

Figure 12.4 (a) Unit cell of the b.c.c. lattice, drawn with (011) in the x-y plane with respect to x-, y- and z-axes parallel to the orthogonal directions $[100]_\beta$, $[01\overline{1}]_\beta$ and $[011]_\beta$, respectively. (b) Distorted hexagonal cell picked out of the b.c.c. lattice, oriented as in (a)

be conveniently represented by a matrix **C**. Suppose $[pqr]$ is a vector in the original lattice referred to the usual crystal axes in the original lattice, and suppose it becomes the vector $[uvw]$ in the transformed lattice referred to the usual crystal axes in the martensite lattice. Then we can write:

$$\begin{bmatrix} u \\ v \\ w \end{bmatrix} = \mathbf{C} \begin{bmatrix} p \\ q \\ r \end{bmatrix} \tag{12.4}$$

where **C** is a 3×3 matrix (see Appendices 1 and 4), known as a correspondence matrix. By substituting for $[pqr]$ the vectors [100], [010] and [001] in turn, it can be seen that the columns of **C** are, respectively, the vectors that these particular vectors become in the transformed lattice. For the correspondence illustrated in Figure 12.4:

$$\mathbf{C} = \begin{pmatrix} 1 & \frac{1}{2} & -\frac{1}{2} \\ 0 & 1 & -1 \\ 0 & \frac{1}{2} & \frac{1}{2} \end{pmatrix} \tag{12.5}$$

where vectors in the hexagonal lattice are written in the three-index system. Thus, for example, the vector $[100]_\beta$ is transformed into the vector $[100]_\alpha$, the vector $\frac{1}{2}[\overline{1}1\overline{1}]_\beta$ is transformed into $[010]_\alpha$, and the vector $\frac{1}{2}[\overline{1}\,\overline{1}1]_\beta$ is transformed into $[\overline{1}\,\overline{1}0]_\alpha$.

Having chosen a lattice correspondence, the next step is to compare the strain that it predicts with the observed strain of a transformed region of crystal. The effect of the pure lattice strain can be illustrated by showing how it would deform a spherical crystal of b.c.c.

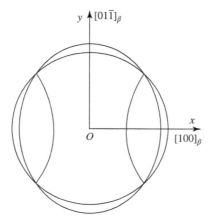

Figure 12.5 A section through a sphere of zirconium and the ellipsoid developed from the sphere by the pure lattice strain. The section is normal to [011]$_\beta$. The curve of intersection is shown in projection on the x–y plane, going through the points common to the circle and ellipse in the x–y plane

zirconium. Any homogeneous strain deforms a sphere into an ellipsoid (Appendix 6). A unit sphere of b.c.c. zirconium oriented as shown in Figure 12.5 and strained in accordance with the lattice correspondence defined by **C** in Eq. (12.5) becomes the ellipsoid:

$$\frac{x^2}{0.90^2} + \frac{y^2}{1.10^2} + \frac{z^2}{1.02^2} = 1 \tag{12.6}$$

All the vectors whose lengths are not changed by the strain are given by the intersection of this ellipsoid with the original sphere; that is, with:

$$x^2 + y^2 + z^2 = 1 \tag{12.7}$$

The curve of intersection is shown in projection on the x–y plane in Figure 12.5. Vectors from the origin to this line define a *cone* of elliptical cross-section and not a plane. However, in the actual transformation it is observed that one plane *does* remain undistorted on the macroscopic scale, namely the habit plane. In the complete theory, the next step is to add formally a slip or twinning deformation, which of course leaves the lattice unchanged, but alters the macroscopic strain, so as to produce a plane that is undistorted on the macroscopic scale.

In the case of zirconium, the 2% strain along [011]$_\beta$ is small enough to make an approximate treatment worthwhile. If this principal strain were zero, the sphere and the ellipsoid would touch at the z-axis, and the unlengthened vectors would lie in a pair of planes, as shown in Figure 12.6. Under these circumstances the two curves of intersection shown in Figure 12.5 would become a simple cross in projection passing through O.

In fact, the necessary and sufficient condition for there to be an undistorted plane is that one of the principal strains should be zero and the other two should be of opposite signs. Because this condition is nearly satisfied for zirconium, the amount of slip or twinning that needs to be added to the pure lattice strain to produce an invariant plane strain is quite small.

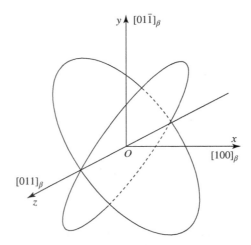

Figure 12.6   Undistorted planes of the strain S′

Reasonable agreement with the observed features of the transformation can be obtained by taking the strain S'_{ij} to be the lattice transformation strain, where:

$$S'_{ij} = \begin{pmatrix} -0.1 & 0 & 0 \\ 0 & 0.1 & 0 \\ 0 & 0 & 0 \end{pmatrix} \qquad (12.8)$$

This strain would be correct if the $[011]_\beta$ lattice vector were exactly equal in length to the c dimension of the hexagonal unit cell. The strains $\varepsilon_x = -0.1$ and $\varepsilon_y = 0.1$ correctly deform the $(011)_\beta$ lattice plane into the basal plane of the hexagonal cell. Since ε_x and ε_y are equal and opposite and small, the strain **S′** is close to being a pure shear (Appendix 6). Now when the correct rotation is added to a pure shear, the net effect is equivalent to a simple shear. The plane on which this simple shear occurs is neither rotated nor distorted, so this plane must be the habit plane.

In the present case, because the magnitude of the (approximate) pure shear is small, the amount of rotation is small and, depending on the sense of the rotation, the plane on which the simple shear occurs is close to one or other of the planes lying at an angle of 45° to $(100)_\beta$ and containing $[011]_\beta$; that is, it is close to the planes whose normals are defined by the unit vectors $\left[\frac{1}{\sqrt{2}}, -\frac{1}{2}, \frac{1}{2}\right]_\beta$ and $\left[\frac{1}{\sqrt{2}}, \frac{1}{2}, -\frac{1}{2}\right]_\beta$.

To be more precise, the rotation must be taken into account. The effect of the strain **S′** on a sphere, viewed along the z-axis, is shown in Figure 12.7; the sphere and ellipsoid here correspond to those in Figure 12.5, without the curve of intersection of the sphere and ellipsoid. Although the planes OQ, OP have not been distorted by the strain, they have been rotated from their initial positions OQ' and OP'. Therefore, to produce an unrotated, as well as undistorted, habit plane, a rotation about the z-axis must be added to the pure lattice strain **S′** in order to return one of these planes to its initial position (OQ to OQ', for example). The *total* lattice strain is then an invariant plane strain that can be identified directly with the observed strain.

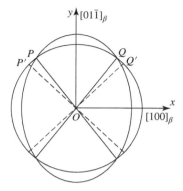

Figure 12.7 Rotation suffered by the undistorted planes of the strain S′

We will choose the plane passing through OQ' as the habit plane (Figure 12.7) and compute the angle between this plane and $[100]_\beta$, which is parallel to the x-axis. Let Q' be the point (x, y). The coordinates of Q are then found by adding the displacement **u** due to the strain **S′**. This is given by the matrix multiplication:

$$\begin{pmatrix} u_x \\ u_y \\ u_z \end{pmatrix} = \mathbf{S'} \begin{pmatrix} x \\ y \\ 0 \end{pmatrix} \tag{12.9}$$

whence:

$$u_x = -0.1x$$
$$u_y = 0.1y \tag{12.10}$$

Therefore, Q is the point $(0.9x, 1.1y)$. The ratio x/y is now given by the condition that $OQ' = OQ$. That is:

$$x^2 + y^2 = (0.9x)^2 + (1.1y)^2$$

Therefore:

$$x/y = 1.05$$

The habit plane therefore makes an angle $\theta = \tan^{-1} 1.05$ ($\approx 46.4°$) with the cube plane of the β phase $(100)_\beta$ and contains the $[011]_\beta$ direction, which becomes the c axis of α, the hexagonal martensite. More specifically, this is the plane whose unit normal is parallel to $[\cos\theta, -\sin\theta/\sqrt{2}, \sin\theta/\sqrt{2}]_\beta$; it is 0.25° away from $(4\bar{3}3)_\beta$. This is not a rational plane of the β lattice.

It is worth noting that, as in this example, even when *no* slip or twinning needs to be added to the lattice strain to produce an invariant plane, the habit plane *still* will not in general be a rational plane. The alternative habit plane, through OP' (Figure 12.7), is crystallographically equivalent to the habit plane through OQ', because the plane normal to the x-axis is a mirror plane; this alternative habit plane is also not a rational plane.

The orientation relationship between the martensite and the matrix can be determined from the lattice correspondence and the *total* lattice strain; that is, the pure strain **S′** followed

(a)

(b)

(c)

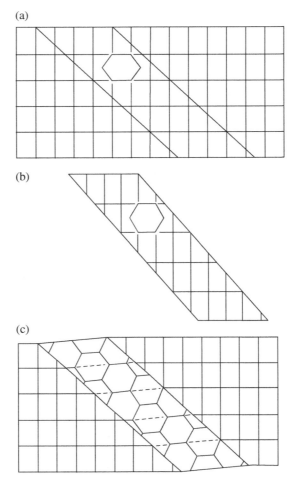

Figure 12.8 Approximate crystallography of a plate of martensite in titanium or zirconium. In (b), the lattice strain is applied to produce the basal plane of the hexagonal structure from the {011} plane of the b.c.c. structure. After a small rotation, the transformed plate fits into the matrix as in (c). In an actual transformation, the pure strain and rotation would occur together, rather than in the two distinct stages shown here

by the relatively small rotation which takes OQ' back to OQ. Since the axis of rotation is normal to $(011)_\beta$, which becomes the basal plane of the martensite, these two planes $(011)_\beta$ and $(0001)_\alpha$ remain exactly parallel to one another. The rotation turns $[100]_\beta$ about 3° from $\langle 2\bar{1}\bar{1}0 \rangle_\alpha$, to which it corresponds. The corresponding close-packed directions, $\langle 111 \rangle_\beta$ and $\langle 2\bar{1}\bar{1}0 \rangle_\alpha$, are about 2.5° apart.

The production of a plate of martensite by the mathematical operation of a lattice strain followed by a subsequent rotation is illustrated schematically in Figure 12.8. In an actual transformation, the mathematical operation of a pure strain followed by a rotation would occur together, rather than in the two distinct operations shown in this figure.

Experimentally, the orientation relationship has been found by using faceted crystals, grown by decomposition of the iodide, ZrI_4 [11, 12]. The external symmetry of these crystals depicts the orientation of the high-temperature phase, β, which cannot be retained at room temperature. Titanium as well as zirconium crystals have been studied in this way [12], and they give essentially the same results. The closest-packed planes $(0001)_\alpha$ and $\{011\}_\beta$ and close-packed directions $\langle 2\bar{1}\bar{1}0 \rangle_\alpha$ and $\langle 111 \rangle_\beta$ are always within a degree or two of being parallel to one another. Groups of parallel striations lie on the surface at room temperature. If presumed to be habit plane traces, these striations show that the habit planes are always nearly normal to the basal plane of the α, as predicted above. Specifically, habit planes close to $\{4\bar{3}3\}_\beta$ have been reported. All these results agree reasonably satisfactorily with the predictions of the approximate theory described above. Further, more detailed, aspects of the crystallography of b.c.c. $\rightarrow$ h.c.p. martensitic transformations in titanium and zirconium alloys, including TEM observations of the martensite, the parent–martensite interface and self-accommodation of martensite variants within a prior b.c.c. grain, can be found in references [13–15]. The principle behind self-accommodation of martensite variants is described in the next section.

12.5   Transformation in Indium–Thallium Alloys

The structure of indium can be described as face-centred tetragonal (f.c.t.) with the axial ratio $c/a = 1.076$ (Section 3.3.1). Alloying with thallium decreases the c/a ratio until at 23 at% Tl the solution becomes c.c.p. The c/a ratio also decreases with temperature, and in the region of 20 at.% Tl a martensitic transformation from c.c.p. (referred to as f.c.c. in this case) to f.c.t. occurs on cooling at $60\,°C$.

This transformation is of interest for two reasons. First, the f.c.t. phase consists of parallel, twin-related lamellae, and it is easy to see how this inhomogeneity produces the macroscopically undistorted habit plane, which could not be produced by the homogeneous lattice strain alone. Second, a single interface between the f.c.c. and the f.c.t. regions can be made to sweep reversibly through a crystal [16]. The theoretical approach of searching for a single macroscopically undistorted, unrotated interface plane can be applied very confidently to this case. A sketch of a partly transformed crystal transforming to a single internally twinned variant of martensite is shown in Figure 12.9.

The f.c.c. and f.c.t. lattices of In–Tl are so nearly alike that the choice of lattice correspondence is obvious. The f.c.t. lattice of a 20.7 at% Tl alloy at the transition temperature is produced from the f.c.c. lattice by expanding one $\langle 001 \rangle$ direction $\frac{2}{3}\%$ and contracting the other two by $\frac{1}{3}\%$. The pure lattice strain referred to the crystal axes of the f.c.c. phase is:

$$S_{ij} = \begin{pmatrix} -\varepsilon & 0 & 0 \\ 0 & 2\varepsilon & 0 \\ 0 & 0 & -\varepsilon \end{pmatrix}, \qquad \varepsilon = 0.0033 \tag{12.11}$$

with the y-axis chosen for the c-axis of the tetragonal lattice. Since the principal strains are so small, a simplified treatment makes it possible to understand the main features of the transformation.

The homogeneous strain $\mathbf{S}$ does not produce an undistorted plane. The intersection of the unit sphere with the ellipsoid which it becomes after applying $\mathbf{S}$ gives a circular cone for

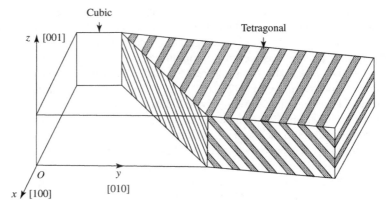

Figure 12.9 Schematic of a partly transformed In–Tl alloy single crystal. The dark and light bands in the tetragonal phase (the martensite) are twin-related lamellae

the locus of unlengthened vectors. However, a macroscopically undistorted plane can exist because the martensite is not in fact homogeneous, but instead consists of lamellae that are twins of one another. The twin relationship is shown in Figure 12.10. It can be produced by transforming two different cube axes into the *c*-axis in the two twin-related regions. If:

$$\mathbf{S}_1 = \begin{pmatrix} -\varepsilon & 0 & 0 \\ 0 & -\varepsilon & 0 \\ 0 & 0 & 2\varepsilon \end{pmatrix} \quad \text{and} \quad \mathbf{S}_2 = \begin{pmatrix} -\varepsilon & 0 & 0 \\ 0 & 2\varepsilon & 0 \\ 0 & 0 & -\varepsilon \end{pmatrix}$$

then $\mathbf{S}_1$ and $\mathbf{S}_2$ produce the two twin-related regions shown in Figure 12.10, apart from a small rotation of one region with respect to the other through the angle $\varphi \approx 3\varepsilon$. Because ε is so small, this rotation can be neglected in the simplified treatment [16]. The macroscopic strain $\overline{\mathbf{S}}$ due to volume fractions x and $1 - x$ of regions 1 and 2 respectively is then:

$$\overline{\mathbf{S}} = x\mathbf{S}_1 + (1 - x)\mathbf{S}_2 \tag{12.12}$$

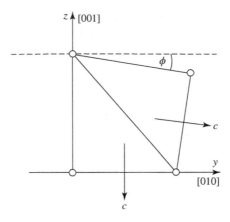

Figure 12.10 The twin relationship of the lamellae shown in Figure 12.9

or:

$$\overline{\mathbf{S}} = \begin{pmatrix} -\varepsilon & 0 & 0 \\ 0 & 2\varepsilon - 3\varepsilon x & 0 \\ 0 & 0 & -\varepsilon + 3\varepsilon x \end{pmatrix} \tag{12.13}$$

By choosing $x = \frac{1}{3}$, $\overline{\mathbf{S}}$ becomes the strain:

$$\overline{\mathbf{S}} = \begin{pmatrix} -\varepsilon & 0 & 0 \\ 0 & \varepsilon & 0 \\ 0 & 0 & 0 \end{pmatrix} \tag{12.14}$$

One of the macroscopic principal strains is now zero, while the other two are opposite in sign, satisfying the condition for there to be an undistorted plane. A similar solution is obtained for $x = 1 - \frac{1}{3} = \frac{2}{3}$. In fact, because ε is so small, the macroscopic strain $\overline{\mathbf{S}}$ defined by Eq. (12.14) is a very close approximation to a pure shear strain, which after adding a small rotation ε is equivalent to a simple shear on a plane at very nearly 45° to the axes Ox and Oy (Appendix 6). The habit plane will be parallel to the plane on which this simple shear occurs; in other words, very close to either (110) or $(1\overline{1}0)$ of the f.c.c. phase. The $(1\overline{1}0)$ habit plane is the one shown in Figure 12.9.

A feature that has important consequences is the fact that a given habit plane may be the invariant plane of either of two opposite macroscopic strains. Thus the plane $(1\overline{1}0)$, as well as serving as the habit plane for a martensite plate internally twinned on the $(011)_{\text{f.c.t.}}$ plane whose macroscopic strain is $\overline{\mathbf{S}}$ as described in Eq. (12.14), may also serve as the habit plane for a plate that is internally twinned on the $(101)_{\text{f.c.t.}}$ plane and therefore has the macroscopic strain

$$\overline{\mathbf{S}} = \begin{pmatrix} \varepsilon & 0 & 0 \\ 0 & -\varepsilon & 0 \\ 0 & 0 & 0 \end{pmatrix}$$

Obviously, by stacking these parallel plates having opposite strains alternately, it is possible to build up a volume of martensite within which the average strain is zero (Figure 12.11). Experimentally, it is observed that the transformation often occurs by the growth of such a stack of martensite plates into the matrix, the plates having alternating types of internal

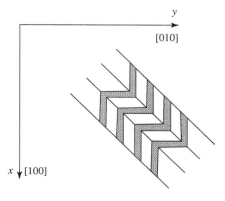

Figure 12.11 Three parallel plates of martensite with alternating shear strains

twinning, as shown in Figure 12.11 [17]. This is a relatively easy to appreciate example of the *self-accommodation* of martensite variants, already referred to in Section 12.4.

Although the theory presented above for In–Tl is inexact, it does show clearly how a shear which preserves the lattice, in this case a twinning shear, can produce a macroscopically undistorted plane to match with the untransformed matrix. Furthermore, this inexact theory produces predictions very close to those from a more precise calculation: since ε is so small, the habit plane as determined by a precise calculation is only 0.5° from {110}, and the tetragonal axes of the f.c.t. martensite are within 2° of being parallel to the cube axes of the f.c.c. parent phase [16].

The absolute thicknesses of the twin lamellae are not specified by the theory. The thinner the twins, the finer the scale in which matching is achieved at the interface. For twins of average thickness h, elastic strains of order ε will spread from the interface to a distance of order h, giving an interfacial energy proportional to $\varepsilon^2 h$. The tendency to reduce interfacial energy by making h small is counterbalanced by the extra volume energy of the martensite due to the twin boundary energy γ, which is of order γ/h per unit volume. The total energy will be a minimum at some finite spacing.

An applied stress has an interesting effect on this transformation, and indeed martensitic transformations in general. For example, it can be seen that a tensile stress along the y-axis will favour a transformation of the particular form shown in Figure 12.9, because this transformation increases the length of the crystal in the Oy direction. The M_s temperature is increased by such an applied tensile stress, so that in a tensile test carried out just above the normal M_s, deformation by transformation occurs at a critical stress. This is the principle behind *superelasticity*, sometimes termed *pseudoelasticity* or *pseudoplasticity*, in which it is possible to accommodate relatively large strains (up to 8%) in suitable shape-memory materials by transforming from austenite to martensite [1]. Superelasticity is reversible; when the stress is removed, the material reverts to austenite. Further crystallographic aspects of superelasticity will be discussed in Section 12.12.3.

The martensitic phase transformation in indium–thallium has a further interesting effect, that of *rubber-like behaviour* [16, 18], in which after being aged for some time in the fully martensitic state, it can be deformed like soft rubber. Thus, for example, after bending, an initially straight rod is able to straighten itself immediately when the bending stress is released [16]. This unusual deformation behaviour is currently interpreted in terms of an atomic rearrangement within the same sub-lattice of the imperfectly ordered alloy during martensite ageing [18].

12.6 Transformations in Steels

Until the recent technological interest in shape-memory materials, almost all of which are non-ferrous alloys and none of which are steels, research on martensitic phenomena was stimulated by the great practical importance of martensite in steels. Martensitic transformations in steel have themselves continued to provide many problems, due to their complexity and their great variety in steels of different compositions.

The one feature that is basic to the theory of all forms of tetragonal martensite in steels is the lattice correspondence, which was proposed by Bain in 1924 [19]. This is depicted in Figure 12.12, in which a b.c.t. lattice is picked out of a c.c.p. lattice.

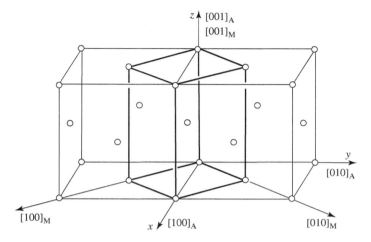

Figure 12.12 The c.c.p. lattice with a b.c.t. cell picked out of it. (After Bain [19])

The correspondence matrix **C** for lattice vectors is given by:

$$\begin{bmatrix} U \\ V \\ W \end{bmatrix}_M = \mathbf{C} \begin{bmatrix} u \\ v \\ w \end{bmatrix}_A \quad \text{where} \quad \mathbf{C} = \begin{pmatrix} 1 & -1 & 0 \\ 1 & 1 & 0 \\ 0 & 0 & 1 \end{pmatrix} \tag{12.15}$$

for the choice of the c-axis shown in Figure 12.12, where subscripts M and A denote martensite (b.c.t.) and austenite (c.c.p.), respectively. When indices for planes are written as column vectors, the correspondence matrix for planes is the transpose of the inverse of **C** (Appendix 4):

$$\begin{pmatrix} H \\ K \\ L \end{pmatrix}_M = \frac{1}{2} \begin{pmatrix} 1 & -1 & 0 \\ 1 & 1 & 0 \\ 0 & 0 & 2 \end{pmatrix} \begin{pmatrix} h \\ k \\ l \end{pmatrix}_A \tag{12.16}$$

This can be confirmed by substituting in turn $(100)_A$, $(010)_A$ and $(001)_A$ for $(hkl)_A$ and inspecting Figure 12.12 to see which martensite planes $(HKL)_M$ these become.

The tetragonality of the b.c.t. lattice as it stands in Figure 12.12 is much greater than that of martensite. To achieve the correct lattice parameters, the [001] direction must be contracted by about 20% and the (001) plane expanded uniformly by about 12%. This pure lattice strain, referred to either A- or M-axes, can be written as:

$$B_{ij} = \begin{pmatrix} \eta_1 & 0 & 0 \\ 0 & \eta_1 & 0 \\ 0 & 0 & \eta_3 \end{pmatrix} \tag{12.17}$$

If the lattice parameter of austenite is a_0 and the lattice parameters of martensite are c and a, then referred to A-axes:

$$\eta_1 = \frac{\sqrt{2}a}{a_0} - 1 \sim 0.12$$

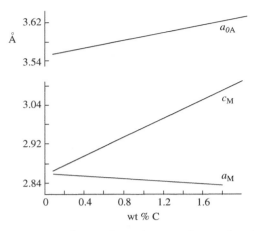

Figure 12.13 Lattice parameters of austenite and martensite as a function of the carbon content of the steel. Source: From [20]; reproduced by permission of the American Institute of Mining, Metallurgical and Petroleum Engineers.

and $\eta_3 = c/a_0-1 \sim -0.20$. The exact values of these principal strains depend upon the carbon content of the steel, which affects the tetragonality of the martensite, as shown in Figure 12.13.

An increase of c/a with carbon content is just what would be expected from the Bain correspondence and the fact that C atoms in austenite occupy octahedral interstices such as $(0, 0, \frac{1}{2})_A$ (Section 3.2.1). If C atoms are trapped in these sites during the transformation, they will oppose the contraction along $[001]_A$. Considering for a moment the Bain correspondence between the c.c.p. lattice and the b.c.c. lattice ($c/a = 1$), it is important to realize that *all* the octahedral interstices in the c.c.p. lattice correspond to *just one of the three* differently oriented squashed octahedral interstices in the b.c.c. lattice; that is, one of the three sites F, A and B in Figure 10.14. Referring to Figure 10.14, if $[001]_M$ corresponds to $[001]_A$ then the c.c.p. octahedral interstices opposing contraction correspond to F-type sites; if $[001]_M$ corresponds to $[010]_A$, the c.c.p. octahedral interstices opposing contraction correspond to B-type sites; and if $[001]_M$ corresponds to $[100]_A$, the c.c.p. octahedral interstices opposing contraction correspond to A-type sites. As a result, the C atoms that are distributed at random in the austenite expand the transformed lattice in one particular direction and produce its tetragonality.

The strain **B** will change a unit sphere into the uniaxial ellipsoid:

$$\frac{x^2}{(1 + \eta_1)^2} + \frac{y^2}{(1 + \eta_1)^2} + \frac{z^2}{(1 + \eta_3)^2} = 1 \qquad (12.18)$$

All the vectors whose length has not been changed by **B** lie on the circular cone defined by the intersection of this ellipsoid with the unit sphere. Since the change in shape of a transforming region is believed to leave one plane macroscopically undistorted and unrotated, to match with the matrix, the problem is to find an additional deformation that will produce such a plane, without changing the lattice. All the principal strains of **B** are quite large, so the approximate arguments used in Sections 12.3–12.5 cannot be applied. The first successful

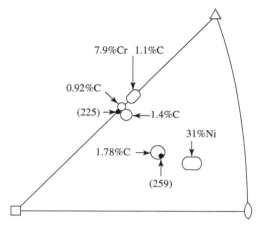

Figure 12.14   Habit plane normals of martensite in various steels plotted on a standard stereographic triangle for the c.c.p. austenite

detailed phenomenological theories for steels were produced in 1953 and were published by Wechsler, Lieberman and Read in 1953 [21] and by Bowles and Mackenzie in 1954 [22]. Although the principles of their theories are quite straightforward, the mathematical details are rather complex, and we shall not reproduce them here.

Experimentally determined habit plane normals for various steels are given in Figure 12.14, plotted as regions rather than points, because the habit plane in a particular steel really does vary a little. The various orientation relationships have in common that one set of closest-packed planes, $\{111\}_A$ and $\{110\}_M$, is roughly parallel. Various directions in these planes may be parallel; in general, the relation is somewhere between $\langle 1\bar{1}0 \rangle_A \parallel \langle 1\bar{1}1 \rangle_M$ (the Kurdjumow–Sachs[1] orientation relationship [23]) and $\langle 2\bar{1}\bar{1} \rangle_A \parallel \langle 1\bar{1}0 \rangle_M$ (the Nishiyama or Nishiyama–Wassermann orientation relationship [24, 25]).

Martensites with a habit near $\{259\}_A$ are well understood. They occur in Fe–Ni–C and Fe–Ni alloys and in Fe–C alloys of relatively high C content. A tolerable prediction of their habit and orientation relationship can be derived from the postulate that the lattice invariant shear strain which is added to the Bain strain is in the $\langle 111 \rangle_M$ direction on a $\{121\}_M$ plane that is derived from a $\{110\}_A$ plane. This shear corresponds to twinning in a b.c.c. crystal, and it was proposed by Wechsler, Lieberman and Read [21] and by Bowles and Mackenzie [22] that the martensite plate consists of a stack of twin-related laths. Under the optical microscope there is little sign that this is the case, but when such steels are examined with the transmission electron microscope, arrays of very thin $\{112\}_M$ twins are indeed found. The confirmation of this by Kelly and Nutting in the early 1960s [26] completed a triumph for the formal crystallographic theory.

However, it is evident from Figure 12.14 that many habit planes in steels are not explicable in terms of a $\{259\}_A$ habit, most notably those with habit planes close to $\{225\}_A$. Bowles and MacKenzie [22] attempted to explain the occurrence of such habit planes through the introduction of a dilatational parameter, δ, allowed to be within a few per cent (typically

---

[1] This spelling of Kurdjumov is that used on the original 1930 paper [23] when Kurdjumov was working in Berlin. Elsewhere, depending on the source, his surname is translated from the Russian either as Kurdjumov or Kurdyumov.

<2%) of unity. Later work showed that δ is not significantly different from unity for the $\{225\}_A$ habit plane steels; that is, for this martensitic transformation the shape strain is an invariant plane strain, rather than one with an additional pure dilatation [27].

Others have attempted to explain the $\{225\}$ habit by advocating multiple lattice invariant shears, rather than one [28, 29]; while these too were initially shown not to be able to account for the experimental data [30], later work examining many more double-shear combinations led to a number of possible alternative combinations to explain the experimental data [31]. Further consideration of the martensitic transformations in steels has involved assessing the role of strain energy [32] and the introduction of a topological model based on interfacial defects [33] (Section 12.13.1), both of which require an advanced understanding of the theory of phase transformations in solids.

12.7   Transformations in Copper Alloys

Several binary and ternary copper alloys, Cu–Al, Cu–Ga, Cu–Sn, Cu–Zn, Cu–Al–Ni, Cu–Zn–Al and Cu–Zn–Ga, behave very similarly within a certain composition range. In each case there exists a high-temperature phase, β, which is b.c.c. If slowly cooled, the b.c.c. phase decomposes into two phases; if cooled rather quickly, it changes structure without changing composition, but also without changing its shape; if cooled very quickly, it undergoes a martensitic transformation. The martensite is a close-packed structure, basically either c.c.p. or, at the low-Cu end of the composition range, h.c.p. The reasons for calling the martensite structures only basically c.c.p. and h.c.p. are rather interesting. The first point is that the alloys usually order before transforming, so that the corresponding pattern of order is built into the martensite (β-brass, Cu–Zn, differs from the remaining alloys in having the CsCl ($B2$) ordered structure instead of the Fe_3Al ($D0_3$) structure; see Section 3.8 for the nomenclature). A second complication to the basically c.c.p. structure is that the plates of martensite are heavily faulted. The stacking faults can be seen under the electron microscope and their effect on diffraction patterns shows that there is one fault on about every third close-packed plane, each produced by the same $\frac{1}{6}\langle 211 \rangle$ displacement [34].

The assertion that the displacements producing the stacking faults supply an undistorted, unrotated plane which forms the face of the martensite plate can be checked by the formal crystallographic theory. The lattice correspondence is the inverse of the Bain correspondence, and a shear on $\{111\}_M$ in the direction $\langle 211 \rangle_M$ is chosen for the complementary strain. Predictions from the phenomenological theory are in good agreement with experiment for Cu-based martensites and lead to the correct habit planes in the range $\{133\}_\beta$–$\{155\}_\beta$, depending on the martensite under consideration [35]. Groups of four martensite variants form from prior β phase in a distinct diamond-shaped morphology in these alloys [35, 36], in such a way that the average shape-deformation matrix arising from each grouping becomes nearly zero; that is, the groups are self-accommodating (Section 12.5).

12.8   Transformations in Ni–Ti-Based Alloys

Other than martensite in steels, Ni–Ti-based alloys are the other major group of alloys in which there has been extensive research into their physical metallurgy as a consequence of its martensitic transformation [37]. This is because the magnitudes of the shape-memory effect and the superelasticity effect in these alloys make them very attractive for a number of applications. In these alloys, the high-temperature equiatomic NiTi phase has the CsCl ordered $B2$ structure (Section 3.8). Depending on the composition of the alloys under consideration, the first martensite to form can either take a monoclinic crystal structure (as in equiatomic NiTi), an orthorhombic crystal structure (as in Ni–Ti–Cu alloys) or a rhombohedral crystal structure (as in Ti–Ni–Fe alloys or aged Ni-rich NiTi alloys). These latter two martensites both have a subsequent tendency to transform to the monoclinic martensite [37].

A number of twinning modes are possible in the monoclinic martensite, designated $B19'$, to provide the necessary lattice invariant shear. The most prevalent twinning mode is a type II twinning mode (Table 11.1), in which K_1 is irrational, but where $\eta_1 = [011]$ in deformation twinning nomenclature [37, 38]. As with the Cu-based alloys, the phenomenological theory of martensitic transformations is able to account for the experimental observations of the crystallography of both the $B2 \rightarrow B19'$ martensitic transformation and the $B2 \rightarrow R \rightarrow B19'$ martensitic transformations [37–39]. Self-accommodation of martensite variants, central to the mechanical properties of these alloys, is on a finer scale in equiatomic and near-equiatomic Ni–Ti alloys than in Cu-based alloys and is more complicated to analyse [37, 40–43].

12.9 Magnetic Shape Memory Alloys

Just as stress can be used to induce reversible elongation in suitable ferromagnetic materials such as Ni–Mn–Ga alloys in their martensitic state, giving rise to shape memory, the crystallography of which is discussed in more detail in Section 12.12, a magnetic field can also induce reversible elongation in these alloys, making such alloys attractive for potential sensor and actuator applications. Ni–Mn–Ga alloys are one of a number of alloys where the high temperature phase has the $L2_1$[2] Heusler alloy crystal structure typified by Cu_2MnAl [44] and named after the discoverer of these ternary alloys, Fritz Heusler (1866–1947). This crystal structure has a cubic F unit cell with 16 atoms arranged on specific sites within the unit cell consistent with the $Fm\bar{3}m$ space group.

On cooling these alloys, a number of possible martensite crystal structures can arise [45]. The most simple of these crystal structures is a non-modulated $L1_0$[2] tetragonal martensite crystal structure with eight atoms per unit cell. The other possible martensite crystal structures in Ni–Mn–Ga alloys can be described as layered structures of a suitable stacking sequence of {110} planes of the high temperature $L2_1$ phase. The crystallography of the martensitic transformation to the $L1_0$ tetragonal martensite and the crystallography of the self-accommodating martensite microstructure in these magnetic shape memory alloys have been the subject of a number of recent studies [46–49].

12.10 Transformations in Non-metals

Martensitic transformations in non-metals are dominated in importance by the martensitic transformation in zirconia, ZrO_2, which provides the basis of *transformation toughening*

---

[2] This notation is a *Strukturbericht* notation (Table 3.5).

[50]. In pure zirconia, a martensitic transformation occurs on cooling from a tetragonal phase above 1100 °C to a monoclinic phase below this temperature. The relatively large positive volume change of 4% which accompanies this phase change on cooling disrupts the zirconia body in which it occurs and thereby limits the usefulness of this otherwise excellent refractory material (melting point of 2680 °C) in its pure state. However, when ZrO_2 is alloyed with other oxides such as CeO_2, MgO and Y_2O_3, the transformation from the tetragonal state to the martensitic state can be prevented if the grain size of the tetragonal phase is sufficiently small, whether it be as separate grains or as precipitates within a cubic matrix [50], so that the tetragonal phase is stabilized down to room temperature. Transformation to the martensitic state can be achieved either by applying a suitable stress (Section 12.5) or by further cooling to sub-zero temperatures. The size of the tetragonal grains is important for transformation toughening: if they are too small, the martensitic transformation cannot be triggered by stress, and if they are too large, they will have already transformed to the martensitic state on cooling. Excellent agreement is found with the predictions of the phenomenological theory of martensitic transformations when applied to zirconia [50]. Other martensitic transformations occur in ceramics, such as the $\alpha' \to \beta$ transformation in dicalcium silicate [51], but have not been shown to be technologically as useful as the one in ZrO_2.

A further class of technically important oxides in which diffusionless phase transformations occur is the ferroelectrics, such as barium titanate, $BaTiO_3$. Above 120 °C $BaTiO_3$ is cubic with the $Pm\bar{3}m$ perovskite crystal structure (Figure 3.22, Section 3.6.1) and is not ferroelectric. On cooling through 120 °C, barium titanate undergoes a phase transition to a tetragonal ferroelectric phase with $P4mm$ space group. Further phase transitions occur at lower temperatures (Section 3.6.1). The tetragonal phase is ferroelectric, with a spontaneous electric polarization in the direction of the c-axis. In this ferroelectric condition, a crystal of $BaTiO_3$ has a domain structure which is reminiscent of the structure of the tetragonal In–Tl phase (Figure 12.9); twinning occurs on $\{101\}$ planes of the tetragonal crystal structure. Stacks of twin lamellae occur in which an individual lamella constitutes a ferroelectric domain. The direction of polarization follows the c-axis and therefore rotates through approximately 90° on passing through a twin boundary.

Mineralogists use the apt word *displacive* to describe changes in structure that do not involve diffusion. Thus, in $BaTiO_3$, the various phase transitions do not involve the interchange of ions, but instead minor displacements ($\ll 1$ Å) of the ions. The volume changes associated with these phase transitions are an order of magnitude less than for martensitic transformations [52], and each transition from one phase to another happens at a well-defined temperature. If, on the other hand, a new structure can only be obtained through breaking and remaking of bonds, such as in the phase transition in ZnS from sphalerite to wurtzite (Section 3.5), the transformation is called *reconstructive*. While displacive transformations have many characteristics in common with martensitic transformations, it is also evident that some reconstructive transformations can also share such characteristics, particularly those which are pressure-induced [53, 54]. It is often then a question of semantics as to whether or not such transformations are regarded as being martensitic.

Martensitic transformations are also able to occur in crystalline polymers. Probably the most well-known example is the orthorhombic $\to$ monoclinic transformation which occurs under stress in oriented high-density polyethylene [55, 56]. A martensitic transformation also occurs under stress between two of the four crystalline forms of syndiotactic polypropylene [57].

12.11 Crystallographic Aspects of Nucleation and Growth

The evolution of the martensitic microstructure from the parent austenite involves nucleation and growth of martensite plates. The process of nucleation of martensite was considered in some detail by Olson and Cohen in a series of papers [58–60]. Olson and Cohen hypothesized that faulting on planes of closest packing in the parent austenite is the first step in the process of producing a martensite embryo. Subsequent steps take place in such a way that they are able to leave the plane of faulting unrotated. As the nucleus grows, its change of shape has to be accommodated by the matrix, and it may be imagined that at some stage the constraint of the matrix forces the nucleus to undergo the heterogeneous deformation by slip, faulting or twinning, which allows the new phase to grow as a plate whose faces match the matrix on a macroscopic scale.

The simplest example to examine in terms of a model of faulting is the c.c.p. → h.c.p. martensitic transformation (Section 12.3). The passing of each partial dislocation on a (111) c.c.p. plane produces two planes of the h.c.p. structure. Olson and Cohen calculated that for spontaneous formation of an h.c.p. embryo, four or five dislocations must be present together in a region of the c.c.p. material [58]. Examples of such a region are grain boundaries and interfaces between the matrix and particles of inclusions (Chapters 13 and 14), with the arrangement of dislocations sufficiently likely to be realistic, but sufficiently rare to make nucleation an uncommon event, in line with the known relatively small density of initial nucleation sites [58].

Similar considerations apply to other transformations, such as the b.c.c. → c.c.p or b.c.c. → h.c.p. martensitic transformations [59]. For example, the transformation of a b.c.c. metal to either a c.c.p. or an h.c.p. structure could conceivably be nucleated by a fault on the {011} plane. Two (011) layers of a b.c.c. crystal are shown in Figure 12.15. If the atoms in the upper layer are shifted by $\frac{1}{8}[0\bar{1}1]$ to positions above the triangular gaps in the layer beneath, the stacking of the two layers becomes more like that of successive close-packed planes of the c.c.p. or h.c.p. structure. It is conceivable that a dislocation with the usual $\frac{1}{2}[1\bar{1}1]$ Burgers vector could provide such a displacement by dissociating on a

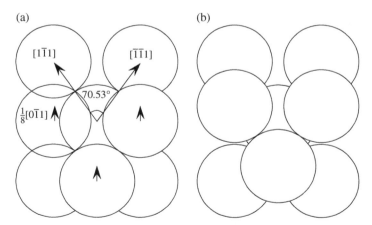

Figure 12.15 The (011) plane of a b.c.c. metal (a) before and (b) after a displacement of the atoms in the upper layer by $\frac{1}{8}[0\bar{1}1]$

(011) plane, according to an equation equivalent to Eq. (9.56):

$$\tfrac{1}{2}[1\bar{1}1] \rightarrow \tfrac{1}{8}[0\bar{1}1] + \tfrac{1}{4}[2\bar{1}1] + \tfrac{1}{8}[0\bar{1}1]$$

Another line of argument leads to the idea that $\{011\}\langle0\bar{1}1\rangle$ shear has a particular significance for the nucleation of a close-packed structure from a b.c.c. structure. Referring to Figure 12.15, it can be seen that a b.c.c. structure made from rigid balls would have no resistance to $\{011\}\langle0\bar{1}1\rangle$ shear. The balls in the upper (011) layer are perched up on saddle points and are unstable because the slightest push in a $\langle0\bar{1}1\rangle$ direction will topple them into the positions shown in Figure 12.15b. The b.c.c. structure is a mechanically unstable one for hard spheres. The stiffness of a real b.c.c. crystal with respect to $\{011\}\langle0\bar{1}1\rangle$ shear is given by the elastic modulus $(c_{11} - c_{12})/2$, and the ratio of $(c_{11} - c_{12})/2$ to c_{44} is a measure of its stability with respect to a close-packed structure. While such considerations enable the critical free energy for embryo nucleation to be lowered, Olson and Cohen concluded that they are not necessary for the nucleation mechanisms they proposed for embryo nucleation [60].

Once nucleated, martensite usually develops as a plate. The edgewise growth of the plate is often very rapid; the velocity may be an appreciable fraction of the velocity of sound. In general, the faces of the plate match the untransformed matrix only on a macroscopic scale. The local distortions can be represented by a set of dislocations in the interface whose long-range stress field vanishes. As the plate thickens, the glide of these dislocations produces the slip, faulting or twinning, which, when added to the lattice distortion, produces an average matching at the interface. Of course, this set of dislocations does not completely solve the problem of accommodating a finite plate in untransformed material. There is a shear on the interface plane which must be accommodated at the edges of the plate. The constraint of the matrix causes the edges of the plate to be tapered, just as in the case of a twin lamella (Figure 11.14). In addition, there is in general an expansion or contraction normal to the faces of the plate which must also be accommodated.

If these shear and tensile strains are small, they may be accommodated by the elastic deformation of the plate and matrix. It is then possible for a plate to be in equilibrium, so that a small drop in temperature will allow the plate to grow by supplying enough chemical free energy to offset the energy of its elastic strain field, whereas a small rise in temperature will allow the elastic strain field to shrink the plate. This situation is analogous to elastic twinning; martensite of this type is called *thermoelastic*, and such martensites will exhibit the shape-memory effect in which an alloy deformed in its fully martensitic state is able to recover its shape prior to deformation by heating it up to the austenite state (Section 12.12).

In steels, the strains are too large to be accommodated elastically and plastic deformation occurs. Since plastic deformation dissipates energy instead of storing it, the growth of a plate is then no longer reversible; consequently, steels do not show shape memory. In addition, the damage that plastic deformation does to the lattice of the matrix may inhibit the further transformation of a partially transformed crystal.

Another way of partly accommodating the strain in one plate is to nucleate another plate of a different strain nearby. This effect leads to the characteristic zigzag arrangement of martensite plates in high-C steels. If one plate nucleates more than one other, a chain reaction develops, which may transform a good part of the whole crystal in one fantastic burst. In the case of alloys in which self-accommodation of martensite variants is readily observed, a group of two or more plates forms a volume with almost no overall strain. Such a volume can have quite arbitrary boundaries, of course.

Advances in computational power have now enabled three-dimensional simulation of the nucleation and growth of martensite from the parent austenite to form internally twinned plates [61, 62]. In such simulations the free energy of the system is modelled using long-range order (lro) parameters, taking into account the strain energy contribution to the total energy of the system. Thus, for example, for a cubic → tetragonal martensitic transformation there are three lros, corresponding to the three possible orientations of the *c*-axis in the tetragonal structure with respect to the cubic structure [61, 62]. The martensitic structure can then be seen to evolve as a function of the three martensite orientation variants. Such simulations are useful in clarifying the ways in which two major factors interact with one another, thereby influencing the way in which martensitic transformations take place. Assuming that it is energetically favourable in terms of total free energy to transform parent crystal to martensite, the first major factor is a tendency to transform by means of the lattice strain, which most simply converts a small unit of the parent lattice into a unit of the martensite lattice. The second major factor is the constraint that is imposed upon a finite transforming region by its surroundings.

12.12 The Shape Memory Effect and Superelasticity

As has already been noted in Sections 12.1, 12.8, 12.9 and 12.11, some, but not all, materials deformed permanently in the martensitic shape are able to 'remember' their physical shape by simply heating them up to the austenitic shape. This is known as the shape memory effect [1]. There are two possible shape memory effects – the one-way shape memory effect and the two-way shape memory effect [1, 37].

12.12.1 The One-Way Shape Memory Effect

In the one-way shape memory effect, a sample annealed in its high temperature austenite phase is cooled through the martensitic phase transformation to a temperature such as room temperature where it is fully martensitic (Figure 12.16a). During the cooling it is also noteworthy that the martensitic phase transformation occurs in a self-accommodating manner, so that any slight change of shape of the cylinder between the high temperature state and the martensitic state is not evident to the naked eye, i.e. it is in essence unchanged macroscopically.

This sample, which in the schematic in Figure 12.16a is a cylinder, is then deformed as in Figure 12.16b. Under these conditions of externally applied stress, i.e. in going from Figure 12.16a to Figure 12.16b, deformation proceeds in the martensite by twin boundary motion within a given group of self-accommodating plates and by growth of more favourably oriented 'habit plane variants' at the expense of less favourable oriented 'habit plane variants' for the particular applied stress, again by twin boundary motion. [1, 37]. In this way, the sample is able to accommodate the stress imposed on it by the ready movement of variant/variant boundaries. The beauty of this process is that it is able to generate relatively large strains of the order of 4–8% without causing permanent plastic deformation of the martensite crystal structure, i.e. without slip taking place in the martensite as a deformation mode. As Saburi et al. were able to demonstrate for single crystals of the austenite phase

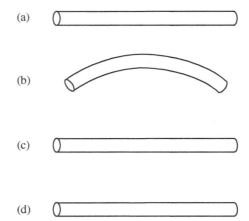

Figure 12.16 The one-way shape memory effect. (a) a sample annealed in its austenitic state is cooled to a temperature where it has fully transformed to the martensitic state; (b) the sample is deformed elastically without inducing permanent plastic strain within the martensite; (c) on heating so that the deformed martensitic state reverts to the austenitic state, the original shape it had in (a) in the austenitic state is recovered; (d) on cooling once again to a temperature where it has fully transformed to the martensitic state, the sample appears unchanged macroscopically

of Cu–Zn–Ga and Cu–Zn–Al alloys [36], when suitably deformed in the martensite state, the 24 orientations of the martensite occurring in six distinct self-accommodating groups are able to convert into a single orientation of martensite of the variant whose shape strain is most favourable with respect to the applied stress.

Removal of this stress at constant ambient temperature will not cause any change in shape of the sample. However, when heated to above its martensitic start temperature, M_s, as in Figure 12.16c, the sample transforms back to the parent phase in the original orientation, because of the lattice correspondence between the martensite and the parent phase [1, 37]. It is this critical aspect of the martensitic transformation which guarantees its reversibility. On cooling once again to a temperature where it is fully martensitic the sample appears unchanged macroscopically, as in Figure 12.16d.

12.12.2 The Two-Way Shape Memory Effect

In the two-way shape memory effect, a sample is 'trained' to remember one shape at high temperature in the fully austenitic phase and another at low temperature in the fully martensitic phase [1, 37]. A schematic of the two-way shape memory effect is shown in Figure 12.17. As with the one-way shape memory effect, a convenient possible starting shape in the martensite state is a cylinder (Figure 12.17a) which has been cooled from a temperature where it is fully austenitic.

One way of achieving the two-way shape memory effect, albeit not the only way, is to deform severely a sample in the fully martensitic phase. This causes irreversible plastic deformation in the martensite, as shown in Figure 12.17b, from the starting configuration in Figure 12.17a. On heating the sample into the high temperature austenitic state, it goes some way towards the cylinder shape, but falls short, as in Figure 12.17c. Cooling the sample once

(a)

(b)

(c)

(d)

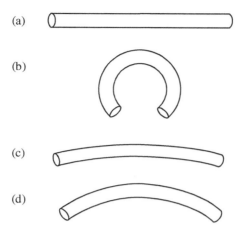

Figure 12.17 The two-way shape memory effect. (a) a sample annealed in its austenitic state is cooled to a temperature where it has fully transformed to the martensitic state; (b) the sample is severely deformed to induce permanent plastic strain within the martensite; (c) on heating so that the deformed martensitic state reverts to the austenitic state, the original shape it had in (a) in the austenitic state is partially, but not fully, recovered; (d) on cooling once again to a temperature where it has fully transformed to the martensitic state, the sample deforms to a new shape in between the shapes shown in (b) and (c). Continued heating and cooling will then cycle the material between the shapes shown in (c) and (d)

again to the martensitic phase causes a change of shape to Figure 12.17d. Continued heating and cooling will then cycle the material between the shapes shown in Figure 12.17c and Figure 12.17d, up to a point – the material can lose the ability to show the two-way shape memory effect because of a gradual change in microstructure as a function of cycling, the so-called *functional fatigue*.

Consideration of the mechanism behind the two-way shape memory effect highlights the need for the material to be able to select repeatedly particular martensite variants on cooling [1, 37, 63]: the crystallography of the two-way shape memory effect is therefore very clearly defined. Internal stress sources within the material such as specific configurations of dislocations or precipitates are required in order to achieve this. It is also evident that slight changes in either the configurations of dislocations or the size of the precipitates as a consequence of thermal cycling, which is in essence a series of mini heat-treatments, can change which variants are selected, and therefore subtly change the shape of the material when it is in the martensitic shape, for example.

12.12.3 Superelasticity

Superelasticity is closely related to shape memory. In this effect, the application of a suitable stress to a fully austenitic state of a material induces the formation of martensite. Further increases of stress can transform the whole material to martensite. As discussed in Section 12.5, the application of a tensile stress will favour those martensite morphologies which will help to elongate the material along the direction of the applied tensile stress; the converse will be true if the applied stress is compressive. On release of such stresses, the material reverts to its initial shape and returns to being fully austenitic. Macroscopic strains similar to those produced for the one-way shape memory effect can be produced before stresses sufficient to cause permanent plastic deformation are required.

A little thought about this effect highlights the fact that, in the austenite phase, there will be a competition between stress-induced martensite and plastic deformation of the austenite by slip [1, 37]. Stress-induced martensite will only be induced in a particular alloy at a particular temperature if the critical stress required for slip of the austenite phase is above that required for martensite formation. Thus, for example, while superelasticity is readily observed in Cu–Al–Ni single crystals, so that these can transform under a suitable applied stress from austenite to single crystals of a particular martensite variant, superelasticity is only usually observed in aged Ni-rich Ni–Ti alloys because the critical stress required for slip in the austenite has been increased in these alloys by precipitation hardening (Section 9.1) [1, 37].

12.13 Modern Theories of Martensitic Transformations

While the phenomenological theory of martensitic transformations has been very successful in predicting the crystallography of these phase transformations, the limitations of this theory have been already been recognized in Section 12.2. In particular, the theory is not mechanistic, in that it does not attempt to predict the paths of atoms during the transformation, and it is also a continuum theory, in the sense that the habit plane that is predicted is an unrelaxed interface which is an invariant plane of the geometrical shape transformation [64]. This recognition of the limitations of the phenomenological theory has led to other models of martensitic transformations being developed, the two most notable being the topological model and the non-linear elasticity model. We shall conclude this Chapter by describing briefly the merits of these two models and highlighting in particular the crystallographic aspects of these models.

12.13.1 The Topological Model

In this model of martensitic transformations, for which, as has already been noted in Section 12.6, it is necessary to have an advanced understanding of the theory of phase transformations in solids, emphasis is placed on the glissile nature of the parent–martensite interface and its structural description in terms of coherent terraces and defects termed disconnections by Pond et al. [33, 64]. These disconnections, or transformation dislocations, exhibit both a dislocation character defined by a suitable Burgers vector $\mathbf{b}$ and a step character defined by a step height h. In general, there are two sets of defects in the parent–martensite interface: one set of disconnections with $(\mathbf{b}_D, h)$ causing the equivalent of the lattice variant shear in the phenomenological theory, and a second defect of the form $(\mathbf{b}_L, 0)$ without a step character causing the equivalent of the lattice invariant shear or lattice invariant deformation (LID) in the phenomenological theory [33]. Hence, in this topological model, these two sets of defects, both further defined by their orientation and spacing within the habit plane, accommodate the strains associated with the martensitic transformation. Lateral motion of the disconnections in the parent–martensite interface then causes transformation from one phase to the other in a diffusionless manner, with additional small shuffles if required

[64], of the same magnitude as those discussed for twinning in Chapter 11. This theory has been applied successfully to a number of martensitic transformations, such as the $\beta \rightarrow \alpha$ transformation in titanium [64], and transformations in steels, Pu–Ga alloys, Cu–Al–Ni and Ni_2MnGa [33, 65–67]. Similar, but not identical, crystallographic predictions are found when this model is compared with the phenomenological model, but the angular differences are relatively small.

12.13.2 Non-linear Elasticity Model

In this continuum mechanics model, a free energy functional is derived from which the crystallography of a particular martensitic phase transformation is produced by energy minimization, with attention being focused on the microstructure as fine mixtures, such as in the partially transformed In–Tl alloy single crystal shown in Figure 12.9 [68]. This model incorporates the phenomenological theory of martensitic transformations as part of its predictive capabilities [69], but is more versatile and advanced, in that is able to predict the detail of microstructures that are observed in materials [70–73]. Therefore, this approach, unlike the topological approach, does not consider the atomic level of the parent–martensite interface and the mechanism of its migration, but instead looks at the fine microstructure of the material as a whole as an energy minimization problem in continuum mechanics. A full appreciation of this theory requires a very advanced understanding of non-linear elasticity theory.

Problems

12.1 A single crystal of c.c.p. cobalt is transformed martensitically to the hexagonal phase. Up to how many orientations of h.c.p. cobalt can be present after the transformation? If the reverse transformation, h.c.p. → c.c.p., is also martensitic, how many orientations of the c.c.p. phase could be produced from a single grain of the hexagonal phase? If the c.c.p. crystal is to transform by means of a pole mechanism (Section 11.5, Figure 11.17), what type of pole dislocation must be available?

12.2 Show that when two different invariant plane strains are applied, one after the other, there always remains a line which is neither rotated nor changed in length and a plane which is not rotated.

12.3 When a crystal of b.c.c. titanium transforms to the h.c.p. phase, how many lattice vectors can transform into the c-axis of the hexagonal unit cell? Is this number equal to the number of different orientations in which the h.c.p. phase originating from a single b.c.c. crystal can be found? Derive the matrix equation which gives the transformation of the indices of planes for the lattice correspondence shown in Figure 12.4 (see Appendix 4).

12.4 The lattice strain S given by Eq. (12.3) produces the correct hexagonal lattice from half the points of the b.c.c. lattice. Describe how the atoms at the remaining points of the b.c.c. lattice must be shuffled in order to place them at the correct positions in the h.c.p. structure.

12.5 Prove that the atoms on the (011) plane of a b.c.c. metal crystal can be brought into a close-packed array by means of a compressive strain of 10% along [100] and a tensile strain of 10% along [0$\bar{1}$1]. What is the c/a ratio of the hexagonal unit cell that is derived from the b.c.c. lattice by means of these two strains alone? Compare this value with the c/a ratio of zirconium.

12.6 Calculate approximately the angle of the tilt that should be observed on the (100) surface of a cubic In–Tl crystal that is transforming by the mode shown in Figure 12.9. Assume that the pure lattice strain is a tensile strain of 0.0066 along that cube axis which transforms into the c-axis of the tetragonal cell and a compressive strain of 0.0033 along the other two cube axes. What is the tensile strain in the Oy direction that can be produced by this mode of transformation? If the twin boundaries within the tetragonal phase are mobile, how can this strain be increased?

12.7 In the transformation of cubic In–Tl to tetragonal In–Tl, interfaces of {110} orientation in the cubic phase divide this phase from the twinned tetragonal phase. How many different interfaces can bound:
(a) A region with a given twin structure?
(b) Any region in which the thicker of the two twin lamellae contains a given c-axis – say parallel to [001] of the cubic phase?
(c) Any region in which either of the twin lamellae contains a given c-axis – say parallel to [001] of the cubic phase?

12.8 Sketch the atomic structure on the planes normal to Ox and Oz in Figure 12.9 at the places where they are intersected by the martensite habit plane. Indicate the form of the local distortions and show how the macroscopic distortion vanishes in both planes.

12.9 The Bain lattice correspondence for the formation of martensite in steels requires a lattice strain of principal components η_1, η_2 and η_3, as defined by Eq. (12.17) and Figure 12.12. Show that the cone of vectors which are not changed in length by this strain has a semi-angle φ_i when the cone is defined by the initial position of these vectors, where:

$$\tan \phi_i = \left(\frac{-2\eta_3 - \eta_3^2}{2\eta_1 + \eta_1^2} \right)^{1/2}$$

Show that, when defined by the position of these vectors after the strain has been applied, the cone has a semi-angle φ_f where:

$$\tan \phi_f = \left(\frac{1 + \eta_1}{1 + \eta_3} \right) \tan \phi_i$$

Calculate approximate values of the angles φ_i and φ_f.

12.10 Show that the lattice correspondence for a b.c.c. → c.c.p. transformation (the Bain correspondence) and the lattice correspondence for the b.c.c. → h.c.p. transformation (e.g. in zirconium) can each be regarded as a special case of a correspondence for a b.c.c. → orthorhombic transformation. Derive a matrix for the transformation of the indices of directions according to this correspondence (see Appendix 4).

Suggestions for Further Reading

Banerjee, S. and Mukhopadhyay, P. (2007). *Phase Transformations: Examples from Titanium and Zirconium Alloys*. Elsevier Science, Oxford: Pergamon Materials Series.
Bhattacharya, K. (2003). *Microstructure of Martensite: Why it Forms and How it Gives Rise to the Shape-Memory Effect*. Oxford: Oxford University Press.
Khachaturyan, A.G. (1983). *Theory of Structural Phase Transformations in Solids*. New York: John Wiley and Sons.
Nishiyama, Z. (1978). *Martensitic Transformations*. New York: Academic Press.
Wayman, C.M. (1964). *Introduction to the Crystallography of Martensitic Transformations*. New York: Macmillan.

References

[1] Otsuka, K. and Wayman, C.M. (eds.) (1998). *Shape Memory Materials*. Cambridge: Cambridge University Press.
[2] Clapp, P.C. (1995). How would we recognize a martensitic transformation if it bumped into us on a dark & austy night? *J. de. Physique IV* 5: C8-11–C8-19.
[3] Massalski, T.B. (1958). The mode and morphology of massive transformations in Cu–Ga, Cu–Zn, Cu–Zn–Ga and Cu–Ga–Ge alloys. *Acta Metall.* 6: 243–253.
[4] Bowles, J.S. and Tegart, W.J.McG. (1955). Crystallographic relationships in aged copper-beryllium alloys. *Acta Metall.* 3: 590–597.
[5] Christian, J.W. (1951). A theory of the transformation in pure cobalt. *Proc. Roy. Soc. Lond. A* 206: 51–64.
[6] Gaunt, P. and Christian, J.W. (1959). The cubic–hexagonal transformation in single crystals of cobalt and cobalt–nickel alloys. *Acta Metall.* 7: 529–533.
[7] Tolédano, P., Krexner, G., Prem, M., Weber, H.-P. and Dmitriev, V.P. (2001). Theory of the martensitic transformation in cobalt. *Phys. Rev. B* 64: 144014-1–144014-17.
[8] Sato, A., Chishima, E., Soma, K. and Mori, T. (1982). Shape memory effect in $\gamma \leftrightarrow \varepsilon$ transformation in Fe–30Mn–1Si alloy single crystals. *Acta Metall.* 30: 1177–1183.
[9] Brooks, J.W., Loretto, M.H. and Smallman, R.E. (1979). *In situ* observations of the formation of martensite in stainless steel. *Acta Metall.* 27: 1829–1838.
[10] Blaschko, O., Dmitriev, V., Krexner, G. and Tolédano, P. (1999). Theory of the martensitic phase transformations in lithium and sodium. *Phys. Rev. B* 59: 9095–9112.
[11] Burgers, W.G. (1934). On the process of transition of the cubic-body-centered modification into the hexagonal-close-packed modification of zirconium. *Physica* 1: 561–586.
[12] Gaunt, P. and Christian, J.W. (1959). The crystallography of the β–α transformation in zirconium and in two titanium–molybdenum alloys. *Acta Metall.* 7: 534–543.
[13] Banerjee, S. and Krishnan, R. (1971). Martensitic transformation in zirconium–niobium alloys. *Acta Metall.* 19: 1317–1326.
[14] Knowles, K.M. and Smith, D.A. (1981). The nature of the parent–martensite interface in titanium–manganese. *Acta Metall.* 29: 1445–1466.
[15] Srivastava, D., Madangopal, K., Banerjee, S. and Ranganathan, S. (1993). Self accommodation morphology of martensite variants in Zr–2.5wt%Nb alloy. *Acta Metall.* 41: 3445–3454.
[16] Burkart, M.W. and Read, T.A. (1953). Diffusionless phase change in the indium-thallium system. *Trans. Am. Inst. Min. Metall. Petrol. Engrs.* 197: 1516–1524.
[17] Basinski, Z.S. and Christian, J.W. (1956). Interpenetrating 'bands' in transformed indium-thallium alloys. *Acta Metall.* 4: 371–378.
[18] Ren, X. and Otsuka, K. (1997). Origin of rubber-like behaviour in metal alloys. *Nature* 389: 579–582.
[19] Bain, E.C. (1924). The nature of martensite. *Trans. Am. Inst. Min. Metall. Petrol. Engrs.* 70: 25–35; discussion 35–46.
[20] Roberts, C.S. (1953). Effect of carbon on the volume fractions and lattice parameters of retained austenite and martensite. *Trans. Am. Inst. Min. Metall. Petrol. Engrs.* 197: 203–204.

[21] Wechsler, M.S., Lieberman, D.S. and Read, T.A. (1953). On the theory of the formation of martensite. *Trans. Am. Inst. Min. Metall. Petrol. Engrs.* 197: 1503–1515.

[22] Bowles, J.S. and Mackenzie, J.K. (1954). The crystallography of martensite transformations. III. Face-centred cubic to body-centred tetragonal transformations. *Acta Metall.* 2: 224–234.

[23] Kurdjumow, G. and Sachs, G. (1930). Über den Mechanismus der Stahlhärtung. *Z. Phys.* 64: 325–343.

[24] Nishiyama, Z. (1934). X-ray investigation of the mechanism of the transformation from face-centred cubic lattice to body-centred cubic. *Sci. Rep. Tôhoku Imp. Univ.* 23: 637–664.

[25] Wassermann, G. (1935). Ueber den Mechanismus der α-γ-Umwandlung des Eisens. *Mitteilungen Kaiser-Wilhelm-Institut für Eisenforschung Düsseldorf* 17: 149–155.

[26] Kelly, P.M. and Nutting, J. (1961). The morphology of martensite. *J. Iron Steel Inst.* 197: 199–211.

[27] Dunne, D.P. and Wayman, C.M. (1969). Measurement of the shape strain for the (225) and (259) martensitic transformations. *Acta Metall.* 17: 201–212.

[28] Acton, A.F. and Bevis, M. (1969/70). A generalised martensite crystallography theory. *Mater. Sci. Eng.* 5: 19–29.

[29] Ross, N.D.H. and Crocker, A.G. (1970). A generalized theory of martensite crystallography and its application to transformations in steels. *Acta Metall.* 18: 405–418.

[30] Dunne, D.P. and Wayman, C.M. (1971). An assessment of the double shear theory as applied to ferrous martensitic transformations. *Acta Metall.* 19: 425–438.

[31] Kelly, P.M. (1993). The crystallography of the (225)$_f$ martensite transformation in steels. In: *ICOMAT-92 Proceedings* (eds. C.M. Wayman and J. Perkins), 293–298.

[32] Kelly, P.M. (2006). Martensite crystallography—The role of the shape strain. *Mater. Sci. Engng.* A 438–440: 43–47.

[33] Pond, R.C., Ma, X. and Hirth, J.P. (2008). Geometrical and physical models of martensitic transformations in ferrous alloys. *J. Mater. Sci.* 43: 3881–3888.

[34] Lovey, F.C. (1987). The fault density in 9R type martensites: a comparison between experimental and calculated results. *Acta Metall.* 35: 1103–1108.

[35] Saburi, T. and Wayman, C.M. (1979). Crystallographic similarities in shape memory martensites. *Acta Metall.* 27: 979–995.

[36] Saburi, T., Wayman, C.M., Takata, K. and Nenno, S. (1980). The shape memory mechanism in 18R martensitic alloys. *Acta Metall.* 28: 15–32.

[37] Otsuka, K. and Ren, X. (2005). Physical metallurgy of Ti–Ni-based shape memory alloys. *Prog. Mater. Sci.* 50: 511–678.

[38] Knowles, K.M. and Smith, D.A. (1981). The crystallography of the martensitic transformation in equiatomic nickel–titanium. *Acta Metall.* 29: 101–110.

[39] Zhang, X. and Sehitoglu, H. (2004). Crystallography of the B2 → R → B19′ phase transformations in NiTi. *Mater. Sci. Engng.* A 374: 292–302.

[40] Madangopal, K. (1997). The self accommodating martensitic microstructure of Ni–Ti shape memory alloys. *Acta Mater.* 45: 5347–5365.

[41] Nishida, M., Nishiura, T., Kawano, H. and Inamura, T. (2012). Self-accommodation of B19′ martensite in Ti–Ni shape memory alloys – Part I. Morphological and crystallographic studies of the variant selection rule. *Phil. Mag.* 92: 2215–2233.

[42] Nishida, M., Okunishi, E., Nishiura, T., Kawano, H., Inamura, T., Ii, S. and Hara, T. (2012). Self-accommodation of B19′ martensite in Ti–Ni shape memory alloys – Part II. Characteristic interface structures between habit plane variants. *Phil. Mag.* 92: 2234–2246.

[43] Inamura, T., Nishiura, T., Kawano, H., Hosoda, H. and Nishida, M. (2012). Self-accommodation of B19′ martensite in Ti–Ni shape memory alloys. Part III. Analysis of habit plane variant clusters by the geometrically nonlinear theory. *Phil. Mag.* 92: 2247–2263.

[44] Bradley, A.J. and Rodgers, J.W. (1934). The crystal structure of the Heusler alloys. *Proc. Roy. Soc. Lond.* A 144: 340–359.

[45] Pons, J., Chernenko, V.A., Santamarta, R. and Cesari, E. (2000). Crystal structure of martensitic phases in Ni–Mn–Ga shape memory alloys. *Acta Mater.* 48: 3027–3038.

[46] Cong, D.Y., Zhang, Y.D., Wang, Y.D., Humbert, M., Zhao, X., Watanabe, T., Zuo, L. and Esling, C. (2007). Experiment and theoretical prediction of martensitic transformation crystallography in a Ni–Mn–Ga ferromagnetic shape memory alloy. *Acta Mater.* 55: 4731–4740.

[47] Pond, R.C., Muntifering, B. and Müllner, P. (2012). Deformation twinning in Ni_2MnGa. *Acta Mater.* 60: 3976–3984.

[48] Muntifering, B., Pond, R.C., Kovarik, L., Browning, N.D. and Müllner, P. (2014). Intra-variant substructure in Ni–Mn–Ga martensite: conjugation boundaries. *Acta Mater.* 71: 255–263.

[49] Muntifering, B., Kovarik, L., Browning, N.D., Pond, R.C., Knowlton, W.B. and Müllner, P. (2016). Stress-assisted removal of conjugation boundaries in non-modulated Ni–Mn–Ga by coordinated secondary twinning. *J. Mater. Sci.* 51: 457–466.

[50] Kelly, P.M. and Rose, L.R.F. (2002). The martensitic transformation in ceramics – its role in transformation toughening. *Prog. Mater. Sci.* 47: 463–557.

[51] Groves, G.W. (1983). Phase transformations in dicalcium silicate. *J. Mater. Sci.* 18: 1615–1624.

[52] Haertling, G.H. (1999). Ferroelectric ceramics: history and technology. *J. Am. Ceram. Soc.* 82: 797–818.

[53] Christy, A.G. (1993). Multistage diffusionless pathways for reconstructive phase transitions: application to binary compounds and calcium carbonate. *Acta Crystall. B* 49: 987–996.

[54] Hu, Q.W., Yan, X.Z., Li, L., Wang, Q.M., Feng, L.H., Qi, L., Zhang, L.L., Peng, F., Ohfuji, H. and He, D.W. (2018). Pressure induced solid-solid reconstructive phase transition in $LiGaO_2$ dominated by elastic strain. *Phys. Rev. B*, 97: 014106-1–014106-6.

[55] Bevis, M. and Crellin, E.B. (1971). The geometry of twinning and phase transformations in crystalline polyethylene. *Polymer* 12: 666–684.

[56] Young, R.J. and Bowden, P.B. (1974). Twinning and martensitic transformations in oriented high-density polyethylene. *Phil. Mag.* 29: 1061–1073.

[57] Auriemma, F. and De Rosa, C. (2003). Time-resolved study of the martensitic phase transition in syndiotactic polypropylene. *Macromolecules* 36: 9396–9410.

[58] Olson, G.B. and Cohen, M. (1976). A general mechanism of martensitic nucleation: Part I. General concepts and the FCC → HCP transformation. *Metall. Trans. A* 7: 1897–1904.

[59] Olson, G.B. and Cohen, M. (1976). A general mechanism of martensitic nucleation: Part II. FCC → BCC and other martensitic transformations. *Metall. Trans. A* 7: 1905–1914.

[60] Olson, G.B. and Cohen, M. (1976). A general mechanism of martensitic nucleation: Part III. Kinetics of martensitic nucleation. *Metall. Trans. A* 7: 1915–1923.

[61] Wang, Y. and Khachaturyan, A.G. (1997). Three-dimensional field model and computer modeling of martensitic transformations. *Acta Mater.* 45: 759–773.

[62] Artemev, A., Jin, Y. and Khachaturyan, A.G. (2001). Three-dimensional phase field model of proper martensitic transformation. *Acta Mater.* 49: 1165–1177.

[63] Müllner, P. and King, A.H. (2010). Deformation of hierarchically twinned martensite. *Acta Mater.* 58: 5242–5261.

[64] Pond, R.C., Celotto, S. and Hirth, J.P. (2003). A comparison of the phenomenological theory of martensitic transformations with a model based on interfacial defects. *Acta Mater.* 51: 5385–5398.

[65] Ma, X. and Pond, R.C. (2011). Martensitic interfaces and transformation crystallography in Pu–Ga alloys. *J. Mater. Sci.* 46: 4236–4242.

[66] Ostapovets, A., Zarubova, N. and Paidar, V. (2012). Topological model of austenite–martensite interfaces in Cu–Al–Ni alloy. *Acta Phys. Polon. A* 122: 493–496.

[67] Ma, X., Wei, Z.Z. and Zhang, X.P. (2014). Theoretical study on the dislocation structure of parent–martensite interface in a magnetic shape memory alloy. *J. Mater. Sci.* 49: 4648–4655.

[68] Ball, J.M. and James, R.D. (1987). Fine phase mixtures as minimizers of energy. *Arch. Ration. Mech. Anal.* 100: 13–52.

[69] Ball, J.M. and James, R.D. (1992). Proposed experimental tests of a theory of fine microstructure and the two-well problem. *Phil. Trans. Roy. Soc. Lond. A* 338: 389–450.

[70] Bhattacharya, K. (1991). Wedge-like microstructure in martensites. *Acta Metall. Mater.* 39: 2431–2444.

[71] Bhattacharya, K. (1992). Self-accommodation in martensite. *Arch. Ration. Mech. Anal.* 120: 201–244.

[72] James, R.D. and Hane, K.F. (2000). Martensitic transformations and shape memory materials. *Acta Mater.* 48: 197–222.

[73] Ball, J.M. (2004). Mathematical models of martensitic microstructure. *Mater. Sci. Eng. A* 378: 61–69.

13

Grain Boundaries

13.1 The Structure of Surfaces and Surface Free Energy

Every real crystal must have at least one imperfection – its surface. It is useful to introduce crystal interfaces by thinking of a surface that is parallel to a prominent crystallographic plane and represents a smooth sheet of atoms, of which the pattern is the same as that of a parallel plane inside the crystal. We shall use this simple clear idea, but recognize it is not strictly true. Surfaces show the phenomenon of *reconstruction*, whereby even in high vacuum the outer layers of atoms rearrange into a more energetically favourable situation. A well-known example is the 7×7 structure found on (111) Si [1]. Similar surface reconstructions occur on clean metal surfaces [2]. These reconstructions can be rationalized in terms of the coordination of the atoms at the surface and the electronic structure at the surface being different from the bulk (which, in practice, typically means distances of >1 nm from the surface).

Rather than such outer surfaces, we are mainly interested in this chapter and the next chapter with *internal* interfaces in crystalline solids. In this chapter we will consider grain boundaries, while in Chapter 14 we will consider interphase boundaries – boundaries between different phases, such as epitaxial interfaces and boundaries between precipitates and the matrix within which they precipitate. The topic of internal interfaces covers a very wide field [3], from which we have attempted in these two chapters to distil the essential aspects relevant to a basic understanding of the crystallography of internal interfaces. We begin in this section with the concept of surface free energy to achieve this. Books listed in the Suggestions for Further Reading in both chapters direct the reader to more advanced treatments of this topic.

We shall formally define a flat surface parallel to any given rational plane by removing all the atoms whose centres lie to one side of a plane of this orientation located within the

Crystallography and Crystal Defects, Third Edition. Anthony Kelly and Kevin M. Knowles.
© 2020 John Wiley & Sons Ltd. Published 2020 by John Wiley & Sons Ltd.
Companion website: www.wiley.com/go/kelly/crystallography3e

bulk crystal. With this assumption, in highly symmetrical crystal structures with one atom per lattice point, such as the c.c.p. and b.c.c. crystal structures, the location of the defining plane will not affect the structure of this idealized surface, because each atom has identical surroundings within the bulk.

In crystal structures with more than one atom per lattice point, the structure of the surface may depend upon the location of the plane. For example, a {111} surface in a NaCl crystal will consist of either a sheet of Na^+ ions or a sheet of Cl^- ions, depending upon the location of the defining plane, or the side from which the atoms are removed. In this structure the same is true for all {hkl} surfaces for which h, k and l are all odd. Similarly, there are two alternative structures for some of the surfaces of an h.c.p. metal. An example is the $(10\bar{1}0)$ surface, shown in Figure 13.1. It is easy to see from Figure 13.2 that in wurtzite there are *four* alternative surfaces parallel to a closest-packed plane. It is interesting to note that any two of these surfaces bounding a crystal must have different structures (a consequence of the absence of a centre of symmetry).

As we have already noted, the atoms at a surface are deprived of some of their neighbours. Since the binding of an atom to its neighbours contributes a favourable negative term to the energy of a crystal, we can attribute some excess energy to the presence of the surface. Imagine two identical surfaces to have been created within a single crystal by breaking the atomic bonds through which a plane passes. The surface energy is then equal to half the energy of the broken bonds. (This equation assumes that the energies of those bonds which are left do not change.) The idea of broken bonds is useful in discussing how the energy of a surface will change as it is rotated out of a low-index orientation.

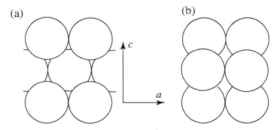

Figure 13.1 The two alternative {$10\bar{1}0$} surfaces of a hexagonal metal. The surface is parallel to the plane of the figure

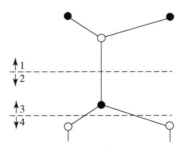

Figure 13.2 The four alternatives for a surface parallel to (0001) in wurtzite. The surface is normal to the plane of the figure, intersecting it at one or other of the dotted lines

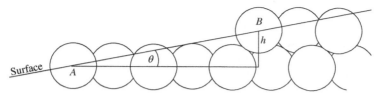

Figure 13.3 Surface at a small angle θ to a {111} plane of a c.c.p. metal. The surface is normal to the plane of the figure

Suppose a {111} surface of a c.c.p. metal is rotated through a small angle θ about a $\langle 1\bar{1}0 \rangle$ direction. It is apparent from Figure 13.3 that it then contains a density of steps ρ per unit length:

$$\rho = \frac{\sin \theta}{h} \tag{13.1}$$

where h is the spacing of the {111} lattice planes. If the angle θ happens to be such that the points A and B in Figure 13.3 lie on $\langle 1\bar{1}0 \rangle$ rows of atoms then the steps will be evenly spaced. For other surfaces of rational orientation, the steps can occur in evenly-spaced groups, but if the surface is irrational, there must be irregularities in the arrangement of steps. An atom on a step lacks more neighbours, and so has more broken bonds, than an atom in the flat {111} surface; consequently, the steps introduce an extra energy which is proportional to the number of steps, as long as they are so far apart that they do not interact with one another. If each step contributes an energy β per unit length, the total energy of unit area of surface is:

$$E = E_0 \cos \theta + \frac{\beta \sin |\theta|}{h} \tag{13.2}$$

where E_0 is the energy of unit area of {111} surface. The surface energy therefore increases as the surface is rotated from its low-index orientation, in either sense. The plot of energy as a function of θ shows a cusp at which $dE/d\theta$ changes discontinuously from $-\beta/h$ to $+\beta/h$ (Figure 13.4).

Similar arguments can be applied to a small rotation of the {111} surface about any axis; therefore a cusp exists in a three-dimensional plot of surface energy against orientation. Such a plot may take the form of a polar diagram in which the energy of a surface is

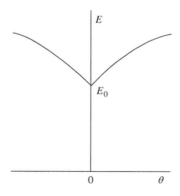

Figure 13.4 A schematic of energy E as a function of angle θ away from a low-index plane

represented by a vector normal to the surface and is of a length in proportion to the energy of the surface. The energy plot can be expected to exhibit a number of cusps at the orientations of various low-index planes.

The energy of a surface is closely related to its surface free energy, defined as the work that can be obtained from the destruction of unit area of surface. The surface free energy is:

$$\gamma = E - TS \tag{13.3}$$

and although the energy E is usually its more important component, the entropy term TS can be significant. For example, the steps upon a surface that is slightly off a low-index plane will introduce a configurational entropy, because their straightness and spacing may vary. Therefore, at a finite temperature, the cusp in the surface free energy plot will not be as sharp as that in the energy plot and it may even disappear for higher-index planes.

A polar diagram of surface free energy is called a γ-plot or Wulff plot. A $(1\bar{1}0)$ section through the γ-plot of a c.c.p. metal is shown in Figure 13.5, as computed from a simple nearest-neighbour bond model.

The meaning of the surface free energy of a solid can be illustrated by describing an experiment that has been used to measure it [5, 6]. A very fine wire of length x and radius r (where $r \approx 50\,\mu m$) is loaded at a temperature close to its melting point. By finding the rate of extension as a function of load, it is possible to find the value of the load W which will just counteract the tendency of the wire to reduce its surface area by shrinking. Considering an infinitesimal increase in length dx at this equilibrium, 'zero-creep' condition, we have [6]:

$$W\,dx = \gamma\,dA + \gamma_b\,dA_b \tag{13.4}$$

In this equation, $\gamma\,dA$ is the increase in free energy due to the creation of an area dA of new surface (Figure 13.6). The term $\gamma_b\,dA_b$ is the decrease in free energy due to a change dA_b in the area of grain boundary within the wire. For the experiments undertaken, each wire was

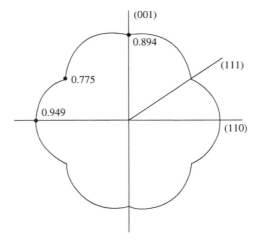

Figure 13.5   Possible $(1\bar{1}0)$ section through the γ-plot of a c.c.p. metal. Values given are in units of the surface energy of a {210} surface, using data from [4]

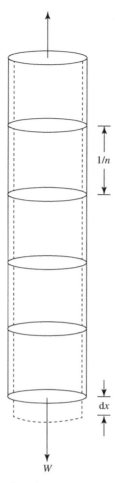

Figure 13.6 A fine wire with a bamboo-like grain structure to which a load *W* is applied to measure surface free energy

carefully annealed to have a bamboo-like set of coarse-sized grains, with grain boundaries transverse to the axis of the wire, as in the schematic in Figure 13.6.

Equation (13.4) is a condition for the free energy of the system to be a minimum. It is important to realize that the wire is conceived to change length by a frictionless flow at constant volume, not by elastic deformation, so that the increase in surface area, d*A*, is obtained by creating a new surface, not by stretching the old one.

For a wire containing *N* transverse grain boundaries, we would have:

$$A = 2\pi xr + 2\pi r^2; \qquad A_b = \pi r^2 N; \qquad V = \pi r^2 x$$

where *V* is the volume of the wire, which remains constant as a result of the frictionless flow process. To a reasonable approximation, the cross-sectional area of the ends of the fine wire can be neglected relative to the area of the cylindrical portion of the wire, so

$A \approx 2\pi xr$. Hence, expressing A and A_b in terms of x and V, we have:

$$A = 2\sqrt{V\pi x}; \quad dA = \sqrt{\frac{V\pi}{x}}dx = \pi r dx; \quad A_b = \frac{VN}{x}; \quad dA_b = -\frac{VN}{x^2}dx = -\frac{\pi r^2 N}{x}dx$$

Substituting these values into Eq. (13.4) and rearranging gives the equation:

$$\gamma = (W/\pi r) + \gamma_b nr \tag{13.5}$$

where $n = N/x$ is the number of transverse grain boundaries per unit length in the specimen.

For sufficiently fine wire and large grain size, the $\gamma_b nr$ term is a small correction term relative to the $W/\pi r$ term. Udin [6] took $\gamma_b = \gamma/3$, a value which is in broad agreement with the surface energies for grain boundaries discussed in Section 13.3. Since the orientation of the surface varies around the circumference of the wire, an average value of γ is obtained from Eq. (13.5).

Some surface free energies obtained by this and other methods are listed in Table 13.1. The data for metallic materials in this table are the selected values from the survey of the available experimental data by Kumikov and Khokonov in 1983 [7]. The surface free energy of the cleavage plane of many brittle crystals has been found from the force needed to split a pre-cracked crystal of convenient shape, such as that shown in Figure 13.7. The basic equation defining the critical force P_c is the balance of energy:

$$2P_c dy = dE + 2\gamma w\, dL \tag{13.6}$$

Table 13.1 Surface free energies of solids

Crystal	Surface	Environment	Temperature (K)	Surface tension (mJ m^{-2})	Method	Ref.
Ag	Average	Vacuum	m.p.[a]	1205	Zero creep	[7]
Au	Average	Vacuum	m.p.	1410	Zero creep	[7]
Cu	Average	Vacuum	m.p.	1520	Zero creep	[7]
CaF$_2$	(111)	Liquid N$_2$	77	450	Cleavage	[8]
Calcite	(10$\bar{1}$0)	Liquid N$_2$	77	230	Cleavage	[8]
Fe (δ)	Average	Argon	m.p.	1910	Zero creep	[7]
Fe 3% Si	(001)	Liquid H$_2$	14	1360	Cleavage	[8]
KCl	(001)	Air	298	110 ± 5	Cleavage	[9]
LiF	(001)	Liquid N$_2$	77	340	Cleavage	[8]
MgO	(001)	Air	298	1150 ± 80	Cleavage	[10]
	(001)	Vacuum	77	1300	Cleavage	[11]
Mo	Average	Vacuum	m.p.	2630	Zero creep	[7]
NaCl	(001)	Vacuum	77	283 ± 30	Cleavage	[11]
	(001)	Liquid N$_2$	77	317 ± 30	Cleavage	[11]
Nb	Average	Vacuum	m.p.	2210	Zero creep	[7]
Ni	Average	Vacuum	m.p.	1940	Zero creep	[7]
Si	(111)	Liquid N$_2$	77	1240	Cleavage	[8]
Sn	Average	Vacuum	m.p.	673	Zero creep	[7]
Zn	Average	Argon	m.p.	868	Zero creep	[7]
Zn	(0001)	Liquid N$_2$	77	575	Cleavage	[7]

[a]m.p. = melting point.

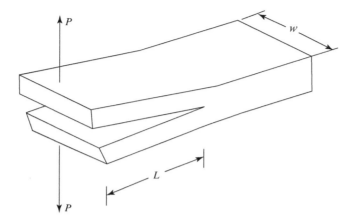

Figure 13.7   Splitting of a crystal of width w with a pre-existing crack of length L by a load P

where dy is the deflection of each of the points of application of P and dE the increase in the elastic energy of the crystal of width w; dy and dE accompany an infinitesimal growth dL of the crack, during which the force P is held constant. To apply this equation, the presumption is made that the applied force P does not have to do work in plastically deforming the crystal as the crack grows, a very reasonable assumption for ceramics at liquid nitrogen temperatures, even for LiF, where at room temperature plastic flow occurs near the tips of slowly moving cracks [8].

13.2   Structure and Energy of Grain Boundaries

A crystalline solid is usually found in the form of a polycrystal; that is, an aggregate of randomly oriented single crystals, called grains. Even so-called single crystals usually contain regions called sub-grains, which have slightly different orientations.

Perhaps the simplest type of grain boundary to visualize is the symmetrical low-angle *tilt boundary*, where the two grains on either side are related by symmetrical rotations about an axis lying in the boundary, and where the rotation is relatively small ($<10°$). A low-angle tilt boundary in a simple cubic structure is shown in Figure 13.8. This boundary could be formed by joining two crystals with unrelaxed stepped surfaces rotated from a cube plane by small angles $+\theta/2$ and $-\theta/2$ about a $\langle 100 \rangle$ direction (Figure 13.8a). When the two surfaces are joined, these steps become edge dislocations, with the Burgers vector equal to the step height, as in the schematic in Figure 13.8b. Setting $h = b$ in Eq. (13.1), the number of dislocations per unit length of boundary becomes:

$$\frac{1}{d} = \frac{2 \sin \theta/2}{b} \tag{13.7}$$

or when θ is small, and expressed in radians:

$$\frac{1}{d} \approx \frac{\theta}{b} \tag{13.8}$$

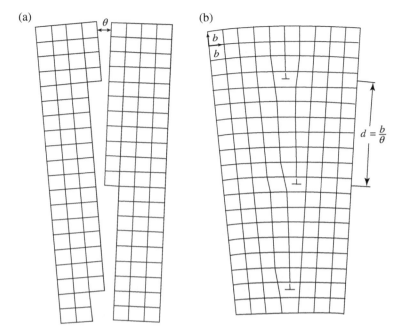

Figure 13.8 Low-angle symmetrical tilt boundary in a simple cubic lattice. The boundary is normal to the plane of the figure [12]. Source: With permission of the American Association of the Advancement of Science

This low-angle symmetrical tilt boundary is, in fact, a 'polygon wall' of dislocations of the type mentioned in Section 8.6. Read and Shockley [13] have calculated the energy of such an array, situated in an infinite medium of shear modulus μ and Poisson's ratio v. They found the energy per unit area to be:

$$E = E_0\theta(A_0 - \ln\theta) \tag{13.9}$$

where:

$$E_0 = \frac{\mu b}{4\pi(1 - v)}$$

and:

$$A_0 = 1 + \ln\left(\frac{b}{2\pi r_0}\right)$$

The length r_0 in Eq. (13.9) is related to the energy of the core of one of the boundary dislocations. Its definition is that an integration of an expression for the elastic strain energy such as Eq. (9.1) down to a radius r_0 gives the *total* energy of the dislocation, including the core energy. A high core energy decreases r_0 and increases A_0. Since from Eq. (13.8) the density of dislocations in the boundary is θ/b, for small θ, the first term in Eq. (13.9) depends on the total core energy in unit area of the boundary. The elastic energy of the boundary enters into the second term. The stresses due to the dislocations largely cancel one another at distances from the boundary that exceed the spacing of the dislocations, b/θ.

The elastic energy within a cylinder of radius b/θ about an edge dislocation of core radius r_c is:

$$E = \frac{\mu b^2}{4\pi(1 - \nu)} \ln \frac{b}{\theta r_c}$$

which accounts for the form of the second term in Eq. (13.9).

Eq. (13.9) can be applied only to boundaries having a small angle of tilt, θ, such that the cores of the dislocations do not overlap. Measurements of the free energy of tilt boundaries in copper show that at angles greater than about 8° the energy is greater than that predicted by fitting Eq. (13.9) to the lower-angle results, to which it properly applies [14]. (The entropy term in the free energy of a low-angle boundary is expected to be small compared to its energy, so that energy and free energy can be identified with one another.)

A plot of Eq. (13.9) is shown in Figure 13.9. The cusp at $\theta = 0°$, where the boundary disappears, is very sharp: because of the logarithm term in Eq. (13.9), $dE/d\theta$ becomes infinite at $\theta = 0$. By comparison, cusps in surface energy plots are relatively blunt (Figure 13.4), because the energy of a surface step is always localized at the step and is limited in amount, whereas a far-reaching stress field is set up when an isolated step is pressed into another crystal to make a grain boundary dislocation. Physically, the reason why the energy of a tilt boundary rises so steeply as its angle increases from zero is that the strain field of each dislocation spreads out to a very large distance when the dislocations are widely separated. The increase in energy becomes less steep as the angle of tilt increases further because the stress fields of the dislocations cancel as they come closer together.

Shallower cusps in the grain boundary energy must exist at angles of tilt at which the dislocations are evenly spaced, as suggested in the early work on computer modelling of grain boundaries by Hasson et al. [15]. This is also consistent with the trends seen in later,

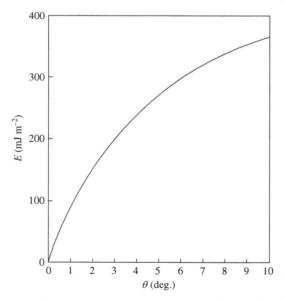

Figure 13.9 Energy of a tilt boundary as a function of the tilt angle θ. Values are given by Eq. (13.9), with $E_0 = 1450 \, \text{mJ m}^{-2}$ (appropriate for Cu) and $A_0 = -0.3$

more detailed, atomistic studies of the structure of symmetrical tilt grain boundaries in c.c.p. metals [16]. A qualitative argument for this can easily be constructed for low angles of tilt. In deriving Eq. (13.9) rigorously, it is assumed that the dislocations are uniformly spaced. It is evident that this is possible in the simple cubic lattice only when θ is such that:

$$\cot \theta/2 = 2n \tag{13.10}$$

where n is an integer. For example, when $\cot \theta/2 = 14$, so that $\theta = 8.17°$, there is one dislocation on every seventh cube plane. To increase the angle of tilt slightly by $\delta\theta$, the separation must occasionally be decreased to six planes. Suppose that this were to be achieved uniformly instead: the corresponding decrease in spacing would be δd ($\ll b$), the original separation being d. Then the spacing, D, of the *actual* disturbances in separation – that is, the spacing between adjacent six-plane separation regions – is:

$$D = \left(\frac{b}{\delta d}\right) d \tag{13.11}$$

That is, d ($= b/\theta$) magnified by the factor ($b/\delta d$). Now:

$$\delta d = \frac{b}{\theta} - \frac{b}{\theta + \delta\theta} \approx \frac{b\delta\theta}{\theta^2} \text{ to } O(\delta\theta)$$

and so:

$$D \approx \frac{b\theta}{\delta\theta} \tag{13.12}$$

If the angle of tilt is decreased, instead of increased, the same equation gives the spacing of the occasional eight-plane separations. This spacing is less than the spacing of dislocations in a tilt boundary of angle $\delta\theta$ would be, by the factor θ ($\theta < 1$ rad). The extra elastic energy due to an increase or decrease in tilt angle of $\delta\theta$ can be guessed, by comparing Eq. (13.12) with Eq. (13.8), to be roughly that of a $\delta\theta$ boundary composed of dislocations of the small Burgers vector $b\theta$ or:

$$\Delta E \sim -E_0\theta\delta\theta \ln \delta\theta \tag{13.13}$$

The form of this expression indicates that a small sharp energy cusp exists at the special angle θ, since $d\Delta E/d(\delta\theta)$ becomes infinite as $\delta\theta \to 0$. Physically, this is because each disturbance in the regular structure of the boundary produces a very far-reaching stress field when the disturbances are widely separated.

Cusps can also be expected to occur at special high angles of tilt. For example, when $\cot \theta/2 = 2$ (or $\theta \approx 53.1°$), a hard sphere model (i.e. a purely geometrical model) of the structure of the boundary is neat and regular, as shown in Figure 13.10. The plane of the boundary is a {210} plane in either grain, and one grain can be described as a twin of the other, with $K_1 = (210)$. (According to the definition of a twin given in Section 11.2, any symmetrical tilt boundary can be described as a twin boundary when it is formed by joining surfaces of rational orientation; that is, in the simple cubic lattice, surfaces for which $\cot \theta/2 = m/n$, where m and n are integers.)

Since the atoms in the twin boundary lie on the lattices of both grains in this purely geometrical model, we may expect the energy to be particularly small when the density of atoms on this boundary is high; that is, when K_1 is a low-index plane. Since (210) is the closest-packed twin plane that can be formed in the simple cubic lattice when the tilt axis

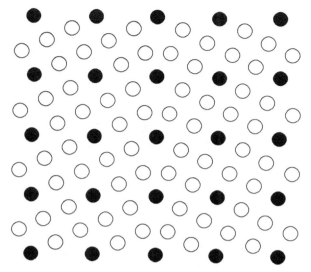

Figure 13.10 Schematic of a high-angle tilt boundary of good fit between one grain on the left and one grain on the right. The boundary is normal to the plane of the figure. The dark circles represent atoms that lie on points of the lattices of both grains [17]

is [001], the deepest of the energy cusps should occur here, at the angle $\theta \approx 53.1°$. Atomistic calculations confirm this qualitative picture: in copper and aluminium, the boundary modelled in Figure 13.10 is a favoured boundary, in that (i) the unit of structure which exists in the boundary is small, and (ii) symmetrical tilt grain boundaries at other angles of rotation about [001] have mixtures of this unit of structure and that of another favoured orientation at $\theta = 90°$ [16].

The symmetrical tilt boundary is a special type of boundary which can be specified by the axis of tilt and the single angle θ. In general, a grain boundary has five degrees of freedom; that is to say, five numbers are required to define it. One way of allocating these numbers is to specify the rotation that brings the lattice of one grain parallel to the lattice of the other – which requires two numbers to define the axis of rotation and one to define the angle of rotation – and then to specify the orientation of the boundary with respect to one of the grains, which uses another two numbers.

We will examine some of the consequences of varying the orientation of the grain boundary in a simple cubic crystal while keeping $\langle 001 \rangle$ as the axis of rotation of one grain with respect to the other.

The tilt boundary of Figure 13.8 can be turned out of its symmetrical orientation by rotating it about the tilt axis, so that it becomes a low-angle asymmetrical tilt grain boundary. The effect of a rotation of χ is shown in Figure 13.11. Edge dislocations with extra planes that are normal to those of the original set are introduced. It can be shown [13] that their density is given by:

$$p_x = \frac{2}{b} \sin |\chi| \sin \frac{\theta}{2} \tag{13.14}$$

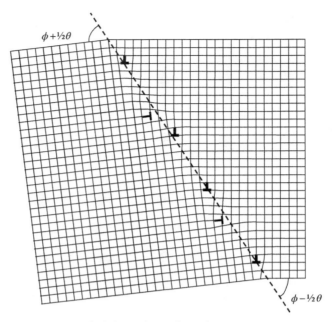

Figure 13.11 An asymmetrical tilt boundary where the misorientation across the boundary is θ. The boundary makes an angle of $\chi = 90° - \varphi$ with the symmetrical tilt orientation

while the density of the original set is reduced to:

$$p_y = \frac{2}{b} \cos \chi \sin \frac{\theta}{2} \tag{13.15}$$

The new dislocations must increase the energy of the boundary sharply as χ increases from zero, because at first they are far apart and therefore have far-reaching strain fields. Read and Shockley [13] showed that the energy of the boundary has the same form as Eq. (13.9); that is:

$$E = E_0'\theta(A - \ln \theta) \tag{13.16}$$

but that now:

$$E_0' = \frac{\mu b}{4\pi(1 - \nu)}(\cos \chi + \sin |\chi|)$$

while:

$$A = A_0 - \frac{1}{2} \sin 2|\chi| - \frac{\sin |\chi| \ln (\sin |\chi|) + \cos \chi \ln (\cos \chi)}{\sin |\chi| + \cos \chi}$$

According to these equations, a sharp energy cusp does indeed exist at the symmetrical orientation $\chi = 0$.

A qualitatively similar cusp should exist for a high-angle tilt boundary which is a twin plane, since rotating such a boundary out of its symmetrical position destroys the good fit of the atoms upon it. Atomistic simulations of asymmetrical high-angle tilt grain boundaries confirm this to be so: cusps occur in the energy against misorientation at both favoured boundary and 'multiple unit reference structure' orientations [18, 19]. A high-angle tilt boundary which is a twin plane is an example of both a favoured boundary and one which has a favoured multiple unit reference structure.

A more drastic change in boundary structure is produced by rotating the tilt boundary of Figure 13.8 through 90° about an axis in the plane of the boundary and normal to the tilt axis. The boundary is then normal to the $\langle 100 \rangle$ axis about which the two grains are rotated, relative to one another, and is called a *twist boundary*. A low-angle twist grain boundary consists of a grid of screw dislocations, as shown in Figure 13.12. The deformation due to one of the two orthogonal sets of screws is such that its pure shear component cancels that of the other set at large distances from the boundary, whereas its rotational component adds to that of the other set and produces the necessary relative rotation of the grains.

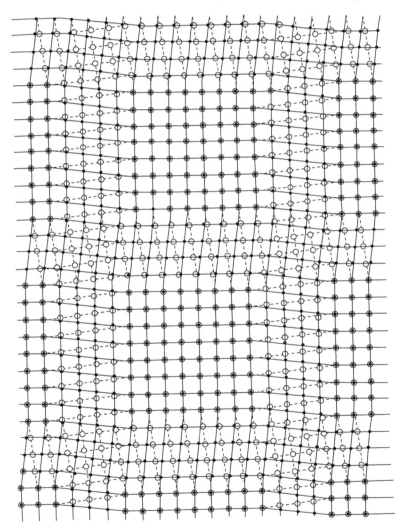

Figure 13.12 A low-angle twist boundary in a simple cubic lattice. The boundary is parallel to the plane of the figure [12]

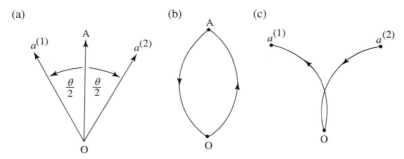

Figure 13.13 (a) Generation of grains 1 and 2 by opposite rotations of $\theta/2$. (b) Burgers circuit through the two grains. (c) Corresponding circuit in the reference lattice

The density of dislocations needed to produce a twist of angle θ can be specified by the following procedure. Choose a mean lattice from which the two grains can be generated by rotations of $\theta/2$ and $-\theta/2$, respectively. In Figure 13.13 the axis of rotation passes through the point O. Choose a vector **OA** which lies in the grain boundary and consists of an array of dislocations in the mean lattice. This array produces the required rotations. Since **OA** cuts through a set of dislocations, a circuit A to O in grain 1 and O to A in grain 2 will surround these dislocations, so that a corresponding circuit taken in a reference lattice will fail to close (Section 8.1).

Choosing the reference lattice to be parallel to the mean lattice, the reference path corresponding to A to O in grain 1 can be found by rotating grain 1 back through $\theta/2$ so as to align it with the reference lattice. This places the starting point at $a^{(2)}$ (Figure 13.13). The back rotation of grain 2 places the finishing point of the circuit at $a^{(1)}$. The closure failure is then $a^{(1)}a^{(2)}$, of magnitude:

$$|\mathbf{B}| = 2|\mathbf{OA}|\sin\theta/2 \tag{13.17}$$

The vector **B** is equal to the sum of the Burgers vectors of all the dislocations intersected by **OA**. Taking **OA** parallel to a $\langle 100 \rangle$ in the mean lattice, the number of $\langle 010 \rangle$ screw dislocations that it intersects, per unit length, is then:

$$p = \frac{2\sin\theta/2}{b} \tag{13.18}$$

This is the same as the density of dislocations in a symmetrical tilt boundary of the same angle, and again for small θ, the dislocation spacing d is:

$$d \approx \frac{b}{\theta} \tag{13.19}$$

The energy of a twist boundary will increase with the angle of twist in the same general way as the energy of a tilt boundary increases with the angle of tilt. The energy should be cusped at those angles of twist at which the atoms fit together well at the boundary. For example, the rotation of 53.1° about [001] at which a symmetrical tilt boundary in a cubic lattice has a good fit also produces a twist boundary, normal to [001], on which the atoms fit together neatly. This is shown by Figure 13.14, in which the atoms lying on the lattice points of both grains are picked out. The net of lattice points common to both

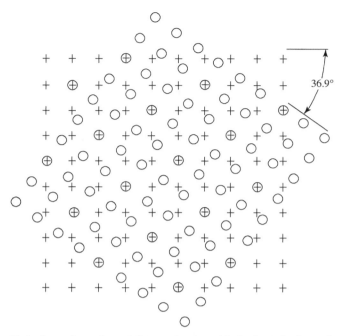

Figure 13.14 Twist boundary of good fit in a simple cubic lattice. The boundary is parallel to the plane of the figure [17]. In this diagram, one in five of the lattice points from the two grains are coincident

grains in Figure 13.14 and the similar net of coincidence points lying on the corresponding tilt boundary (Figure 13.10) are each one of the planes of a single *coincidence site lattice* (CSL). The reciprocal of the density of common points is termed Σ [20], so for the example in Figure 13.14, $\Sigma = 5$. Detailed atomistic calculations on the structure of high-angle twist boundaries in c.c.p. metals confirm that this geometrical picture is useful, even if the actual relaxed structure computed for such boundaries is more complicated, involving small volume increases at the boundaries, relaxations parallel to the boundaries, and also translations parallel to the boundaries [21].

A CSL is a lattice formed by those lattice points that lie on top of one another after two initially superimposed lattices have been rotated with respect to each other. A rotation about [001] of approximately 53.1°, or equivalently of $90° - 53.1° = 36.9°$, causes one-fifth of the lattice points of a cubic crystal to coincide. This is true for b.c.c. and c.c.p. lattices, as well as for simple cubic. This is a geometrically equivalent description of the CSL shown in Figure 13.14. The b.c.t. lattice of coincident points produced from a c.c.p. lattice by this rotation of 36.9°, about [001] is shown in Figure 13.15. The plane that has the highest density of coincident lattice points is the {110} in the coincidence lattice, which corresponds to a {210} plane in one of the c.c.p. lattices.

A number of CSLs formed from cubic lattices by rotations about prominent crystallographic directions are shown in Table 13.2 [20]. Because of symmetry, each coincidence lattice can be produced by several different rotations (Section 13.3). A grain boundary that contains some points common to the lattices of both grains can be formed by choosing

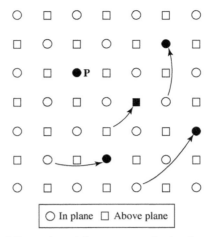

Figure 13.15 Part of the CSL produced from a c.c.p. lattice by a rotation of 36.9° about a [001] at P, the plane of the figure being (001)

Table 13.2 Some coincidence lattices for c.c.p. and b.c.c. crystals [20]

Fraction of lattice points on coincidence lattice	Rotations producing coincidence lattice		Closest-packed plane of coincidence lattice written as plane of parent lattice	
	Axis	Angle (°)	c.c.p.	b.c.c.
1 : 3	110	70.5	111	112
	111	60		
	210	131.8		
	211	180		
	311	146.4		
1 : 5	100	36.9		
	210	180	210	310
	211	101.6		
	221	143.1		
	310	180		
	311	95.7		
	331	154.2		
1 : 7	111	38.2		
	210	73.4	123	123
	211	135.6		
	310	115.4		
	320	149		
	321	180		
	331	110.9		

any plane of a coincidence lattice and removing one of the component lattices on one side of this plane and the other on the other side. The coincidence lattice therefore provides a useful way of *grouping together* various boundaries of good fit between grains with the various equivalent orientation relationships that produce the same coincidence lattice. The degree of fit is striking only when the coincidence lattice has a high density of points and the boundary orientation is that of a prominent plane in this lattice.

The possibility of grouping together boundaries of good fit on the basis of coincidence lattices is really a consequence of crystal symmetry. It should be noted that such groupings are purely geometrical and descriptive for perfect lattices; nothing quantitative can be said about the relative energies of boundaries in real crystalline materials. In less symmetrical crystals, such a treatment is not usually possible. For example, a twin boundary of excellent fit in a hypothetical monoclinic crystal is shown in Figure 13.16. The orientation relationship can be described as a rotation of π about the normal to the twin plane *OA*. In general, *OA* is irrational and no lattice of coincidence sites exists.

It is evident from such considerations that orientation relationships between any two different lattices can be described geometrically in terms of either coincident cells or *near-coincident* cells; algorithms for determining orientations at which there are near-coincidence cells for any two lattices in either real space or reciprocal space have been widely discussed in the scientific literature [22–26], the presumption being that such orientations are more likely to be favoured than other orientations. However, Sutton and Balluffi [27] conclude that there is a lack of quantitative evidence that a geometric criterion of 'good fit' between lattices, however described, is actually useful for making predictions about low-energy interfaces. Instead, they suggest that details of the interfacial structure and the nature of the atomic bonding in a particular situation have to be taken into consideration when attempting to predict low interfacial energies.

The alternative viewpoint is that, notwithstanding the conclusions of Sutton and Balluffi, energy minimization does correlate with geometrical features, such as good atom matching across interfaces. This viewpoint is inspired by work such as that of Shiflet and van der Merwe [28] on interfaces between c.c.p. and b.c.c. crystals in which (111) interfaces of c.c.p. crystals are forced to be parallel to (110) b.c.c. interfaces. In such circumstances, geometrical features such as row matching are found to correspond to energy minimization, albeit not through an atomistic simulation analysis, but through instead an analysis of the energy savings to be made from the introduction of structural ledges as misfit-compensating

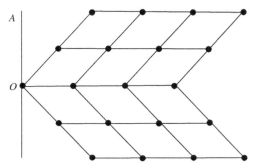

Figure 13.16   Twin boundary in a monoclinic lattice. The boundary is normal to the plane of the figure

defects. It is likely that the two viewpoints will only be able to be reconciled once there are robust atomistic simulation procedures capable of determining with confidence interfacial energies of grain boundaries and interphase boundaries with complex geometry; at the time of writing, such procedures have yet to be developed.

The method employed above to determine the dislocation content of a twist boundary can be generalized to cover any type of boundary. We can produce any grain boundary by means of an array of dislocations in a lattice, which we may call the source lattice. The total Burgers vector of the dislocations intersected by a vector $\mathbf{R}$ which lies in the boundary is given by:

$$\mathbf{B} = \mathbf{r}_2 - \mathbf{r}_1 \tag{13.20}$$

The vectors $\mathbf{r}_1$ and $\mathbf{r}_2$ are vectors generated from $\mathbf{R}$ by rotations equal in magnitude but opposite in sense to those required to generate grains 1 and 2 respectively from the source lattice. (The reversal of sense arises from the use of a closure failure in a reference lattice that is parallel to the source lattice.) In a real crystal, the vector $\mathbf{B}$ must then be allotted to possible crystal dislocations.

For a small rotation, expressed in magnitude and direction by the vector $\boldsymbol{\theta}$, so that the direction of $\boldsymbol{\theta}$ is parallel to the axis of rotation and the magnitude of $\boldsymbol{\theta}$, θ, is the angle of rotation in radians:

$$\mathbf{B} = \mathbf{R} \times \boldsymbol{\theta} \tag{13.21}$$

as is easily seen from Figure 13.17. When θ is large, the rotation cannot be simply described by a vector and Eq. (13.21) becomes:

$$\mathbf{B} = (\mathbf{R} \times \mathbf{l})2 \sin \theta/2$$

where $\mathbf{l}$ is a unit vector in the mean lattice (source lattice) such that grain 1 is generated from the source lattice by a rotation of $-\theta/2$ and grain 2 by a rotation of $+\theta/2$ about $\mathbf{l}$. $\mathbf{B}$ and $\mathbf{R}$ have the same meanings as for very small θ and so $\mathbf{B}$ and $\mathbf{R}$ are defined in the source lattice [29, 30].

While it is possible to determine a dislocation array that produces the correct orientation relationship for any grain boundary, such an array has little physical significance unless the

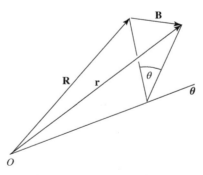

Figure 13.17 Graphical representation of the total Burgers vector $\mathbf{B}$ of the dislocations intersected by a vector $\mathbf{R}$ for a small rotation θ between two grains. The vector $\boldsymbol{\theta}$ is parallel to the axis of rotation, with a magnitude equal to the angle of rotation in radians

dislocations within it are so far apart that their core structures do not overlap. Only then is it possible to relate the properties of the grain boundary to the properties of individual dislocations and the local atomic structure of the grain boundaries [3].

13.3 Equivalent Geometrical Descriptions of High-Angle Grain Boundaries

Aspects of the geometry of high-angle grain boundaries have already been considered in Section 13.2, where it was recognized that two grains either side of a grain boundary are related to one another by an axis of rotation and an angle of rotation. For example, two crystals forming symmetrical low-angle tilt grain boundaries in a simple cubic crystal are related to one another by a rotation θ ($< 10°$) about a $\langle 100 \rangle$ direction. In general, however, the rotation will be round an axis with direction cosines n_1, n_2 and n_3. Under these circumstances, for a rotation of an angle θ anticlockwise about this axis, a rotation matrix $\mathbf{R}$ rotates a vector $\mathbf{r}$ to $\mathbf{r}'$:

$$\mathbf{r}' = \mathbf{R}\mathbf{r} \tag{13.22}$$

where

$$\mathbf{R} = \begin{bmatrix} \cos\theta + n_1^2(1 - \cos\theta) & n_1 n_2(1 - \cos\theta) - n_3 \sin\theta & n_3 n_1(1 - \cos\theta) + n_2 \sin\theta \\ n_1 n_2(1 - \cos\theta) + n_3 \sin\theta & \cos\theta + n_2^2(1 - \cos\theta) & n_2 n_3(1 - \cos\theta) - n_1 \sin\theta \\ n_3 n_1(1 - \cos\theta) - n_2 \sin\theta & n_2 n_3(1 - \cos\theta) + n_1 \sin\theta & \cos\theta + n_3^2(1 - \cos\theta) \end{bmatrix} \tag{13.23}$$

(Section A.1.4). Hence, for the example shown in Figure 13.14, a rotation of 53.1° about [001], for which one-fifth of the lattice points coincide, so that $\Sigma = 5$,

$$\mathbf{R} = \frac{1}{5} \begin{bmatrix} 3 & -4 & 0 \\ 4 & 3 & 0 \\ 0 & 0 & 5 \end{bmatrix} \tag{13.24}$$

A little thought about this rotation shows that there are equivalent descriptions of this rotation, as has already been noted in Section 13.2. To formalize this, on one side of a grain boundary we can identify a clockwise set of crystal axes in the grain under consideration, grain A, which we can term $\mathbf{x}_A$, $\mathbf{y}_A$ and $\mathbf{z}_A$. On the other side of the grain boundary we can also identify a clockwise set of crystal axes in the grain, grain B, which we can term $\mathbf{x}_B$, $\mathbf{y}_B$ and $\mathbf{z}_B$ into which $\mathbf{x}_A$, $\mathbf{y}_A$ and $\mathbf{z}_A$ are rotated by $\mathbf{R}$. However, having identified the directions in grain B along which $\langle 100 \rangle$ directions occur, it is evident that we can choose other settings of the axes which are symmetrically equivalent to the choice we first made for grain B.

Thus, for example, for the most symmetric cubic crystals with $m\bar{3}m$ point group symmetry, i.e. the symmetry of the cubic P, I and F Bravais lattices, a rotation of 90° about the direction we have chosen to be [100] in grain B will rotate the shape of a cube into a symmetrically equivalent orientation. When all such symmetrically equivalent rotations are taken into account there are 24 symmetrically equivalent descriptions through the following:

1. The identity rotation of 0° about any axis;
2. 90°, 180° and 270° rotations about [100], [001] and [001];
3. 120° and 240° rotations about [111], [$\bar{1}$11], [1$\bar{1}$1] and [11$\bar{1}$];
4. 180° rotations about [110], [011], [101], [1$\bar{1}$0], [01$\bar{1}$] and [10$\bar{1}$].

giving $1 + 9 + 8 + 6 = 24$ symmetrically equivalent descriptions. If we consider equivalent descriptions to a rotation of $53.1°$ about $[001]$ relating the crystal axes of A and B, so that

$$x_B = R_{11}x_A + R_{12}y_A + R_{13}z_A$$

$$y_B = R_{21}x_A + R_{22}y_A + R_{23}z_A$$

$$z_B = R_{31}x_A + R_{32}y_A + R_{33}z_A \qquad (13.25)$$

using Eq. (A4.2), for the elements R_{ij} of the rotation matrix under consideration, then if a particular direction written in terms of the cell-edge vectors x_A, y_A and z_A is $r_A = [uvw]_A$ and is $r_B = [UVW]_B$ when written in terms of the cell-edge vectors x_B, y_B and z_B,

$$\begin{pmatrix} U \\ V \\ W \end{pmatrix}_B = \begin{pmatrix} R_{11} & R_{21} & R_{31} \\ R_{12} & R_{22} & R_{32} \\ R_{13} & R_{23} & R_{33} \end{pmatrix}^{-1} \begin{pmatrix} u \\ v \\ w \end{pmatrix}_A \equiv \begin{pmatrix} R_{11} & R_{12} & R_{13} \\ R_{21} & R_{22} & R_{23} \\ R_{31} & R_{32} & R_{33} \end{pmatrix} \begin{pmatrix} u \\ v \\ w \end{pmatrix}_A \qquad (13.26)$$

using Eq. (A4.5) and the property of a rotation matrix that its inverse is identical to its transpose (Section A1.4). Therefore, for a rotation of $53.1°$ about $[001]$, Eq. (13.26) can be written in the form

$$r_B = Rr_A$$

with R given by Eq. (13.24). It is evident that this is the same form as Eq. (13.22). In general, therefore, Eq. (13.22) can be interpreted either as (i) a rotation of a vector to a new vector referred to the same crystal axes (e.g. x_A, y_A and z_A), or as (ii) the same vector expressed in terms of different crystal axes rotated with respect to one another, e.g. the cell-edge vectors x_A, y_A and z_A and the cell-edge vectors x_B, y_B and z_B. For grain boundary geometry, it is evident that it is the second of these interpretations which is the appropriate one to use.

There are 23 equivalent descriptions obtained by rotating the crystal axes of B to a new set B'. Of particular interest are the *disorientation* description and any twin orientation descriptions. The disorientation description is the description for which the magnitude of the rotation angle is the smallest, while twin orientation descriptions are ones for which the rotation angle is $180°$, so that is evident that the CSL description is equivalent to a twinning operation (Chapter 11).

The disorientation description for a rotation of $53.1°$ about $[001]$ is obtained by rotating the crystal axes of B to B' by $270°$ about $[001]$, so that

$$x_{B'} = R'x_B = R'Rx_A = {}_{B'}R_A x_A \qquad (13.27)$$

with

$$R' = \begin{bmatrix} 0 & 1 & 0 \\ -1 & 0 & 0 \\ 0 & 0 & 1 \end{bmatrix} \text{ and } {}_{B'}R_A = \frac{1}{5}\begin{bmatrix} 4 & 3 & 0 \\ -3 & 4 & 0 \\ 0 & 0 & 5 \end{bmatrix} \qquad (13.28)$$

giving the disorientation description as a rotation of $36.9°$ about $[00\bar{5}]$. By contrast, if

$$R' = \begin{bmatrix} -1 & 0 & 0 \\ 0 & 1 & 0 \\ 0 & 0 & -1 \end{bmatrix}, {}_{B'}R_A = \frac{1}{5}\begin{bmatrix} 4 & 3 & 0 \\ -3 & 4 & 0 \\ 0 & 0 & 5 \end{bmatrix} \qquad (13.29)$$

which describes a 180° rotation about [120], i.e. a twin description. Here, $\mathbf{R}'$ describes a symmetry rotation of 180° about [010].

A second twin description occurs if

$$\mathbf{R}' = \begin{bmatrix} 0 & 1 & 0 \\ 1 & 0 & 0 \\ 0 & 0 & -1 \end{bmatrix} \text{when } {}_{B'}\mathbf{R}_A = \frac{1}{5}\begin{bmatrix} 4 & 3 & 0 \\ 3 & -4 & 0 \\ 0 & 0 & -5 \end{bmatrix} \tag{13.30}$$

which describes a 180° rotation about [310]. In this second example, $\mathbf{R}'$ describes a symmetry rotation of 180° about [110]. Further equivalent descriptions can be found by appropriate matrix manipulation to confirm the data in Table 13.2 for when one-fifth of the lattice points coincide.

An alternative approach to determining equivalent geometrical descriptions of high-angle grain boundaries is to use quaternion geometry (Sections 1.6 and A1.4), making use of the multiplication law for two quaternions $\mathbf{p}$ and $\mathbf{q}$ (Problem 13.12). Using quaternion geometry it is straightforward to show that the lowest value of Σ for which a twin description is not possible for CSLs in cubic crystals occurs when $\Sigma = 39$ [31, 32], for which the disorientation description is 50.13° about a $\langle 321 \rangle$ axis. A suitable quaternion description for this disorientation is the unit quaternion $(1/\sqrt{78})\{8, 3, 2, 1\}$.

Grimmer [33] has used quaternion algebra to establish the maximum rotation angle for any disorientation angle/axis combination in the cubic crystal system. He has shown that the maximum disorientation angle θ_m about an axis [$mn1$] with $m \leq n \leq 1$ within the standard stereographic triangle is determined by either

$$\tan(\theta_m/2) = [(3 - 2\sqrt{2})(1 + m^2 + n^2)]^{1/2}$$

or

$$\tan(\theta_m/2) = (1 + m^2 + n^2)^{1/2}(1 + m + n)^{-1}$$

whichever gives the smaller value of θ_m. These values for selected rotation axes in cubic lattices are shown in Figure 13.18. The maximum possible angle of 62.8° $(= 2\tan^{-1}(23 - 16\sqrt{2})^{1/2})$ for any disorientation description in a cubic crystal within the

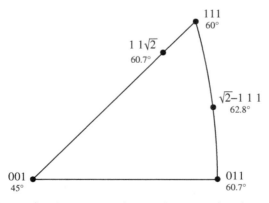

Figure 13.18 Maximum disorientation angles as a function of angle/axis descriptions within the standard stereographic triangle. Source: After Figure 1 of [33]

standard stereographic triangle occurs when the rotation axis is parallel to $[\sqrt{2} - 1, 1, 1]$. As Frank has noted [34], this disorientation description is more simply described as a rotation about a $\langle 110 \rangle$ axis of 90°.

Disorientation descriptions have also found to be very useful in the context of twinning in h.c.p. metals. Deformation twinning arises on $\{10\bar{1}2\}$ planes in all h.c.p. metals (Chapter 11). In addition, depending on the particular h.c.p. metal, twinning can occur on other planes such as $\{11\bar{2}1\}$, $\{11\bar{2}2\}$, $\{11\bar{2}4\}$ and $\{10\bar{1}1\}$ (Table 11.1). For electron back-scattered diffraction work (EBSD) on h.c.p. metals such as titanium, disorientation descriptions of these possible compound twinning operations are used instead of descriptions in terms of 180° rotations about either the twin plane normals or the shear direction within the twin plane. Practitioners of EBSD of deformed h.c.p. metals become adept at recognizing these twins from their disorientation descriptions.

As an example, deformation twinning on the $(10\bar{1}2)$ plane in an h.c.p. metal can be described either as a rotation of 180° about the normal to the $(10\bar{1}2)$ plane or as a rotation of 180° about the $[\bar{1}011]$ vector in the twin plane (Table 11.1). The disorientation descriptions has an angle dependent on the specific c/a ratio of the metal under consideration; the disorientation axis is the vector lying in $(10\bar{1}2)$ perpendicular to $[\bar{1}011]$. Therefore, for titanium for which $\gamma = c/a = 1.58595$, a suitable misorientation description for $\{10\bar{1}2\}$ twinning is 84.96° about $\langle \bar{1}2\bar{1}0 \rangle$. Formulae for disorientation descriptions for values of $\gamma < \sqrt{3}$ for compound twins in h.c.p. metals are shown in Table 13.3.

The formula for the disorientation angle will change as a function of γ. Typically, for h.c.p. metals the value of γ varies between 1.57 and 1.9 (Table 3.1). Therefore, for all practical values of γ, the formulae in Table 13.3 all produce acute values of the disorientation angle apart from the formula for the disorientation angle for twinning on $\{10\bar{1}2\}$ planes. Here, the formula produces an angle of 90° for $\gamma = \sqrt{3}$. For $\gamma > \sqrt{3}$, such as Zn and Cd, the disorientation angle for $\{10\bar{1}2\}$ twinning is $2 \tan^{-1}\left(\frac{\sqrt{3}}{\gamma}\right)$ rather than that given in Table 13.3.

Table 13.3 Disorientation descriptions for compound twins in h.c.p. metals

Twin plane	Disorientation axis	Disorientation angle for a general $\gamma < \sqrt{3}$	Disorientation angle for $\gamma = 1.58595$
$\{10\bar{1}2\}$	$\langle \bar{1}2\bar{1}0 \rangle$	$180° - 2\tan^{-1}\left(\dfrac{\sqrt{3}}{\gamma}\right)$	84.96°
$\{11\bar{2}1\}$	$\langle 1\bar{1}00 \rangle$	$2\tan^{-1}\left(\dfrac{1}{2\gamma}\right)$	35.00°
$\{11\bar{2}2\}$	$\langle 1\bar{1}00 \rangle$	$2\tan^{-1}\left(\dfrac{1}{\gamma}\right)$	64.46°
$\{11\bar{2}4\}$	$\langle 1\bar{1}00 \rangle$	$180° - 2\tan^{-1}\left(\dfrac{2}{\gamma}\right)$	76.82°
$\{10\bar{1}1\}$	$\langle \bar{1}2\bar{1}0 \rangle$	$2\tan^{-1}\left(\dfrac{\sqrt{3}}{2\gamma}\right)$	57.27°

13.4 Interface Junctions

Interface junctions occur when a grain boundary comes to a surface, for example, or when three grains are in contact with one another inside a polycrystal. From the equilibrium positions of the interfaces at such junctions, it is possible to deduce something about the relative magnitudes of their interfacial free energies, or about the variation of the energies with interface orientation.

When the energy of each intersecting interface is independent of orientation, the equilibrium configuration is easily deduced. In effect, each boundary, by trying to reduce its area, exerts a force upon the junction and the total force must vanish. When a grain boundary meets a surface, it pulls the surface down into a groove until its own tension is balanced by the surface tensions, provided that the temperature is high enough to permit atoms to move and thereby accomplish the equilibrium configuration – to form the groove, material must move from the intersection out to the sides, where it forms hills, as shown in Figure 13.19.

The equilibrium configuration for the symmetrical case shown in Figure 13.19 can be deduced as follows. Consider an infinitesimal displacement of the junction in a direction parallel to the grain boundary. The total free energy change accompanying this displacement from equilibrium must be zero. If unit length of junction is displaced by dx, the area of grain boundary decreases by dx, while the area of surface increases by $2dx \sin \theta/2$. Therefore:

$$-\gamma_b dx + 2\gamma_s dx \sin \theta/2 = 0$$

or:

$$\gamma_b = 2\gamma_s \sin \theta/2 \tag{13.31}$$

where γ_b and γ_s are the grain boundary and surface free energies. Eq. (13.31) can also be obtained by balancing the tensions that are, in effect, exerted by the interfaces, and this shows that the magnitudes of these tensions are γ_b and γ_s. By measuring the angles of grain boundary grooves, the ratio γ_b/γ_s has been found for several materials. Some values are listed in Table 13.4, along with values of γ_b deduced from the ratio γ_b/γ_s and independent measurements of γ_s. It is evident from this table that for metals $\gamma_b/\gamma_s \approx 1/3$.

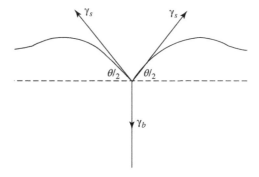

Figure 13.19   Grain boundary groove, seen in cross-section

Table 13.4 Experimentally determined energies of high-angle grain boundaries

Material	Temp. (K)	γ_b/γ_s	γ_b mJ m^{-2}	Method	Ref.
Au	1300	0.27	390	Grooving	[35]
Co	1728	0.39	1010	Grooving	[36]
Cu	1338	0.32	530	Grooving	[35]
Fe (γ)	1373		790	Polyphase equilibria	[35]
Fe (δ)	1753	0.39	985	Grooving	[36]
Nb	2523	0.36	760	Grooving	[35]
Ni	1643	0.37	930	Grooving	[36]
Sn	486	0.235	160	Grooving	[35]
NaCl	~873	—	270 ± 50	Dihedral angles	[37]
LiF	~873	—	400 ± 80	Dihedral angles	[37]

Unfortunately, the assumption that the free energies of the interfaces are independent of their orientations is not usually correct. If a piece of interface can lower its free energy by turning into a different orientation, then in order to prevent it from doing so, a couple must be applied to it. This results in the addition of so-called *torque terms* to the balance of forces at a junction.

The forces that would have to be applied to the edges of a segment of mobile interface in order to hold it in equilibrium are given by the condition for the free energy to be a minimum, or:

$$dG = 0$$

where dG is the change in free energy of the system which accompanies an infinitesimal displacement from equilibrium. By considering the creation of an extra length dx of the interface of length l shown in Figure 13.20:

$$-F_x dx + \gamma dx = 0$$

where F_x is a force per unit length. Therefore:

$$F_x = \gamma \tag{13.32}$$

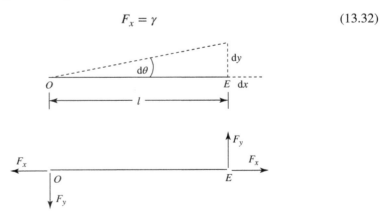

Figure 13.20 A segment of an interface, *OE*, held in equilibrium by forces F_x, F_y

By considering the displacement dy, the force per unit length F_y has to satisfy the equation:

$$-F_y \mathrm{d}y + l\frac{\mathrm{d}y}{\mathrm{d}\theta}\mathrm{d}\theta = 0$$

Since $\mathrm{d}y \approx l\mathrm{d}\theta$:

$$F_y = \frac{\mathrm{d}\gamma}{\mathrm{d}\theta} \tag{13.33}$$

The force F_x must be applied to ends of the boundary to stop it from shrinking and the force F_y must be applied to stop it from turning into an orientation of lower energy. If the point E, for example, is an interface junction then the boundary OE exerts forces upon the junction which are equal and *opposite* to F_x and F_y.

If the interface happens to be at the orientation of a cusp in free energy, where $\mathrm{d}\gamma/\mathrm{d}\theta$ changes sharply from, say, $-\alpha$ to $+\alpha$, then the situation becomes indeterminate. No force F_y is then needed to stop the boundary from turning, since any rotation of the boundary out of its cusp orientation increases its energy. In fact, a force F_y of any value between the limits $-\alpha$ and $+\alpha$ could be applied without causing the boundary to rotate.

The existence of energy cusps at special grain boundary orientations can lead to odd-looking configurations. For example, two boundaries of the same twin joining at right angles are shown schematically in Figure 13.21. It is supposed that both are special boundaries; for example, boundary 1 may be the K_1 plane of the twin, so that the atoms upon it are common to the lattices of both twin and matrix, and boundary 2 may be some other plane of good fit. All that can be deduced from this configuration is that the grain boundary energy of boundary 2 is not great enough to pull boundary 1 out of its special orientation:

$$\gamma_2 < \frac{\mathrm{d}\gamma_1}{\mathrm{d}\theta} \tag{13.34}$$

and similarly:

$$\gamma_1 < \frac{\mathrm{d}\gamma_2}{\mathrm{d}\theta} \tag{13.35}$$

where, in each case, $\mathrm{d}\gamma/\mathrm{d}\theta$ is the magnitude of the slope at the cusp in the plot of boundary free energy against the angle of rotation θ about an axis parallel to the line of intersection of boundaries 1 and 2.

The case of a triple interface junction is shown in Figure 13.22. If it is assumed that none of the intersecting interfaces lies in a special orientation, two equations of equilibrium can be derived by allowing the junction to be infinitesimally displaced in two orthogonal

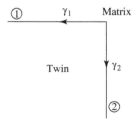

Figure 13.21 Two boundaries of the same twin joining at right angles to one another

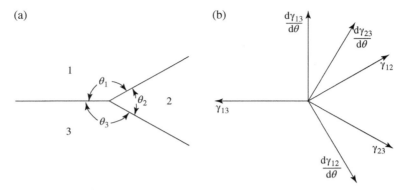

Figure 13.22 The junction of the interfaces between three grains. Each interface is normal to the plane of the figure. (a) Geometry of junction. (b) Force diagram. The positive sense of $d\theta$ is taken to be a clockwise rotation of the interface concerned

directions and by developing the condition that the resultant change in free energy must be zero. The same equations can be easily written down from the force diagram of Figure 13.22b. The forces per unit length in Figure 13.22b are forces per unit length that the boundaries exert upon the junction (i.e. they are opposite in sense to the forces per unit length F_x and F_y shown in Figure 13.20, which are forces per unit length exerted upon the boundary). It is often convenient to use the three symmetrical equations, only two of which are independent, that are obtained by setting the sum of the components of force per unit length in the directions of the three interfaces equal to zero. These are:

$$\gamma_{13} + \gamma_{12}\cos\theta_1 + \gamma_{23}\cos\theta_3 - \frac{d\gamma_{12}}{d\theta}\sin\theta_1 + \frac{d\gamma_{23}}{d\theta}\sin\theta_3 = 0$$

$$\gamma_{12} + \gamma_{23}\cos\theta_2 + \gamma_{13}\cos\theta_1 - \frac{d\gamma_{23}}{d\theta}\sin\theta_2 + \frac{d\gamma_{13}}{d\theta}\sin\theta_1 = 0 \qquad (13.36)$$

$$\gamma_{23} + \gamma_{13}\cos\theta_3 + \gamma_{12}\cos\theta_2 - \frac{d\gamma_{13}}{d\theta}\sin\theta_3 + \frac{d\gamma_{12}}{d\theta}\sin\theta_2 = 0$$

These equations were first derived by Herring [38].

Returning to the case of a grain boundary intersecting a surface, we see that if there are important, unknown, torque terms $d\gamma/d\theta$, it is generally impossible to find the ratio of grain boundary energy to surface energy from the angles at a single junction. However, in the special case of the intersection of a twin lamella with the surface of a c.c.p. crystal, it has been possible to simplify the set of three equations in Eq. (13.36) and apply them in an elegant way.

The simplification follows from the relative magnitudes of the various terms. The energy of the {111} twin boundary is expected to be small; indeed, it may well be even smaller than the magnitude of the torque term $d\gamma_s/d\theta$ arising from the anisotropy of the surface free energy. As a result, the torque terms acting at a junction between a twin boundary and the surface can overcome the tension of the twin boundary and pull the surface up into a ridge, as shown in Figure 13.23.

The occurrence of a ridge, instead of a groove, demonstrates most strikingly the existence of the torque arising from the tendency of the surface to rotate into a more favourable

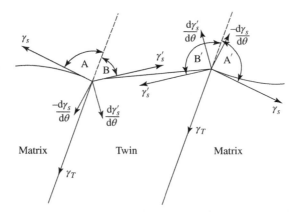

Figure 13.23 Twin boundary grooving, seen in cross-section

orientation. Because the twin boundary is itself at a special orientation where there is a cusp in free energy, the only one of the set of three equations in Eq. (13.36) that can be applied is the one that does not contain a torque term for the twin boundary. In other words, we can only balance the forces acting parallel to the twin boundary. At the two junctions shown in Figure 13.23, we have, respectively:

$$\gamma_T = \gamma_s \cos A + \gamma'_s \cos B + \frac{d\gamma_s}{d\theta} \sin A - \frac{d\gamma'_s}{d\theta} \sin B$$

and : $\qquad\qquad\qquad\qquad\qquad\qquad\qquad\qquad\qquad\qquad\qquad\qquad$ (13.37)

$$\gamma_T = \gamma_s \cos A' + \gamma'_s \cos B' - \frac{d\gamma_s}{d\theta} \sin A' + \frac{d\gamma'_s}{d\theta} \sin B'$$

Adding the two equations in Eq. (13.37) together and assuming that $\gamma_s \approx \gamma'_s$, we obtain:

$$2\gamma_T = \gamma_s(\cos A + \cos A' + \cos B + \cos B')$$
$$+ \frac{d\gamma_s}{d\theta}(\sin A - \sin A') + \frac{d\gamma_s}{d\theta}(\sin B' - \sin B)$$

Now by observation, $A + A' \approx 180°$, $B + B' \approx 180°$, so that $\sin A - \sin A'$ and $\sin B - \sin B'$ are small quantities. Since $d\gamma_s/d\theta$ and $d\gamma'_s/d\theta$, although larger than γ_T, are themselves expected to be much smaller than γ_s, we can write:

$$2\gamma_T \approx \gamma_s(\cos A + \cos A' + \cos B + \cos B') \qquad\qquad (13.38)$$

Hence the ratio of the free energies of the twin boundary and surface, γ_T/γ_s, can be found. This technique was first used by Mykura [39] to measure twin boundary free energies (knowing the surface energy) and to build up a picture of the anisotropy of the surface free energy in c.c.p. metals. The main conclusions are that the free energy of a {111} boundary of a twin in a c.c.p. metal is very small compared with a surface free energy and that the surface free energy does not vary by more than 10% with change in orientation. Provided that the surfaces are clean, it is found that the close-packed {111} surface has the smallest free energy, as might be expected. Some values of twin boundary free energies for c.c.p. metals are listed in Table 13.5.

Table 13.5 Twin boundary free energies, γ_T, relative to surface energies, γ_s, for some c.c.p. metals

Metal	Temp. (K)	Atmosphere	γ_T/γ_s	Ref.
Ag	1123	H$_2$	0.016 ± 0.001	[40]
Co	1273	Vacuum	0.0158 ± 0.002	[40]
Cu	1173	Vacuum	0.016 ± 0.002	[40]
γ-Fe	1323	Vacuum	0.016 ± 0.0027	[40]
Ni	1273	Vacuum	0.0133 ± 0.0025	[40]
Pt	1773	Vacuum	0.028 ± 0.03	[41]

13.5 The Shapes of Crystals and Grains

Although the shape of a crystal is usually a consequence of the way in which it grew, or perhaps of its cleavage, there must nevertheless be some equilibrium shape that might be reached in practice by an unconstrained small crystal or by a small void within a crystal. This shape is determined by the surface free energy γ and is such as to minimize the total free energy. That is:

$$\int_A \gamma \, dA = a \, minimum \tag{13.39}$$

If the surface free energy is isotropic, the equilibrium shape is a sphere; that is, the surface area of a given volume is minimized. If the surface free energy varies with orientation, the equilibrium shape can be derived from the γ-plot by means of a theorem due to Wulff [42]. The equilibrium shape, often referred to as the Wulff form, is that of the inner envelope of planes normal to, and passing through the ends of, the vectors representing surface free energies on the γ-plot ('Wulff planes'). The proof of Wulff's theorem for a body of fixed volume is non-trivial – for discussion of the proof, see Herring [43] and Johnson and Chakerian [44].

If the γ-plot is deeply cusped, the equilibrium shape will be a polyhedron. This is illustrated in two dimensions in Figure 13.24 for the hypothetical γ-plot of a c.c.p. metal first shown in Figure 13.5. The equilibrium shape may contain rounded regions if the cusps are less pronounced, although it must still possess flat surfaces at the cusp orientations. The question of whether a surface of given orientation will appear on the equilibrium shape, or 'Wulff form', can be studied by means of a construction due to Herring [43].

Consider some orientation, OA in Figure 13.25, O being the origin of the γ-plot, which is smoothly curved at A. If the surface that is normal to OA is to appear on the Wulff form then it must do so in a region of the body that touches the Wulff plane AB, normal to OA. Where on the plane AB does this happen? This question is answered by considering points in the neighbourhood of A, such as A′. These points represent the free energies of surfaces of slightly different orientations to the surface whose free energy is represented by A, and so these surfaces must appear on the Wulff form close to the point at which the surface is normal to OA. As the points are brought closer and closer to A, their Wulff planes must touch the Wulff form at points closer and closer to the point at which the Wulff plane at A touches it.

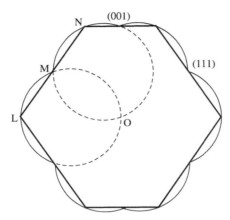

Figure 13.24   The γ-plot of Figure 13.5, showing the equilibrium shape (or Wulff form) of the crystal

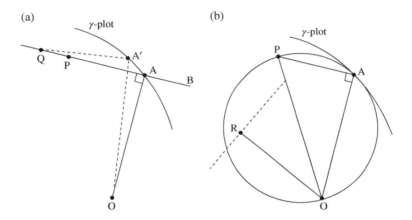

Figure 13.25   Construction due to Herring [43]

The plane that is normal to OA$'$ intersects the line AB at Q. As various points A$'$, A$''$ … around A are brought in towards A, the lines of intersection of their Wulff planes tend towards an intersection at a single point, P. If the Wulff form exhibits the surface that is normal to OA then it does so at the point P. Now, in two dimensions, the locus of points defined by the property that the normals to the radius vector OA$'$ at each point intersect in a common point P$'$ is a circle through the origin O and with OP$'$ as diameter, since the diameter of a circle subtends an angle of $90°$ at all points on the circumference.

There is a corresponding result in three dimensions, from which it follows that the γ-plot at A touches a sphere with OP as diameter, as shown in Figure 13.25b. An important result follows immediately. Consider any other direction through O. If it meets the γ-plot inside the sphere of diameter OP, as at R, then its Wulff plane will cut off the point P, which cannot then lie on the *inner envelope* of Wulff planes.

It follows that the surface that is normal to OA cannot appear if the γ-plot passes anywhere inside the sphere through O which is tangent to the γ-plot at A. If, instead, the γ-plot is everywhere outside this tangent sphere, then access to P cannot be cut off and the surface that is normal to OA must appear on the Wulff form. By similar reasoning, it follows that an orientation at which there is an energy cusp will appear on the Wulff form if and only if a sphere having the line from O to the cusp point as diameter lies entirely inside the γ-plot.

An interesting special case arises when the γ-plot coincides with a portion of the tangent sphere. This is illustrated in two dimensions in Figure 13.24. Here, the section of the γ-plot consists of circles through the origin, two of which are shown. OL is a diameter of the circle LMO, so that all Wulff planes to points on the γ-plot in the section L $\rightarrow$ M touch the Wulff form at the corner L. Similarly, ON is a diameter of the circle OMN and all Wulff planes to points on the γ-plot from M to N touch the Wulff form at the corner N. In fact, every Wulff plane from the γ-plot touches a corner of the two-dimensional Wulff form, so that, strictly speaking, surfaces of all orientations appear on it, although only surfaces of the cusp orientations such as the {111} surface LN have any extension, other surfaces being crowded into a corner such as N.

If the cusp points of Figure 13.24 are kept fixed, but the rest of the γ-plot is expanded, as in Figure 13.26, the Wulff form remains the same, but now Wulff planes from general points do not touch it, even at a corner. This distinction has some significance in connection with thermal faceting, which is defined below.

An ordinary-sized crystal of arbitrary shape never reaches its equilibrium shape in practice, because of the large redistribution of material required and the small amount of energy to be gained thereby. It is more likely that the surface of a crystal will be able to reduce its energy by means of the relatively small atom movements needed to break up the surface into small sections, one or more of which has an orientation of lower energy. This process is called thermal faceting (Figure 13.27) and is particularly relevant to an understanding of the shapes of nanoparticles.

Herring has shown that a surface of an orientation not represented on the Wulff form can always reduce its total surface free energy by faceting. A surface parallel to a Wulff plane which touches the Wulff form cannot reduce its energy by faceting, even when the plane touches only a corner of the Wulff form, as in Figure 13.24.

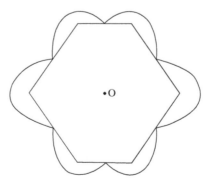

Figure 13.26 The same equilibrium shape as shown in Figure 13.24, arising from a different γ-plot

Figure 13.27 A surface that has reduced its energy by breaking up into facets, one of which is a low-index plane with a small surface tension γ_0. The surface is normal to the plane of the figure

Crystals seldom appear in their equilibrium shape; it is much more common for their shape to be the result of the way in which they grew. When the growth rate of a crystal is determined only by the rate at which its surfaces can accept atoms, the rate of advance of a surface will depend only on its crystallographic orientation. In this way, the shape of a salt crystal grown from aqueous solution may be determined simply by the relative growth rates at its various surfaces. The faster-growing orientations will tend to grow themselves out, leaving the crystal with its more slowly growing surfaces. The steady-state shape of the growing crystal can be derived from a polar diagram of the growth rate by means of the same construction used to derive the equilibrium shape from the γ-plot. The analogue of the Wulff theorem is that the crystal becomes trapped in that shape which grows at the slowest possible rate.

Often, the rate of advance of the surface of a growing crystal does not depend solely on its orientation. Instead, the growth rate may be controlled by the rate at which material can be transported to the surface or by the rate at which heat can be removed. Under these conditions an unstable form of growth sometimes occurs, which produces a spiky or 'dendritic' morphology. Dendrites are observed frequently when a crystal has grown from a liquid, and occasionally in solid-state transformations.

To illustrate how growth instabilities can occur, consider for example a plane surface of a crystal growing into a supersaturated solution. Suppose that the growth rate is limited only by the rate of diffusion of solute atoms to the surface. Then there must be a concentration gradient from the supersaturated value in the solution to the concentration in equilibrium with the crystal at the surface. If now, by chance, a bump starts to develop, the concentration gradient at its nose must become steeper, because it is thrust in towards the region of high solute concentration. The steepening of the gradient increases the rate of diffusion to the bump and tends to amplify it. Opposing this is the action of surface tension, which tends to flatten the bump and sets up a flux of solute from the bump out to the sides. Provided that the bump is not too sharp, its amplitude will show a net increase. The growth of the bump into a spike, followed by the development of bumps upon the spike, and so on, produces a dendrite. Often crystallographic factors combine with conditions of unstable growth to produce a dendritic crystal whose dendrite arms have crystallographic orientations.

The arrangement of the grains within a polycrystal is observed to be quite similar to that of bubbles within a froth. This suggests that the pattern of grains is governed by grain boundary tensions, just as the pattern of bubbles in a froth is governed by the surface tension of the liquid between the bubbles. In drawing this analogy, any anisotropy of the grain boundary free energies in a crystalline material is neglected. The two principles that govern the shapes of the grains are then, first, that they must fill space and, second, that grain boundary tensions must balance at grain edges and corners.

The only edges that should occur are those along which three boundaries meet, and the only corners those at which four edges meet, because other junctions can reduce the

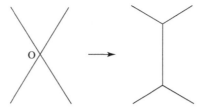

Figure 13.28   The instability of four interfaces meeting along a line through O, normal to the figure

grain boundary area by decomposing after the manner shown in Figure 13.28. (In fact, polycrystals, unlike froths, have been found to contain a small proportion of abnormal junctions, perhaps because of some anisotropy in the boundary free energy.) The polyhedron which can be stacked to fill space under this constraint is the truncated octahedron, shown in Figure 13.29a. This shape is also known as either an orthic tetrakaidecahedron or the Wigner–Seitz cell of the b.c.c. crystal structure [45]. However, this polyhedron, when stacked, does not fulfil the requirement that the boundary tensions should balance at its edges and corners. Kelvin [46] showed that to make the three faces meet at 120° and the four edges at 109.47°, each hexagonal face must acquire complex curvatures, as shown in Figure 13.29b [45].

It turns out that a three-dimensional stacking of the shape proposed by Kelvin is not after all the structure in three dimensions in which the surface energy is minimized, and where each shape has the same volume. This would appear to be the Weaire–Phelan structure (Figure 13.30), in which the periodic repeating unit is a simple cubic unit cell consisting of six 14-sided tetrakaidecahedra and two 12-sided irregular pentagonal dodecahedra [47]. This minimizes surface area subject to a constraint of fixed volumes of the individual foam cells. It has a surface energy that is approximately 0.3% less than the Kelvin solution.

Of course, neither polycrystals nor froths ever attain the ideal regularity of an array of cells with equal volumes. Grains vary in size, in the number of faces that they possess and in the shapes of their faces. Nevertheless, their average properties are found to correspond closely to those of the truncated octahedron. This has 14 faces, 36 edges and 24 vertices, so that the average number of edges per face is 36/7, each edge being shared by two faces. A face with five edges occurs in assemblies of grains in crystals more frequently than any other [48].

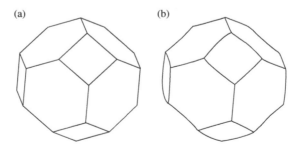

Figure 13.29   (a) Truncated octahedron and (b) distorted truncated octahedron, which, when stacked, meets the requirement that the surface tensions balance at all junctions. Source: Diagram (b) reproduced from [45] with permission

(a)   (b)

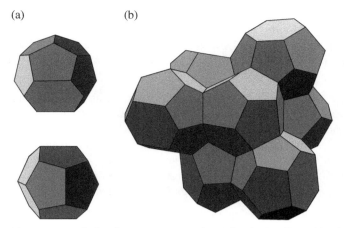

Figure 13.30 The Weaire–Phelan foam structure. The individual cells within the foam structure are shown in (a), and the stacking together of these cells is shown in (b)

In a froth, the faces of bubbles with fewer than average faces have to bow out so as to increase the angles at which they meet to 120°. To maintain this convexity, the gas in the bubble must be at a higher pressure than the gas in neighbouring bubbles. If the gas can diffuse through the bubble wall, it will leak out and the bubble will shrink. The ultimate loss of such bubbles increases the average bubble size. Grain growth also occurs in polycrystals at high temperatures [49]. The tendency of a curved grain boundary to reduce its area while at the same time maintaining a balance of tensions at its edges causes it to migrate towards its centre of curvature.

Problems

13.1 Assuming no atomic relaxations on or near the surfaces, sketch the structure (or structures) of surfaces of the following orientations in an h.c.p. metal: (a) $(10\bar{1}2)$, (b) $(11\bar{2}2)$.

13.2 Show that any dislocation line of Burgers vector **b** emerging through a surface of unit normal **n** produces a step whose height, measured normal to the surface, is **n** · **b**. Show that the step cannot be removed by evaporation of the crystal except in one special case. Write down an expression for the height of the step in units of the spacing of the lattice planes parallel to the surface. Apply the expression to the case of a dislocation with a $\frac{1}{2}[110]$ Burgers vector in a crystal with a c.c.p. lattice when the dislocation emerges through the following surfaces: (111), (211), (100), (110).

13.3 According to Mackenzie, Moore and Nicholas, when the energy of a surface in a c.c.p. metal is estimated on the basis of the number of bonds between nearest neighbours which are broken, the energy of a surface of orientation (*hkl*) is related to the energy of the {210} surface by the equation:

$$\gamma_{hkl} = \gamma_{210} \cos \theta$$

where θ is the angle between the surface $\{hkl\}$ and the nearest $\{210\}$ surface. Show that the γ-plot for this model consists of portions of spheres whose diameters are the vectors representing $\{210\}$ energies on the plot and sketch the section of the plot for surfaces in the $[001]$ zone, giving the values of the energies of low-index surfaces in terms of γ_{210}.

13.4 A silver wire of radius $65\,\mu m$ stressed in tension at an elevated temperature was observed to contract slowly when the stress was $14\,kPa$ and to extend at about the same rate when the stress was increased to $17\,kPa$. If the wire contained 6000 grain boundaries running across it, per metre length of wire, estimate the surface free energy of the wire. Assume that the grain boundary free energy is one-third of the surface free energy. Determine the error incurred if the effect of the grain boundaries is neglected and comment on the significance of this error.

13.5 Derive a simple expression for the energy of a small-angle tilt boundary between two crystals in terms of the angle of tilt and the (isotropic) elastic properties of the material. Use your expression to investigate:
(a) Whether two parallel boundaries of the same kind attract or repel one another.
(b) How the energy of a boundary of a given angle changes as the number of dislocations in it is reduced and their Burgers vectors are increased.

13.6 Consider the low-angle tilt boundary produced by a wall of edge dislocations corresponding to one of the predominant slip systems in each of the following materials: (a) NaCl, (b) Al, (c) Zn. Sketch the boundary structure on a plane normal to the edge dislocations, give the indices of the boundary plane and of the axis of tilt, and give an expression for the angle of tilt in terms of the separation of the dislocations in the boundary.

13.7 A low-angle symmetric tilt grain boundary made up of edge dislocations is formed in a c.c.p. bicrystal where the lattice parameter of each crystal is $4\,\text{Å}$. The rotation axis is $[11\bar{2}]$.
(a) Specify the boundary plane.
(b) Determine the length of dislocation line per unit area of boundary if the two crystals are rotated by an angle of $0.5°$ about $[11\bar{2}]$ in the bicrystal.
(c) Show that the edge dislocations in the bicrystal in (b) are stable with respect to displacement of any one of them on its slip plane.
(d) At what angle of rotation about $[11\bar{2}]$ would the individual edge dislocations making up a low-angle symmetrical tilt grain boundary become indistinguishable for these c.c.p. crystals?

13.8 In graphite, twinned regions are found in which the twin and matrix are related by a tilt of approximately $20.5°$ about an axis of the type $\langle 1\bar{1}00\rangle$. The axial ratio of graphite, $c/a = 2.72$. Investigate the dislocation structure of the boundary and find the simplest structures employing (a) total dislocations and (b) partial dislocations.

13.9 Sketch and determine the structure of the coincidence lattice produced in a b.c.c. lattice by a rotation of $36.9°$ about a $\langle 100\rangle$ axis.

13.10 A misorientation between two grains in a c.c.p. material is described by the rotation matrix

$$\mathbf{R}_1 = \frac{1}{3}\begin{pmatrix} 2 & -2 & -1 \\ 2 & 1 & 2 \\ -1 & -2 & 2 \end{pmatrix}$$

Determine the angle and axis associated with this rotation. Is this a possible disorientation description?

13.11 Determine the rotation matrix associated with a rotation of 180° about [111] for a c.c.p. crystal. Call this $\mathbf{R}_2$. Determine the rotation matrix $\mathbf{R}_3$ for which $\mathbf{R}_1 = \mathbf{R}_2\mathbf{R}_3$ (with $\mathbf{R}_1$ given in Question 13.10) and determine its angle and axis of rotation. Comment on this result.

Specify the axis and angle for a disorientation description symmetrically equivalent to the rotation of 180° about [111].

13.12 Determine the angles and axes of rotation associated with the unit quaternions

$$\mathbf{p} = \frac{1}{\sqrt{51}}\{0,1,1,7\} \text{ and } \mathbf{q} = \frac{1}{\sqrt{51}}\{7,-1,1,0\}$$

Using the formula for the product of two quaternions given Appendix A, Section A1.4, determine the quaternions $\mathbf{p}\cdot\mathbf{q}$ and $\mathbf{q}\cdot\mathbf{p}$ and the angle and axes of rotation which these new quaternions represent. Comment on your results.

13.13 A triple junction shown below has three grains A, B and C of a c.c.p. metal meeting at the junction.

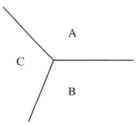

Figure for Problem 13.13

Electron back-scattered diffraction patterns show that grains A and B are related to one another by a rotation of 60° about a ⟨111⟩ axis. Grains A and C are also related to one another by a rotation of 60° about a ⟨111⟩ axis, but a different ⟨111⟩ axis to that which relates grain A to grain B.

Taking the ⟨111⟩ axis for the rotation from A to B to be [111] and the ⟨111⟩ axis for the rotation from A to C to be [11$\bar{1}$], rotation matrices $\mathbf{R}_1$ and $\mathbf{R}_2$ can be defined so that the components of a vector $\mathbf{r}$ defined with respect to basis A are related to the components defined with respect to the bases B and C are such that

$$\mathbf{r}_B = \mathbf{R}_1\mathbf{r}_A \text{ and } \mathbf{r}_C = \mathbf{R}_2\mathbf{r}_A$$

Determine the rotation $\mathbf{R}$ which relates $\mathbf{r}_B$ to $\mathbf{r}_C$:

$$\mathbf{r}_C = \mathbf{R}\mathbf{r}_B$$

and determine the angle and axis of rotation of $\mathbf{R}$.

You suspect that $\mathbf{R}$ might define a coincidence site lattice orientation. A colleague suggests to you that it is a $\Sigma = 9$ CSL defined by a rotation angle of $\cos^{-1}(7/9)$ about [011] if a suitable symmetry operation of the holosymmetric cube is applied to the crystal axes of the B grain. Show that this is a correct suggestion.

13.14 Confirm that the disorientation descriptions in Table 13.3 for h.c.p. metals with $\gamma < \sqrt{3}$, where $\gamma = c/a$, are correct.

13.15 A twin lamella in a c.c.p. metal intersects the surface of the metal at 90°. After annealing, a ridge is observed along one twin boundary–surface junction and a groove along the other. If the twin boundary free energy is one-twentieth of the surface free energy, what difference do you expect to find between the angle at the base of the notch and the angle at the tip of the ridge? Carefully point out any assumptions or approximations in your calculation.

13.16 If a bubble is blown in a liquid, its radius of curvature r is related to the excess pressure p inside the bubble by the well-known equation:

$$p = \frac{2\gamma}{r}$$

where γ is the surface free energy of the liquid. If this equation is applied without modification to a material whose surface energy is anisotropic (e.g. to a gas bubble in a crystal at an elevated temperature), it appears to predict that the orientations of highest surface free energy on the bubble will have the largest radius of curvature and therefore the largest extent. This conflicts with the requirement that the total free energy should be a minimum; that is, it predicts a shape quite different from that of the Wulff form. Resolve this apparent paradox.

13.17 The surface of a crystal lies at 10° to a low-index plane. After annealing, it is observed that the surface has broken up into facets, one of which is the low-index surface and the other a surface of no special orientation. If the free energies of the original surface and of the non-special facet are both γ and the angle that the non-special facet makes with the low-index plane is 26°, determine the free energy of the low-index surface and the effective free energy of the faceted surface (i.e. the free energy per unit area of the original surface). Assume that the free energy of the non-special facet does not vary with rotation about its present orientation.

Suggestions for Further Reading

Christian, J.W. (2002). *The Theory of Transformations in Metals and Alloys (Part I + II)*. Oxford: Pergamon Press, Elsevier Science.

Flewitt, P.E.J. and Wild, R.K. (2001). *Grain Boundaries: Their Microstructure and Chemistry*. Chichester: Wiley.

Howe, J.M. (1997). *Interfaces in Materials: Atomic Structure, Kinetics and Thermodynamics of Solid–Vapor, Solid–Liquid and Solid–Solid Interfaces*. New York: Wiley.

Tiller, W.A. (1991). *The Science of Crystallization: Macroscopic Phenomena and Defect Generation*. Cambridge: Cambridge University Press.

Tiller, W.A. (1991). *The Science of Crystallization: Microscopic Interfacial Phenomena*. Cambridge: Cambridge University Press.

References

[1] Haneman, D. (1987). Surfaces of silicon. *Rep. Prog. Phys.* 50: 1045–1086.

[2] Besenbacher, F. (1996). Scanning tunnelling microscopy studies of metal surfaces. *Rep. Prog. Phys.* 59: 1737–1802.

[3] Sutton, A.P. and Balluffi, R.W. (1995). *Interfaces in Crystalline Materials*. Oxford: Clarendon Press.

[4] Mackenzie, J.K., Moore, A.J.W. and Nicholas, J.F. (1962). Bonds broken at atomically flat crystal surfaces–I. Face-centred and body-centred cubic crystals. *J. Phys. Chem. Solids* 23: 185–196.

[5] Udin, H., Shaler, A.J. and Wulff, J. (1949). The surface tension of solid copper. *Trans. Am. Inst. Min. Metall. Petrol. Engrs.* 185: 186–190.

[6] Udin, H. (1951). Grain boundary effect in surface tension measurement. *Trans. Am. Inst. Min. Metall. Petrol. Engrs.* 189: 63.

[7] Kumikov, V.K. and Khokonov, K.B. (1983). On the measurement of surface free energy and surface tension of solid metals. *J. Appl. Phys.* 54: 1346–1350.

[8] Gilman, J.J. (1960). Direct measurements of the surface energies of crystals. *J. Appl. Phys.* 31: 2208–2218.

[9] Westwood, A.R.C. and Hitch, T.T. (1963). Surface energy of {100} potassium chloride. *J. Appl. Phys.* 34: 3085–3089.

[10] Westwood, A.R.C. and Goldheim, D.L. (1963). Cleavage surface energy of {100} magnesium oxide. *J. Appl. Phys.* 34: 3335–3339.

[11] Gutshall, P.L. and Gross, G.E. (1965). Cleavage surface energy of NaCl and MgO in vacuum. *J. Appl. Phys.* 36: 2459–2460.

[12] Read, W.T. (1953). *Dislocations in Crystals*. New York: McGraw-Hill.

[13] Read, W.T. and Shockley, W. (1950). Dislocation models of crystal grain boundaries. *Phys. Rev.* 78: 275–289.

[14] Gjostein, N.A. and Rhines, F.N. (1959). Absolute interfacial energies of [001] tilt and twist grain boundaries in copper. *Acta Metall.* 7: 319–330.

[15] Hasson, G., Boos, J.-Y., Herbeuval, I., Biscondi, M. and Goux, C. (1972). Theoretical and experimental determinations of grain boundary structures and energies: correlations with various experimental results. *Surf. Sci.* 31: 115–137.

[16] Sutton, A.P. and Vitek, V. (1983). On the structure of tilt grain boundaries in cubic metals. I. Symmetrical tilt boundaries. *Phil. Trans. Roy. Soc. Lond.* A 309: 1–36.

[17] Shewmon, P.G. (1966). Energy and structure of grain boundaries. In: *Recrystallization, Grain Growth and Textures*, pp. 165–199. Metals Park, Ohio: American Society for Metals.

[18] Sutton, A.P. and Vitek, V. (1983). On the structure of tilt grain boundaries in cubic metals. II. Asymmetrical tilt boundaries. *Phil. Trans. Roy. Soc. Lond.* A 309: 37–54.

[19] Sutton, A.P. and Vitek, V. (1983). On the structure of tilt grain boundaries in cubic metals. III. Generalizations of the structural study and implications for the properties of grain boundaries. *Phil. Trans. Roy. Soc. Lond.* A 309: 55–68.

[20] Brandon, D.G., Ralph, B., Ranganathan, S. and Wald, M.S. (1964). A field ion microscope study of atomic configuration at grain boundaries. *Acta Metall.* 12: 813–821.

[21] Bristowe, P.D. and Crocker, A.G. (1978). The structure of high-angle (001) CSL twist boundaries in f.c.c. metals. *Phil. Mag.* A 38: 487–502.

[22] Bollmann, W. (1970). *Crystal Defects and Crystalline Interfaces*. New York: Springer-Verlag.

[23] Bonnet, R. and Cousineau, E. (1977). Computation of coincident and near-coincident cells for any two lattices – related DSC-1 and DSC-2 lattices. *Acta Crystallogr.* A 33: 850–856.

[24] Zhang, W.-Z. and Weatherly, G.C. (2005). On the crystallography of precipitation. *Prog. Mater. Sci.* 50: 181–292.

[25] Zhang, M.-X. and Kelly, P.M. (2009). Crystallographic features of phase transformations in solids. *Prog. Mater. Sci.* 54: 1101–1170.

[26] Gautam, A.R.S. and Howe, J.M. (2011). A method to predict the orientation relationship, interface planes and morphology between a crystalline precipitate and matrix. Part I. Approach. *Phil. Mag.* 91: 3203–3327.

[27] Sutton, A.P. and Balluffi, R.W. (1987). On geometric criteria for low interfacial energy. *Acta Metall.* 35: 2177–2201.

[28] Shiflet, G.J. and van der Merwe, J.H. (1994). The role of structural ledges as misfit-compensating defects: fcc-bcc interphase boundaries. *Metall. Mater. Trans. A* 25: 1895–1903.

[29] Frank, F.C. (1950). The resultant content of dislocations in an arbitrary intercrystalline boundary. In: *A Symposium on the Plastic Deformation of Crystalline Solids*, pp. 150–151. Pittsburgh, PA: Mellon Institute/US Office of Naval Research; discussion pp. 151–154.

[30] Bollmann, W. (1964). Some basic problems concerning subgrain boundaries. *Discuss. Faraday Soc.* 38: 26–34.

[31] Woirgard, J. and de Fouquet, J. (1972). Réponse a "axis/angle pair description of coincidence site lattice grain boundaries". *Scripta Metall.* 6: 21–30.

[32] Grimmer, H. (1973). Coincidence rotations for cubic lattices. *Scripta Metall.* 7: 1295–1300.

[33] Grimmer, H. (1974). Disorientations and coincidence rotations for cubic lattices. *Acta Crystallogr. A* 30: 685–688.

[34] Frank, F.C. (1988). Orientation mapping. *Metall. Trans. A.* 19: 403–408.

[35] Hondros, E.D. (1969). Energetics of solid-solid interfaces. In: *Interfaces Conference*, Melbourne (ed. R.C. Gifkins), pp. 77–100. London: Butterworths.

[36] Roth, T.A. (1975). The surface and grain boundary energies of iron, cobalt and nickel. *Mater. Sci. Eng.* 18: 183–192.

[37] Spitzer, D.P. (1962). Intercrystalline energies in the alkali halides. *J. Phys. Chem.* 66: 31–38.

[38] Herring, C. (1951). Surface tension as a motivation for sintering. In: *The Physics of Powder Metallurgy* (ed. W.E. Kingston), pp. 143–178. New York: McGraw-Hill; discussion pp. 178–179.

[39] Mykura, H. (1961). The variation of the surface tension of nickel with crystallographic orientation. *Acta Metall.* 9: 570–576.

[40] Kudrman, J. and Čadek, J. (1969). The energy of coherent twin boundaries of some metals and alloys. *Phil. Mag.* 20: 413–420.

[41] McLean, M. and Mykura, H. (1966). Temperature coefficient of twin boundary energy of platinum and cobalt. *Phil. Mag.* 14: 1191–1197.

[42] Wulff, G. (1901). Zur Frage der Geschwindigkeit des Wachsthums und der Auflösung der Krystallflächen. *Z. Kristallogr.* 34: 449–530.

[43] Herring, C. (1951). Some theorems on the free energies of crystal surfaces. *Phys. Rev.* 82: 87–93.

[44] Johnson, C.A. and Chakerian, G.D. (1965). On the proof and uniqueness of Wulff's construction of the shape of minimum surface free energy. *J. Math. Phys.* 6: 1403–1404.

[45] Weaire, D. (2008). Kelvin's foam structure: a commentary. *Phil. Mag. Lett.* 88: 91–102.

[46] Thomson, W. (Lord Kelvin) (1887). On the division of space with minimum partitional area. *Phil. Mag.* 24: 503–514.

[47] Weaire, D. and Phelan, R. (1994). A counter-example to Kelvin's conjecture on minimal surfaces. *Phil. Mag. Lett.* 69: 107–110.

[48] Smith, C.S. (1952). Grain shapes and other metallurgical applications of topology. In: *Metal Interfaces*, pp. 65–108. Cleveland, Ohio: American Society for Metals; discussion pp. 108–113.

[49] Atkinson, H.V. (1988). Theories of normal grain growth in pure single phase systems. *Acta Metall.* 36: 469–491.

14

Interphase Boundaries

14.1   Boundaries Between Different Phases

Like a grain boundary, a boundary between two different crystalline phases, an *interphase* boundary, can be specified by describing the orientation relationship between the lattices of the two crystals and the orientation of the boundary itself. Corresponding to grain boundaries of special orientation, such as twin boundaries, there are special boundaries between two different crystals in a specific orientation relationship; corresponding to high-angle grain boundaries, there are interfaces between randomly oriented crystals of different kinds. Examples of special orientation relationships between two phases are the Kurdjumov[1]–Sachs and Nishiyama–Wassermann orientation relationships found between austenite and martensite considered in Chapter 12. These orientation relationships, and variants of these orientation relationships, are also found in diffusional phase transformations between c.c.p. and b.c.c. phases.

The free energy of a random interphase boundary relative to that of a grain boundary can be obtained from the angles between interfaces in a two-phase mixture. A particle of a phase B situated at a grain boundary of the phase A is shown in Figure 14.1. Assuming that the interfacial energies do not vary with orientation (i.e. neglecting torque terms), the 'dihedral angle' θ is given by an equation similar to Eq. (13.31):

$$\cos \frac{\theta}{2} = \frac{\gamma_{AA}}{2\gamma_{AB}} \tag{14.1}$$

Some results derived from the measurement of dihedral angles are shown in Table 14.1. It can be seen that the interfacial energies of random interphase boundaries do not differ

---

[1] See footnote 1 of Chapter 12, p. 391. In this Chapter, Chapter 14, the spelling of Kurdjumov will be used.

Crystallography and Crystal Defects, Third Edition. Anthony Kelly and Kevin M. Knowles.
© 2020 John Wiley & Sons Ltd. Published 2020 by John Wiley & Sons Ltd.
Companion website: www.wiley.com/go/kelly/crystallography3e

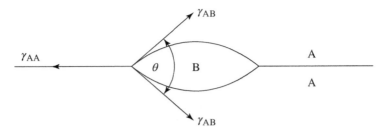

Figure 14.1 A particle of a phase B situated at a grain boundary of the phase A. γ_{AA} is the surface energy of the grain boundary of A and γ_{AB} is the surface energy of the AB interphase boundary. θ is the dihedral angle

Table 14.1 Relative energies of random interphase boundaries

System	Temp. (°C)	Phase A	Phase B	γ_{AB}/γ_{AA}	γ_{AB}/γ_{BB}	Ref.
Fe–C	910	α b.c.c. 0.01 wt% C	γ c.c.p. 0.22 wt% C	0.93 ± 0.02	0.90 ± 0.02	[1]
Fe–Cu	825	α b.c.c.	Cu	0.74	0.86	[2]
	1000	γ c.c.p.	Cu	0.61	0.87	[2]
Cu–Zn	700	α c.c.p.	β b.c.c.	0.83	1.00	[2]
Cu–Sn	750	α c.c.p.	β b.c.c.	0.76	0.93	[2]
NaCl–NaF	600	NaCl	NaF	0.78	0.90	[3]
NaCl–LiF	600	NaCl	LiF	0.76	1.13	[3]
NaCl–NaI	550	NaCl	NaI	1.08	0.78	[3]
LiF–CsCl	550	LiF	CsCl	1.38	0.65	[3]

greatly from the interfacial energies of high-angle grain boundaries in the same system, and are often slightly less.

At the opposite extreme to the random interphase boundary is a boundary between two crystals that have different atoms but identical structures, so that boundaries of any orientation between parallel crystals are fully coherent. This situation occurs when a precipitate is formed by ageing a supersaturated solid solution of Ag in Al. The Ag atoms collect into small spherical clusters on a continuous c.c.p. lattice. Another possibility is that the crystal structures of a precipitate and matrix differ, but possess a common sublattice. For example, magnesioferrite, $MgFe_2O_4$, contains a c.c.p. sublattice of oxygen ions that is almost identical to that of MgO (Section 3.6). Small $MgFe_2O_4$ particles precipitating within MgO preserve the oxygen ion sublattice, and an interface passing through oxygen ions, such as a suitably located $\{111\}$, is common to both structures. A third possibility is that the two crystal structures possess only a plane in common. For example, the close-packed planes of a c.c.p. and a h.c.p. crystal match exactly if the interatomic distances are equal; the martensitic transformation in cobalt illustrates this case (Chapter 12). In general, of course, it is impossible to find identical planes in two different crystal structures, and arbitrarily chosen crystals cannot be joined by a coherent boundary.

Usually, even the corresponding planes of two specially related crystals match only approximately. An exact fit could then be achieved by means of a uniform elastic strain, or alternatively, the misfit could be localized at dislocation lines. A boundary in which a small misfit is taken up by an array of dislocations is called a partially coherent boundary; it is analogous to a low-angle grain boundary. A difference is that an array of dislocations which constitutes a low-angle boundary is always such that the strain vanishes at large distances, whereas a partially coherent boundary array would produce a long-range strain if it were located within a single homogeneous medium.

The *actual* strain within the pair of different crystals can be thought of as being the result of a homogeneous elastic strain, which gives an exact fit, and a strain due to the array of dislocations at the interface, the two strains cancelling at large distances from the interface. For example, the case of a misfit in one direction in an interface between two orthorhombic crystals is shown in Figure 14.2. The junction can be thought of as having been made by first stretching the upper crystal horizontally in the *a* direction to give a perfect fit, and then introducing parallel edge dislocations that cancel this strain at large distances, but at the expense of producing local regions of mismatch at the boundary.

The number of extra planes to be accommodated in the upper crystal of Figure 14.2, in unit distance, is:

$$\rho = \frac{1}{a_1} - \frac{1}{a_2} \tag{14.2}$$

The dislocation spacing $p = \rho^{-1}$ is therefore:

$$p = \frac{a_1 a_2}{a_2 - a_1} \tag{14.3}$$

The dimension of a small particle precipitating from solid solution might be much less than the dislocation spacing given by Eq. (14.3). In this case, that particle would be fully coherent, the misfit being taken up by elastic 'coherency strain'.

A quantity such as $(a_2 - a_1)/a_1$ may be defined as the *misfit*. The definition of misfit clearly depends upon the choice of reference lattice; if the reference lattice has a lattice

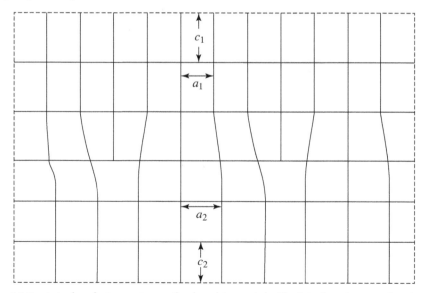

Figure 14.2 Interface between two orthorhombic crystals. The interface is normal to the plane of the figure

parameter a_1, the misfit is $(a_2 - a_1)/a_1$; likewise, if the reference lattice had a lattice parameter a_2, the misfit would be referred to as $(a_2 - a_1)/a_2$. A third option would be to define the misfit relative to a median reference lattice. There is no 'correct' definition of misfit.

Defining the misfit as $(a_2 - a_1)/a_1$, when this reaches 10%, the edge dislocations in Figure 14.2 are separated by only 10 atom spacings; that is, $10a_1$. The core regions of the dislocations are then on the point of overlapping and the dislocation array becomes useless for the purposes of calculating atom positions and the energy of the boundary by elasticity theory.

When the misfit is small, the energy of the boundary can be calculated by elasticity theory. Its dependence on the misfit is very similar to the dependence of the energy of a low-angle grain boundary on its angle of tilt or twist. For a boundary between phases with identical elastic properties, Brooks [4] obtained the expression:

$$E = \frac{\mu b \delta}{4\pi(1 - \nu)}(A_0 - \ln \delta) \tag{14.4}$$

where δ is the misfit and A_0 has the same meaning as in Eq. (13.9). A plot of a similar result obtained by van der Merwe [5] is shown in Figure 14.3. Van der Merwe took explicit account of the dislocation core energy by using a Peierls model; that is, by assuming that a sinusoidal force law operates between atoms on either side of the boundary (Section 8.5). For misfits greater than about 0.1, the energy stored in the bonds between these atoms exceeds the energy stored in the two elastic media.

The simple case of a misfit in one direction is less likely to be encountered in real materials than are cases of a two-dimensional misfit, where more than one set of dislocations is required to take up the mismatch. For example, a misfit in the b dimensions of the orthorhombic crystals of Figure 14.2 could be taken up by a set of edge dislocations orthogonal to that which matches the a dimensions. Square networks of this type have been seen at the faces of plates of UC_2 precipitated within a matrix of UC [6]. The structure of UC is the same as that of NaCl; the square pattern of U atoms in the {100} planes almost matches that of the U atoms in the (001) of UC_2, which has a b.c.t. structure. A square net of edge dislocations takes up the small mismatch at the faces of the (001) plates of UC_2, which precipitate on the cube planes of UC.

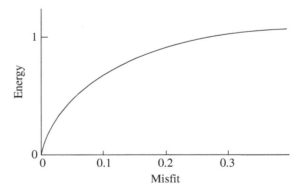

Figure 14.3 Energy of a boundary of the type shown in Figure 14.2, between two large crystals having the same elastic constants. (After [5].)

Apart from its application to phase transformations (see Section 14.2 and also the Suggestions for Further Reading), the theory of interphase boundaries is of interest in connection with the phenomenon of *epitaxial growth* – the growth of one crystal upon the surface of another in some definite orientation relationship. The earliest investigations were of growth from solution, where, for example, one alkali halide will grow upon the surfaces of another, in the same orientation. In this case it was found that epitaxy was limited to crystals whose lattice parameters differed by less than about 15%. More frequently studied today, particularly in the growth of thin-film semiconductor heterostructures by molecular beam epitaxy (Section 14.3), is the epitaxial growth of a thin film of crystal by condensation from a vapour. When the film and substrate crystal structures differ only by having a slightly different lattice parameter in one direction within the interface, a dislocation array like that of Figure 14.2 may take up the mismatch. The elastic energy of the array will be relatively unimportant in comparison with the misfit energy at the interface when the film is very thin, but a model such as that of van der Merwe, which assumes a sinusoidal force law between the atoms at the interface, can still be used to study the energy of the array.

Although the results of early studies had suggested that epitaxy was restricted to crystals of the same type of bonding and with a small mismatch in the interface plane, many contrary examples were soon found. Some examples of epitaxial growth of c.c.p. metals on inorganic substrates are shown in Table 14.2. It will be seen that c.c.p. metals grow epitaxially on cleaved mica and that they can be condensed epitaxially on to heated rock salt with magnitudes of misfit of up to 38%. The uncertainty of predictions made solely on the basis of geometrical fit is demonstrated by the epitaxy of both Ag and Au on {001} surfaces of NaCl. With (001) Ag or Au parallel to (001) NaCl, a misfit of atom sites of only -3% can be found by orienting $[011]_{Ag} \parallel [001]_{NaCl}$. However, the commonly *observed* orientation relationship is $[001]_{Ag} \parallel [001]_{NaCl}$ (Figure 14.4).

More recent work on this well-studied system has shown that this picture of the epitaxy of silver on rock salt is actually too simplistic: more than one possible orientation can occur,

Table 14.2 Epitaxial deposition of c.c.p. metals. (After [7].)

Substrate	Metal	Temp. (°C)	Orientation relationship				Misfit (%)
			Parallel planes		Parallel directions		
			Substrate	Metal	Substrate	Metal	
NaCl	Ni	370	(001)	(001)	[100]	[100]	−38
	Cu	300	(001)	(001)	[100]	[100]	−36
	Ag	150	(001)	(001)	[100]	[100]	−28
	Au	400	(001)	(001)	[100]	[100]	−28
MoS$_2$	Ni	≥ 120	(0001)	(111)	[10$\bar{1}$0]	[1$\bar{1}$0]	−6
	Cu	≥ 50	(0001)	(111)	[10$\bar{1}$0]	[1$\bar{1}$0]	−3
	Ag	≥ 20	(0001)	(111)	[10$\bar{1}$0]	[1$\bar{1}$0]	+9
	Au	≥ 80	(0001)	(111)	[10$\bar{1}$0]	[1$\bar{1}$0]	+9
Mica	Ag	150	(001)	(111)	[100]	[1$\bar{1}$0] or [2$\bar{1}\bar{1}$]	−44
		250	(001)	(001)	[100]	[100]	−21
	Au	450	(001)	(111)	[100]	[1$\bar{1}$0] or [2$\bar{1}\bar{1}$]	−44

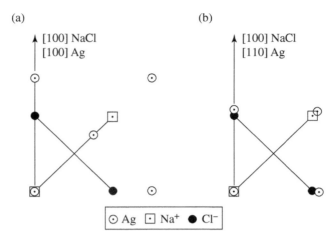

Figure 14.4 Epitaxy of Ag deposited on (001) of NaCl: (a) observed orientation relationship and (b) orientation relationship of small misfit

depending upon the quality of the vacuum during the deposition process, the cleanliness of the rock salt surface, the thickness of the film deposited, the temperature of deposition and the rate of deposition [8]. Thus, while defining the misfit as $(a_2 - a_1)/a_1$ can be useful in predicting orientation relationships adopted between two phases where one is deposited on the other by condensation from a vapour, other microstructural factors can dominate considerations of the way mismatch between two crystal structures is accommodated at interphase boundaries.

14.2 Interphase Boundaries Between C.C.P. and B.C.C. Phases

In a number of materials different phases are stable at different temperatures. Iron is a good example. At room temperature the stable phase is ferrite (α; b.c.c), while at temperatures above 910 °C the stable phase is austenite (γ; c.c.p.). Under slow cooling from above 910 °C the phase transformation from γ to α is one in which diffusion occurs, so that there is no correspondence between atom positions in the γ phase and their corresponding positions in the α phase. If austenite is quenched to room temperature, there is a diffusionless phase change to martensite (α'), a hard, brittle, body-centred tetragonal (b.c.t.) phase (Chapter 12). The martensite structure is a very slightly distorted version of the crystal structure of α caused by the presence of the carbon atoms which are unable to move (diffuse) during the quenching operation.

Irrespective of whether the transformation between γ and α (or α') is one where diffusion does, or does not, occur, it is found that there are favoured orientation relationships which occur between the c.c.p. and the b.c.c. (b.c.t.) phases. The same phenomenon is found in thin film deposition: in experimental investigations of the deposition formed by condensing vapour in vacuum of c.c.p. metals on (110) b.c.c. metal substrates and of b.c.c. metals deposited on (111) c.c.p. metal substrates, it is also found that the metals deposit in favoured orientation relationships [9]. Therefore, an understanding of the possible favoured orientation relationships between c.c.p. and b.c.c. phases, and why they might be favoured, is a subject of wide interest in materials science. The principles established when considering the crystallography of interphase boundaries between c.c.p. and b.c.c. phases are also clearly relevant for other boundaries between dissimilar phases.

Two orientation relationships in particular are commonly encountered when considering the crystallography of c.c.p. – b.c.c. interfaces in metallic materials. These are the Nishiyama-Wassermann (N–W) and the Kurdjumov–Sachs (K–S) orientation relationships (Chapter 12) [10–12]. In the N–W orientation relationship, {111} close-packed planes in the c.c.p. phase are found to be parallel to {110} planes in the b.c.c. phase and the bisectors of close-packed directions within these pairs of planes are found to be parallel, e.g.

$$[11\bar{2}]_F \parallel [\bar{1}10]_B$$
$$[\bar{1}10]_F \parallel [001]_B \tag{14.5}$$
$$[111]_F \parallel [110]_B$$

where the subscripts F and B denote c.c.p. and b.c.c. crystal structures (or the *F*ace-centred and *B*ody-centred Bravais lattices with one atom per lattice point), respectively. In the K–S orientation relationship, {111} close-packed planes in the c.c.p. phase are found to be parallel to {110} planes in the b.c.c. phase and two close-packed directions within these pairs of planes are found to be parallel, e.g.

$$[\bar{1}01]_F \parallel [1\bar{1}1]_B$$
$$[1\bar{2}1]_F \parallel [1\bar{1}\bar{2}]_B \tag{14.6}$$
$$[111]_F \parallel [110]_B$$

Other orientation relationships have also been reported, related to the K–S and N–W orientation relationships by relatively small rotations [13], such as the Bain orientation relationship and the inverse Nishiyama–Wassermann orientation relationship. One possible variant of the Bain orientation relationship [14] takes the form

$$[100]_F \parallel [10\bar{1}]_B$$
$$[010]_F \parallel [010]_B \tag{14.7}$$
$$[001]_F \parallel [101]_B$$

and one variant of the inverse Nishiyama–Wassermann orientation relationship (also known as the Pitsch orientation [15]) takes the form

$$[1\bar{1}0]_F \parallel [1\bar{1}\bar{1}]_B$$
$$[110]_F \parallel [12\bar{1}]_B \tag{14.8}$$
$$[001]_F \parallel [101]_B$$

It is termed the *inverse* N–W because the indices of the {011} and {111} planes parallel are the *inverse* of that found in the classic N–W orientation relationship, i.e. an {011}$_F$ plane parallel to a {111}$_B$ plane. An understanding of the rotations required to carry one of these orientation relationships into another (Problems 14.2–14.5) is therefore useful in the understanding of 'favoured' orientation relationships found experimentally between c.c.p. and b.c.c. phases.

In geometrical theories of interphase boundaries, the approach used to explain the occurrence of such 'favoured' orientation relationships is first to identify a correspondence between multiples of unit cells in the two phases either side of the interphase boundary, such as the approach used for grain boundaries discussed in Section 13.2 [16–20]. This approach therefore generalizes the concept of 'misfit' discussed in Section 14.1. It has also to be understood that the correspondence is one in which strains have to be invoked to quantify the degree to which these multiples of unit cells differ in their shape. Depending on whether the situation under consideration is a three-dimensional problem, such as the precipitation of one phase in another, or whether it is a two-dimensional problem, such as the condensation from the vapour of one phase onto the surface of another, different approaches can be used to predict how two phases 'fit' together at their boundary or boundaries.

As an example of this methodology, it is instructive to analyse the Nishiyama–Wassermann (N–W) orientation relationship found in the condensation from the vapour of Ni and Cu onto (110) W surfaces [9]. Bruce and Jaeger [9] superimposed nets of discs representing atoms on (110) b.c.c. planes and (111) c.c.p. planes to demonstrate misfit between these planes in the N–W orientation relationship as a function of the ratio of the bulk interplanar spacing of these two planes:

$$\rho = d_0 \, (110) \, \text{b.c.c.} / d_0 \, (111) \, \text{c.c.p.} = \sqrt{3} a_B / \sqrt{2} a_F \qquad (14.9)$$

where a_B and a_F are the lattice parameters of the b.c.c. and c.c.p. phases, respectively. This approach can be expressed mathematically in terms of a Bain correspondence between these two phases (Chapter 12) and a suitable matrix which specifies a strain which formally strains non-primitive unit cells of the b.c.c. and c.c.p. phases into one another [21]. To do this, the Bain correspondence between the b.c.c. and c.c.p. phases can be used (Chapter 12), e.g.

$$_B\mathbf{C}_F = \begin{pmatrix} 0 & 0 & 1 \\ 1 & 1 & 0 \\ -1 & 1 & 0 \end{pmatrix} \qquad (14.10)$$

so that the vectors $[001]_F$, $\frac{1}{2}[110]_F$ and $\frac{1}{2}[\bar{1}10]_F$ correspond to the vectors $[100]_B$, $[010]_B$ and $[001]_B$, respectively.

If we now choose an orthonormal axis system, O, within which we define the epitaxial interface misfit in the N–W orientation so that

$$\mathbf{x}_O \parallel [11\bar{2}]_F \parallel [\bar{1}10]_B$$

$$\mathbf{y}_O \parallel [\bar{1}10]_F \parallel [001]_B \qquad (14.11)$$

$$\mathbf{z}_O \parallel [111]_F \parallel [110]_B$$

then following the methodology in Appendix 4, a matrix $_F\mathbf{T}_O$ can be determined which relates vectors defined in the orthonormal axis system to those defined in the c.c.p. phase:

$$_F\mathbf{T}_O = \frac{1}{a_F} \begin{pmatrix} 1/\sqrt{6} & -1/\sqrt{2} & 1/\sqrt{3} \\ 1/\sqrt{6} & 1/\sqrt{2} & 1/\sqrt{3} \\ -2/\sqrt{6} & 0 & 1/\sqrt{3} \end{pmatrix} \tag{14.12}$$

This is a rotation matrix premultiplied by the factor $1/a_F$. Similarly, a matrix $_B\mathbf{T}_O$ can be determined which relates vectors defined in the orthonormal axis system to those defined in the b.c.c. phase:

$$_B\mathbf{T}_O = \frac{1}{a_B} \begin{pmatrix} -1/\sqrt{2} & 0 & 1/\sqrt{2} \\ 1/\sqrt{2} & 0 & 1/\sqrt{2} \\ 0 & 1 & 0 \end{pmatrix} \tag{14.13}$$

Within this orthonormal frame of reference, in which the epitaxial interface is the (001) plane, the strain matrix in this orthonormal basis system, $(_O\mathbf{S}_O)$, defining the deformation which formally carries the c.c.p. lattice into the b.c.c. lattice is

$$(_O\mathbf{S}_O) = (_O\mathbf{T}_B)(_B\mathbf{C}_F)(_F\mathbf{T}_O) \tag{14.14}$$

where $_O\mathbf{T}_B$ is the inverse of $_B\mathbf{T}_O$, i.e. a_B multiplied by the transpose of the rotation matrix in the formula for $_B\mathbf{T}_O$ in Eq. (14.13). Hence,

$$(_O\mathbf{S}_O) = \frac{a_B}{a_F} \begin{pmatrix} 2/\sqrt{3} & 0 & 1/\sqrt{6} \\ 0 & \sqrt{2} & 0 \\ 0 & 0 & \sqrt{6}/2 \end{pmatrix} \tag{14.15}$$

where from the nature of the problem being considered, that of condensation of c.c.p. metals from the vapour onto {110} b.c.c. metal substrates, it is evident that there is no mechanistic relationship between the two phases, in the sense that there is no requirement that the interphase boundary under consideration is one whose movement causes the phase on one side of the boundary to be consumed by the other, as would happen in either a diffusional or a diffusionless phase transformation.

The total Burgers vector content $\mathbf{B}$ of the dislocations intersected by a vector $\mathbf{p}$ lying in an interface is then the generalization of Eq. (13.20) for interphase boundaries:

$$\mathbf{B}_o = [\mathbf{I} - (_O\mathbf{S}_O)^{-1}]\mathbf{p}_o \tag{14.16}$$

where $\mathbf{I}$ is the unit 3×3 matrix and $\mathbf{B}$ and $\mathbf{p}$ are defined in the orthonormal reference frame. Therefore, in the epitaxial (110) b.c.c. ∥ (111) c.c.p. interface, for a vector $\mathbf{p}_o = [uv]$ lying in this interface,

$$\mathbf{B}_o = \begin{pmatrix} 1 - \frac{\sqrt{3}}{2}\frac{a_F}{a_B} & 0 \\ 0 & 1 - \frac{1}{\sqrt{2}}\frac{a_F}{a_B} \end{pmatrix} \begin{bmatrix} u \\ v \end{bmatrix} \tag{14.17}$$

Therefore, as the nets considered by Bruce and Jaeger demonstrate, if a_F/a_B is either $2/\sqrt{3} = 1.155$ or $\sqrt{2} = 1.414$, the nets will have perfect fit, i.e. zero misfit, along one direction within the interface. Intriguingly, for Ni deposited onto W (110) surfaces, $a_F/a_B = 1.113$, and when Au is deposited onto Fe (110) surfaces, $a_F/a_B = 1.422$. For both these situations, the N–W orientation relationship is the one observed. By contrast, when Ag is deposited onto W (110) surfaces, for which $a_F/a_B = 1.291$, the K–S orientation relationship is the one observed, and at this relative orientation of the two phases, there is a direction along which there is almost perfect fit [9, 21]. Nets representing (110) b.c.c. || (111) c.c.p. planes are shown in the N–W and K–S relationships in Figure 14.5 for $a_F/a_B = 1.291$.

$[1\bar{1}2]_F \parallel [1\bar{1}0]_B$

$[1\bar{1}0]_F \parallel [001]_B$

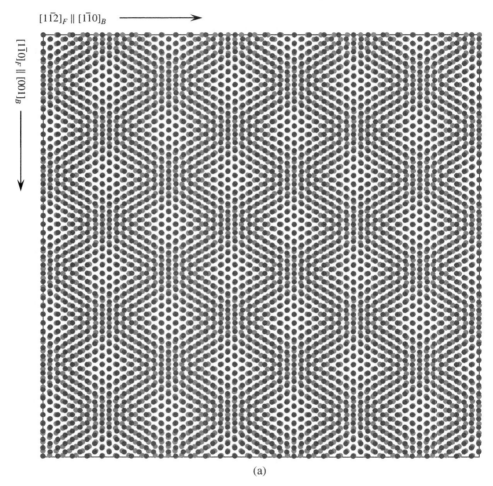

(a)

Figure 14.5 Superposition of nets representing atoms in unrelaxed (110) b.c.c. || (111) c.c.p. planes shown in (a) a N–W and (b) a K–S relationship for $a_F/a_B = 1.291$. The atoms in the b.c.c. planes are represented by light grey discs and those in the c.c.p. planes by dark grey discs (after Bruce and Jaeger [9]). In the K–S relationship, $[1\bar{1}1]_B \parallel [10\bar{1}]_F$, so that the c.c.p. net in (b) is rotated 5.26° clockwise with respect to its orientation in (a)

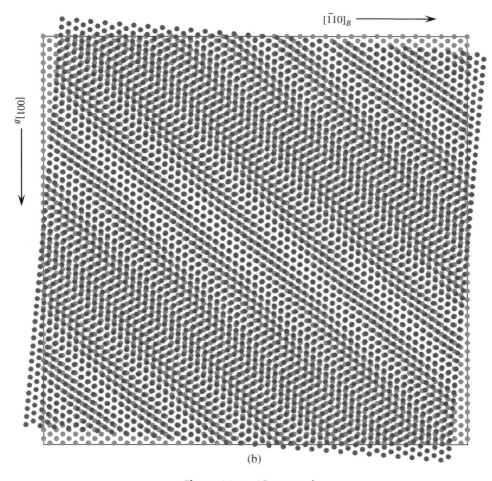

$[\bar{1}10]_B \longrightarrow$

$[001]_B \downarrow$

(b)

Figure 14.5 *(Continued)*

In Figure 14.5 areas where the discs representing the (110) b.c.c. ‖ (111) c.c.p. planes coincide are unfavourable – (110) b.c.c. planes have a staggered arrangement of atoms on adjacent (110) planes, as do (111) c.c.p. planes. It is evident that areas where these discs coincide are more elongated in the K–S arrangement, Figure 14.5(b), than in the N–W arrangement, Figure 14.5(a). In addition, the area fraction of such areas is quantifiably lower in (b) than (a) for this ratio of $a_F/a_B = 1.291$ [21].

Consideration of these principles have led to the concept of the 'invariant line' as being a key parameter to rationalize observed 'favoured' orientation relationships in systems where one phase precipitates within another, rather than a criterion based on planar mismatch [13]. With the recent advances in electron back-scattered electron diffraction, where data can be gathered from large numbers of interphase boundaries, it is therefore interesting that a recent study of ferrite-austenite interfaces in duplex stainless steels has shown that only 12% of the orientation relationships found were K–S and 4% were N–W (within a 2° tolerance angle), so that ~84% of the orientation relationships were 'other'. [22]. Clearly, therefore, even in the apparently simple case of c.c.p.–b.c.c. interphase boundaries, there is still significant room for improvement in our understanding.

14.3 Strained Layer Epitaxy of Semiconductors

In contrast to more complex geometries of interphase boundaries, aspects of which were considered in Section 14.2, almost perfect epitaxy is obtained in some semiconductor crystals. The high purity of the materials, their similar crystal structures, and the precision of modern growth techniques, such as molecular beam epitaxy (MBE), can lead to epitaxial layers grown on single-crystal substrates with no detectable defects at the interfaces. The interfaces are completely coherent in the sense defined in Section 14.1. The resulting layers can be in a state of biaxial strain. For example, for an InGaAs alloy layer grown on a (001)-oriented GaAs substrate wafer, the biaxial strain in the plane of the layer changes the crystal structure from cubic to tetragonal because of the Poisson effect (Section 6.5), with the result that interesting electronic properties are produced in the layer, or quantum well; such quantum wells have uses as lasers. Similarly, InGaAs alloy layers grown on the (111) surface of binary alloys such as GaAs undergo a trigonal distortion of the unit cell, which results in a strain-induced piezoelectric effect.

Growth of perfect *pseudomorphic* strained layers is generally possible for layer misfits below 2% and for layer thicknesses below a critical thickness. Above this critical thickness it is energetically favourable for some or all of the layer strain to be reduced or *relaxed* through the introduction of interfacial misfit dislocations, as discussed in Section 14.1. It is useful to be able to obtain the appropriate elasticity equations for layers grown in any orientation. Dunstan [23] has provided the general method for achieving this. The special case of a (001)-oriented strained layer is shown in Figure 14.6.

Suppose that the substrate and the layer have cubic lattice parameters a_s and a_ℓ respectively when in their bulk forms. In order for the layer to grow coherently on the substrate, it will be strained to adopt the in-plane lattice parameter of the substrate. This will result in a distortion of the lattice dimension perpendicular to the interface plane. The strained layer lattice parameters then become a_1, a_2 and a_3.

For a fully strained layer, $a_1 = a_2 = a_3$. For a relaxed layer, a_1 and a_2 need not be equal and may take on values between a_ℓ and a_s. In general, whether fully strained or partially relaxed, the principal strains in the layer can be defined as follows:

$$\varepsilon_1 = \frac{a_1 - a_\ell}{a_\ell}, \quad \varepsilon_2 = \frac{a_2 - a_\ell}{a_\ell}, \quad \varepsilon_3 = \frac{a_3 - a_\ell}{a_\ell} \tag{14.18}$$

These strains are measurable, by using X-ray diffraction techniques, for example. In order to understand the relationships between these strains, or to obtain the stress or strain energy in the layer, it is necessary to use the equations developed in Chapter 6.

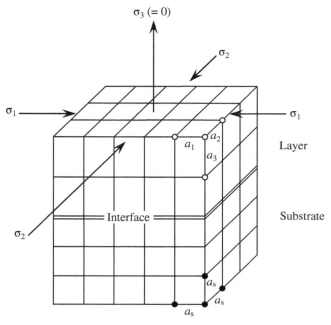

Figure 14.6   A strained epitaxial layer in the (001) orientation. The in-plane principal stresses and strains are taken to be compressive

Using the contracted two-suffix notation described in Chapter 6, the stress–strain relationship for a layer in the (001) orientation is:

$$
\begin{pmatrix} \sigma_1 \\ \sigma_2 \\ 0 \\ 0 \\ 0 \\ 0 \end{pmatrix} = \begin{pmatrix} c_{11} & c_{12} & c_{12} & 0 & 0 & 0 \\ c_{12} & c_{11} & c_{12} & 0 & 0 & 0 \\ c_{12} & c_{12} & c_{11} & 0 & 0 & 0 \\ 0 & 0 & 0 & c_{44} & 0 & 0 \\ 0 & 0 & 0 & 0 & c_{44} & 0 \\ 0 & 0 & 0 & 0 & 0 & c_{44} \end{pmatrix} \begin{pmatrix} \varepsilon_1 \\ \varepsilon_2 \\ \varepsilon_3 \\ 0 \\ 0 \\ 0 \end{pmatrix}
\tag{14.19}
$$

The free surface requires σ_3 to vanish. Shear stresses are ruled out because of the translational symmetry in the layer. In pseudomorphic structures, the in-plane strains ε_1 and ε_2 will normally be equal and have the value of the misfit, ε_0. The strain perpendicular to the interface is obtained from the third line of Eq. (14.19):

$$
0 = c_{12}\varepsilon_1 + c_{12}\varepsilon_2 + c_{11}\varepsilon_3
\tag{14.20}
$$

whence:

$$
\varepsilon_3 = -\frac{2c_{12}}{c_{11}} \frac{\varepsilon_1 + \varepsilon_2}{2}
\tag{14.21}
$$

This equation is also used for the case where there is some relaxation of the strain via plastic relaxation. However, in the case of plastic relaxation the in-plane strains ε_1 and ε_2 will not necessarily be equal. The strain is also commonly expressed in the form:

$$\varepsilon_3 = -\nu_2 \varepsilon_{ave} \tag{14.22}$$

where:

$$\varepsilon_{ave} = \frac{\varepsilon_1 + \varepsilon_2}{2}$$

This defines a two-dimensional Poisson's ratio, ν_2, which turns out to be close to unity [23]. The first two lines of Eq. (14.19) give the in-plane stress–strain relationships:

$$\begin{aligned} \sigma_1 &= c_{11}\varepsilon_1 + c_{12}\varepsilon_2 + c_{12}\varepsilon_3 \\ \sigma_2 &= c_{12}\varepsilon_1 + c_{11}\varepsilon_2 + c_{12}\varepsilon_3 \end{aligned} \tag{14.23}$$

Thus, adding these together and using Eq. (14.21) we have:

$$\begin{aligned} \sigma_{ave} &= (c_{11} + c_{12} - 2c_{12}^2/c_{11})\varepsilon_{ave} \\ &= M\varepsilon_{ave} \end{aligned} \tag{14.24}$$

which defines a two-dimensional Young's modulus denoted by M. Values of M for some common semiconductors are given in Table 14.3. For a pseudomorphic layer, σ_1, σ_2 and σ_{ave} are all equal and can be written as σ_0

The elastic strain energy of the layer can be written in terms of the stress and strain tensors as:

$$E = \int_V \frac{1}{2} \sum \sigma_{ij}\varepsilon_{ij} dV \tag{14.25}$$

Table 14.3 Lattice parameters and stiffness constants for some common semiconductors

Compound	Lattice parameter (Å)	c_{11} (GPa)	c_{12} (GPa)	c_{44} (GPa)	M_{001} (GPa)
Si	5.4310	166	64	79.5	181
Ge	5.6537	124	41	68	138
AlP	5.4580	160	75	40	165
AlAs	5.6610	120	57	59	123
AlSb	6.1353	88	43	41	89
GaP	5.4505	140	62	70	147
GaAs	5.6535	118	53.5	59	123
GaSb	6.0954	88	40	43	92
InP	5.8688	101	56	45.5	95
InAs	6.0585	83	45	39.5	79
InSb	6.4788	67	38	31	62
$In_xGa_{1-x}As$[a]	$5.6535 + 0.4050x$	$118 - 35x$	$53.5 - 8.5x$	$59 - 19.5x$	$123 - 44x$

[a]Where ternary alloys such as InGaAs exist as single-phase solutions throughout the entire composition range, Vegard's law can be used to interpolate for the lattice parameter and stiffness constant between the binary compound values.
Source: The c_{ij} and M_{001} values are taken from [23], with permission.

or per unit area for a pseudomorphic layer of thickness h, as:

$$E = 2 \times \frac{1}{2} h \sigma_0 \varepsilon_0 = h M \varepsilon_0^2 \tag{14.26}$$

The elastic energy of the layer therefore depends on the square of the misfit.

If dislocations are introduced into the interface between the layer and substrate, the elastic strain energy can be reduced. The situation is a little different from that shown in Figure 14.2 since the layer is of a finite thickness h. The introduction of dislocations will lower the elastic energy because the coherency strain is relieved close to the dislocation. However, the dislocations possess self-energy. If they are spaced p apart and are close to the interface, the additional energy per unit area of the interface due to the dislocations will be approximately:

$$E_\perp = \frac{Gb^2}{4\pi(1-\nu)} \frac{1}{p} \ln\left(\frac{\beta h}{b}\right) \tag{14.27}$$

where β is a constant (see Eq. 9.8). If the strain were completely relieved, the value of p would be equal to a_s/ε_0 from Eq. (14.3).

The elastic strain energy of the layer increases as $h\varepsilon_0^2$ from Eq. (14.26), while the energy of the dislocation array will increase as ε_0 from Eq. (14.27). These equations therefore predict that for small values of h the pseudomorphic layer is stable with no dislocations present, but that above a certain critical thickness it should be energetically favourable to introduce dislocations to relieve the strain (see Problem 14.6).

The critical thickness is inversely proportional to ε_0. Dunstan, Young and Dixon [24] have given a simple geometrical argument as to why this should be so. This is illustrated in Figure 14.7. The strain field of a dislocation at a depth h from a surface must decay laterally to zero within a distance of approximately h. Consequently, a dislocation can only relieve the strain within the layer over a lateral distance mh, where m is a small number around one or two. Outside this region the relaxation is zero, and within it averages b/mh, where b is the Burgers vector. A dislocation will not be produced if the strain induced, b/mh, exceeds the strain in the layer.

Thus, the layer is prevented from relaxing at small thicknesses because the relaxation (b/mh) exceeds the strain ε_0. It may relax when $(b/mh) = \varepsilon_0$. Rearranging, we obtain a

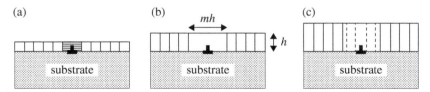

Figure 14.7 The relaxed region with lateral dimension *mh* around a misfit dislocation is shown for three layer thicknesses, *h*. Vertical hatching corresponds to strain as grown-in. Horizontal hatching indicates strain of the opposite sign. In a thin layer below the critical thickness, (a), relaxation induces an opposite strain in the region around the dislocation. When the layer is at critical thickness, (b), the strain is fully relieved around the dislocation. In the thickest layer, (c), the strain is only partly relieved

prediction for the critical thickness of:

$$h_c = \frac{b}{m\varepsilon_0} \tag{14.28}$$

If we take b as the atomic separation and $m = 1$, the equation becomes:

$$h_c = \frac{b}{\varepsilon_0} \tag{14.29}$$

where h_c is measured in units of monolayers. This equation is in agreement with experiment and is consistent with treatments based on the change of energy.

Layers may be grown which are considerably thicker than the critical thickness and remain strained for long periods at high temperature. One reason for this is the difficulty of producing dislocations with the correct Burgers vector to relieve the strain. This difficulty arises in face-centred cubic lattices strained biaxially on (001), as in Figure 14.6, because the Burgers vectors of common glide dislocations and their slip planes are quite inconvenient for relieving this particular pattern of strain (see Problem 14.7).

Problems

14.1 The cleavage plane of mica contains potassium ions at the corners of a network of equilateral triangles (i.e. similar to a close-packed array). If the K^+ ion separation is $5.2\,\text{Å}$, sketch the atomic fit obtained by epitaxially depositing silver with the close-packed plane of the silver parallel to the cleavage plane of mica and the $\langle 112 \rangle$ of the silver parallel to a closest-packed direction of the K^+ ions. Calculate the misfit in this direction. Show that, with this orientation relationship, the silver may grow epitaxially in either of two twin-related forms. If the silver deposits as isolated islands, having $\{111\}$ surfaces, sketch the shape of islands of both forms of deposit.

14.2 Show that the Nishiyama–Wassermann orientation, written in the form

$$[\bar{1}2\bar{1}]_F \parallel [\bar{1}10]_B$$

$$[\bar{1}01]_F \parallel [001]_B$$

$$[111]_F \parallel [110]_B$$

where F and B refer to the c.c.p. and b.c.c. crystal structures, respectively, is a rotation of $5.26°$ away from two Kurdjumov–Sachs orientation relationships of the form

$$\langle \bar{1}01 \rangle_F \parallel \langle 1\bar{1}1 \rangle_B$$

$$\langle 1\bar{2}1 \rangle_F \parallel \langle 1\bar{1}2 \rangle_B$$

$$[111]_F \parallel [110]_B$$

Specify both possible forms (variants) of the Kurdjumov–Sachs orientation relationship.

14.3 Given the Bain orientation relationship in the form

$$[100]_F \parallel [10\bar{1}]_B$$

$$[010]_F \parallel [010]_B$$

$$[001]_F \parallel [101]_B$$

in the nomenclature used in Problem 14.2, show that a rotation of 9.74° about $[001]_B$ is required to rotate this into the Nishiyama–Wassermann orientation given in Problem 14.2.

14.4 Show that a rotation of 9.74° about $[101]_B$ is required to rotate the Bain orientation relationship given in Problem 14.3 into the *inverse* Nishiyama–Wassermann orientation (also known as the Pitsch orientation), written in the form

$$[1\bar{1}0]_F \parallel [1\bar{1}\bar{1}]_B$$

$$[110]_F \parallel [12\bar{1}]_B$$

$$[001]_F \parallel [101]_B$$

using the nomenclature used in Problem 14.2.

14.5 Without detailed mathematics (i.e. without actually carrying out the calculation), explain how the angle with respect to one of the lattices required to bring the Bain orientation relationship into a nearby Kurdjumov–Sachs orientation relationship can be established.

14.6 Use Eqs. (14.26) and (14.27) to obtain an estimate of the critical thickness of a pseudomorphic layer below which it will not relax. Take M in Eq. (14.26) to be $2G(1 + v)/(1 - v)$, where G is the shear modulus. Hence show the relationship is of the same form as Eq. (14.28).

14.7 Take the layer in Figure 14.6 to represent the lattice of a c.c.p. crystal and consider how $a/2\langle 110 \rangle$ dislocations moving on $\{111\}$ can relieve the biaxial strain. What is the minimum number of slip systems that must operate in order to relieve the strain?

Suggestions for Further Reading

Christian, J.W. (2002). *The Theory of Transformations in Metals and Alloys (Part I + II)*. Oxford: Pergamon Press, Elsevier Science..

Howe, J.M. (1997). *Interfaces in Materials: Atomic Structure, Kinetics and Thermodynamics of Solid–Vapor, Solid–Liquid and Solid–Solid Interfaces*. New York: John Wiley and Sons.

Nix, W.D. (1989). Mechanical properties of thin films. *Metall. Trans. A* 20: 2217–2245.

Sutton, A.P. and Balluffi, R.W. (1995). *Interfaces in Crystalline Materials*. Oxford: Clarendon Press.

Tiller, W.A. (1991). *The Science of Crystallization: Macroscopic Phenomena and Defect Generation*. Cambridge: Cambridge University Press.

Tiller, W.A. (1991). *The Science of Crystallization: Microscopic Interfacial Phenomena*. Cambridge: Cambridge University Press.

References

[1] Gjostein, N.A., Domian, H.A., Aaronson, H.I. and Eichen, E. (1966). Relative interfacial energies in Fe–C alloys. *Acta Metall.* 14: 1637–1644.

[2] Smith, C.S. (1964). Some elementary principles of polycrystalline microstructure. *Metall. Rev.* 9: 1–48.

[3] Spitzer, D.P. (1962). Intercrystalline energies in the alkali halides. *J. Phys. Chem.* 66: 31–38.

[4] Brooks, H. (1952). Theory of internal boundaries, in *Metal Interfaces*, American Society for Metals, Cleveland, Ohio, pp. 20–64; discussion p. 64.

[5] van der Merwe, J.H. (1963). Crystal interfaces. Part II: finite overgrowths. *J. Appl. Phys.* 34: 123–127; erratum 3420.

[6] Whitton, J.L. (1964). Transmission electron microscopy of uranium monocarbide. *J. Nucl. Mater.* 12: 115–119.

[7] Pashley, D.W. (1956). The study of epitaxy in thin surface films. *Adv. Phys.* 5: 173–240.

[8] Yang, F.-L., Somekh, R.E. and Greer, A.L. (1998). UHV magnetron sputtering of silver films on rocksalt: quantitative X-ray texture analysis of substrate-temperature-dependent microstructure. *Thin Solid Films* 322: 46–55.

[9] Bruce, L.A. and Jaeger, H. (1978). Geometric factors in f.c.c. and b.c.c. metal-on-metal epitaxy III. The alignments of (111) f.c.c. – (110) b.c.c. epitaxed metal pairs. *Phil. Mag. A* 38: 223–240.

[10] Nishiyama, Z. (1934). X-ray investigation of the mechanism of the transformation from face-centred cubic lattice to body-centred cubic. *Sci. Rep. Tôhoku Imp. Univ.* 23: 637–664.

[11] Wassermann, G. (1935). Ueber den Mechanismus der α-γ-Umwandlung des Eisens. *Mitteilungen Kaiser-Wilhelm-Institut für Eisenforschung Düsseldorf* 17: 149–155.

[12] Kurdjumow, G. and Sachs, G. (1930). Über den Mechanismus der Stahlhärtung. *Z. Phys.* 64: 325–343.

[13] Dahmen, U. (1982). Orientation relationships in precipitation systems. *Acta Metall.* 30: 63–73.

[14] Bain, E.C. (1924). The nature of martensite. *Trans. Am. Inst. Min. Metall. Petrol. Engrs.* 70: 25–35; discussion 35–46.

[15] Pitsch, W. (1959). The martensite transformation in thin foils of iron-nitrogen alloys. *Phil. Mag.* 4: 577–584.

[16] Bollmann, W. (1970). *Crystal Defects and Crystalline Interfaces*. New York: Springer-Verlag.

[17] Bonnet, R. and Cousineau, E. (1977). Computation of coincident and near-coincident cells for any two lattices – related DSC-1 and DSC-2 lattices. *Acta Crystall. A* 33: 850–856.

[18] Zhang, W.-Z. and Weatherly, G.C. (2005). On the crystallography of precipitation. *Prog. Mater. Sci.* 50: 181–292.

[19] Zhang, M.-X. and Kelly, P.M. (2009). Crystallographic features of phase transformations in solids. *Prog. Mater. Sci.* 54: 1101–1170.

[20] Gautam, A.R.S. and Howe, J.M. (2011). A method to predict the orientation relationship, interface planes and morphology between a crystalline precipitate and matrix. Part I. Approach. *Phil. Mag.* 91: 3203–3327.

[21] Knowles, K.M. and Smith, D.A. (1982). The application of surface dislocation theory to the f.c.c.–b.c.c. interface. *Acta Crystall. A* 38: 34–40.

[22] Haghdadi, N., Cizek, P., Hodgson, P.D., Tari, V., Rohrer, G.S. and Beladi, H. (2018). Five-parameter crystallographic characteristics of the interfaces formed during ferrite to austenite transformation in a duplex stainless steel. *Phil. Mag.* 98: 1284–1306.

[23] Dunstan, D.J. (1997). Strain and strain relaxation in semiconductors. *J. Mater. Sci.: Mater. Electron.* 8: 337–375.

[24] Dunstan, D.J., Young, S. and Dixon, R.H. (1991). Geometrical theory of critical thickness and relaxation in strained-layer growth. *J. Appl. Phys.* 70: 3038–3045.

15

Texture

15.1 Texture

Because large strains can cause a reorientation of the lattice of a crystal, it follows that, in a polycrystalline material, continued plastic flow, such as that which occurs during the working of metals, tends to develop a *texture* or a *preferred orientation* of the lattice within the grains, as well as a preferred change of shape of the grains. Working involves very large plastic strains, often hundreds of per cent. The development of a texture changes the crystallographic properties of the material significantly. A fine-grained material in which the grains have random lattice orientations will have isotropic properties, provided that other crystal defects such as inclusions and boundaries are uniformly distributed. However, a specimen with a preferred orientation will have anisotropic properties, which may be desirable or undesirable, depending upon the intended use of the material.

The final texture that develops may resemble a single orientation or may comprise crystals distributed between two or more preferred orientations. During plastic flow the process of reorientation is gradual. In theory, the orientation change proceeds until a texture is reached that is stable against indefinitely continued flow of a given type. The theoretical end distribution of orientations and the manner in which it develops are a characteristic of the material, its crystal structure, its microstructure and the nature of the forces arising from the deformation process. Whether a material actually reaches its stable orientation depends upon many variables, such as the rate of working and the temperature at which it is carried out. Textural changes may correspond with marked changes in the ductility of a material. Thus, there is an interplay between the role of deformation in creating texture and the role of texture in facilitating deformation. While the basic principles underlying texture development for single crystals have been presented in Section 7.4, the prediction of actual textures during forming processes is necessarily very complicated [1–3].

Crystallography and Crystal Defects, Third Edition. Anthony Kelly and Kevin M. Knowles.
© 2020 John Wiley & Sons Ltd. Published 2020 by John Wiley & Sons Ltd.
Companion website: www.wiley.com/go/kelly/crystallography3e

The various processes of texture development have been categorized according to their final state and to the kind of working involved, for example wire and fibre textures, compression textures, sheet and rolling textures, torsion textures and deep drawing textures. Some examples of ideal end textures are presented in Table 15.1. In addition to deformation textures, anisotropic properties of a polycrystalline material can also arise as a result of recrystallization following working (see Chapter 11, Footnote 1) or as a result of material fabrication processes that do not involve plastic deformation. For example, the vapour deposition of layers on substrates often results in a layer having a polycrystalline texture with a preferred orientation in the growth direction and random orientations perpendicular to the growth direction. Some examples of deposition textures are included in Table 15.1.

There is usually a statistical distribution of orientations of the various crystals in a sample. This distribution can be measured and represented on a pole figure, such as in Figure 15.1. A pole figure is similar to a stereographic projection. On it is presented the statistical distribution of the orientations of the plane normals from a particular set of planes. A pole figure is usually obtained using diffraction from X-rays, neutrons or electrons[1]. In the typical case of X-ray diffraction in the laboratory, the detector position is fixed to collect the diffraction from one particular set of planes. The sample is rotated about two axes so as to scan over all possible orientations.

The corresponding pole figure is usually presented as a contour map, where the contours are labelled according to the relative strength of the diffraction signal. For example, Figure 15.1 is a pole figure showing the relative intensities of the reflections from {111} planes in a cold-rolled copper sample. The relative populations of planes in the various orientations are proportional to the numbers against the contour lines. The pole figure is interpreted in terms of an ideal end orientation. In uniaxial textures, such as those obtained in tension, compression and wire- and rod-forming processes, it is often sufficient to specify which crystallographic direction (or directions) lies parallel to the unique axis.

For sheet textures, ideal orientations are given by a plane or planes lying parallel to the plane of the sheet and a direction or directions in the plane lying parallel to a significant direction in the sheet, such as the rolling direction (RD). For example, to identify {110}⟨001⟩ and {110}⟨1$\bar{1}$2⟩ textures the procedures illustrated in Figure 15.2 might be followed. First, identify the orientations of the {111} poles relative to the {110} plane and the ⟨001⟩ and ⟨112⟩ directions. A stereographic projection is shown in Figure 15.2a with the (110) plane normal perpendicular to the plane of the paper. All of the {111} poles are shown, along with those ⟨001⟩ and ⟨112⟩ directions that are in the (110) plane and thus could lie parallel to the RD. The orientations of the {111} poles when the RD corresponds to ⟨001⟩ and ⟨112⟩ are shown in Figures 15.2b and c, respectively. Note that there are two possible ⟨112⟩ orientations, which results in there being 10 {111} pole positions shown on this diagram. The case where the end texture contains both {110}⟨001⟩ and {110}⟨1$\bar{1}$2⟩ orientations is illustrated in Figure 15.2d.

A pole figure from an ideal end texture will have the appearance of a pole plot from one or a few single crystals, depending upon how many orientations comprise the end texture.

---

[1] For X-ray and neutron diffraction, the experiment obtains an average texture over volumes of the order of cubic millimetres. Using electron diffraction, textures from different regions of the sample can be discriminated, giving rise to the terms *microtexture* and *mesotexture* used in the study of textural variations within a sample.

Table 15.1 Some common end or final textures

Crystal structure	Description	Example material	Texture	Comments	References
c.c.p.	Rolling texture {plane}/⟨axis⟩	Al	{110} ⟨11̄2⟩ then {112} ⟨111̄⟩	Known as 'pure metal' texture at high temperatures and large deformations	[4]
		Al single crystals	{111} ⟨112̄⟩ {001} ⟨110⟩	Up to 90% reduction After 90% reduction	[5]
		Cu	{110} ⟨11̄2̄⟩ and {112} ⟨111̄⟩ {123} ⟨412̄⟩ + {146} ⟨211̄⟩	Orientations between these extremes or spread around these ideal end textures	[4]
		α-brass (70Cu/30Zn)	{110} ⟨11̄2⟩ and {110} ⟨001⟩	Known as 'alloy' or 'brass' texture; changes to 'pure metal' texture at high temperatures	[4]
	Tension (fibre) texture ⟨axis⟩	Al, Cu, Au, Ag, Ni	⟨111⟩ and ⟨100⟩	Relative proportions dependent upon temperature, strain rate, percentage reduction, stacking fault energy and twinning	[6]
		Some single crystals	⟨211⟩	Except for starting orientation ⟨111⟩	[5]
			⟨111⟩	For starting orientation ⟨111⟩	[6]
	Compression texture ⟨axis⟩	Single crystal and polycrystalline Al	Range from ⟨110⟩ to ⟨311⟩	Tendency to rotate away from ⟨111⟩ and ⟨100⟩	[5]
		α-brass (70Cu/30Zn)	Range from ⟨110⟩ to ⟨311⟩; some ⟨111⟩	Tendency to rotate away from ⟨100⟩	[6]
	Deposition texture ⟨growth direction⟩	Au on cleaved NaCl by thermal evaporation	⟨111⟩	Random orientation in the plane of growth	[7]

Table 15.1 *(continued)*

Crystal structure	Description	Example material	Texture	Comments	References
b.c.c.	Rolling texture {plane}⟨axis⟩	Fe single crystal and polycrystalline	Between {001}⟨110⟩ and {112}⟨110⟩; {111}⟨112⟩	Often both components together Occasionally	[4]
	Tension (fibre) texture ⟨axis⟩	Mo, Nb, Ta, V, W, ferrosilicon, β-brass	⟨110⟩		[5]
	Compression texture ⟨axis⟩	Fe	⟨111⟩, some ⟨100⟩	High growth rates	[5]
	Deposition texture ⟨growth direction⟩	Mo on (001) Si by magnetron sputtering	⟨111⟩		[8, 9]
			⟨110⟩	Low growth rates and low substrate temperatures Some in-plane texture in both cases	
h.c.p.	Rolling texture {plane}⟨axis⟩	Mg	{0001}⟨11$\bar{2}$0⟩		[6]
		Zn, Cd	20–25° off {0001}⟨11$\bar{2}$0⟩		[6]
		Zr, Ti	30–40° off {0001}⟨10$\bar{1}$0⟩	Deviation is in the transverse sense, i.e. about the rolling direction	[6]
	Tension (fibre) texture ⟨axis⟩	Mg	⟨10$\bar{1}$0⟩	Low temperatures	[6]
		Zn	⟨$\bar{2}$110⟩	High temperatures	
			⟨0001⟩	Small reductions	[6]
			70° from ⟨0001⟩ spiral texture	Large reductions	
	Compression texture ⟨axis⟩	Zr, Ti	⟨10$\bar{1}$0⟩		[6]
		Mg	⟨0001⟩		[6]
	Deposition texture ⟨growth direction⟩	Ti on oxidized Si (001) by magnetron sputtering	⟨0001⟩	Random in-plane orientation	[10]

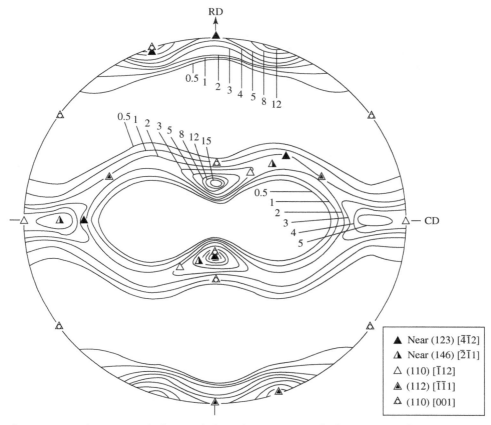

Figure 15.1 The {111} pole figure of electrolytic copper rolled to 96.6% reduction at room temperature, where RD is the rolling direction and CD is the cross direction. The numbers represent relative intensities of the diffraction signal [11]

A real textured material will show a broader distribution of intensities about the ideal orientation. How close the actual texture is to the ideal texture is a qualitative judgement. Actual textures are described as varying from 'weak' (a small proportion of grains having rotated to the endpoint) to 'strong' (a large volume fraction of grains having rotated to or close to the endpoint).

A more computationally intensive, but more representative, measure of texture is in the *orientation distribution function* (ODF). Using data from pole figures taken for several reflections, the orientations of three primary axes of the crystallites are plotted with respect to three standard axes. ODFs can be displayed as three-dimensional plots or as a series of two-dimensional sections. An ODF relevant for both heavily rolled α-brass and heavily rolled silver is shown in Figure 15.3, where ϕ_1, Φ and ϕ_2 are the angles between principal directions in the crystallites and significant directions in the forming process and are known as the Euler angles (Section 15.2). In this diagram the surface shown represents the half-maximum density of the orientation distribution corresponding to a $(011)[2\bar{1}1]$ texture.

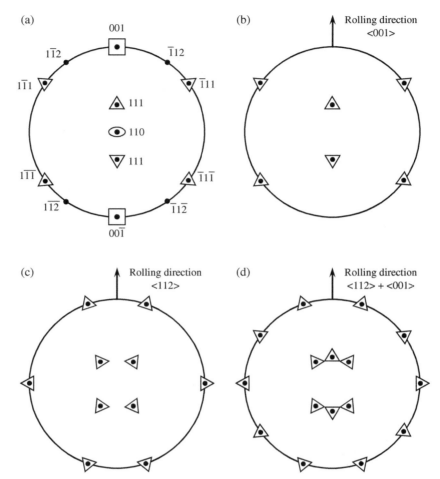

Figure 15.2   (a) Standard stereographic projection in the (110) orientation showing the poles for the {001}, {112} and {111} plane normals, (b) orientations of the {111} plane normals for the (110)⟨001⟩ rolling texture, (c) orientations of the {111} plane normals for the (110)⟨1$\bar{1}$2⟩ rolling texture, and (d) a combination of (b) and (c)

Typical Euler angles for rolled sheet are illustrated in Figure 15.4: Φ is the angle between the crystallite [001] direction and the sheet normal (ND), ϕ_2 is the angle between the crystallite [100] direction and the nodal direction, and ϕ_1 is the angle between the nodal direction and RD. The nodal direction lies perpendicular to ND and the crystallite [001] direction and is parallel to the direction defined by the vector product of these two directions. We see in Figure 15.3 that the spread in ϕ_1 is about 10°, the spread in Φ is about 10° and the spread in ϕ_2 is about 20°, indicating a pronounced texture with only a small spread in orientation. Using the mathematical analysis of the ODF, a near-quantitative description of texture is possible.

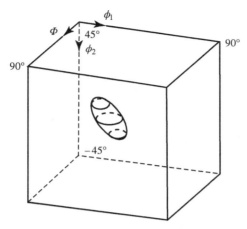

Figure 15.3 Spatial representation of the half-maximum density of the ODF relevant for both heavily cold-rolled heavily rolled α-brass and heavily rolled silver [1, 2, 12]

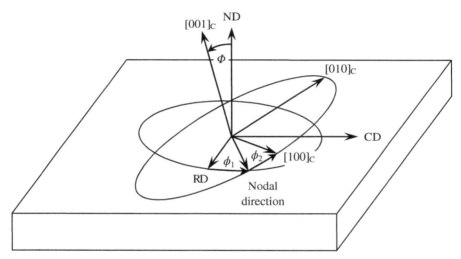

Figure 15.4 Definition of the orientation of a crystallite in rolled sheet by means of three Euler angles, ϕ_1, Φ and ϕ_2. The nodal direction is parallel to the vector product of ND and the crystallite [001] direction. RD is the rolling direction, CD the cross direction (or transverse direction, TD) and ND is the sheet normal

15.2 Euler Angles

There are a number of ways in which the Euler angles can be defined [1, 13, 14]. Here, the most commonly used convention due to Bunge [1] shown in Figure 15.4 will be described in some detail. As shown in Figure 15.4, three Euler angles, ϕ_1, Φ and ϕ_2 define a set of rotations which rotate a set of axes defined by the normal direction, ND, and the rolling direction RD into a set of orthonormal crystal directions [100], [010] and [001]. It is evident from Figure 15.4 that the sequence of rotation operations and the sense of the rotation operations are both very important.

Suppose that initially a right-hand set of orthornomal directions **i**, **j**, **k** are defined so that RD is parallel to **i**, CD is parallel to **j** and ND is parallel to **k**. The first rotation, ϕ_1, is an anticlockwise rotation around **k** which takes **i**, **j**, **k** into the set of orthonormal directions **i'**, **j'**, **k'** where **k** is parallel to **k'** and where **i'** is parallel to the nodal direction, i.e., the direction perpendicular to ND and the crystallite [001] direction, as shown in Figure 15.4. Hence, following the methodology in Appendix 4,

$$\begin{pmatrix} \mathbf{i'} \\ \mathbf{j'} \\ \mathbf{k'} \end{pmatrix} = \begin{pmatrix} \cos\phi_1 & \sin\phi_1 & 0 \\ -\sin\phi_1 & \cos\phi_1 & 0 \\ 0 & 0 & 1 \end{pmatrix} \begin{pmatrix} \mathbf{i} \\ \mathbf{j} \\ \mathbf{k} \end{pmatrix} = \mathbf{R}_{\phi_1} \begin{pmatrix} \mathbf{i} \\ \mathbf{j} \\ \mathbf{k} \end{pmatrix} \tag{15.1}$$

It is important to appreciate that $\mathbf{R}_{\phi_1}$ here is not the same as the rotation matrix which defines how a vector **x** is rotated into a vector **x'** *with respect to a fixed set of axes* **i**, **j**, **k** (Section A1.4): $\mathbf{R}_{\phi_1}$ is the transpose of such a matrix (Problem 15.1).

A second rotation of Φ about the nodal direction, **i'**, rotates **i'**, **j'**, **k'** into the set of orthonormal directions **i''**, **j''**, **k''** where **k''** is now parallel to the crystallite [001]$_C$ direction, **k**$_C$, so that

$$\begin{pmatrix} \mathbf{i''} \\ \mathbf{j''} \\ \mathbf{k''} \end{pmatrix} = \begin{pmatrix} 1 & 0 & 0 \\ 0 & \cos\Phi & \sin\Phi \\ 0 & -\sin\Phi & \cos\Phi \end{pmatrix} \begin{pmatrix} \mathbf{i'} \\ \mathbf{j'} \\ \mathbf{k'} \end{pmatrix} = \mathbf{R}_{\Phi} \begin{pmatrix} \mathbf{i'} \\ \mathbf{j'} \\ \mathbf{k'} \end{pmatrix} \tag{15.2}$$

Finally, a third rotation of ϕ_2 about **k''** ∥ **k**$_C$ rotates **i''** into the crystallite [100]$_C$ direction, **i**$_C$ and **j''** into the crystallite [010]$_C$ direction, **j**$_C$:

$$\begin{pmatrix} \mathbf{i}_C \\ \mathbf{j}_C \\ \mathbf{k}_C \end{pmatrix} = \begin{pmatrix} \cos\phi_2 & \sin\phi_2 & 0 \\ -\sin\phi_2 & \cos\phi_2 & 0 \\ 0 & 0 & 1 \end{pmatrix} \begin{pmatrix} \mathbf{i''} \\ \mathbf{j''} \\ \mathbf{k''} \end{pmatrix} = \mathbf{R}_{\phi_2} \begin{pmatrix} \mathbf{i''} \\ \mathbf{j''} \\ \mathbf{k''} \end{pmatrix} \tag{15.3}$$

Therefore, overall,

$$\begin{pmatrix} \mathbf{i}_C \\ \mathbf{j}_C \\ \mathbf{k}_C \end{pmatrix} = \mathbf{R}_{\phi_2}\mathbf{R}_{\Phi}\mathbf{R}_{\phi_1} \begin{pmatrix} \mathbf{i} \\ \mathbf{j} \\ \mathbf{k} \end{pmatrix} = \mathbf{R} \begin{pmatrix} \mathbf{i} \\ \mathbf{j} \\ \mathbf{k} \end{pmatrix} \tag{15.4}$$

where the elements of **R** are:

$$\mathbf{R}_{11} = \cos\phi_1 \cos\phi_2 - \sin\phi_1 \sin\phi_2 \cos\Phi$$

$$\mathbf{R}_{12} = \sin\phi_1 \cos\phi_2 + \cos\phi_1 \sin\phi_2 \cos\Phi$$

$$\mathbf{R}_{13} = \sin\phi_2 \sin\Phi$$

$$\mathbf{R}_{21} = -\cos\phi_1 \sin\phi_2 - \sin\phi_1 \cos\phi_2 \cos\Phi$$

$$\mathbf{R}_{22} = -\sin\phi_1 \sin\phi_2 + \cos\phi_1 \cos\phi_2 \cos\Phi \qquad (15.5)$$

$$\mathbf{R}_{23} = \cos\phi_2 \sin\Phi$$

$$\mathbf{R}_{31} = \sin\phi_1 \sin\Phi$$

$$\mathbf{R}_{32} = -\cos\phi_1 \sin\Phi$$

$$\mathbf{R}_{33} = \cos\Phi$$

[1, 13]. A rolling texture represented as $(hkl)[uvw]$ in which the normal to (hkl) is ND and $[uvw]$ is RD can be represented as a set of three Euler angles ϕ_1, Φ and ϕ_2, so that

$$\phi_1 = \arcsin\left[\frac{w}{\sqrt{u^2 + v^2 + w^2}} \cdot \frac{\sqrt{h^2 + k^2 + l^2}}{\sqrt{h^2 + k^2}}\right]$$

$$\Phi = \arccos\frac{l}{\sqrt{h^2 + k^2 + l^2}} \qquad (15.6)$$

$$\phi_2 = \arccos\frac{k}{\sqrt{h^2 + k^2}} = \arcsin\frac{h}{\sqrt{h^2 + k^2}}$$

[1]. Conversely, given a set of three Euler angles ϕ_1, Φ and ϕ_2, the texture $(hkl)[uvw]$ is determined from the conditions

$$h : k : l = \sin\Phi \sin\phi_2 : \sin\Phi \cos\phi_2 : \cos\Phi \qquad (15.7)$$

and $u : v : w = P : Q : R$ where

$$P = \cos\phi_1 \cos\phi_2 - \sin\phi_1 \sin\phi_2 \cos\Phi$$

$$Q = -\cos\phi_1 \sin\phi_2 - \sin\phi_1 \cos\phi_2 \cos\Phi \qquad (15.8)$$

$$R = \sin\phi_1 \sin\Phi$$

Thus, for example, the centre of the ellipsoid in Figure 15.3 representing the point in Euler space with the texture $(011)[2\bar{1}1]$ has $\phi_1 = 35.26°$, $\Phi = 45°$ and $\phi_2 = 0°$.

15.3 Microtexture

It is evident from Section 15.2 that Euler angles can also be used to describe the orientations of individual crystals in a polycrystalline aggregate. The technique of choice, described in [13], is that of electron back-scattered diffraction (EBSD), in which the orientation of crystals as small as 200 nm in diameter can be determined under favourable circumstances.

For cubic crystals, there will of course be the ambiguity that for each crystal there 24 different ways in which the crystal axes $[100]_C$, $[010]_C$ and $[001]_C$ can be defined (Section 13.3).

Commercial automated EBSD systems enable large numbers of sampling points to be assigned orientations from which information can be extracted on the spatial distribution of the components of microtexture, such as the identification of newly recrystallised grains in a deformed matrix, or on misorientations across interfaces (Chapters 13 and 14). In such applications, use is made of Rodrigues vectors to represent both the orientations of individual crystals and also the misorientations across grain boundaries.

15.3.1   Rodrigues Vectors and Rodrigues Space

Like quaternions (Section A1.4), Rodrigues vectors have the advantage of describing rotations in a convenient compact notation. Rather than describing a rotation between two sets of orthonormal axes through the nine elements of the rotation matrix **R** defined by Eq. (15.5), the angle θ and the unit vector **n** describing the axis of rotation can be extracted from Eq. (15.5) using the methodology described in Section A1.4, with the caveat that **R** here is the transpose of the matrix specified in Eq. (A.1.54) (Section 15.2). The Rodrigues vector λ is then defined by the equation

$$\lambda = \tan \frac{\theta}{2}\, \mathbf{n} \tag{15.9}$$

[13, 15–17]. A set of λ vectors each representing the orientation of individual crystals can then be represented in a three-dimensional space known as Rodrigues space or Rodrigues–Frank space [15, 16]. For each individual crystal, it is convenient to choose its representation in terms of λ so that has the lowest possible value of the angle θ from the various crystallographically equivalent descriptions possible, depending on symmetry of the crystal structure under consideration. In this way, the set of λ vectors can be made to lie in a polyhedron known as the fundamental zone of Rodrigues space [13]. For cubic crystals this fundamental zone is a cube with truncated {111} faces, the apices of which define disorientations of 62.8° about $< \sqrt{2} - 1, 1, 1 >$ orientations (Section 13.3) representing the maximum possible disorientation descriptions for any axis/angle combination. These disorientations correspond to $< 3 - 2\sqrt{2}, \sqrt{2} - 1, \sqrt{2} - 1 > = $ <0.1716, 0.4142, 0.4142> λ vectors. Furthermore, the fundamental zone can be divided further into symmetrically equivalent subvolumes for the display of preferred misorientation [18]. In cubic crystals there are 48 such subvolumes. Further, more detailed, aspects of the representation of microtexture using the Rodrigues vector formalism can be found in [13, 18, 19].

Problems

15.1 (a) Consider the simple case of a rotation of a vector **x** defined with respect to orthonormal axes by an angle θ anticlockwise about [001] to a vector **x'**. Determine the elements of the rotation matrix $\mathbf{R}_1$ when the components of **x'** and **x** are referred to the *same* set of orthonormal axes.

(b) Determine the elements of the rotation matrix $\mathbf{R}_2$ when $\mathbf{x}'$ and $\mathbf{x}$ refer to the same vector, but when they refer to *different* sets of orthonormal axes with a common [001] where the axes defining the vector as $\mathbf{x}'$ have been rotated angle θ anti-clockwise about [001] with respect to those defining the vector as $\mathbf{x}$. Specify the relationship between $\mathbf{R}_1$ and $\mathbf{R}_2$.

(c) Without detailed mathematics, deduce the result between $\mathbf{R}_1$ and $\mathbf{R}_2$ in the general case where the rotation θ under consideration is anticlockwise about a unit vector $\mathbf{n}$, rather than the special case of [001].

15.2 At room temperature, the lattice parameter of silicon is 0.54306 nm and the lattice parameter of $NiSi_2$ is 0.5406 nm. $NiSi_2$ has the CaF_2 structure. When a layer of $NiSi_2$ is deposited on the (111) plane of single crystal silicon, it can take one of two orientations with respect to the silicon. In orientation A all planes and directions are coincident in the two crystals. In orientation B the two crystals have the (111) plane and [111] direction in common, but the [110] direction in the silicon crystal corresponds to a [114] direction in the $NiSi_2$ layer.

Draw an accurate stereogram centred on (001) including the [111], [110] and [114] directions of silicon. By constructing a suitable small circle show that if the stereogram is rotated by 180° about [111] the new position of the [110] pole coincides with the original position of the [114] pole. Assuming that the $NiSi_2$ layer nucleates by the formation of a close-packed monolayer of Ni atoms on the Si (111) plane, suggest reasons for the occurrence of the two growth orientations A and B.

15.3 A thin film of silver is deposited on to a single-crystal ceramic substrate. Two pole figures are determined at different values of 2θ. The stereogram below is a super-position of the two figures, in the correct relative orientation. One shows the poles, using the 111 reflection, from a thin film of silver. The other shows poles from the 110 reflection from the ceramic substrate (also cubic).

Label the poles. What is the orientation relationship between the silver and the substrate? The silver lattice parameter is 4.086 Å, while that of the substrate material is 5.760 Å. Why might this orientation relationship be adopted?

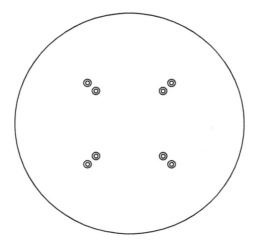

Figure for Problem 15.3

15.4 Pole figures using the 200 reflections from heavily cold-rolled and annealed copper show very strong reflections consistent with the $\{100\}\langle001\rangle$ cube texture formed in such a material, but also weak reflections arising from an additional $\{221\}\langle21\bar{2}\rangle$ texture. This weaker texture is produced by annealing twins nucleated during the annealing process which form as $\{111\}$ twins in this c.c.p. metal.

(a) Confirm that reflection of $\langle100\rangle$ directions in $\{111\}$ plane produces directions parallel to $\langle221\rangle$ directions.

(b) Draw a $\{200\}$ pole figure with the normal direction, ND, at the centre of the stereogram and the rolling direction, RD, in the position normally occupied by $[\bar{1}00]$ to show these $\{100\}\langle001\rangle$ and $\{221\}\langle21\bar{2}\rangle$ textures and their relative strengths.

15.5 On cold rolling commercial 70–30 brass strip (Cu-rich, c.c.p.), a strong texture is developed in which it is suggested that (110) is parallel to the plane of the sheet and $[1\bar{1}2]$ is parallel to the rolling direction. Macroscopically, the strip has *mmm* point symmetry as a consequence of the cold rolling.

Draw a pole figure as in Problem 15.4 so that the $1\bar{1}2$ pole is then in the position normally occupied by $[\bar{1}00]$ and 110 is then in the position normally occupied by 001. Mark the positions of the 100, 010 and 001 poles on this figure consistent with the *mmm* point symmetry produced in the strip by the rolling process.

Suggestions for Further Reading

Bunge, H.-J. (1982). *Texture Analysis in Materials Science, Mathematical Methods*. London: Butterworths.

Engler, O. and Randle, V. (2010). *Introduction to Texture Analysis: Macrotexture, Microtexture and Orientation Mapping*, 2nd Edition. Boca Raton, FL: CRC Press.

Suwas, S. and Ray, R.K. (2014). *Crystallographic Texture of Materials*. London: Springer-Verlag.

References

[1] Bunge, H.-J. (1982). *Texture Analysis in Materials Science, Mathematical Methods*. London: Butterworths.

[2] Cahn, R.W. (1991). Measurement and control of texture, in volume 15. In: *Processing of Metals and Alloys, Materials Science and Technology* (series editors R.W. Cahn, P. Haasen and E.J. Kramer), (ed. R.W. Cahn), Chapter 10, pp. 429–480. Chichester/Weinheim: Wiley–VCH.

[3] Kocks, U.F., Tomé, C.N. and Wenk, H.-R. (1998). *Texture and Anisotropy: Preferred Orientations in Polycrystals and their Effect on Materials Properties*. Cambridge: Cambridge University Press.

[4] Dillamore, I.L. and Roberts, W.T. (1965). Preferred orientation in wrought and annealed metals. *Metall. Rev.* 10: 271–380.

[5] Hu, H., Cline, R.S. and Goodman, S.R. (1965). Deformation textures in metals. In: *Recrystallization, Grain Growth and Textures*, pp. 295–367. Metals Park, OH: American Society for Metals.

[6] Barrett, C.S. and Massalski, T.B. (1980). *Structure of Metals: Crystallographic Methods, Principles and Data*, pp. 541–583. Oxford: Pergamon.

[7] Harris, K.E. and King, A.H. (1994). Localized texture formation and its detection in polycrystalline thin films of gold. In: *Mechanisms of Thin Film Evolution, Volume 317* (eds. S.M. Yalisove, C.V. Thompson and D.J. Eaglesham), pp. 425–430. Pennsylvania: Materials Research Society.

[8] Vill, M., Malhotra, S.G., Rek, Z., Yalisove, S.M. and Bilello, J.C. (1994). White-beam transmission characterization of texture in Mo thin films and Mo/W multilayers. In: *Mechanisms of Thin Film Evolution, Volume 317* (eds. S.M. Yalisove, C.V. Thompson and D.J. Eaglesham), pp. 413–418. Pennsylvania: Materials Research Society.

[9] Karpenko, O.P., Vill, M., Malhotra, S.G. Bilello, J.C. and Yalisove, S.M. (1994). Texture in sputtered Mo films. In: *Mechanisms of Thin Film Evolution, Volume 317* (eds. S.M. Yalisove, C.V. Thompson and D.J. Eaglesham), pp. 467–472. Pennsylvania: Materials Research Society.

[10] Ahuja, R. and Fraser, H.L. (1994). Structural transitions in titanium-aluminum thin film multilayers. In: *Mechanisms of Thin Film Evolution, Volume 317* (eds. S.M. Yalisove, C.V. Thompson and D.J. Eaglesham), pp. 479–484. Pennsylvania: Materials Research Society.

[11] Hu, H. and Goodman, S.R. (1963). Texture transition in copper. *Trans. Am. Inst. Min. Metall. Petrol. Engrs.* 227: 627–639.

[12] Bunge, H.J. (1987). Three-dimensional texture analysis. *Int. Mater. Rev.* 32: 265–291.

[13] Engler, O. and Randle, V. (2010). *Introduction to Texture Analysis: Macrotexture, Microtexture and Orientation Mapping*, 2nd Edition. Boca Raton, FL: CRC Press.

[14] Suwas, S. and Ray, R.K. (2014). *Crystallographic Texture of Materials*. London: Springer-Verlag.

[15] Rodrigues, O. (1840). Des lois géométriques qui régissent les déplacements d'un système solide dans l'espace, et de la variation des coordonnées provenant de ces déplacements considérés indépendamment des causes qui peuvent les produire. *J. Math. Pures Appl.* 5: 380–440.

[16] Frank, F.C. (1988). Orientation mapping. *Metall. Trans. A* 19: 403–408.

[17] Piña, E. (2011). Rotations with Rodrigues' vector. *Eur. J. Phys.* 32: 1171–1178.

[18] Randle, V. (1990). Representation of grain misorientations (mesotexture) in Rodrigues–Frank space. *Proc. Roy. Soc. Lond. A.* 431: 61–69.

[19] Krakow, R., Bennett, R.J., Johnstone, D.N., Vukmanovic, Z., Solano-Alvarez, W., Lainé, S.J., Einsle, J.F., Midgley, P.A., Rae, C.M.F. and Hielscher, R. (2017). On three-dimensional misorientation spaces. *Proc. Roy. Soc. Lond. A* 473, 20170274-1–20170274-28.

Appendix 1

Crystallographic Calculations

While a number of basic crystallographic calculations can be undertaken using three-dimensional coordinate geometry and spherical trigonometry, more detailed crystallographic calculations make extensive use of vector algebra and matrix algebra. In addition, quaternion algebra is relevant both for describing allowed combinations of rotations for crystals and for describing rotations between grains within polycrystalline materials. This appendix contains a summary of definitions and relationships in vector algebra, matrix algebra and quaternion algebra relevant to crystallographic calculations. Many of the topics covered here are covered in mathematics textbooks for science and engineering students, such as [1–4].

A1.1   Vector Algebra

A vector is a quantity possessing both magnitude and direction. Both must be given to specify the vector. To add two vectors $\mathbf{a}$ and $\mathbf{b}$[1] we can represent them both in magnitude and direction by lines. Each is then represented by a certain displacement (Figure A1.1) and their sum is the resultant displacement. Thus $\mathbf{c}$ in Figure A1.1 is equivalent to the displacements $\mathbf{a}$ and $\mathbf{b}$ applied successively. We may therefore write:

$$\mathbf{c} = \mathbf{a} + \mathbf{b} \tag{A1.1}$$

and:

$$\mathbf{b} = \mathbf{c} - \mathbf{a} \tag{A1.2}$$

---

[1] In Section A1.1, $\mathbf{a}$, $\mathbf{b}$ and $\mathbf{c}$ represent general vectors, rather than lattice vectors in crystals. We use this nomenclature because it is the nomenclature most often encountered in other textbooks where vectors are introduced. The context in which vectors are used in this book will make it clear whether or not vectors under consideration are lattice vectors.

Crystallography and Crystal Defects, Third Edition. Anthony Kelly and Kevin M. Knowles.
© 2020 John Wiley & Sons Ltd. Published 2020 by John Wiley & Sons Ltd.
Companion website: www.wiley.com/go/kelly/crystallography3e

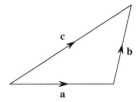

Figure A1.1 The addition of two vectors, **a** and **b**, to produce a third resultant vector, **c**

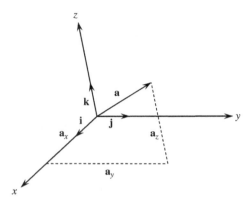

Figure A1.2 The components a_x, a_y and a_z of a vector **a** referred to three axes x, y and z, not necessarily orthogonal

Therefore, the reversal of sign of a vector is equivalent to a reversal of its direction. Note that in applying an equation such as (A1.2), any change in origin is disregarded. A vector is represented by heavy type, thus **a**. Usually, vectors are represented by small letters, but for a quantity such as an electric field, capital letters are used, for example **E**. The magnitude alone of a vector, without the idea of direction, is represented by ordinary type a or by |a|.

A unit vector is a vector of unit magnitude in a given direction. Therefore, if **i** is a vector of unit magnitude in the direction of **a**:

$$\mathbf{a} = \mathbf{i}|\mathbf{a}| = \mathbf{i}a \tag{A1.3}$$

If **i**, **j**, **k** are *unit* vectors in the directions of three coordinate axes, x, y, z, not necessarily orthogonal, and if a_x, a_y, a_z are the components of the vector **a** referred to the three axes:

$$\mathbf{a} = a_x\mathbf{i} + a_y\mathbf{j} + a_z\mathbf{k} \tag{A1.4}$$

(see Figure A1.2).

A1.1.1 The Scalar Product

The scalar product of two vectors **a** and **b** is a scalar equal in magnitude to the product of the magnitudes of **a** and **b** and the cosine of the angle between the directions of **a** and **b**. It is denoted $\mathbf{a} \cdot \mathbf{b}$ (read as '**a** dot **b**'). Thus:

$$\mathbf{a} \cdot \mathbf{b} = ab\cos\theta \tag{A1.5}$$

where θ is the angle between the vectors. Since $\cos(2\pi - \theta) = \cos\theta$ it does not matter whether θ is measured from **a** to **b** or from **b** to **a**. It follows from Eq. (A1.5) that the scalar product is commutative; that is, $\mathbf{a} \cdot \mathbf{b} = \mathbf{b} \cdot \mathbf{a}$. If θ is the angle between the directions of **a** and **b**:

$$\cos\theta = \frac{\mathbf{a} \cdot \mathbf{b}}{|\mathbf{a}||\mathbf{b}|} \tag{A1.6}$$

The scalar product $\mathbf{a} \cdot \mathbf{b}$ is equal to the projection of **a** on the direction of **b** multiplied by the magnitude of **b**, or equivalently, the projection of **b** on the direction of **a** multiplied by the magnitude of **a**. If **n** is a unit vector, $\mathbf{a} \cdot \mathbf{n}$ is the projection of **a** on the direction of **n**. If **a** is a unit vector then its projections on the axes are equal to its direction cosines.

The scalar product of a vector with itself is always equal to the square of its magnitude. For *orthogonal axes* where the vectors defining the axes are also each of the same length – that is, they are *orthonormal*,[2] this gives:

$$|\mathbf{a}|^2 = a_x^2 + a_y^2 + a_z^2 \tag{A1.7}$$

Let a_x, a_y, a_z and b_x, b_y, b_z be the components of the vectors **a** and **b** referred to a set of axes. Then from Eq. (A1.4):

$$\mathbf{a} \cdot \mathbf{b} = (a_x\mathbf{i} + a_y\mathbf{j} + a_z\mathbf{k}) \cdot (b_x\mathbf{i} + b_y\mathbf{j} + b_z\mathbf{k}) \tag{A1.8}$$

Scalar multiplication of vectors is distributive, for multiplication over addition. If **i, j, k** are a set of mutually orthonormal unit vectors, then:

$$\mathbf{i} \cdot \mathbf{i} = \mathbf{j} \cdot \mathbf{j} = \mathbf{k} \cdot \mathbf{k} = 1 \tag{A1.9}$$

$$\mathbf{i} \cdot \mathbf{j} = \mathbf{j} \cdot \mathbf{k} = \mathbf{k} \cdot \mathbf{i} = 0 \tag{A1.10}$$

and so Eq. (A1.8) becomes:

$$\mathbf{a} \cdot \mathbf{b} = a_x b_x + a_y b_y + a_z b_z \tag{A1.11}$$

If $\alpha_1, \beta_1, \gamma_1$ and $\alpha_2, \beta_2, \gamma_2$ are the angles which the directions of **a** and **b** make with the axes, respectively, then from Eqs. (A1.6), (A1.7) and (A1.10), the cosine of the angle between **a** and **b** is given by:

$$\cos\theta = \cos\alpha_1 \cos\alpha_2 + \cos\beta_1 \cos\beta_2 + \cos\gamma_1 \cos\gamma_2 \tag{A1.12}$$

for both *orthogonal* and *orthonormal* axes.

If orthonormal axes are used and the axes are termed 1, 2 and 3, rather than x, y and z, Eq. (A1.11) can be written in the form:

$$\mathbf{a} \cdot \mathbf{b} = a_1 b_1 + a_2 b_2 + a_3 b_3 \tag{A1.13}$$

This can be written in the shortened form:

$$\mathbf{a} \cdot \mathbf{b} = a_i b_i \tag{A1.14}$$

using the Einstein summation convention (Section 5.3), in which it is implicit that the summation is over the dummy subscript i, where i can take the values 1, 2 and 3.

---

[2] For orthorhombic and tetragonal unit cells, the axes are orthogonal, but care must be taken when determining angles between vectors when they are expressed in the form [uvw] – the magnitude of the 'unit' along each axis is the relevant lattice parameter. For this reason, the term 'orthonormal' is to be preferred to 'orthogonal' here to eliminate any possible ambiguity about the definition of the 'unit' under consideration.

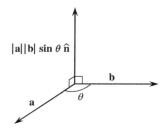

Figure A1.3 The vector product $\mathbf{a} \times \mathbf{b} = (|\mathbf{a}||\mathbf{b}| \sin\theta) \, \hat{\mathbf{n}}$ of two vectors $\mathbf{a}$ and $\mathbf{b}$, where θ is the angle less than 180° between $\mathbf{a}$ and $\mathbf{b}$

A1.1.2 The Vector Product

The vector product of two vectors $\mathbf{a}$ and $\mathbf{b}$ is itself a vector equal to $(|\mathbf{a}||\mathbf{b}| \sin\theta) \, \hat{\mathbf{n}}$, where $\hat{\mathbf{n}}$ is a unit vector perpendicular to the plane containing both $\mathbf{a}$ and $\mathbf{b}$ and in such a direction that a right-handed screw driven in the direction of $\mathbf{n}$ would carry $\mathbf{a}$ into $\mathbf{b}$ through the angle θ, where θ is the angle less than 180° between $\mathbf{a}$ and $\mathbf{b}$ (Figure A1.3). The symbol $\mathbf{a} \times \mathbf{b}$ (read as '$\mathbf{a}$ cross $\mathbf{b}$') is the vector product of $\mathbf{a}$ and $\mathbf{b}$, sometimes written as $\mathbf{a} \wedge \mathbf{b}$. It is clear from the definition of 'vector product' that the magnitude of $\mathbf{a} \times \mathbf{b}$ is the area of the parallelogram of which $\mathbf{a}$ and $\mathbf{b}$ are adjacent sides, and that:

$$\mathbf{a} \times \mathbf{b} = -\mathbf{b} \times \mathbf{a} \tag{A1.15}$$

If two vectors are parallel, their vector product vanishes ($\theta = 0°$). If they are perpendicular to one another, the magnitude of the vector product equals the product of their magnitudes. Therefore, if the axes are orthonormal:

$$\mathbf{i} \times \mathbf{i} = \mathbf{j} \times \mathbf{j} = \mathbf{k} \times \mathbf{k} = 0$$

$$\mathbf{i} \times \mathbf{j} = \mathbf{k}, \qquad \mathbf{j} \times \mathbf{k} = \mathbf{i}, \qquad \mathbf{k} \times \mathbf{i} = \mathbf{j} \tag{A1.16}$$

$$\mathbf{j} \times \mathbf{i} = -\mathbf{k}, \qquad \mathbf{k} \times \mathbf{j} = -\mathbf{i}, \qquad \mathbf{i} \times \mathbf{k} = -\mathbf{j}$$

Using these relations, we find that the components of the vector product of $\mathbf{a}$ and $\mathbf{b}$ are

$$\mathbf{a} \times \mathbf{b} = (a_x\mathbf{i} + a_y\mathbf{j} + a_z\mathbf{k}) \times (b_x\mathbf{i} + b_y\mathbf{j} + b_z\mathbf{k})$$
$$= (a_y b_z - a_z b_y)\mathbf{i} + (a_z b_x - a_x b_z)\mathbf{j} + (a_x b_y - a_y b_x)\mathbf{k} \tag{A1.17}$$

The quantities multiplying $\mathbf{i}$, $\mathbf{j}$ and $\mathbf{k}$ are the components of the vector product of $\mathbf{a}$ and $\mathbf{b}$. The relationship (A1.17) may be memorized easily by writing it in the form of a determinant (Section (A1.3)):

$$\mathbf{a} \times \mathbf{b} = \begin{vmatrix} \mathbf{i} & \mathbf{j} & \mathbf{k} \\ a_x & a_y & a_z \\ b_x & b_y & b_z \end{vmatrix} \tag{A1.18}$$

The expression $\mathbf{a} \cdot [\mathbf{b} \times \mathbf{c}]$ is a scalar and is called the scalar triple product. If the components of $[\mathbf{b} \times \mathbf{c}]$ are written out using Eq. (A1.17) and then Eq. (A1.18) is employed, we have

$\mathbf{a} \cdot [\mathbf{b} \times \mathbf{c}] = a_x(b_y c_z - b_z c_y) + a_y(b_z c_x - b_x c_z) + a_z(b_x c_y - b_y c_x)$, so:

$$\mathbf{a} \cdot [\mathbf{b} \times \mathbf{c}] = \begin{vmatrix} a_x & a_y & a_z \\ b_x & b_y & b_z \\ c_x & c_y & c_z \end{vmatrix} \tag{A1.19}$$

$\mathbf{a} \cdot [\mathbf{b} \times \mathbf{c}]$ is the volume of a parallelepiped of which $\mathbf{a}, \mathbf{b}, \mathbf{c}$ are three concurrent edges in cyclic order. It should be noted that a change in the cyclic order of the vectors in the scalar triple product results in a change of sign of the product. Hence:

$$\mathbf{a} \cdot [\mathbf{b} \times \mathbf{c}] = \mathbf{c} \cdot [\mathbf{a} \times \mathbf{b}] = \mathbf{b} \cdot [\mathbf{c} \times \mathbf{a}]$$

$$= -\mathbf{a} \cdot [\mathbf{c} \times \mathbf{b}] = -\mathbf{c} \cdot [\mathbf{b} \times \mathbf{a}] = -\mathbf{b} \cdot [\mathbf{a} \times \mathbf{c}]$$

The expression $\mathbf{a} \times [\mathbf{b} \times \mathbf{c}]$ is a vector called the vector triple product. It is a vector perpendicular to $\mathbf{a}$ in the plane containing $\mathbf{b}$ and $\mathbf{c}$. It should be noted that $\mathbf{a} \times [\mathbf{b} \times \mathbf{c}]$ is not equal to $[\mathbf{a} \times \mathbf{b}] \times \mathbf{c}$, so the order of multiplication must be shown by the brackets.

Using the Einstein summation convention and the three-subscript Levi-Civita symbol ε_{ijk}, the value of which is given by:

$\varepsilon_{ijk} = +1$ if i, j, k is a cyclic permutation of $1, 2$ and 3, i.e.$123, 231$ or 312

$\varepsilon_{ijk} = -1$ if i, j, k is an anticyclic permutation of $1, 2$ and 3, i.e.$132, 213$ or 321 (A1.20)

$\varepsilon_{ijk} = 0$ otherwise,

Equation (A1.17) can be written in the more convenient form:

$$\mathbf{a} \times \mathbf{b} = \varepsilon_{ijk} a_j b_k \tag{A1.21}$$

Likewise, the scalar triple product is:

$$\mathbf{a} \cdot [\mathbf{b} \times \mathbf{c}] = \varepsilon_{ijk} a_i b_j c_k \tag{A1.22}$$

and the vector triple product is:

$$\mathbf{a} \times [\mathbf{b} \times \mathbf{c}] = \varepsilon_{ijk}\varepsilon_{klm} a_j b_l c_m = \varepsilon_{ijk}\varepsilon_{lmk} a_j b_l c_m = (\delta_{il}\delta_{jm} - \delta_{im}\delta_{jl}) a_j b_l c_m$$

$$= b_i a_j c_j - c_i a_j b_j \tag{A1.23}$$

making use of the identity:

$$\varepsilon_{ijk}\varepsilon_{lmk} = \delta_{il}\delta_{jm} - \delta_{im}\delta_{jl} \tag{A1.24}$$

where δ_{ij} is the Kronecker delta. This is equal to 1 if $i = j$ and is 0 otherwise. It follows from Eq. (A1.23) that:

$$\mathbf{a} \times [\mathbf{b} \times \mathbf{c}] = \mathbf{b}(\mathbf{a} \cdot \mathbf{c}) - \mathbf{c}(\mathbf{a} \cdot \mathbf{b}) \tag{A1.25}$$

A further useful result which we will use in Appendix 3 is the identity:

$$[\mathbf{a} \times \mathbf{b}] \cdot [\mathbf{c} \times \mathbf{d}] = (\mathbf{a} \cdot \mathbf{c})(\mathbf{b} \cdot \mathbf{d}) - (\mathbf{a} \cdot \mathbf{d})(\mathbf{b} \cdot \mathbf{c}) \tag{A1.26}$$

which can be proved using the identities in Eqs. (A1.22) and (A1.24).

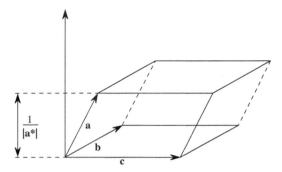

Figure A1.4 The definition of the reciprocal lattice vector **a***: the direction of **a*** is normal to the plane containing **b** and **c**, while the magnitude of **a*** is equal to the reciprocal of the spacing of the (100) planes in the real lattice

A1.2 The Reciprocal Lattice

When calculating the angle between two lattice planes of given indices, or when transforming the indices of a given plane or direction from those corresponding to one unit cell to those appropriate to another, it is often extremely useful to use a device called the reciprocal lattice. This was first developed for carrying out lattice sums and then became invaluable in diffraction problems and in the theory of the behaviour of electrons in crystals. We will introduce it in a formal way in this appendix and then illustrate its utility by deriving some formulae of general use.

Suppose we have a crystal lattice and we select a primitive unit cell with translation vectors **a**, **b** and **c**. We define three vectors reciprocal to **a**, **b** and **c**, and call these **a***, **b*** and **c***. The defining relations are:

$$\mathbf{a}^* = \frac{\mathbf{b} \times \mathbf{c}}{\mathbf{a} \cdot [\mathbf{b} \times \mathbf{c}]}, \quad \mathbf{b}^* = \frac{\mathbf{c} \times \mathbf{a}}{\mathbf{b} \cdot [\mathbf{c} \times \mathbf{a}]} \quad \text{and} \quad \mathbf{c}^* = \frac{\mathbf{a} \times \mathbf{b}}{\mathbf{c} \cdot [\mathbf{a} \times \mathbf{b}]} \tag{A1.27}$$

From these definitions it is immediately apparent that **a*** is normal to the plane containing **b** and **c**. Its magnitude is equal to the reciprocal of the spacing of the (100) planes in the real lattice (Figure A1.4). **a***, **b*** and **c*** are taken to define the primitive unit cell of the lattice which is reciprocal to the real crystal lattice with cell edge vectors **a**, **b** and **c**. Once **a***, **b*** and **c*** have been found from relations in ((A1.27)), this reciprocal lattice can be constructed. The utility of the reciprocal lattice depends upon the following properties:

1. A lattice vector of the reciprocal lattice such as:

$$\mathbf{r}^* = h\mathbf{a}^* + k\mathbf{b}^* + l\mathbf{c}^*$$

is normal to the planes of Miller indices (*hkl*) in the real lattice. Thus if *h*, *k*, *l* are given small integral values, the reciprocal lattice vectors **r***(*hkl*) represent the normals to low index planes.

2. The magnitude of **r***, that is, |**r***| is equal to the reciprocal of the spacing of the planes of the real lattice of Miller indices (*hkl*).

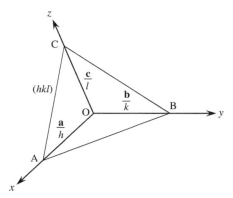

Figure A1.5 The plane (*hkl*) in the real crystal making intercepts of **a**/*h*, **b**/*k* and **c**/*l* along the *x*-, *y*- and *z*-axes, respectively

To prove these two propositions, we note from the definition of **a***, **b*** and **c*** that **a*** · **a** = 1, **b*** · **b** = 1 and **c*** · **c** = 1, but that **a*** · **b** = **a*** · **c** = **b*** · **a** = **b*** · **c** = **c*** · **a** = **c*** · **b** = 0. The two properties of **r***(*hkl*) then follow quite simply. In Figure A1.5, ABC is the plane of indices (*hkl*) in the real crystal. The vector (**a**/*h* − **b**/*k*) represents the vector **BA** in this plane and (**a**/*h* − **c**/*l*) represents the vector **CA** in the same plane. If **r*** is normal to the plane ABC, it must be normal to both **BA** and **CA**, which are nonparallel directions in ABC. The dot product of **r*** with these two vectors must be equal to zero. This is clearly so, since, for example:

$$\mathbf{r}^* \cdot \left(\frac{\mathbf{a}}{h} - \frac{\mathbf{b}}{k} \right) = (h\mathbf{a}^* + k\mathbf{b}^* + l\mathbf{c}^*) \cdot \left(\frac{\mathbf{a}}{h} - \frac{\mathbf{b}}{k} \right) = \frac{h}{h}\mathbf{a} \cdot \mathbf{a}^* - \frac{k}{k}\mathbf{b} \cdot \mathbf{b}^* = 0$$

and a similar calculation shows that **r***·(**a**/*h* − **c**/*l*) = 0. This proves proposition (1).

We prove proposition (2) by noting that the spacing of the lattice planes of indices (*hkl*) equals the projection of **a**/*h* onto their normal. By proposition (1), a unit vector along the normal to this plane is **r***/|**r***|. Then the spacing of the planes d_{hkl} is given by:

$$d_{hkl} = \frac{\mathbf{a}}{h} \cdot \frac{\mathbf{r}^*}{|\mathbf{r}^*|} = \frac{h\mathbf{a} \cdot \mathbf{a}^*}{h|\mathbf{r}^*|} = \frac{1}{|\mathbf{r}^*|}$$

It follows from the definition of the reciprocal lattice vectors that the plane (*hkl*) is in the zone [*uvw*] if and only if:

$$hu + kv + lw = 0 \tag{A1.28}$$

This is so because if (*hkl*) is in the zone we must have **r***(*hkl*) normal to [*uvw*]. That is:

$$(h\mathbf{a}^* + k\mathbf{b}^* + l\mathbf{c}^*) \cdot (u\mathbf{a} + v\mathbf{b} + w\mathbf{c}) = 0$$

i.e.:

$$hu + kv + lw = 0$$

The indices of the zone axis [*uvw*] common to the two planes of indices $(h_1 k_1 l_1)$, $(h_2 k_2 l_2)$ are derived in Section 1.3. The condition for three planes $(h_1 k_1 l_1)$, $(h_2 k_2 l_2)$, $(h_3 k_3 l_3)$ to be in the same zone is that the corresponding reciprocal lattice vectors are coplanar. If the corresponding reciprocal lattice vectors are r_1^*, r_2^* and r_3^*, this implies that they define a parallelepiped of zero volume; that is, $r_1^* \cdot [r_2^* \times r_3^*]$ equals zero. From Eq. (A1.19), this implies:

$$\begin{vmatrix} h_1 & k_1 & l_1 \\ h_2 & k_2 & l_2 \\ h_3 & k_3 & l_3 \end{vmatrix} = 0 \tag{A1.29}$$

The condition that a given lattice point, say $r (= u\mathbf{a} + v\mathbf{b} + w\mathbf{c})$ from the origin, should lie in a given lattice plane (hkl) is useful to know in drawing arrangements of lattice points and/or atoms in lattice planes. If r lies in the pth plane from the origin of indices (hkl) then the projection of r on to the normal to the lattice planes (hkl) must equal p times the interplanar spacing. That is:

$$r \cdot \frac{r^*(h\,k\,l)}{|\,r^*(h\,k\,l)\,|} = \frac{p}{|r^*(h\,k\,l)|}$$

Therefore, multiplying out:

$$hu + kv + lw = p \tag{A1.30}$$

The angle between two sets of planes, with Miller indices $(h_1 k_1 l_1)$ and $(h_2 k_2 l_2)$, must be equal to that between the two reciprocal vectors:

$$r_1^* = h_1 \mathbf{a}^* + k_1 \mathbf{b}^* + l_1 \mathbf{c}^*$$

and:

$$r_2^* = h_2 \mathbf{a}^* + k_2 \mathbf{b}^* + l_2 \mathbf{c}^*$$

Therefore:

$$r_1^* \cdot r_2^* = |r_1^*||r_2^*| \cos \phi$$

where ϕ is the required angle. Hence:

$$\cos \phi = d_1 d_2 (h_1 \mathbf{a}^* + k_1 \mathbf{b}^* + l_1 \mathbf{c}^*) \cdot (h_2 \mathbf{a}^* + k_2 \mathbf{b}^* + l_2 \mathbf{c}^*) \tag{A1.31}$$

where d_1 and d_2 are the corresponding spacings of the planes.

Spacings of lattice planes can be derived from the relation:

$$\frac{1}{d_{hkl}} = |r^*(hkl)| = |h\mathbf{a}^* + k\mathbf{b}^* + l\mathbf{c}^*| \tag{A1.32}$$

Expanded versions of Eqs. (A1.31) and (A1.32) are given in Appendix 3 for the various crystal systems.

A1.3   Matrices

A matrix is an array of quantities that can be helpful in writing down a set of equations in a compact form and in manipulating them. For example, the equations giving the components (j_1, j_2, j_3) of the current density $\mathbf{j}$ referred to three mutually perpendicular axes

Ox_1, Ox_2, Ox_3 as a function of applied electric field, represented by the vector **E** with components (E_1, E_2, E_3) referred to the same set of axes, are Eqs. (5.1):

$$j_1 = \sigma_{11}E_1 + \sigma_{12}E_2 + \sigma_{13}E_3$$
$$j_2 = \sigma_{21}E_1 + \sigma_{22}E_2 + \sigma_{23}E_3$$
$$j_3 = \sigma_{31}E_1 + \sigma_{32}E_2 + \sigma_{33}E_3$$

We define the matrix σ (which in this case is a tensor of the second rank) as the following array of coefficients:

$$\sigma = \begin{pmatrix} \sigma_{11} & \sigma_{12} & \sigma_{13} \\ \sigma_{21} & \sigma_{22} & \sigma_{23} \\ \sigma_{31} & \sigma_{32} & \sigma_{33} \end{pmatrix} \tag{A1.33}$$

The equations in Eq. (5.1) can now be written in the form:

$$\begin{pmatrix} j_1 \\ j_2 \\ j_3 \end{pmatrix} = \sigma \begin{pmatrix} E_1 \\ E_2 \\ E_3 \end{pmatrix} \tag{A1.34}$$

provided that the following meaning is attached to the multiplication of the column (E_1, E_2, E_3) by the matrix σ. The first step is to multiply each term in the first row of the matrix by the corresponding term in the column (E_1, E_2, E_3) and to add the products. This is easy to remember as 'first of row times first of column plus second of row times second of column plus third of row times third of column'. The sum is set equal to the first term in the resulting column, j_1, which gives the first of Eqs. (5.1). The second and third of Eqs. (5.1) are obtained in a similar way from the second and third rows of the matrix. Eqs. (5.1) can now be written still more compactly as:

$$\mathbf{j} = \sigma\mathbf{E} \tag{A1.35}$$

where the vectors **j** and **E** are understood to be represented by columns for the purpose of the multiplication. The reason for writing the components of a vector in a column will become clear when we consider what meaning should be attached to the multiplication of one matrix by another.

To take a specific example, suppose that two rotations are applied to a body, one after the other, as in Section 1.6, in which possible combinations of rotational symmetries for crystals are considered. If a first rotation carries the point **x** to **x**′ then:

$$\mathbf{x}' = \mathbf{R}_1\mathbf{x} \tag{A1.36}$$

where $\mathbf{R}_1$ is the matrix representing the first rotation. A second rotation carries the point **x**′ to the point **x**″, where:

$$\mathbf{x}'' = \mathbf{R}_2\mathbf{x}' \tag{A1.37}$$

where $\mathbf{R}_2$ is the matrix representing the second rotation. From Eq. (A1.36):

$$\mathbf{x}'' = \mathbf{R}_2\mathbf{R}_1\mathbf{x} \tag{A1.38}$$

When Eqs. (A1.36) and (A1.37) are written out in full, it will be seen that Eq. (A1.38), describing the net effect of the two rotations, can be written as:

$$\mathbf{x}'' = \mathbf{Rx} \tag{A1.39}$$

where $\mathbf{R}$ $(= \mathbf{R}_2\mathbf{R}_1)$ is a matrix whose terms can be obtained as follows. The term in the ith row and jth column of $\mathbf{R}$ is obtained by multiplying each term in the ith row of the left-hand matrix $\mathbf{R}_2$ by the corresponding term in the jth column of the right-hand matrix $\mathbf{R}_1$ and adding the products. For instance, if:

$$\mathbf{R}_2 = \begin{pmatrix} a_{11} & a_{12} & a_{13} \\ a_{21} & a_{22} & a_{23} \\ a_{31} & a_{32} & a_{33} \end{pmatrix}, \quad \mathbf{R}_1 = \begin{pmatrix} b_{11} & b_{12} & b_{13} \\ b_{21} & b_{22} & b_{23} \\ b_{31} & b_{32} & b_{33} \end{pmatrix}$$

and:

$$\mathbf{R} = \mathbf{R}_2\mathbf{R}_1 = \begin{pmatrix} c_{11} & c_{12} & c_{13} \\ c_{21} & c_{22} & c_{23} \\ c_{31} & c_{32} & c_{33} \end{pmatrix}$$

then:

$$c_{11} = a_{11}b_{11} + a_{12}b_{21} + a_{13}b_{31}$$

or, in general, the ijth element is:

$$c_{ij} = a_{i1}b_{1j} + a_{i2}b_{2j} + a_{i3}b_{3j} = a_{ik}b_{kj} \tag{A1.40}$$

using the Einstein summation convention.

Compounding the rows of the left-hand matrix with the columns of the right-hand matrix in this way gives a general definition of the multiplication of two matrices. The operation can be performed with any two arrays of numbers *provided* that the number of columns of the left-hand array equals the number of rows of the right-hand array. The multiplication of a vector by a matrix, as in Eq. (A1.34), can be regarded as a special case of this rule where the vector is represented by a matrix with three rows and one column (i.e. a 3×1 matrix). Generally, therefore, the product of an $(m \times n)$ matrix with one of order $(n \times p)$ is a matrix of order $(m \times p)$.

It should be noted that in the definition of multiplication, the left-hand and right-hand matrices are treated differently, so that in general for two matrices $\mathbf{D}_1$ and $\mathbf{D}_2$:

$$\mathbf{D}_1\mathbf{D}_2 \neq \mathbf{D}_2\mathbf{D}_1$$

In our example, the rotation produced by two successive rotations depends in general on the order in which the two rotations are applied.

If we examine the equations representing homogeneous strain of a body in Eq. (A6.3) we see that they can be written in the form:

$$\mathbf{x}' = \mathbf{Ix} + \mathbf{Sx} \tag{A1.41}$$

where:

$$\mathbf{I} = \begin{pmatrix} 1 & 0 & 0 \\ 0 & 1 & 0 \\ 0 & 0 & 1 \end{pmatrix} \quad \text{and} \quad \mathbf{S} = \begin{pmatrix} e_{11} & e_{12} & e_{13} \\ e_{21} & e_{22} & e_{23} \\ e_{31} & e_{32} & e_{33} \end{pmatrix}$$

Writing Eq. (A1.41) as:

$$\mathbf{x}' = (\mathbf{I} + \mathbf{S})\mathbf{x}$$

and comparing with an equivalent matrix form:

$$\mathbf{x}' = \mathbf{D}\mathbf{x}$$

we see that $\mathbf{D} = \mathbf{I} + \mathbf{S}$. In this case, and in general, one matrix is added to another merely by adding together corresponding terms in the two arrays. It is said that the two matrices are conformable for addition, when each has the same number of rows and each has the same number of columns. The 3×3 matrix $\mathbf{I}$, defined above, has the special property that:

$$\mathbf{IA} = \mathbf{A} = \mathbf{AI} \tag{A1.42}$$

where $\mathbf{A}$ is any 3×3 matrix. It is therefore called the unit matrix. A unit $n \times n$ matrix is defined in exactly the same way. Given a square $(n \times n)$ matrix $\mathbf{B}$, if we can find another $n \times n$ matrix that produces the unit matrix when it multiplies $\mathbf{B}$, we say that we have found the inverse or reciprocal of $\mathbf{B}$, written $\mathbf{B}^{-1}$. Thus:

$$\mathbf{B}^{-1}\mathbf{B} = \mathbf{I} = \mathbf{BB}^{-1} \tag{A1.43}$$

The inverse of a matrix representing a homogeneous strain, for example, is the matrix representing the 'opposite' strain that will return the body to its undistorted state. Likewise, the inverse of a matrix representing a rotation is the matrix representing the 'opposite' rotation that will return the body to its unrotated state. As a third example, consider Hooke's law in its contracted form (Eq. (6.53)). The six equations giving the stress in terms of the strain can be written in matrix form as:

$$\begin{pmatrix} \sigma_1 \\ \sigma_2 \\ \sigma_3 \\ \sigma_4 \\ \sigma_5 \\ \sigma_6 \end{pmatrix} = \mathbf{C} \begin{pmatrix} \varepsilon_1 \\ \varepsilon_2 \\ \varepsilon_3 \\ \varepsilon_4 \\ \varepsilon_5 \\ \varepsilon_6 \end{pmatrix} \tag{A1.44}$$

where $\mathbf{C}$ is a 6×6 matrix of stiffness constants. Eq. (A1.44) can be written more compactly as:

$$\sigma = \mathbf{C}\varepsilon \tag{A1.45}$$

with the understanding that σ and ε are 6×1 matrices. Hooke's law can also be written in a contracted form giving the strain in terms of the stress:

$$\varepsilon = \mathbf{S}\sigma \tag{A1.46}$$

where $\mathbf{S}$ is a 6×6 matrix of compliances. The matrix $\mathbf{S}$ is in fact the inverse of the matrix $\mathbf{C}$, since by multiplying both sides of Eq. (A1.45) by $\mathbf{C}^{-1}$ we have:

$$\mathbf{C}^{-1}\sigma = \mathbf{C}^{-1}\mathbf{C}\varepsilon = \mathbf{I}\varepsilon$$

or:

$$\mathbf{C}^{-1}\sigma = \varepsilon$$

Therefore:

$$\mathbf{C}^{-1} = \mathbf{S} \tag{A1.47}$$

It is apparent from these equations that solving the six equations represented by Eq. (A1.44) simultaneously to obtain ε in terms of σ is equivalent to finding the inverse of the matrix $\mathbf{C}$.

In order for a matrix to possess an inverse it must be both square and nonsingular. In order to find the inverse of a matrix $\mathbf{A}$ we have to find the determinant of the matrix, $|\mathbf{A}|$, and we have to be able to write down the adjoint matrix to $\mathbf{A}$, adj $\mathbf{A}$. If the matrix has elements a_{ij}, so that:

$$\mathbf{A} = \begin{bmatrix} a_{11} & a_{12} & a_{13} & - & - & - & a_{1n} \\ a_{21} & a_{22} & a_{23} & - & - & - & a_{2n} \\ a_{31} & a_{32} & a_{33} & - & - & - & a_{3n} \\ - & - & - & - & - & - & - \\ - & - & - & - & - & - & - \\ - & - & - & - & - & - & - \\ a_{n1} & a_{n2} & a_{n3} & - & - & - & a_{nn} \end{bmatrix} \tag{A1.48}$$

then the cofactor of the element a_{ik} in $\mathbf{A}$ is defined as $(-1)^{i+k}$ times the value of the determinant formed by deleting the row and column in which a_{ik} occurs.

The matrix of cofactors, A_{ik}, so formed, which is clearly also a square matrix of the same order as $\mathbf{A}$, is:

$$\begin{bmatrix} A_{11} & A_{12} & A_{13} & - & - & - & A_{1n} \\ A_{21} & A_{22} & A_{23} & - & - & - & A_{2n} \\ A_{31} & A_{32} & A_{33} & - & - & - & A_{3n} \\ - & - & - & - & - & - & - \\ - & - & - & - & - & - & - \\ - & - & - & - & - & - & - \\ A_{n1} & A_{n2} & A_{n3} & - & - & - & A_{nn} \end{bmatrix}$$

The adjoint matrix, adj $\mathbf{A}$, is then defined as the transposed matrix $\tilde{\mathbf{A}}$ (i.e. the matrix obtained by interchanging rows and columns) of the cofactors of $\mathbf{A}$. Thus:

$$\text{adj } \mathbf{A} = \begin{bmatrix} A_{11} & A_{21} & A_{31} & - & - & - & A_{n1} \\ A_{12} & A_{22} & A_{32} & - & - & - & A_{n2} \\ A_{13} & A_{23} & A_{33} & - & - & - & A_{n3} \\ - & - & - & - & - & - & - \\ - & - & - & - & - & - & - \\ - & - & - & - & - & - & - \\ A_{1n} & A_{2n} & A_{3n} & - & - & - & A_{nn} \end{bmatrix} \tag{A1.49}$$

Having formed the adjoint matrix, adj $\mathbf{A}$, and having evaluated $|\mathbf{A}|$, the inverse of $\mathbf{A}$, $\mathbf{A}^{-1}$, is given by:

$$\mathbf{A}^{-1} = \frac{\text{adj } \mathbf{A}}{|\mathbf{A}|} \tag{A1.50}$$

A set of n simultaneous equations in n unknowns can generally be solved, and so the inverse of a square matrix can generally be found. However, if it happens that one of the equations is merely a linear combination of some of the others, then there are only $n - 1$ essentially different equations and a solution is not possible. It follows that a matrix has no inverse if one of its rows is a linear combination of others, and it can be shown that the same is true if there is a linear relationship amongst its columns. Such a matrix is called a *singular* matrix; it has a zero determinant.

A simple illustration of these ideas is to consider the condition for three lattice vectors to be coplanar. Since any two directions $[u_1 v_1 w_1]$ and $[u_2 v_2 w_2]$ in a crystal define a plane (hkl) in which these two vectors lie, a third vector $[u_3 v_3 w_3]$ will be coplanar with these two vectors if it can be written as a linear combination of these two vectors; that is, if:

$$\begin{vmatrix} u_1 & v_1 & w_1 \\ u_2 & v_2 & w_2 \\ u_3 & v_3 & w_3 \end{vmatrix} = 0 \tag{A1.51}$$

A1.4 Rotation Matrices and Unit Quaternions

Rotations can be described either using the algebra of matrices or by using the concepts of quaternions. Both approaches are used in this book, depending on the context.

If a rotation carries the point **x** to **x'** then, following Eq. (A1.36):

$$\mathbf{x'} = \mathbf{Rx} \tag{A1.52}$$

where **R** is the matrix representing the rotation. Suppose this is an anticlockwise rotation of angle θ about an axis **n** where $\mathbf{n.n} = 1$ with respect to an orthonormal axis system (Figure A1.6). Then $\mathbf{n} \times \mathbf{x}$ is a vector into the plane of the paper normal to **n** and **x** with a magnitude equal to the radius of the circle shown in Figure A1.6; $\mathbf{n} \times \mathbf{n} \times \mathbf{x}$ is a vector perpendicular to **n** and $\mathbf{n} \times \mathbf{x}$, also with a magnitude equal to the radius of this circle. Hence:

$$\mathbf{x'} = \mathbf{x} + (1 - \cos\theta)\,\mathbf{n} \times \mathbf{n} \times \mathbf{x} + \sin\theta\ \mathbf{n} \times \mathbf{x} \tag{A1.53}$$

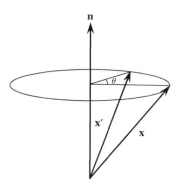

Figure A1.6 An anticlockwise rotation of θ about **n** carrying the point **x** to **x'**

The elements of $\mathbf{R}$ can then be determined by comparing Eqs. (A1.52) and (A1.53):

$$\mathbf{R} = \begin{bmatrix} \cos\theta + n_1^2(1-\cos\theta) & n_1 n_2(1-\cos\theta) - n_3\sin\theta & n_3 n_1(1-\cos\theta) + n_2\sin\theta \\ n_1 n_2(1-\cos\theta) + n_3\sin\theta & \cos\theta + n_2^2(1-\cos\theta) & n_2 n_3(1-\cos\theta) - n_1\sin\theta \\ n_3 n_1(1-\cos\theta) - n_2\sin\theta & n_2 n_3(1-\cos\theta) + n_1\sin\theta & \cos\theta + n_3^2(1-\cos\theta) \end{bmatrix}$$

$$(A1.54)$$

from which it follows that:

$$\mathbf{R}_{ii} = (2\cos\theta + 1) \tag{A1.55}$$

$$\varepsilon_{ijk}\mathbf{R}_{jk} = -2n_i\sin\theta \tag{A1.56}$$

using the Einstein summation convention. Equations (A1.55) and (A1.56) therefore enable an angle and axis to be obtained if $\mathbf{R}$ is known, perhaps as a consequence of the multiplication of two other rotation matrices. A rotation matrix $\mathbf{R}$ has the property that its inverse is identical to its transpose; that is, the matrix formed by interchanging the rows and columns of $\mathbf{R}$. Such matrices are said to be *orthogonal*.

Rotations can also be described using quaternion algebra. Quaternions are sets of four numbers originally developed by Sir William Rowan Hamilton which obey particular rules of addition and multiplication [5–8]. If a particular quaternion $\mathbf{q}$ consists of the scalar quantities q_0, q_1, q_2 and q_3 then $\mathbf{q}$ can be written in the form:

$$\mathbf{q} = \{q_0, q_1, q_2, q_3\} = q_0 + q_1\mathbf{i} + q_2\mathbf{j} + q_3\mathbf{k} \tag{A1.57}$$

where $\mathbf{i}, \mathbf{j}$ and $\mathbf{k}$ are vector quantities obeying the multiplication properties:

$$\mathbf{i}\cdot\mathbf{i} = \mathbf{j}\cdot\mathbf{j} = \mathbf{k}\cdot\mathbf{k} = \mathbf{i}\cdot\mathbf{j}\cdot\mathbf{k} = -1$$
$$\mathbf{i}\cdot\mathbf{j} = \mathbf{k} = -\mathbf{j}\cdot\mathbf{i}$$
$$\mathbf{j}\cdot\mathbf{k} = \mathbf{i} = -\mathbf{k}\cdot\mathbf{j} \tag{A1.58}$$
$$\mathbf{k}\cdot\mathbf{i} = \mathbf{j} = -\mathbf{i}\cdot\mathbf{k}$$

Using these properties, it is a straightforward exercise to show that the product of two quaternions $\mathbf{p} = \{p_0, p_1, p_2, p_3\}$ and $\mathbf{q} = \{q_0, q_1, q_2, q_3\}$ is:

$$\begin{aligned}\mathbf{p}\cdot\mathbf{q} = \{&p_0 q_0 - p_1 q_1 - p_2 q_2 - p_3 q_3, \\ &p_0 q_1 + p_1 q_0 + p_2 q_3 - p_3 q_2, \\ &p_0 q_2 - p_1 q_3 + p_2 q_0 + p_3 q_1, \\ &p_0 q_3 + p_1 q_2 - p_2 q_1 + p_3 q_0\}\end{aligned} \tag{A1.59}$$

The complex conjugate $\mathbf{q}^*$ of a quaternion $\mathbf{q}$ is the set of four numbers $\mathbf{q}^* = \{q_0, -q_1, -q_2, -q_3\}$. Hence it follows that $\mathbf{q}+\mathbf{q}^*$ is $\{2q_0, 0, 0, 0\}$; that is, a scalar. The product $\mathbf{q}\cdot\mathbf{q}^* = \mathbf{q}^*\cdot\mathbf{q}$ is also a scalar quantity:

$$\mathbf{q}\cdot\mathbf{q}^* = \mathbf{q}^*\cdot\mathbf{q} = q_0^2 + q_1^2 + q_2^2 + q_3^2 \tag{A1.60}$$

If $\mathbf{q} \cdot \mathbf{q}^* = 1$, the quaternion is said to be a *unit quaternion*. Unit quaternions are particularly useful when describing rotations. The rotation matrix described by Eq. (A1.54) is homomorphic to the unit quaternion:

$$\mathbf{q} = \{q_0, q_1, q_2, q_3\} = \left\{\cos \tfrac{1}{2}\theta, \ n_1 \sin \tfrac{1}{2}\theta, \ n_2 \sin \tfrac{1}{2}\theta, \ n_3 \sin \tfrac{1}{2}\theta\right\} \tag{A1.61}$$

Multiplying two unit quaternions produces a resultant quaternion which is also a unit quaternion. It is apparent from Eq. (A1.59) that the quaternion product is noncommutative: in general $\mathbf{p} \cdot \mathbf{q} \neq \mathbf{q} \cdot \mathbf{p}$. This is a statement in quaternion algebra terms that the order of rotations is important in defining the resultant of two successive rotations.

References

[1] Boas, M.L. (2006). *Mathematical Methods in the Physical Sciences*, 3rd Edition. New York: Wiley.
[2] Riley, K.F., Hobson, M.P. and Bence, S.J. (2006). *Mathematical Methods for Physics and Engineering*, 3rd Edition. Cambridge: Cambridge University Press.
[3] Kreysig, E. (2010). *Advanced Engineering Mathematics*, 10th Edition. New York: Wiley.
[4] Arfken, G.B., Weber, H.J. and Harris, F.E. (2012). *Mathematical Methods for Physicists: A Comprehensive Guide*, 7th Edition. Waltham, Massachusetts: Academic Press.
[5] Hamilton, W.R. (1844). On a new species of imaginary quantities connected with a theory of quaternions. *Proc. R. Irish Acad.* 2: 424–434.
[6] Kuipers, J.B. (1999). *Quaternions and Rotation Sequences*. Princeton: Princeton University Press.
[7] Altmann, S.L. (1986). *Rotations, Quaternions and Double Groups*. Oxford: Clarendon Press.
[8] Grimmer, H. (1974). Disorientations and coincidence rotations for cubic lattices. *Acta Crystall. A* 30: 685–688.

Appendix 2

The Stereographic Projection

A2.1 Principles

In the study of crystallography it is often useful to be able to represent crystal planes and crystal directions on a diagram in two dimensions so that angular relationships and the symmetrical arrangements of crystal faces can be discussed upon a flat piece of paper, and if required, *measured*. Clearly, the most useful type of diagram will be one in which the angular relationships in three dimensions in the crystal are faithfully reproduced in a plane in some form of projectional geometry. Mathematically, a projection from three dimensions to two dimensions in which angular relationships are faithfully reproduced is known as a *conformal projection*. The conformal projection used in crystallography is the *stereographic projection*.

To picture how a stereographic projection is used in crystallography, imagine a crystal to be positioned with its centre at the centre of a sphere, which we call the *sphere of projection* (Figure A2.1a), and draw normals to crystal planes through the centre of the sphere to intersect the surface of the sphere, say at P. P is called the *pole* of the plane of which OP is the normal. A direction is similarly represented by a point on the surface of the sphere, defined as the point where the line parallel to the given direction, passing through the centre of the sphere, strikes the surface of the sphere. A crystal plane can also be represented by drawing the parallel plane through the centre of the sphere and extending it until it strikes the sphere (Figure A2.1a). Since the plane passes through the centre of the sphere, it is a *diametral plane*, and the line of intersection of the sphere with such a plane is called a *great circle*. A great circle is a circle on the surface of a sphere with a radius equal to the radius of the sphere.

At this stage we have represented directions in the crystal – that is, normals to lattice planes or lattice directions – by points (poles) on the surface of the sphere. We have a *spherical projection* of the crystal. The angle between two planes of which the normals are OP and OQ (Figure A2.1b) is equal to the angle between these normals, which is the angle subtended at the centre of the sphere of projection by the arc of the great circle drawn

Crystallography and Crystal Defects, Third Edition. Anthony Kelly and Kevin M. Knowles.
© 2020 John Wiley & Sons Ltd. Published 2020 by John Wiley & Sons Ltd.
Companion website: www.wiley.com/go/kelly/crystallography3e

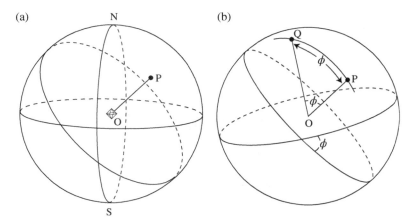

Figure A2.1   (a) Sphere of projection. (b) The angle between two planes is equal to the angle ϕ between the two poles

through the poles P and Q. To make a drawing in two dimensions in which angular relationships are preserved, we now project the poles onto a suitable two-dimensional plane such as a piece of paper.

The spherical projection is like a terrestrial globe. Let us define north and south poles, N and S in Figure A2.1a, by analogy with the north and south poles of a globe. The equatorial plane passes through the centre of the sphere normal to the line NS and cuts the sphere in a great circle called the equator. There are various ways of projecting points on the sphere onto a two-dimensional plane. A number of ways are shown in Figure A2.2.

In the *orthographic* projection a pole P is projected from a point at infinity onto a plane parallel to the equatorial plane to form P'_O on a plane parallel to the equatorial plane passing through N. In the *gnomonic* projection the point of projection is the centre of the sphere, giving the projected pole at P'_G on a plane parallel to the equatorial plane passing

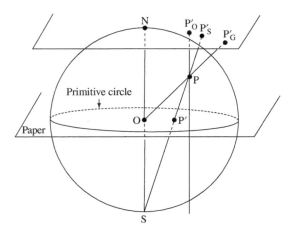

Figure A2.2   Projections of poles on the surface of a sphere onto a flat piece of paper

through N. Both these projections have their uses in crystallography – the orthographic projection is useful for visualizing crystal shapes and the gnomonic projection is relevant for labelling electron back-scattered diffraction patterns in scanning electron microscopes. However, neither of these projections is conformal, and so angles are distorted in these projections.

In the *stereographic* projection the pole P is projected from a point on the surface of the sphere, say S, called the pole of projection, onto a plane normal to OS. This plane can pass through any point on NS. If it passes through N, the point P projects to P'_S. The most convenient plane for our purpose is the equatorial plane normal to SO. If we project the point P from S onto this plane we define the point P' so produced as the *stereographic projection* of P. In what follows we shall always take the plane of projection as the equatorial plane. The line of intersection of the plane of projection with the sphere of projection is a great circle called the primitive circle, or, for brevity, the primitive. The method of projection we shall adopt is shown in Figure A2.3a. A pole P_1 in the northern hemisphere projects to P'_1, inside the primitive, and is marked with a dot on the paper. All poles in the northern hemisphere project inside the primitive. Poles in the southern hemisphere, say P_2, give a projection P'_2 outside the primitive. The point P'_2 is the *true* projection of P_2. It is often inconvenient to work with projected poles outside the primitive, and to avoid this a pole P_2, in the southern hemisphere, may be projected from the north pole N (diametrically opposite S) to give the projected pole at P''_2. The projected pole P''_2 is then distinguished from the *true projection* of P_2 (at P'_2) by marking the point P''_2 with a ring instead of with a dot.

In addition to being angle true, the stereographic projection has a second very useful property: all circles (great or small) on the surface of the sphere of projection project as circles. This is illustrated for a small circle in Figure A2.3b. A proof of both these properties is given at the end of this appendix in Section A2.4.

We can now proceed to draw the stereographic representation or stereogram of the poles of crystal planes in a cubic crystal. In cubic crystals the normal to a plane (*hkl*) is parallel to the vector [*hkl*]; therefore in stereographic projections of such crystals the pole *hkl*

(a)   (b)

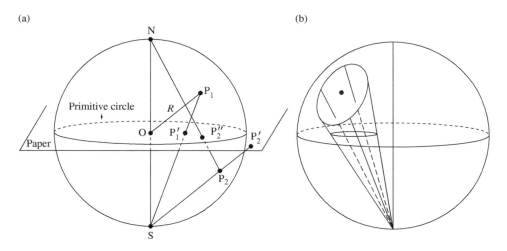

Figure A2.3   (a) Stereographic projection. (b) A small circle projects as a circle

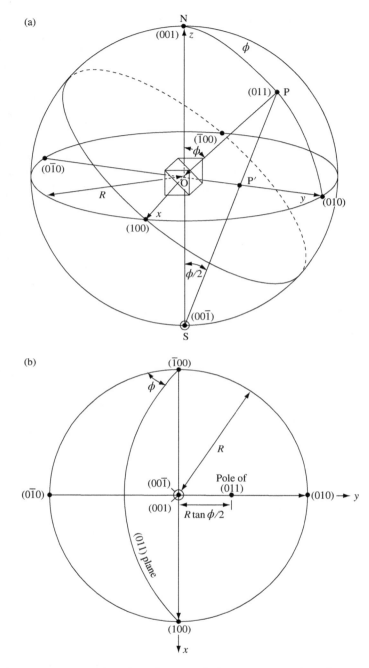

Figure A2.4 (a) Poles of a cubic crystal. (b) Stereogram of a cubic crystal

can represent either the normal to the plane (*hkl*) or the [*hkl*] zone. The crystal axes are positioned with respect to the pole and plane of projection as in Figure A2.4a. The three axes are orthogonal and of equal length (Table 1.3). In the *standard* projection shown in Figure A2.4b, the *z*-axis of the crystal is taken normal to the plane of projection and, since the axes are orthogonal, the *x*- and *y*-axes then lie in the plane of projection at 90° to one another. The pole of the (001) planes coincides with N and projects to the centre of the primitive (Figure A2.4b). The poles of (100), (010), ($\bar{1}$00) and (0$\bar{1}$0) lie on the primitive equally spaced at angles of 90°. The (00$\bar{1}$) planes would lie at infinity if projected from S, so we project it from N and denote it by the ring. The (011) planes are represented by the pole P; (011) lies in the zone of which the *x*-axis is the zone axis; that is, [100]. The poles of all planes in the [100] zone lie on the great circle defined by the locus of all points 90° from the (100) pole. This great circle projects as the *line* on the stereogram joining (0$\bar{1}$0), (001) and (010). Therefore, P projects somewhere between (001) and (010). The angle ϕ in Figure A2.4a is the angle between (001) and (011); for the cubic crystal, $\phi = 45°$. From Figure A2.4a, the distance OP′ is given by:

$$OP' = R \tan \phi/2 \qquad (A2.1)$$

where *R* is the radius of the sphere of projection. This follows since S, O, N, P and P′ all lie in the same plane, and the angle OSP is equal to $\phi/2$ because OSP is the angle at the circumference standing on the same arc NP as the angle NOP at the centre. We can therefore insert the (011) pole on the stereogram at a distance $R \tan \phi/2$ (in this case, $R \tan 45°/2 = R \tan 22.5°$ from the (001) pole along the radius of the primitive joining (001) and (010)).

The plane (011) itself can be drawn upon the stereogram instead of just the pole of (011) by drawing the projection of the great circle which is the locus of points 90° from the pole (011). This is drawn in Figure A2.4b.

Drawing a great circle of which the pole is given can be accomplished either by construction or by using graphical aids. We will first deal with some constructions on the stereogram and then with the use of the graphical aid known as the Wulff net.

A2.2   Constructions

To obtain a thorough understanding of the stereogram, it is wise for the beginner to carry out a number of constructions accurately and without any graphical aid. We will now describe some of these. However, since all of these constructions can be accomplished with graphical aids or software packages, this section can be omitted without prejudice to the rest of the appendix.

A2.2.1   To Construct a Small Circle

A2.2.1.1   About the Centre of the Primitive
The stereographic projection of the required angular radius ϕ of the small circle is plotted on either side of N, at X′Y′, so that NY′ = NX′ = $R \tan \phi/2$, where *R* is the radius of the primitive. A circle is then described with N as centre, and NX′ (or NY′) as the radius. This is the *only* case where the centre of the small circle in projection coincides with the stereographic projection of the centre of the small circle.

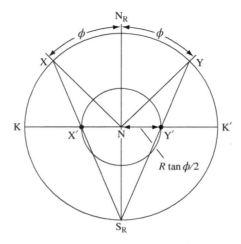

Figure A2.5   Construction of a small circle about the centre of the primitive

Alternatively, we could locate the points X′ and Y′ in projection solely by construction as follows. Draw the diameter of the primitive upon which we wish to locate X′ and Y′ and then draw the diameter of the primitive normal to this, $N_R N S_R$ (Figure A2.5). Find the point X, on the primitive, such that the arc $X N N_R$ subtends an angle ϕ at the centre of the primitive. If we join $X S_R$ then X′ is located where $X S_R$ cuts the first diameter of the primitive. The justification for this construction is shown in detail in Figure A2.6, where it is seen that if we imagine keeping the plane of projection fixed but rotate the sphere of projection through 90° about KK′ (a line lying in the plane of projection), the pole of projection S comes to lie on the primitive at S_R. Similarly, N lies at N_R and the validity of the construction follows. This useful trick of imagining the whole sphere of projection rotated through 90° is often used to establish constructions on the stereogram.

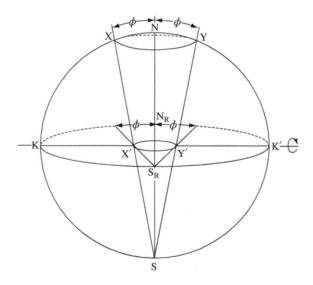

Figure A2.6   Rotation of the sphere of projection in Figure A2.5 by 90° about the axis KK′

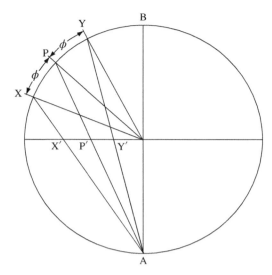

Figure A2.7   Construction of a small circle about a pole within the primitive

A2.2.1.2   About a Pole within the Primitive – say about P (Figure A2.7)
Draw the diameter through P′ and the diameter normal to this to locate A and B, as in
Figure A2.7. Project P′ from A to the primitive to give P, say. Measure off ϕ (the required
radius of the small circle) along the primitive on either side of P to obtain the points X,
Y on the primitive (Figure A2.7), noting that ϕ is the angle subtended at the centre of the
primitive by the arc XP. Reproject X and Y from A to obtain the points X′, Y′, which
are opposite ends of the diameter of the required small circle. It should be noted that ϕ
is the angle subtended at the centre of the primitive and that P′ is not halfway between
X′ and Y′. This is a demonstration of the fact that while small circles project as small
circles, in general, the centre of the small circle in projection does not coincide with the
stereographic projection of the centre of the small circle. It is a useful exercise to draw a
diagram similar to Figure A2.6 to justify this construction.

A2.2.1.3   About a Pole on the Primitive – say about P (Figure A2.8)
Draw the radius NP and from P measure off the angle ϕ (equal to the angular radius
of the required circle) subtended at the centre of the primitive to locate the point M, as in
Figure A2.8. At M, draw the tangent to the primitive to meet NP produced in S. S is the *centre* of the required small circle and SM its radius. This construction is a little quicker than
that described previously in Section A2.2.1.2 for the case of a small circle about a pole
within the primitive. It is a useful exercise in trigonometry to verify this construction.

A2.2.2   To Find the Opposite of a Pole

The opposite of a given pole is the pole 180° away from the particular pole under consideration; that is, the other end of a diameter of the sphere of projection passing through the pole.
Suppose that we wish to find the stereographic projection of the opposite of the pole V in
Figure A2.9. Draw the diameter of the primitive through V. Then the opposite of V, say V″,
clearly lies on this diameter and 180° from V. Draw the diameter of the primitive normal to
the diameter through V and mark opposite ends of this, S_R and N_R. Project V from S_R to

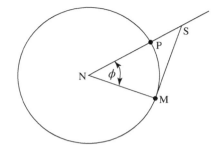

Figure A2.8 Construction of a small circle about a pole on the primitive

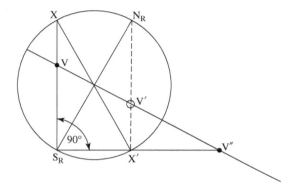

Figure A2.9 To find the opposite of a given pole

the primitive to find the point X. Draw the diameter of the primitive through X and mark the other end of this diameter at X′. X′ is projected from S_R onto the diameter through V to give the required opposite of V at V″ (Figure A2.9). The justification for this construction is easily seen from Figure A2.10, where we use the same device as in Figure A2.6 and imagine the sphere of projection rotated 90° about the diameter of the primitive containing the projection of V, so that S comes to lie on the primitive at S_R but V and V″, lying on the axis of rotation, do not move. The angle $XS_RX′$ in Figure A2.9 is clearly 90°, so in practice V″ is easily located by putting a 90° set square at S_R with one edge running through V and using the other edge to locate V″.

If the given pole V lies inside the primitive in projection then the true opposite V″ lies outside the primitive circle. Therefore, it is usually more convenient to work with the opposite V′ obtained by projecting not from S but from N on the sphere of projection (Figure A2.3a). Clearly, V′ is always found on the diameter of the primitive through V at an equal distance from the centre on the opposite side (Figure A2.9).

A2.2.3 To Draw a Great Circle through Two Poles

Find the true opposite of one of them using the construction in A2.2.2, and then construct the circle passing through the two given points and this opposite. This is the required great circle.

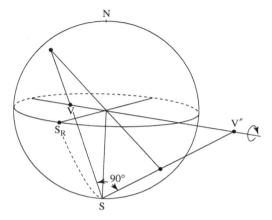

Figure A2.10 An alternative view of the construction in Figure A2.9 to show how to find the opposite of a given pole

A2.2.4 To Find the Pole of a Great Circle

Suppose we wish to find the pole of ACB in Figure A2.11. Draw the diameter AB of the primitive, which is a chord of the great circle, and draw the diameter of the primitive normal to AB. Let this diameter intersect the great circle at T. Project T from B to the primitive at X and measure 90° from X over the pole A to find the pole Y. Project Y from B onto the diameter of the primitive through T to find the pole D, which is the required pole. Since the arc XY subtends an angle of 90° at the centre of the primitive, clearly the angle TBD is 45° and so D may be rapidly located once T is found by placing the 45° angle of a set square as shown in Figure A2.11. This construction is easily justified by drawing a diagram similar to Figure A2.10.

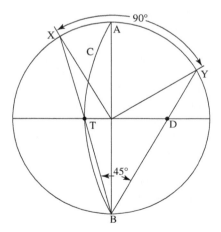

Figure A2.11 To find the pole of a great circle

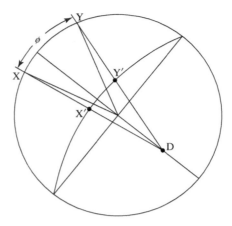

Figure A2.12   To find the angle between two poles

A2.2.5   To Measure the Angle Between Two Poles on an Inclined Great Circle

Let X′Y′ be the two projected poles (in Figure A2.12). Locate the pole D of the great circle on which they lie by the construction in A2.2.4. From D, project X′ and Y′ to the primitive to locate X and Y, respectively. The angle ϕ subtended by the arc *XY at the centre of the primitive* is the angle between the two poles. It is a useful exercise to draw a diagram similar to Figure A2.10 to prove this.

A2.3   Constructions with the Wulff Net

A graphical aid for the construction of stereograms, which is also very useful for taking angular measurements upon them, is called a Wulff net [1].[1] A net is shown in Figure A2.13b. A downloadable Excel program for generating a Wulff net is available on the Wiley Web page accompanying this book at www.wiley.com. The net is the projection of one-half of the terrestrial globe with lines of latitude and longitude marked upon its surface and with the north and south poles lying in the plane of projection. The relationship between the lines of latitude (which are all small circles, except the equator) and the lines of longitude (all great circles) on the surface of the sphere of projection and their representation on the net is shown in Figures A2.13a and b. The radius of the net and that of the sphere of the projection are of course equal, and equal to the radius of the primitive circle of the projection. The net is used by placing it under the stereogram, which is drawn on transparent tracing paper, and the centres of the two are located by a pin. The stereogram can be rotated above the net. The angle between poles within the primitive is measured by rotating the stereogram until the two poles lie on the same great circle and the angle is measured by counting the small circles between the poles (Figures A2.14a and b). Angles between poles on the primitive are measured directly.

---

[1] Phillips [2], p. 33, makes a compelling argument for Wulff's full name to be transcribed more correctly as Yurii (Georgii) Viktorovich Vulf, but the German transcription of the original Russian is the one in usage outside Russia, even if it is often Anglicized in pronunciation.

(a)

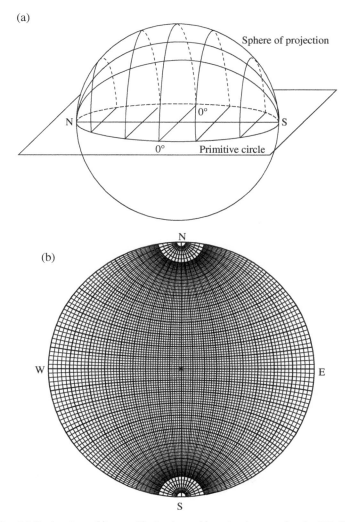

(b)

Figure A2.13 (a) Projection of lines of latitude and longitude to make the Wulff net. (b) Wulff net drawn to 2° intervals

The great circle corresponding to the locus of points 90° from a given pole is found by placing the pole on the equator of the net (the line marked 0°0° in Figures A2.13a and A2.15) and tracing out the great circle 90° from the given pole, as indicated in Figure A2.15. This great circle is the trace of the plane of which P_2' is the pole.

It is often useful to be able to rotate a stereogram about a given axis. To rotate any pole, say A_1 in Figure A2.16, about the pole B lying on the primitive, the net is rotated until the axis NS of the net lies along the diameter of the primitive through B. A_1 is then rotated the required number of degrees about B by moving A_1 to A_2 along the small circle, as shown in Figure A2.16. A pole which will pass outside the primitive upon rotation is also shown in Figure A2.16. The true projection of C_2, as distinct from its opposite (shown), would be on the same small circle as C_2.

(a)

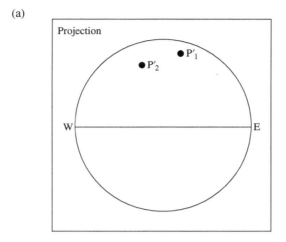

(b)

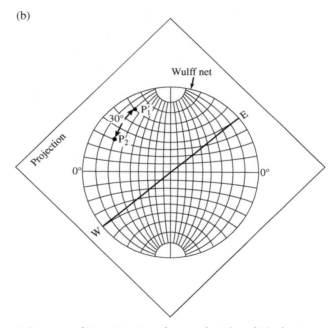

Figure A2.14 (a) Stereographic projection of two poles P'_1 and P'_2. (b) Rotation of the projection to put both poles on the same great circle of the Wulff net. Angle between poles = 30°. (Based on Cullity [3] with permission of Pearson Education Inc.)

If B does not lie on the primitive, the procedure shown in Figure A2.17 can be followed. The net is rotated until NS lies perpendicular to the radius of the primitive through B. B and A are then both rotated about the axis NS until B lies at the centre of the primitive (B′ in Figure A2.17) and A moves to A′. If it is required to rotate A, say, 40° clockwise about B, we now rotate A′ 40° clockwise about B′ to A″. This is easily done, as shown in

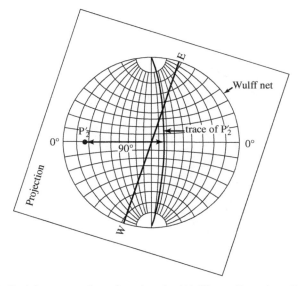

Figure A2.15   To find the trace of a pole using the Wulff net. (Based on Cullity [3] with permission of Pearson Education Inc.)

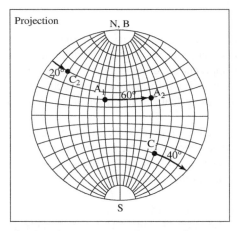

Figure A2.16   Rotation of poles about an axis B lying on the primitive. (Based on Cullity [3] with permission of Pearson Education Inc.)

Figure A2.17. We now rotate B′ back to B by rotation about NS and rotate A″ about NS by the same amount in the same sense to obtain A‴, the required rotated position of A.

A2.3.1   Two-Surface Analysis

In the identification and subsequent study of planar imperfections in crystals using optical microscopy or scanning electron microscopy, it is often necessary to identify a crystal

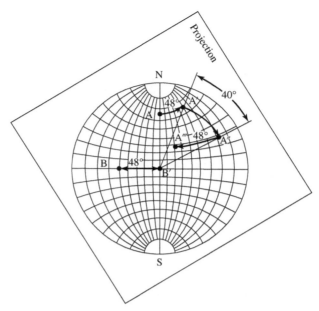

Figure A2.17   Rotation of poles about an inclined axis. (Based on Cullity [3] with permission of Pearson Education Inc.)

plane from the linear traces which it makes in two (or more) other non-parallel planes. Even when identifying features such as faults or twins in crystalline materials using transmission electron microscopy, an appreciation of the geometry behind two-surface analysis is useful for identifying such features using information from electron diffraction patterns and combining this with images of these features taken when the electron beam is along different directions within the crystals.

The procedure is illustrated in Figure A2.18. Suppose the planes A and B (which are flat surfaces of a crystal) intersect along the line PQ and their (*hkl*) indices are known. Let us draw a stereogram with the plane of projection parallel to B so that the pole of B lies at the centre of the primitive. The angle between A and B is ϕ (the angle between the outward normals to A and B) and so the pole of A lies on the stereogram as shown in Figure 2.18b. The planes A and B intersect along PQ, and so this intersection can be marked upon the stereogram. The plane we are interested in is MNT, which makes an angle θ_A with PQ in the face A and θ_B with PQ in the face B. Consider the trace TT′ in the face B. The direction TT′ lies in plane B and at angle θ_B from PQ measured clockwise from P, as shown in Figure A2.18a. We can therefore plot the direction TT′ on the stereogram in Figure A2.18b at an angle θ_B to PQ. The plane MNT when drawn on the stereogram must project as a great circle which passes through the points T and T′. An infinite number of great circles pass through TT′, corresponding to all the possible planes that intersect the plane B in a direction parallel to TT′.

To locate another point on the stereogram through which the projection of the plane MNT must pass, we consider the trace MM′ of the plane MNT in the face A. The trace MM′ makes a planar angle in the face A of θ_A with PQ. We have already plotted the direction of PQ in

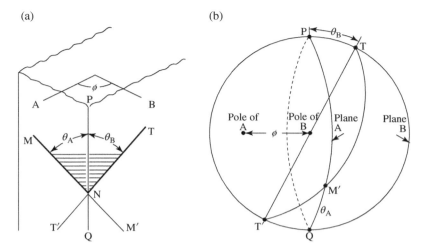

Figure A2.18   Two-surface analysis

the stereogram, so θ_A must be measured in the appropriate sense from PQ and *in the plane* A. The point M′ giving the direction of the trace of MM′ in A is at an angle θ_A measured anticlockwise from Q along the great circle representing the plane A in Figure A2.18b. The angle θ_A can be measured on the stereogram either by using the method of construction to measure the angle between two poles on an inclined great circle (see Section A2.2) or more simply by using the Wulff net so that NS in Figure A2.13b lies along the line PQ in Figure A2.18b and θ_A is measured off by counting the lines of latitude along the great circle coinciding with the trace of the plane A.

We have now located T, M′ and T′ as poles on the stereogram through which the projection of the plane MNT must pass. We can therefore draw in MNT as the great circle passing through these poles, and use the Wulff net or a construction to find the pole of MNT and hence the indices of MNT if those of A and B are known.

A2.4   Proof of the Properties of the Stereographic Projection

Suppose R is the radius of the sphere of projection in Figure A2.19. S is the point of projection and N is diametrically opposite S. P is the pole on the sphere of any direction in space. Let the angle NOP equal ϕ. The angle NPS is a right angle, since NS is a diameter of the sphere. The distance SP is therefore equal to $2R \cos \phi/2$, since $\hat{OSP} = \frac{1}{2} \hat{NOP}$. P′ is the stereographic projection of P. The triangle SOP′ has a right angle at O. Therefore, the length SP′ is equal to $R \sec \phi/2$. The product of the lengths SP and SP′ is:

$$SP \cdot SP' = 2R^2 \tag{A2.2}$$

This product is independent of the angle ϕ and hence it is constant for all poles P on the surface of the sphere of projection. Any pole P and its stereographic projection P′ are said to be the *inverse of one another* because their distances from a fixed point (S) are related by Eq. (A2.2).

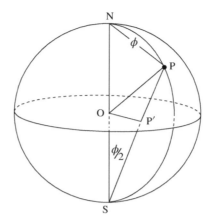

Figure A2.19 The fundamental projectional geometry of the stereographic projection: P′ is the stereographic projection of P, while S is the point of projection

The geometry of *inversion* was invented by Ludwig Immanuel Magnus in 1832 [4].[2] The formal definition of the geometry of inversion is as follows. Given a fixed sphere of radius k and centre O, we define the inverse of any point P (distinct from O) to be the point P′ *on the ray* OP whose distance from O satisfies the equation:

$$OP \cdot OP' = k^2 \tag{A2.3}$$

P is clearly the inverse of P′ itself and the point O is singular. Comparing this definition with our definition of the stereographic projection, we see that all poles P on the sphere of projection and their stereographic projections P′ are such that P and P′ are the inverse of one another in a sphere of centre S in Figure A2.19 and of radius $\sqrt{2}R$. The sphere of inversion is *not* drawn in Figure A2.19.

The importance of the stereographic projection in crystallography arises because of the following features of the geometry of inversion. All spheres invert into spheres (or planes) and therefore all circles, being intersections of spheres, invert into circles or straight lines. The transformation inverts *angles into equal angles*. The stereographic projection is therefore said to be angle true. These properties are proved below.

The mathematician defines the stereographic projection as that inversion which puts points of a plane into a one-to-one correspondence with the points on the surface of a sphere. In Figure A2.19 we have chosen a particular example in which the centre of the sphere of inversion, S, and the radius of the sphere of inversion are so chosen that all points on the surface of the *sphere of projection* invert into points that lie in a diametral plane of the sphere of projection. This plane is the plane of projection as we use it in this book.

A proof that any sphere inverts into a sphere is as follows. Let P be any point on a sphere σ of centre C, as in Figure A2.20. It is required to show that the inverse of σ is another sphere. Let O in Figure A2.20 be the centre of the sphere of inversion and k its radius (the sphere of inversion is not shown in Figure A2.20). Join PO and let the second intersection of PO with the sphere σ be Q.

---

[2] In the original 1832 paper his name is given as L.J. Magnus.

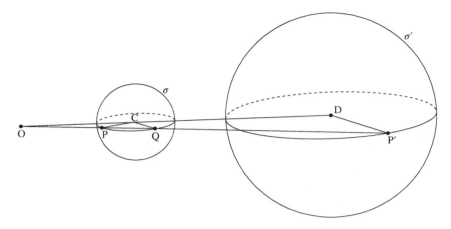

Figure A2.20 Geometry to show that any sphere inverts into a sphere

The product OP · OQ is independent of the position of P on the sphere.[3] Let the product OP · OQ equal p. Join OC and find the point D on OC such that $OD/OC = k^2/p$. Draw the sphere σ' of centre D and radius $(k^2/p) \cdot CP$; that is, a sphere of radius k^2/p times the radius of σ. Draw the radius of this sphere, DP', in the plane of OCD and P, parallel to CQ. Join OP'. Since:

$$\frac{DP'}{QC} = \frac{k^2}{p} \frac{CP}{QC} = \frac{k^2}{p} = \frac{OD}{OC} \tag{A2.4}$$

triangles OQC, OP'D are similar and hence OP and OP' coincide. It follows immediately that P' is the inverse of P, since:

$$\frac{OP'}{OQ} = \frac{OD}{OC} = \frac{k^2}{p} \tag{A2.5}$$

and

$$OQ = \frac{p}{OP} \tag{A2.6}$$

Therefore:

$$OP \cdot OP' = k^2 \tag{A2.7}$$

Since P was any point on σ, we have proved that the sphere σ inverts to the sphere σ' and, incidentally, since any circle can be regarded as the intersection of two spheres, we have also shown that all circles invert to circles. It should be noted that, if the sphere σ passes through O, σ' is a plane (a sphere of infinite radius) and a circle through O inverts into a line. It is also worth calling attention to the fact that D is not necessarily the inverse of C.

That the stereographic projection is angle true – that is to say, that the angle of intersection of two curves on the sphere of projection is the same as the angle between the two projected curves – is also easily shown from the geometry of inversion. To prove that the angle at which two curves cut is equal to the angle at which their inverse curves cut at the

---

[3] This property follows from Euclid III.37 for a circle [5]: if from a point outside a circle a secant and a tangent are drawn, the rectangle contained by the whole secant and the part outside the circle is equal to the square on the tangent.

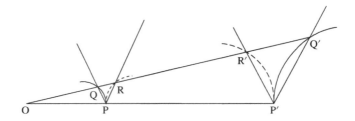

Figure A2.21 Geometry to show that a stereographic projection is angle true

corresponding point of intersection, we proceed as follows. O is the centre of inversion (Figure A2.21). PP′ are the points of intersection of the two pairs of curves. Through O, draw any line OQRR′Q′ cutting the two given curves at Q, R and their inverses at Q′, R′.

Since OQ·OQ′ = OP·OP′ = OR·OR′, PQQ′P′ and PRR′P′ are cyclic quadrilaterals. Therefore:

$$O\hat{P}Q = O\hat{Q}'P' \quad \text{and} \quad O\hat{P}R = O\hat{R}'P' \tag{A2.8}$$

and so:

$$Q\hat{P}R = Q'\hat{P}'R' \tag{A2.9}$$

By taking the line OQ sufficiently close to OP, the difference between these two angles and the angles between the two pairs of curves may be made as small as we please, thus proving the theorem.

References

[1] Wulff, G. (1902). Untersuchungen im Gebiete der optischen Eigenschaften isomorpher Krystalle. *Z. Krystall.* 36: 1–28.
[2] Phillips, F.C. (1971). *Introduction to Crystallography*, 4th Edition. Edinburgh: Oliver and Boyd.
[3] Cullity, B.D. (1956). *Elements of X-ray Diffraction*, 1st Edition. Reading, Massachusetts: Addison-Wesley.
[4] Magnus, L.J. (1832). Nouvelle méthode pour découvrir des théorèmes de géométrie. *J. Reine Angew. Math. (Crelle's Journal)* 8: 51–63.
[5] Densmore, D. (ed.) (2002). *Euclid's Elements* (authored by Euclid and translated by T.L. Heath). Santa Fe, NM: Green Lion Press.

Appendix 3

Interplanar Spacings and Interplanar Angles

A3.1 Interplanar Spacings

By definition, two successive planes of indices (hkl) make intercepts on the crystal axes of na/h, nb/k, nc/l and $(n+1)a/h$, $(n+1)b/k$, $(n+1)c/l$, respectively, where n is an integer. The perpendicular distance between successive planes, or interplanar spacing d_{hkl}, has been shown to be given by the inverse of the magnitude of the corresponding reciprocal lattice vector (Eq. (A1.32)); that is:

$$d_{hkl} = \frac{1}{|h\mathbf{a}^* + k\mathbf{b}^* + l\mathbf{c}^*|} \tag{A3.1}$$

Now:

$$
\begin{aligned}
|h\mathbf{a}^* + k\mathbf{b}^* + l\mathbf{c}^*|^2 &= (h\mathbf{a}^* + k\mathbf{b}^* + l\mathbf{c}^*) \cdot (h\mathbf{a}^* + k\mathbf{b}^* + l\mathbf{c}^*) \\
&= h^2 \mathbf{a}^{*2} + k^2 \mathbf{b}^{*2} + l^2 \mathbf{c}^{*2} + 2kl\mathbf{b}^* \cdot \mathbf{c}^* + 2lh\mathbf{c}^* \cdot \mathbf{a}^* + 2hk\mathbf{a}^* \cdot \mathbf{b}^* \\
&= h^2 a^{*2} + k^2 b^{*2} + l^2 c^{*2} + 2klb^* c^* \cos \alpha^* + 2lhc^* a^* \cos \beta^* \\
&\quad + 2hka^* b^* \cos \gamma^*
\end{aligned}
\tag{A3.2}
$$

where α^* is the angle between the reciprocal lattice vectors $\mathbf{b}^*$ and $\mathbf{c}^*$, β^* is the angle between $\mathbf{c}^*$ and $\mathbf{a}^*$ and γ^* is the angle between $\mathbf{a}^*$ and $\mathbf{b}^*$.

Crystallography and Crystal Defects, Third Edition. Anthony Kelly and Kevin M. Knowles.
© 2020 John Wiley & Sons Ltd. Published 2020 by John Wiley & Sons Ltd.
Companion website: www.wiley.com/go/kelly/crystallography3e

The interplanar spacing can be evaluated in terms of the parameters of the direct lattice by substituting for the magnitudes a^*, b^* and c^* and for $\cos \alpha^*$, $\cos \beta^*$ and $\cos \gamma^*$ in these two equations. Formulae for these are now given for each crystal system.

A3.1.1 Triclinic

Substituting in Eq. (A1.27) for $\mathbf{a}^*$, $\mathbf{b}^*$ and $\mathbf{c}^*$, it is apparent for a triclinic crystal that:

$$a^* = \frac{bc \sin \alpha}{V}, \quad b^* = \frac{ca \sin \beta}{V} \quad \text{and} \quad c^* = \frac{ab \sin \gamma}{V} \tag{A3.3}$$

where:

$$V = abc\sqrt{1 - \cos^2\alpha - \cos^2\beta - \cos^2\gamma + 2\cos\alpha \cos\beta \cos\gamma} \tag{A3.4}$$

(Eq. (1.42)). Using Eqs. (A1.26) and (A1.27), it is also apparent that:

$$\cos \alpha^* = \frac{\cos \beta \cos \gamma - \cos \alpha}{\sin \beta \sin \gamma}$$

$$\cos \beta^* = \frac{\cos \alpha \cos \gamma - \cos \beta}{\sin \alpha \sin \gamma} \tag{A3.5}$$

$$\cos \gamma^* = \frac{\cos \alpha \cos \beta - \cos \gamma}{\sin \alpha \sin \beta}$$

Together with Eqs. (A3.1) and (A3.2), these formulae can be used in spreadsheets and computer programs to evaluate interplanar spacings for crystals of any crystal systems. A downloadable Excel program for calculating interplanar spacings based on Eqs. (A3.1–A3.5) is available on the Web site accompanying this book.

Useful simplifications of these general formulae occur as the symmetry shown by the crystal system under consideration increases.

A3.1.2 Monoclinic

Since $\alpha^* = \alpha = 90°$ and $\gamma^* = \gamma = 90°$,

$$d_{hkl}^2 = \frac{1}{h^2 a^{*2} + k^2 b^{*2} + l^2 c^{*2} + 2lhc^* a^* \cos \beta^*} \tag{A3.6}$$

where:

$$\beta^* = 180° - \beta, \quad a^* = \frac{1}{a \sin \beta}, \quad b^* = \frac{1}{b}, \quad c^* = \frac{1}{c \sin \beta} \tag{A3.7}$$

A3.1.3 Orthorhombic

Since $\alpha^* = \beta^* = \gamma^* = 90°$,

$$d_{hkl}^2 = \frac{1}{h^2 a^{*2} + k^2 b^{*2} + l^2 c^{*2}} \tag{A3.8}$$

where:

$$a^* = \frac{1}{a}, \quad b^* = \frac{1}{b}, \quad c^* = \frac{1}{c} \tag{A3.9}$$

A3.1.4 Trigonal

Since $\alpha^* = \beta^* = \gamma^*$ and $a^* = b^* = c^*$,

$$d_{hkl}^2 = \frac{1}{[h^2 + k^2 + l^2 + 2(kl + lh + hk)\cos\alpha^*]\,a^{*2}} \tag{A3.10}$$

where:

$$\cos(\alpha^*/2) = \frac{1}{2\cos(\alpha/2)}, \quad a^* = \frac{1}{a\sin\alpha\sin\alpha^*} \tag{A3.11}$$

A3.1.5 Tetragonal

Since $\alpha^* = \beta^* = \gamma^* = 90°$ and $a^* = b^*$,

$$d_{hkl}^2 = \frac{1}{(h^2 + k^2)\,a^{*2} + l^2 c^{*2}} \tag{A3.12}$$

where:

$$a^* = \frac{1}{a}, \quad c^* = \frac{1}{c} \tag{A3.13}$$

A3.1.6 Hexagonal

Since $\alpha^* = \beta^* = 90°, \gamma^* = 180° - \gamma = 60°$ and $a^* = b^*$,

$$d_{hkl}^2 = \frac{1}{(h^2 + k^2 + hk)\,a^{*2} + l^2 c^{*2}} \tag{A3.14}$$

where:

$$a^* = \frac{2}{a\sqrt{3}}, \quad c^* = \frac{1}{c} \tag{A3.15}$$

A3.1.7 Cubic

Since $\alpha^* = \beta^* = \gamma^* = 90°$ and $a^* = b^* = c^*$,

$$d_{hkl}^2 = \frac{1}{(h^2 + k^2 + l^2)\,a^{*2}} \tag{A3.16}$$

where:

$$a^* = \frac{1}{a} \tag{A3.17}$$

It should be appreciated that the value of d_{hkl}, given by Eq. (A3.1), is the perpendicular distance between planes making intercepts of na/h, nb/k, nc/l and $(n+1)a/h$, $(n+1)b/k$, $(n+1)$ c/l on the crystal axes, whether or not these planes coincide with sheets of lattice points. If h, k and l are integers having no common factor, they are the Miller indices of a lattice plane.

The spacing of successive lattice planes of Miller indices (hkl) in the stack of these planes which builds up the complete lattice is given by Eq. (A3.1), provided that the Bravais lattice is primitive; that is, provided that the vectors **a**, **b** and **c** define a cell that contains a single lattice point. If the unit cell of the Bravais lattice is body-centred, face-centred or base-centred, Eq. (A3.1) can give twice the spacing of successive lattice planes, depending on the values of h, k and l.

If the unit cell is body-centred, the formula in Eq. (A3.1) gives twice the lattice plane spacing whenever $(h + k + l)$ is an odd number. For example, the spacing of lattice planes of Miller indices (100) in a b.c.c. lattice is easily seen to be one-half the value of d_{100} given by Eq. (A3.1). The rule for obtaining the true lattice plane spacing from Eq. (A3.1) in a body-centred (I) lattice is therefore to double the Miller indices whenever $(h + k + l)$ is odd. In a face-centred (F) lattice, the rule is to double the Miller indices if either h or k or l is even, since the spacing of all lattice planes except those with h, k and l all odd numbers is one-half d_{hkl}. Zero counts as an even number. For a cell which is centred on, say, the (001) face, the rule is to double the Miller indices when $(h + k)$ is an odd number.

A3.2   Interplanar Angles

It has been shown that the angle between two lattice planes is equal to the angle between the corresponding reciprocal lattice vectors (see Eq. (A1.31)). This leads to the following general formula for the angle ϕ between the planes (hkl) and ($h'k'l'$):

$$\cos \phi = d_{hkl}d_{h'k'l'}[hh'a^{*2} + kk'b^{*2} + ll'c^{*2} + (kl' + lk')b^*c^* \cos \alpha^*$$
$$+ (hl' + lh')a^*c^* \cos \beta^* + (hk' + kh')a^*b^* \cos \gamma^*] \tag{A3.18}$$

where d_{hkl} is given by Eq. (A3.1).

With the extensive use of scientific calculators, spreadsheets and computer programs, tabulations of interplanar angles are, not surprisingly, no longer necessary. Values of some interplanar angles in cubic crystals are, however, listed in Table A3.1 in the first edition of this book and on the Web site accompanying this third edition. In addition, there is a downloadable Excel program available on the Web site for calculating interplanar angles based on Eq. (A3.18).

In crystal systems other than the triclinic, some simplification of Eq. (A3.18) occurs. Expressions for determining interplanar angles in the orthorhombic, hexagonal and cubic systems are shown below.

A3.2.1   Orthorhombic

$$\cos \phi = d_{hkl}d_{h'k'l'}[hh'a^{*2} + kk'b^{*2} + ll'c^{*2}] \tag{A3.19}$$

where a^*, b^* and c^* are defined in Eq. (A3.9).

A3.2.2 Hexagonal

$$\cos \phi = d_{hkl} d_{h'k'l'} \left[\left\{ hh' + kk' + \tfrac{1}{2}(hk' + kh') \right\} a^{*2} + ll'c^{*2} \right]$$ (A3.20)

where a^* and c^* are defined in Eq. (A3.15).

A3.2.3 Cubic

$$\cos \phi = \frac{hh' + kk' + ll'}{\sqrt{h^2 + k^2 + l^2} \sqrt{h'^2 + k'^2 + l'^2}}$$ (A3.21)

Appendix 4

Transformation of Indices Following a Change of Unit Cell

When studying crystals different choices for the unit cell of a given crystal may be convenient for different considerations, and so it is often necessary to know how the indices of directions and the Miller indices of planes alter when the choice of unit cell is altered. In this appendix we show how to obtain the transformation formulae in Sections A4.1 and A4.2 and give examples of their use in Sections A4.3 and A4.4.

A4.1 Change of Indices of Directions

Let $[uvw]$ be the indices of a given direction in terms of the old unit cell of cell-edge vectors $\mathbf{a}$, $\mathbf{b}$, $\mathbf{c}$ and $[UVW]$ be those of the same direction in terms of the cell of cell-edge vectors $\mathbf{A}$, $\mathbf{B}$, $\mathbf{C}$. It then follows that:

$$U\mathbf{A} + V\mathbf{B} + W\mathbf{C} = u\mathbf{a} + v\mathbf{b} + w\mathbf{c} \tag{A4.1}$$

We write $\mathbf{A}$, $\mathbf{B}$ and $\mathbf{C}$ in terms of $\mathbf{a}$, $\mathbf{b}$ and $\mathbf{c}$ so that:

$$\mathbf{A} = s_{11}\mathbf{a} + s_{12}\mathbf{b} + s_{13}\mathbf{c}$$
$$\mathbf{B} = s_{21}\mathbf{a} + s_{22}\mathbf{b} + s_{23}\mathbf{c} \tag{A4.2}$$
$$\mathbf{C} = s_{31}\mathbf{a} + s_{32}\mathbf{b} + s_{33}\mathbf{c}$$

Crystallography and Crystal Defects, Third Edition. Anthony Kelly and Kevin M. Knowles.
© 2020 John Wiley & Sons Ltd. Published 2020 by John Wiley & Sons Ltd.
Companion website: www.wiley.com/go/kelly/crystallography3e

It is evident from Eqs. (A4.1) and (A4.2) that:

$$u = s_{11}U + s_{21}V + s_{31}W$$
$$v = s_{12}U + s_{22}V + s_{32}W \tag{A4.3}$$
$$w = s_{13}U + s_{23}V + s_{33}W$$

This can be written out in matrix form:

$$\begin{pmatrix} u \\ v \\ w \end{pmatrix} = \begin{pmatrix} s_{11} & s_{21} & s_{31} \\ s_{12} & s_{22} & s_{32} \\ s_{13} & s_{23} & s_{33} \end{pmatrix} \begin{pmatrix} U \\ V \\ W \end{pmatrix} \tag{A4.4}$$

Therefore:

$$\begin{pmatrix} U \\ V \\ W \end{pmatrix} = \begin{pmatrix} s_{11} & s_{21} & s_{31} \\ s_{12} & s_{22} & s_{32} \\ s_{13} & s_{23} & s_{33} \end{pmatrix}^{-1} \begin{pmatrix} u \\ v \\ w \end{pmatrix} \tag{A4.5}$$

Alternatively, if we write **a**, **b** and **c** in terms of **A**, **B** and **C** so that:

$$\mathbf{a} = t_{11}\mathbf{A} + t_{12}\mathbf{B} + t_{13}\mathbf{C}$$
$$\mathbf{b} = t_{21}\mathbf{A} + t_{22}\mathbf{B} + t_{23}\mathbf{C} \tag{A4.6}$$
$$\mathbf{c} = t_{31}\mathbf{A} + t_{32}\mathbf{B} + t_{33}\mathbf{C}$$

then it follows that:

$$\begin{pmatrix} U \\ V \\ W \end{pmatrix} = \begin{pmatrix} t_{11} & t_{21} & t_{31} \\ t_{12} & t_{22} & t_{32} \\ t_{13} & t_{23} & t_{33} \end{pmatrix} \begin{pmatrix} u \\ v \\ w \end{pmatrix} \tag{A4.7}$$

The t matrix in Eq. (A4.7) is the inverse of the s matrix in (A4.4).

As well as being used to transform vectors – that is, directions (or zone axis symbols) – Eqs. (A4.4) and (A4.7) are the necessary relations for transforming the coordinates of atom or ion positions in the two unit cells.

The positions of the s's and t's in Eqs. (A4.2) and (A4.6) and their positions in (A4.4) and in (A4.7) should be very carefully noted: the rows and columns are interchanged in going from (A4.2) to (A4.4) and in going from (A4.6) to (A4.7).

If V_1 is the volume of the cell **a**, **b**, **c** and V_2 that of **A**, **B**, **C** then it follows from Eq. (A1.19) that the determinants of the arrays of s's and t's relate the volumes of the two unit cells:

$$V_1 : V_2 = \begin{vmatrix} 1 & 0 & 0 \\ 0 & 1 & 0 \\ 0 & 0 & 1 \end{vmatrix} : \begin{vmatrix} s_{11} & s_{12} & s_{13} \\ s_{21} & s_{22} & s_{23} \\ s_{31} & s_{32} & s_{33} \end{vmatrix} = \begin{vmatrix} t_{11} & t_{12} & t_{13} \\ t_{21} & t_{22} & t_{23} \\ t_{31} & t_{32} & t_{33} \end{vmatrix} : \begin{vmatrix} 1 & 0 & 0 \\ 0 & 1 & 0 \\ 0 & 0 & 1 \end{vmatrix} \tag{A4.8}$$

That is:

$$V_1 : V_2 = 1 : \begin{vmatrix} s_{11} & s_{12} & s_{13} \\ s_{21} & s_{22} & s_{23} \\ s_{31} & s_{32} & s_{33} \end{vmatrix} = \begin{vmatrix} t_{11} & t_{12} & t_{13} \\ t_{21} & t_{22} & t_{23} \\ t_{31} & t_{32} & t_{33} \end{vmatrix} : 1 \tag{A4.9}$$

A4.2   Change of Indices of Planes

Let (hkl) be the indices of a set of lattice planes referred to the unit cell of translation vectors **a, b, c**, which we will call the 'old' cell. We wish to find the indices of the same set of planes referred to the 'new' cell, which has translation vectors **A, B, C** given the relationships between **a, b, c** and **A, B, C** established through Eqs. (A4.2) and (A4.6). Referred to the 'new' cell, the indices of the set of lattice planes are (HKL).

The result is that, in matrix form, (HKL) are given in terms of (hkl) as follows:

$$\begin{pmatrix} H \\ K \\ L \end{pmatrix} = \begin{pmatrix} s_{11} & s_{12} & s_{13} \\ s_{21} & s_{22} & s_{23} \\ s_{31} & s_{32} & s_{33} \end{pmatrix} \begin{pmatrix} h \\ k \\ l \end{pmatrix} \tag{A4.10}$$

Note that the positions of the s's in this equation and that of Eq. (A4.2) are the same. (hkl) are given in terms of (HKL) through the equation:

$$\begin{pmatrix} h \\ k \\ l \end{pmatrix} = \begin{pmatrix} t_{11} & t_{12} & t_{13} \\ t_{21} & t_{22} & t_{23} \\ t_{31} & t_{32} & t_{33} \end{pmatrix} \begin{pmatrix} H \\ K \\ L \end{pmatrix} \tag{A4.11}$$

and the positions of the t's in this equation and that in Eq. (A4.6) are also the same.

The proof that these are the correct equations is straightforward. The cell with translation vectors **a, b, c** has a reciprocal cell with translation vectors **a*, b*, c***, where **a*, b*, c*** are related to **a, b, c** by equations such as Eq. (A1.27). The cell with translation vectors **A, B, C** has a corresponding reciprocal cell with vectors **A*, B*, C***.

A given lattice point of the reciprocal lattice represents a plane in real space and so must be the same distance from the origin regardless of the choice of unit cell. Therefore:

$$h\mathbf{a}^* + k\mathbf{b}^* + l\mathbf{c}^* = H\mathbf{A}^* + K\mathbf{B}^* + L\mathbf{C}^* \tag{A4.12}$$

If we form the dot product of both sides of this equation first with **A**, then with **B** and then with **C**, we obtain:

$$\mathbf{A} \cdot (h\mathbf{a}^* + k\mathbf{b}^* + l\mathbf{c}^*) = H$$
$$\mathbf{B} \cdot (h\mathbf{a}^* + k\mathbf{b}^* + l\mathbf{c}^*) = K \tag{A4.13}$$
$$\mathbf{C} \cdot (h\mathbf{a}^* + k\mathbf{b}^* + l\mathbf{c}^*) = L$$

If we now substitute for **A** in the first part of Eq. (A4.13) by using Eq. (A4.2) and multiply out the dot product, we have:

$$H = s_{11}h + s_{12}k + s_{13}l$$

Expressions for K and L are obtained by substituting for **B** and **C**, thus obtaining Eq. (A4.10). To obtain Eq. (A4.11), we form the dot product of both sides of Eq. (A4.12) first with **a**, then with **b** and **c** to generate equations for h, k and l, and proceed similarly, substituting for **a, b** and **c** respectively using Eq. (A4.6) to generate the equations for h, k and l in terms of the t_{ij} and H, K and L.

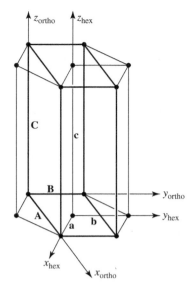

Figure A4.1   Relationship between the conventional hexagonal unit cell with cell-edge vectors **a**, **b** and **c** and the orthorhombic unit cell with cell-edge vectors **A**, **B** and **C**.

A4.3   Example 1: Interchange of Hexagonal and Orthorhombic Indices for Hexagonal Crystals

In dealing with hexagonal crystals it is sometimes convenient to use the orthorhombic cell shown in Figure A4.1, for example in the phenomenological theory of martensitic transformations if either the parent phase or the martensitic phase is hexagonal. This cell is sometimes called an orthohexagonal cell. The conventional hexagonal cell has cell-edge vectors **a**, **b**, **c** as shown and the *new* orthorhombic cell has edges **A**, **B**, **C**. It is apparent from Figure A4.1 that:

$$\mathbf{A} = 2\cdot\mathbf{a} + 1\cdot\mathbf{b} + 0\cdot\mathbf{c}$$
$$\mathbf{B} = 0\cdot\mathbf{a} + 1\cdot\mathbf{b} + 0\cdot\mathbf{c} \qquad (A4.14)$$
$$\mathbf{C} = 0\cdot\mathbf{a} + 0\cdot\mathbf{b} + 1\cdot\mathbf{c}$$

Hence, if (*hkl*) are the Miller indices of a plane referred to the hexagonal cell and (*HKL*) are the indices of the same plane referred to the orthorhombic cell, then, comparing Eqs. (A4.2), (A4.10) and (A4.14):

$$H = 2h + k$$
$$K = k \qquad (A4.15)$$
$$L = l$$

That is:

$$\begin{pmatrix} H \\ K \\ L \end{pmatrix} = \begin{pmatrix} 2 & 1 & 0 \\ 0 & 1 & 0 \\ 0 & 0 & 1 \end{pmatrix} \begin{pmatrix} h \\ k \\ l \end{pmatrix} \tag{A4.16}$$

whence:

$$\begin{pmatrix} h \\ k \\ l \end{pmatrix} = \begin{pmatrix} 2 & 1 & 0 \\ 0 & 1 & 0 \\ 0 & 0 & 1 \end{pmatrix}^{-1} \begin{pmatrix} H \\ K \\ L \end{pmatrix} \tag{A4.17}$$

That is:

$$\begin{pmatrix} h \\ k \\ l \end{pmatrix} = \begin{pmatrix} \frac{1}{2} & -\frac{1}{2} & 0 \\ 0 & 1 & 0 \\ 0 & 0 & 1 \end{pmatrix} \begin{pmatrix} H \\ K \\ L \end{pmatrix} \tag{A4.18}$$

confirming that:

$$\begin{aligned}
\mathbf{a} &= \tfrac{1}{2} \cdot \mathbf{A} - \tfrac{1}{2} \cdot \mathbf{B} + 0 \cdot \mathbf{C} \\
\mathbf{b} &= 0 \cdot \mathbf{A} + 1 \cdot \mathbf{B} + 0 \cdot \mathbf{C} \\
\mathbf{c} &= 0 \cdot \mathbf{A} + 0 \cdot \mathbf{B} + 1 \cdot \mathbf{C}
\end{aligned} \tag{A4.19}$$

and:

$$\begin{aligned}
h &= \tfrac{1}{2}(H - K) \\
k &= K \\
l &= L
\end{aligned} \tag{A4.20}$$

It also follows from Eqs. (A4.2), (A4.4) and (A4.14) that:

$$\begin{pmatrix} u \\ v \\ w \end{pmatrix} = \begin{pmatrix} 2 & 0 & 0 \\ 1 & 1 & 0 \\ 0 & 0 & 1 \end{pmatrix} \begin{pmatrix} U \\ V \\ W \end{pmatrix} \tag{A4.21}$$

and from Eqs. (A4.6), (A4.7) and (A4.19) that:

$$\begin{pmatrix} U \\ V \\ W \end{pmatrix} = \begin{pmatrix} \frac{1}{2} & 0 & 0 \\ -\frac{1}{2} & 1 & 0 \\ 0 & 0 & 1 \end{pmatrix} \begin{pmatrix} u \\ v \\ w \end{pmatrix} \tag{A4.22}$$

A4.4   Example 2: Interchange of Rhombohedral and Hexagonal Indices

As a second example of the transformation of indices, we will consider the relationship between rhombohedral and hexagonal cells in the trigonal crystal system. A trigonal crystal may possess either a rhombohedral primitive unit cell (Figure 1.19k) or else a cell of the shape of Figure 1.19j. A hexagonal crystal must possess a primitive unit cell of the type shown in Figure 1.19j. When hexagonal axes are used with a crystal that has a true rhombohedral Bravais lattice, then if the smallest unit cells are chosen, the smallest hexagonal cell is triply-primitive. We will consider this case.

The rhombohedral axis can be oriented in two different ways relative to the corresponding hexagonal axes. These are shown in Figure A4.2. Consider the *obverse* setting in Figures A4.2a and A4.2b with the rhombohedral cell-edge vectors $\mathbf{a}_R$, $\mathbf{b}_R$, $\mathbf{c}_R$ and let the hexagonal cell-edge vectors be $\mathbf{a}_H$, $\mathbf{b}_H$, $\mathbf{c}_H$. Let the indices of a vector referred to the hexagonal cell be $[UVW]$ and those of the same vector referred to the rhombohedral cell be $[uvw]$. Likewise, let the indices of a plane referred to the hexagonal cell be (HKL) and those of the same plane referred to the rhombohedral cell be (hkl).

Writing the hexagonal cell vectors in terms of the rhombohedral cell vectors, we have:

$$\mathbf{a}_H = 1 \cdot \mathbf{a}_R + (-1) \cdot \mathbf{b}_R + 0 \cdot \mathbf{c}_R$$

$$\mathbf{b}_H = 0 \cdot \mathbf{a}_R + 1 \cdot \mathbf{b}_R + (-1) \cdot \mathbf{c}_R \qquad \text{(A4.23)}$$

$$\mathbf{c}_H = 1 \cdot \mathbf{a}_R + 1 \cdot \mathbf{b}_R + 1 \cdot \mathbf{c}_R$$

Hence, using Eq. (A4.10):

$$\begin{pmatrix} H \\ K \\ L \end{pmatrix} = \begin{pmatrix} 1 & \bar{1} & 0 \\ 0 & 1 & \bar{1} \\ 1 & 1 & 1 \end{pmatrix} \begin{pmatrix} h \\ k \\ l \end{pmatrix} \qquad \text{(A4.24)}$$

and so $(10\bar{1})_R = (110)_H$ and $(20\bar{1})_R = (211)_H$. An interesting example is $(41\bar{2})_R = (333)_H$: the $(41\bar{2})_R$ plane has an interplanar spacing one-third of the $(111)_H$ plane in the triply primitive hexagonal unit cell.

It follows from Eqs. (A4.23) and (A4.24) that (hkl) are derived from (HKL) through the equation:

$$\begin{pmatrix} h \\ k \\ l \end{pmatrix} = \begin{pmatrix} \frac{2}{3} & \frac{1}{3} & \frac{1}{3} \\ -\frac{1}{3} & \frac{1}{3} & \frac{1}{3} \\ -\frac{1}{3} & -\frac{2}{3} & \frac{1}{3} \end{pmatrix} \begin{pmatrix} H \\ K \\ L \end{pmatrix} \qquad \text{(A4.25)}$$

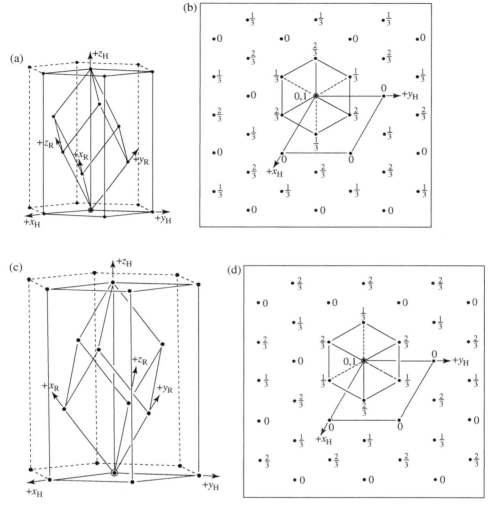

Figure A4.2 Unit cells in the rhombohedral lattice. (a) Obverse setting of rhombohedron and corresponding hexagonal non-primitive unit cell. (b) Plan of obverse setting (- - - - - lower edges, ——— upper edges of rhombohedron). (c) Reverse setting of rhombohedron and corresponding hexagonal non-primitive unit cell. (d) Plan of reverse setting (- - - - - lower edges, ——— upper edges of rhombohedron).

confirming that $(110)_H = (10\bar{1})_R$, $(211)_H = (20\bar{1})_R$ and $(333)_H = (41\bar{2})_R$. It is a useful exercise to show that $\mathbf{a}_R$, $\mathbf{b}_R$, $\mathbf{c}_R$ can therefore be written in terms of $\mathbf{a}_H$, $\mathbf{b}_H$, $\mathbf{c}_H$ as:

$$\mathbf{a}_R = \tfrac{2}{3}\cdot\mathbf{a}_H + \tfrac{1}{3}\cdot\mathbf{b}_H + \tfrac{1}{3}\cdot\mathbf{c}_H$$

$$\mathbf{b}_R = -\tfrac{1}{3}\cdot\mathbf{a}_H + \tfrac{1}{3}\cdot\mathbf{b}_H + \tfrac{1}{3}\cdot\mathbf{c}_H \qquad\qquad (A4.26)$$

$$\mathbf{c}_R = -\tfrac{1}{3}\cdot\mathbf{a}_H - \tfrac{2}{3}\cdot\mathbf{b}_H + \tfrac{1}{3}\cdot\mathbf{c}_H$$

It also follows from Eqs. (A4.2), (A4.4) and (A4.23) that:

$$
\begin{pmatrix} u \\ v \\ w \end{pmatrix} = \begin{pmatrix} 1 & 0 & 1 \\ \bar{1} & 1 & 1 \\ 0 & \bar{1} & 1 \end{pmatrix} \begin{pmatrix} U \\ V \\ W \end{pmatrix}
\tag{A4.27}
$$

and from Eqs. (A4.6), (A4.7) and (A4.26) that:

$$
\begin{pmatrix} U \\ V \\ W \end{pmatrix} = \begin{pmatrix} \frac{2}{3} & -\frac{1}{3} & -\frac{1}{3} \\ \frac{1}{3} & \frac{1}{3} & -\frac{2}{3} \\ \frac{1}{3} & \frac{1}{3} & \frac{1}{3} \end{pmatrix} \begin{pmatrix} u \\ v \\ w \end{pmatrix}
\tag{A4.28}
$$

Thus, for example, $[111]_R = [001]_H$ and $[330]_R = [122]_H$.

The matrices for accomplishing the transformation of indices of planes and of directions between the hexagonal cell and the *reverse* rhombohedral cell can be deduced from Figures A4.2c and d.

Appendix 5

Slip Systems in C.C.P. and B.C.C. Crystals

Here we use the methodology introduced in Section 7.3 to show that c.c.p. crystals, in which slip occurs in $\langle 1\bar{1}0 \rangle$ directions on $\{111\}$ planes, and b.c.c. crystals, in which slip occurs in $\langle 1\bar{1}1 \rangle$ directions on $\{110\}$ planes, both have five independent slip systems. We also give a proof of how to determine the slip system on which the resolved shear stress is the greatest for a particular tensile or compressive axis along which a load is applied.

A5.1   Independent Glide Systems in C.C.P. Metals

There are 12 slip systems of the form $\langle 1\bar{1}0 \rangle\{111\}$. Considering each of these in turn, we can choose to look at slip directions within a given slip plane, and the strain tensors they produce.

A5.1.1   Example: Slip Along [$1\bar{1}0$] on the (111) Slip Plane

If the amount of slip is of magnitude γ, where γ is the angle of shear due to glide, then from Eq. (7.14):

$$\varepsilon = \frac{\gamma}{2\sqrt{6}} \begin{pmatrix} 2 & 0 & 1 \\ 0 & -2 & -1 \\ 1 & -1 & 0 \end{pmatrix}$$

since:

$$\mathbf{n} = \left[\frac{1}{\sqrt{3}}, \frac{1}{\sqrt{3}}, \frac{1}{\sqrt{3}} \right] \quad \text{and} \quad \boldsymbol{\beta} = \left[\frac{1}{\sqrt{2}}, -\frac{1}{\sqrt{2}}, 0 \right] \quad .$$

Crystallography and Crystal Defects, Third Edition. Anthony Kelly and Kevin M. Knowles.
© 2020 John Wiley & Sons Ltd. Published 2020 by John Wiley & Sons Ltd.
Companion website: www.wiley.com/go/kelly/crystallography3e

Table A5.1   The twelve slip systems in c.c.p. metal crystals, labelled here for convenience A–L.

Label	Slip system
A	$[1\bar{1}0]\,(111)$
B	$[01\bar{1}]\,(111)$
C	$[10\bar{1}]\,(111)$
D	$[1\bar{1}0]\,(11\bar{1})$
E	$[011]\,(11\bar{1})$
F	$[101]\,(11\bar{1})$
G	$[0\bar{1}\bar{1}]\,(1\bar{1}1)$
H	$[110]\,(1\bar{1}1)$
I	$[10\bar{1}]\,(1\bar{1}1)$
J	$[110]\,(\bar{1}11)$
K	$[\bar{1}0\bar{1}]\,(\bar{1}11)$
L	$[01\bar{1}]\,(\bar{1}11)$

A5.1.2   Number of Independent Glide Systems

Following the example in (A5.1.1), we can determine the 12 strain tensors produced by the 12 slip systems for slip of magnitude γ, designating the slip systems A–L (Table A5.1).

The strain tensors are therefore the following matrices:

$$\varepsilon_A = \frac{\gamma}{2\sqrt{6}}\begin{pmatrix} 2 & 0 & 1 \\ 0 & -2 & -1 \\ 1 & -1 & 0 \end{pmatrix}; \quad \varepsilon_B = \frac{\gamma}{2\sqrt{6}}\begin{pmatrix} 0 & 1 & -1 \\ 1 & 2 & 0 \\ -1 & 0 & -2 \end{pmatrix}; \quad \varepsilon_C = \frac{\gamma}{2\sqrt{6}}\begin{pmatrix} 2 & 1 & 0 \\ 1 & 0 & -1 \\ 0 & -1 & -2 \end{pmatrix}$$

$$\varepsilon_D = \frac{\gamma}{2\sqrt{6}}\begin{pmatrix} 2 & 0 & -1 \\ 0 & -2 & 1 \\ -1 & 1 & 0 \end{pmatrix}; \quad \varepsilon_E = \frac{\gamma}{2\sqrt{6}}\begin{pmatrix} 0 & 1 & 1 \\ 1 & 2 & 0 \\ 1 & 0 & -2 \end{pmatrix}; \quad \varepsilon_F = \frac{\gamma}{2\sqrt{6}}\begin{pmatrix} 2 & 1 & 0 \\ 1 & 0 & 1 \\ 0 & 1 & -2 \end{pmatrix}$$

$$\varepsilon_G = \frac{\gamma}{2\sqrt{6}}\begin{pmatrix} 0 & -1 & -1 \\ -1 & 2 & 0 \\ -1 & 0 & -2 \end{pmatrix}; \quad \varepsilon_H = \frac{\gamma}{2\sqrt{6}}\begin{pmatrix} 2 & 0 & 1 \\ 0 & -2 & 1 \\ 1 & 1 & 0 \end{pmatrix}; \quad \varepsilon_I = \frac{\gamma}{2\sqrt{6}}\begin{pmatrix} 2 & -1 & 0 \\ -1 & 0 & 1 \\ 0 & 1 & -2 \end{pmatrix}$$

$$\varepsilon_J = \frac{\gamma}{2\sqrt{6}}\begin{pmatrix} -2 & 0 & 1 \\ 0 & 2 & 1 \\ 1 & 1 & 0 \end{pmatrix}; \quad \varepsilon_K = \frac{\gamma}{2\sqrt{6}}\begin{pmatrix} 2 & -1 & 0 \\ -1 & 0 & -1 \\ 0 & -1 & -2 \end{pmatrix}; \quad \varepsilon_L = \frac{\gamma}{2\sqrt{6}}\begin{pmatrix} 0 & -1 & 1 \\ -1 & 2 & 0 \\ 1 & 0 & -2 \end{pmatrix}$$

It is readily apparent that there can only be two independent slip systems on each $\{111\}$ slip plane, so that, for example, in the above:

$$\varepsilon_C = \varepsilon_A + \varepsilon_B$$

$$\varepsilon_F = \varepsilon_D + \varepsilon_E$$

$$\varepsilon_I = \varepsilon_G + \varepsilon_H$$

$$\varepsilon_L = \varepsilon_J + \varepsilon_K$$

and so we are not totally unrestricted in the choice of the slip systems that we identify as candidate independent slip systems for c.c.p. crystals. Therefore, we need only look at slip systems A, B, D, E, G, H, J and K.

If out of these we choose our independent slip systems to be A, B, D, E and G, the strain tensors of the other seven slip systems can all be written in terms of ε_A, ε_B, ε_D, ε_E and ε_G, demonstrating that there are five independent slip systems for c.c.p. crystals.

(An examination of the forms of the strain tensor for A, B, D, E and G shows that they are all independent of one another. Thus, for example, the strain tensor for G cannot be constructed from a suitable combination of the strain tensors of A, B, D and E to reduce the number of independent slip systems still further.)

Thus:

$$\varepsilon_C = \varepsilon_A + \varepsilon_B$$

$$\varepsilon_F = \varepsilon_D + \varepsilon_E$$

$$\varepsilon_H = \varepsilon_D + \varepsilon_E - \varepsilon_B$$

$$\varepsilon_I = \varepsilon_D + \varepsilon_E + \varepsilon_G - \varepsilon_B$$

$$\varepsilon_J = \varepsilon_E - \varepsilon_A - \varepsilon_B$$

$$\varepsilon_K = \varepsilon_A + \varepsilon_G$$

$$\varepsilon_L = \varepsilon_E + \varepsilon_G - \varepsilon_B$$

and so we have shown that there are five independent slip systems in c.c.p. crystals, the number required for a material to exhibit general plasticity. It is also immediately apparent from this analysis that there have to be five independent slip systems in b.c.c. crystals in which slip occurs in $\langle 1\bar{1}1 \rangle$ directions on $\{110\}$ planes.

A5.2 Diehl's Rule and the OILS Rule

Once the slip systems characteristic of a particular crystal structure have been established, the general approach to establishing which slip system will operate in a tensile test on a single crystal is to compute the Schmid factor $\cos\phi\cos\lambda$ for each distinct slip system, and for each orientation of the tensile axis as deformation proceeds. As the applied stress is increased, the system with the largest Schmid factor – that is, the one experiencing the greatest resolved shear stress – will be that on which slip first occurs as the critical resolved shear stress is exceeded.

Diehl's rule [1] is a simple stereographic method for finding the slip system with the highest Schmid factor, without calculation, and can be used for:

1. c.c.p. crystals slipping on $\{111\}\langle 1\bar{1}0 \rangle$
2. b.c.c. crystals slipping on $\{1\bar{1}0\}\langle 111 \rangle$

A5.2.1 Use of Diehl's Rule for $\{111\}\langle 1\bar{1}0 \rangle$ Slip (such as C.C.P. Metals)

1. Sketch a cubic stereogram displaying standard triangles – that is, showing all poles of the forms $\{100\}$, $\{110\}$ and $\{111\}$ for the holosymmetric cubic class and the great circles connecting them to form 48 right-angled spherical triangles, as in Figure 7.13. (Each of the small spherical triangles in the figure is bounded by mirror planes and its contents can, in turn, be reflected into all the other 47 standard triangles covering the surface of the projection sphere; that is, any standard triangle comprises the smallest angular region necessary for considering any property variation with angle in the holosymmetric cube.)
2. Locate the standard triangle containing the pole of the tensile axis $[uvw]$, for example in the spherical triangle 001–011–111.
3. The identity of the slip plane is found by taking the $\{111\}$-type pole in this standard triangle and forming its reflection in the side of the triangle opposite to it. For example, $(\bar{1}11)$ is the reflection of (111) in the great circle 001–011–010.
4. Similarly, the identity of the slip direction is found by forming the mirror image of the $\langle 110 \rangle$ – type pole of the triangle in the side opposite to it. For example, [101] is the reflection of [011] in the great circle 001–111–110.
5. $(\bar{1}11)$[101] is then the slip system with the highest Schmid factor, A III in the nomenclature used in Figure 7.13.

Note:

1. Care is needed when considering triangles bordering the primitive, since one of the poles calculated using this procedure may be in the lower hemisphere of the stereographic projection.
2. If $[uvw]$ lies at the boundary (junction plane) of two (or more) standard triangles, the Schmid factors are equal on the corresponding two (or more) systems.

A5.2.2 Use of Diehl's Rule for $\{1\bar{1}0\}\langle 111 \rangle$ Slip (such as B.C.C. Metals)

For b.c.c. metals slipping on $\{1\bar{1}0\}\langle 111 \rangle$, the slip plane/slip direction indices are interchanged. Thus, for a tensile axis $[uvw]$ in the spherical triangle 001–011–111, the slip system would be (101)$[\bar{1}11]$.

Once deformation has been initiated by slip starting on the slip system with the highest critical resolved shear stress, the tensile axis always rotates towards the slip direction. Thus, for c.c.p. crystals, a tensile axis initially in the spherical triangle 001–011–111 would move towards [101], while for b.c.c. crystals it would move towards $[\bar{1}11]$.

A5.2.3 The OILS Rule

This rule is exactly equivalent to Diehl's rule. To determine the operative slip system (the one with the highest Schmid factor) in a specimen under uniaxial tension or compression, the procedure is as follows [2]:

1. Ignoring the signs, identify the highest (**H**), intermediate (**I**) and lowest (**L**) valued indices of the loading axis $[uvw]$.
2. The $\langle 110 \rangle$ slip direction or $\{110\}$ slip plane (whichever is appropriate) is the one with zer**O** in the position of the **I** index and the signs of the other two indices preserved.
3. The $\{111\}$ slip plane or $\langle 111 \rangle$ slip direction (whichever is appropriate) is the one with the **L** index **S**ign reversed and the signs of the other two indices preserved.

Thus, if $[uvw]$ is in the spherical triangle 001–011–111, u, v and w are all positive and $u < v < w$. Hence, for a c.c.p. crystal, the slip direction is [101] and the slip plane ($\bar{1}11$); that is, the same result as that obtained by Diehl's rule.

A5.3 Proof of Diehl's Rule and the OILS Rule

The proof is straightforward. For simplicity, we can choose $[uvw]$ to lie within the standard stereographic triangle defined by the 001, 011 and 111 poles. Under these circumstances, $0 < u < v < w$, as we have just noted in Section A5.2. For c.c.p. crystals we have the 12 possible slip systems specified in Section A5.1. For each of these, we can specify the Schmid factor $\cos \phi \cos \lambda$, as in Table A5.2, where for convenience we have chosen the indices of the slip directions and slip planes for each of the 12 possibilities to conform to the labels used in Figure 7.13.

We now have to identify the row in the far right-hand column of Table A5.2, for which $\cos \phi \cos \lambda$ is a maximum.

A comparison of the Schmid factors for A II, A III and A VI (which have a common factor $(v + w - u)$) shows that the one for A III is the largest of the three for $0 < u < v < w$. Likewise, for $0 < u < v < w$, (i) the Schmid factor for B IV is greater than B II or B III, (ii) the Schmid factor for C I is greater than C III or C V, and (iii) the Schmid factor for D I is greater than D IV or D VI.

Therefore, we have only have to determine which of A III, B IV, C I and D I has the largest Schmid factor. Comparing C I and D I, which have a common factor $(v + w)$, it is apparent that D I has the greater Schmid factor when $u > 0$. Looking more closely at the Schmid factors for A III and B IV, we have:

$$\text{A III}: \ |\cos \phi \cos \lambda|(\times \sqrt{6}(u^2 + v^2 + w^2)) = w^2 - u^2 + v(w + u)$$

$$\text{B IV}: \ |\cos \phi \cos \lambda|(\times \sqrt{6}(u^2 + v^2 + w^2)) = w^2 - u^2 + v(w - u)$$

and so the Schmid factor for A III is greater than that for B IV when $u > 0$.

Finally, comparing the Schmid factors for A III and D I, we have:

$$\text{A III}: \ |\cos \phi \cos \lambda|(\times \sqrt{6}(u^2 + v^2 + w^2)) = w^2 + vu + vw - u^2$$

$$\text{D I}: \ |\cos \phi \cos \lambda|(\times \sqrt{6}(u^2 + v^2 + w^2)) = w^2 + vu + uw - v^2$$

Table A5.2 Slip systems and their corresponding Schmid factors in c.c.p. metal crystals for a tensile axis [*uvw*] within the standard stereographic triangle. The nomenclature for the four slip planes (A–D) and six slip directions (I–VI) is that used in Figure 7.13.

Slip system	Slip plane and slip direction	$\mid \cos\phi \cos\lambda \mid (\times \sqrt{6}(u^2 + v^2 + w^2))$
A II	$(\bar{1}11)[0\bar{1}1]$	$(v + w - u)(w - v)$
A III	$(\bar{1}11)[101]$	$(v + w - u)(u + w)$
A VI	$(\bar{1}11)[110]$	$(v + w - u)(u + v)$
B II	$(111)[0\bar{1}1]$	$(u + v + w)(w - v)$
B IV	$(111)[\bar{1}01]$	$(u + v + w)(w - u)$
B V	$(111)[1\bar{1}0]$	$(u + v + w)\mid v - u \mid$
C I	$(\bar{1}\bar{1}1)[011]$	$\mid w - u - v \mid (v + w)$
C III	$(\bar{1}\bar{1}1)[101]$	$\mid w - u - v \mid (u + w)$
C V	$(\bar{1}\bar{1}1)[1\bar{1}0]$	$\mid w - u - v \mid \mid v - u \mid$
D I	$(1\bar{1}1)[011]$	$(u + w - v)(v + w)$
D IV	$(1\bar{1}1)[\bar{1}01]$	$(u + w - v)(w - u)$
D VI	$(1\bar{1}1)[110]$	$(u + w - v)(u + v)$

and since:

$$vw + v^2 > uw + u^2$$

for $0 < u < v < w$, it follows that A III is the slip system with the greatest Schmid factor; that is, the slip system $(\bar{1}11)$ [101]. This is of course the slip system found using either Diehl's rule or OILS rule for a pole within the 001–011–111 standard stereographic triangle in c.c.p. crystals. Thus, Diehl's rule and OILS rule are both proved using this methodology. The extension of this analysis to other stereographic triangles can be achieved by suitable permutations of the indices defining the poles at the corners of the standard stereographic triangle.

A further simple extension of this methodology enables special cases, such as where $u = v > 0$, to be considered when two or more slip systems have the same Schmid factor.

References

[1] Diehl, J. personal communication to A. Seeger, acknowledged on p. 24 of A. Seeger (1958) *Kristallplastizität*, *Handbuch der Physik*, Band VII.2: Kristallphysik II (ed. S. Flügge), Springer-Verlag, Berlin, pp. 1–210.
[2] Hutchings, I.M. (1993). Quick non-graphical method for deducing slip systems in cubic close packed metals in tension or compression. *Mater. Sci. Technol.* 9: 929–930.

Appendix 6

Homogeneous Strain

A body is said to be homogeneously strained if the distortion is the same everywhere. Under these circumstances, the e_{ij} in Eqs. (6.5) and (6.9) are constants, independent of position in the body. The equation:

$$du_i = \frac{\partial u_i}{\partial x_j} dx_j = e_{ij} \, dxj \tag{A6.1}$$

can then be integrated immediately to obtain the displacements:

$$u_i = (u_0)_i + e_{ij}x_j \quad (i,j = 1,2,3) \tag{A6.2}$$

It is apparent from Eq. (A6.2) that in homogeneous strain, the displacements are linear functions of the position coordinates.

The constants $(u_0)_i$ represent the translation of the body as a whole, and are of no further interest. Subtracting this translation from the displacement, we obtain the residual displacement:

$$x_i' - x_i = e_{ij}x_j \tag{A6.3}$$

where x_i' are the new coordinates (referred to axes that have been translated by $(u_0)_i$ but not rotated) of the point which was originally at x_i. Equation (A6.3) written out in full is:

$$x_1' = (1 + e_{11}) x_1 + e_{12}x_2 + e_{13}x_3$$
$$x_2' = e_{21}x_1 + (1 + e_{22}) x_2 + e_{23}x_3$$
$$x_3' = e_{31}x_1 + e_{32}x_2 + (1 + e_{33}) x_3$$

The shape into which a line or surface whose general equation is:

$$f(x_1, x_2, x_3) = 0 \tag{A6.4}$$

Crystallography and Crystal Defects, Third Edition. Anthony Kelly and Kevin M. Knowles.
© 2020 John Wiley & Sons Ltd. Published 2020 by John Wiley & Sons Ltd.
Companion website: www.wiley.com/go/kelly/crystallography3e

is deformed can be determined by solving Eq. (A6.3) for x_1, x_2 and x_3 in terms of x_1', x_2' and x_3', and substituting these values into Eq. (A6.4). In this way it can be shown that any homogeneous strain, large or small, has the following four properties:

1. Straight lines remain straight lines, and in general are rotated and stretched or contracted, all straight lines in the same direction being rotated through the same angle and stretched or contracted in the same ratio. Similarly, planes are deformed into planes.
2. A sphere is deformed into an ellipsoid. The ellipsoid into which a sphere of unit radius is deformed is called the *strain ellipsoid*.
3. The axes of the strain ellipsoid are derived from three mutually perpendicular diameters of the unit sphere, which are called the principal axes. These axes form the only set of three orthogonal directions which remain orthogonal after the strain.
4. In the unstrained state there is a particular ellipsoid, called the *reciprocal strain ellipsoid*, which deforms into the unit sphere. Its axes are the principal axes.

In general, the principal axes are rotated by the strain. A deformation which leaves the principal axes unrotated is called a pure strain. A general homogeneous strain can be accomplished in two stages: a pure strain in which the principal axes receive their correct extensions, followed by a rotation in which they are brought to their final position. The above results are applied in the discussion of the geometry of martensitic transformations in Chapter 12.

We shall now mention three very simple types of homogeneous strain.

A6.1   Simple Extension

$$
\begin{aligned}
x_1' &= kx_1, \\
x_2' &= x_2, \\
x_3' &= x_3,
\end{aligned}
\qquad
\left(
\begin{array}{lll}
e_{11} = k - 1 & e_{12} = 0 & e_{13} = 0 \\
e_{21} = 0 & e_{22} = 0 & e_{23} = 0 \\
e_{31} = 0 & e_{32} = 0 & e_{33} = 0
\end{array}
\right)
\tag{A6.5}
$$

The principal axes Ox_1, Ox_2 and Ox_3 are not rotated; therefore, simple extension is a pure strain.

A6.2   Simple Shear

$$
\begin{aligned}
x_1' &= x_1 + gx_2, \\
x_2' &= x_2, \\
x_3' &= x_3,
\end{aligned}
\qquad
\left(
\begin{array}{lll}
e_{11} = 0 & e_{12} = g & e_{13} = 0 \\
e_{21} = 0 & e_{22} = 0 & e_{23} = 0 \\
e_{31} = 0 & e_{32} = 0 & e_{33} = 0
\end{array}
\right)
\tag{A6.6}
$$

where g is the magnitude of the shear strain, as defined in Eq. (6.2). A more detailed description of simple shear is developed in Chapters 7 and 11, where it is required for the description of slip and twinning. Simple shear is not a pure strain; its non-rotational part is described below.

A6.3   Pure Shear

$$
\begin{aligned}
x_1' &= kx_1, \\
x_2' &= k^{-1}x_2, \\
x_3' &= x_3,
\end{aligned}
\qquad
\left(
\begin{array}{lll}
e_{11} = k - 1 & e_{12} = 0 & e_{13} = 0 \\
e_{21} = 0 & e_{22} = (1/k) - 1 & e_{23} = 0 \\
e_{31} = 0 & e_{32} = 0 & e_{33} = 0
\end{array}
\right)
\tag{A6.7}
$$

In this strain, the principal axes are Ox_1, Ox_2 and Ox_3, and they are not rotated.

A6.4 The Relationship Between Pure Shear and Simple Shear

The relationship between a pure shear and a simple shear is illustrated by the diagram in Figure A6.1. A simple shear of strength g, referred to axes Ox_1°, Ox_2°, distorts a circle into the ellipse shown in Figure A6.1. The same ellipse can be produced by a strain which is a pure shear, referred to axes Ox_1, Ox_2, with Ox_1 at an angle of $\left((\pi/4) - \frac{1}{2}\tan^{-1}(g/2)\right)$ to Ox_1°, as shown in Figure A6.1. The required magnitude of the pure shear is given by:

$$k - (1/k) = g \tag{A6.8}$$

In order for the *displacements* produced by this pure shear to be identical to those produced by the simple shear, it must be preceded by a rotation of $\alpha = \tan^{-1} g/2$. Thus point P, for example, is rotated to Q, and the pure shear then carries Q to P'. Alternatively, the displacements of the simple shear can be produced by a pure shear referred to axes at $\pm(\pi/4) + \frac{1}{2}\tan^{-1}(g/2)$ to Ox_1° followed by a rotation of $\tan^{-1} g/2$. These axes are principal axes of the simple shear; Ox_1, Ox_2 and Ox_3 (which remains unchanged) are the principal axes of the simple shear in their final, rotated position.

When the simple shear is of very small magnitude ($g = \gamma$, where $\gamma << 1$), it is equivalent to a pure shear referred to axes rotated through an angle of $\pi/4$ together with a rotation of $\gamma/2$. This can also be seen by carefully studying Figure 7.9.

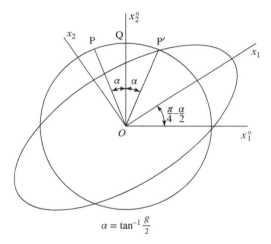

$$\alpha = \tan^{-1} \frac{g}{2}$$

Figure A6.1 Illustration of the relationship between a pure shear and a simple shear

Appendix 7

Crystal Structure Data

A7.1 Crystal Structures of the Elements, Interatomic Distances and Six-Fold Coordination-Number Ionic Radii

Atomic number	Element	Crystal structure[a]	Atomic radius (Å)[b]	Ion	Ionic radius[c] (Å)
1	H	—	—	H^-	1.40; 1.53[d]
2	He	—	—	—	—
3	Li	b.c.c.	1.52	Li^+	0.76
4	Be	h.c.p.	1.11	Be^{2+}	0.45
5	B	trigonal	0.83	B^{3+}	0.27
6	C (graphite)	hexagonal	0.71	C^{4+}	0.16
	C (diamond)	cubic	0.77	—	—
7	N	—	—	N^{3-}	1.46[e]
8	O	—	—	O^{2-}	1.40
9	F	—	—	F^-	1.33
10	Ne	c.c.p.	1.60	—	—
11	Na	b.c.c.	1.86	Na^+	1.02
12	Mg	h.c.p.	1.60	Mg^{2+}	0.72
13	Al	c.c.p.	1.43	Al^{3+}	0.535
14	Si	cubic (diamond)	1.17	Si^{4+}	0.400

Crystallography and Crystal Defects, Third Edition. Anthony Kelly and Kevin M. Knowles.
© 2020 John Wiley & Sons Ltd. Published 2020 by John Wiley & Sons Ltd.
Companion website: www.wiley.com/go/kelly/crystallography3e

Atomic number	Element	Crystal structure[a]	Atomic radius (Å)[b]	Ion	Ionic radius[c] (Å)
15	P (black)	orthorhombic	1.11	P^{3+}	0.44
16	S	orthorhombic	1.02	S^{2-}	1.84
17	Cl	orthorhombic	0.99	Cl^-	1.81
18	Ar	c.c.p	1.86	—	—
19	K	b.c.c.	2.31	K^+	1.38
20	Ca	c.c.p.	1.98	Ca^{2+}	1.00
21	Sc	h.c.p.	1.62	Sc^{3+}	0.745
22	Ti	h.c.p.	1.45	Ti^{3+}	0.67
				Ti^{4+}	0.605
23	V	b.c.c.	1.31	V^{2+}	0.79
				V^{3+}	0.64
				V^{4+}	0.58
				V^{5+}	0.54
24	Cr	b.c.c.	1.25	Cr^{2+}	0.80
				Cr^{3+}	0.615
				Cr^{4+}	0.55
				Cr^{6+}	0.44
25	Mn	cubic	1.29	Mn^{2+}	0.83
				Mn^{3+}	0.645
				Mn^{4+}	0.53
26	Fe	b.c.c.	1.24	Fe^{2+}	0.780
				Fe^{3+}	0.645
27	Co	h.c.p.	1.25	Co^{2+}	0.745
				Co^{3+}	0.61
28	Ni	c.c.p.	1.25	Ni^{2+}	0.69
				Ni^{3+}	0.60
29	Cu	c.c.p.	1.28	Cu^+	0.77
				Cu^{2+}	0.73
30	Zn	h.c.p.	1.33	Zn^{2+}	0.74
31	Ga	orthorhombic	1.41	Ga^{3+}	0.62
32	Ge	cubic (diamond)	1.22	Ge^{4+}	0.53
33	As	trigonal	1.26	As^{3+}	0.58
				As^{5+}	0.46
34	Se	trigonal	1.16	Se^{2-}	1.98
35	Br	orthorhombic	1.14	Br^-	1.96
36	Kr	c.c.p	1.99	—	—
37	Rb	b.c.c.	2.42	Rb^+	1.52
38	Sr	c.c.p.	2.15	Sr^{2+}	1.18
39	Y	h.c.p.	1.78	Y^{3+}	0.90
40	Zr	h.c.p.	1.59	Zr^{4+}	0.72
41	Nb[f]	b.c.c.	1.43	Nb^{3+}	0.72
				Nb^{4+}	0.68
				Nb^{5+}	0.64

Atomic number	Element	Crystal structure[a]	Atomic radius (Å)[b]	Ion	Ionic radius[c] (Å)
42	Mo	b.c.c.	1.36	Mo^{3+}	0.69
				Mo^{4+}	0.65
				Mo^{5+}	0.61
				Mo^{6+}	0.59
43	Tc	h.c.p.	1.35	Tc^{4+}	0.645
				Tc^{5+}	0.60
44	Ru	h.c.p.	1.32	Ru^{3+}	0.68
				Ru^{4+}	0.62
				Ru^{5+}	0.565
45	Rh	c.c.p.	1.34	Rh^{3+}	0.665
				Rh^{4+}	0.60
				Rh^{5+}	0.55
46	Pd	c.c.p.	1.38	Pd^{2+}	0.86
				Pd^{3+}	0.76
				Pd^{4+}	0.615
47	Ag	c.c.p.	1.44	Ag^{+}	1.15
				Ag^{2+}	0.94
48	Cd	h.c.p.	1.49	Cd^{2+}	0.95
49	In	tetragonal	1.69	In^{3+}	0.80
50	Sn (white)	tetragonal	1.51	Sn^{4+}	0.69
51	Sb	rhombohedral	1.45	Sb^{3+}	0.76
				Sb^{5+}	0.60
52	Te	trigonal	1.42	Te^{2-}	2.21
53	I	orthorhombic	1.34	I^{-}	2.20
54	Xe	c.c.p.	2.17	—	—
55	Cs	b.c.c.	2.62	Cs^{+}	1.67
56	Ba	b.c.c.	2.18	Ba^{2+}	1.35
57	La	double h.c.p.	1.87	La^{3+}	1.032
58	Ce	c.c.p.	1.82	Ce^{3+}	1.01
				Ce^{4+}	0.87
59	Pr	double h.c.p.	1.82	Pr^{3+}	0.99
				Pr^{4+}	0.85
60	Nd	double h.c.p.	1.81	Nd^{3+}	0.83
61	Pm	double h.c.p.	1.80	Pm^{3+}	0.97
62	Sm	rhombohedral	1.79	Sm^{3+}	0.958
63	Eu	b.c.c.	1.98	Eu^{3+}	0.947
64	Gd	h.c.p.	1.78	Gd^{3+}	0.938
65	Tb	h.c.p.	1.76	Tb^{3+}	0.923
				Tb^{4+}	0.76
66	Dy	h.c.p.	1.75	Dy^{3+}	0.912
67	Ho	h.c.p.	1.74	Ho^{3+}	0.901
68	Er	h.c.p.	1.73	Er^{3+}	0.890
69	Tm	h.c.p.	1.72	Tm^{3+}	0.88

Atomic number	Element	Crystal structure[a]	Atomic radius (Å)[b]	Ion	Ionic radius[c] (Å)
70	Yb	c.c.p.	1.94	Yb^{2+}	1.02
				Yb^{3+}	0.868
71	Lu	h.c.p.	1.72	Lu^{3+}	0.861
72	Hf	h.c.p.	1.56	Hf^{4+}	0.71
73	Ta	b.c.c.	1.43	Ta^{3+}	0.72
				Ta^{4+}	0.68
				Ta^{5+}	0.64
74	W	b.c.c.	1.37	W^{4+}	0.66
				W^{5+}	0.62
				W^{6+}	0.60
75	Re	h.c.p.	1.37	Re^{4+}	0.63
				Re^{5+}	0.58
				Re^{6+}	0.55
				Re^{7+}	0.53
76	Os	h.c.p.	1.34	Os^{4+}	0.63
				Os^{5+}	0.575
				Os^{6+}	0.545
				Os^{7+}	0.525
77	Ir	c.c.p.	1.35	Ir^{3+}	0.68
				Ir^{4+}	0.625
				Ir^{5+}	0.57
78	Pt	c.c.p.	1.39	Pt^{2+}	0.80
				Pt^{4+}	0.625
				Pt^{5+}	0.57
79	Au	c.c.p.	1.44	Au^{+}	1.37
80	Hg	rhombohedral	1.50	Hg^{+}	1.19
				Hg^{2+}	1.02
81	Tl	h.c.p.	1.71	Tl^{+}	1.50
				Tl^{3+}	0.885
82	Pb	c.c.p.	1.75	Pb^{2+}	1.19
				Pb^{4+}	0.775
83	Bi	rhombohedral	1.54	Bi^{3+}	1.03
				Bi^{5+}	0.76
84	Po	simple cubic	1.68	Po^{4+}	0.94
				Po^{6+}	0.67
85	At	—	—	—	—
86	Rn	—	—	—	—
87	Fr	—	2.70	Fr^{+}	1.80
88	Ra	b.c.c	2.23	—	—
89	Ac	c.c.p.	1.88	Ac^{3+}	1.12
90	Th	c.c.p.	1.80	Th^{4+}	0.94

Atomic number	Element	Crystal structure[a]	Atomic radius (Å)[b]	Ion	Ionic radius[c] (Å)
91	Pa	b.c. tetragonal	1.61	Pa^{3+}	1.04
				Pa^{4+}	0.90
				Pa^{5+}	0.78
92	U	orthorhombic	1.54	U^{3+}	1.025
				U^{4+}	0.89
				U^{5+}	0.76
				U^{6+}	0.73
93	Np	orthorhombic	1.46	Np^{3+}	1.01
				Np^{4+}	0.87
				Np^{5+}	0.75
				Np^{6+}	0.72
94	Pu	monoclinic	1.58	Pu^{3+}	1.00
				Pu^{4+}	0.86
				Pu^{5+}	0.74
				Pu^{6+}	0.71
95	Am	double h.c.p.	1.73	Am^{3+}	0.975
				Am^{4+}	0.85
96	Cm	double h.c.p.	1.74	Cm^{3+}	0.97
				Cm^{4+}	0.85
97	Bk	c.c.p.	1.77	Bk^{3+}	0.96
		double h.c.p.	1.70	Bk^{4+}	0.83
98	Cf	double h.c.p.	1.69	Cf^{3+}	0.95
				Cf^{4+}	0.821

Note: Low-atomic-number elements such as H, He, O and N which are gases at room temperature and pressure crystallize under high pressures; however, data relating to atomic distances in these crystalline forms have been omitted from this table.

[a] Data are for room temperature and ambient pressure, apart from those for elements which are gaseous or liquid at room temperature.

[b] Defined as half the shortest distance between atoms in b.c.c. and c.c.p. crystals, half the shortest distance between neighbouring atoms in h.c.p. and double h.c.p. crystals, half the shortest distance between adjacent atoms in As, Bi, P, Sb, Hg, Br, Cl and I, and as an average metallic radius specified for more complex crystal structures. The atomic radius of an element increases as its coordination number increases – see, for example, Pauling, L. (1947). Atomic radii and interatomic distances in metals. *J. Am. Chem. Soc.* 69: 542–553. Atomic radii quoted here for b.c.c. metals should be increased by 2% for a coordination number of 12.

[c] Values taken, with permission, from Table 1 of Shannon, R.D. (1976). Revised effective ionic radii and systematic studies of interatomic distances in halides and chalcogenides. *Acta Crystall. A* 32: 751–767. In crystal structures where the coordination number is 4, the data in Figure 2 of Shannon suggest that the cation radius should be decreased by about 21%; with coordination number 8 the cation radius should be increased by 12–22%; and with coordination number 12 the cation radius should be increased by 30–34%.

[d] Shannon suggests that the value of 1.40 Å is appropriate for the more electronegative metals, while the value of 1.53 Å is appropriate for the large alkali halides.

[e] Value for fourfold coordination.

[f] Formerly known as columbium (Cb), and still occasionally referred to as such, particularly in North America.

A7.2 Crystals with the Sodium Chloride Structure

The data in Sections A7.2 to A7.8 are taken from more extensive tables in R.G. Wyckoff (1963) *Crystal Structures*, Interscience, New York.

Crystal	a_0 (Å)	Crystal	a_0 (Å)
Antimonides		*Chlorides*	
CeSb	6.40	AgCl	5.55
LaSb	6.47	KCl	6.29
ScSb	5.86	LiCl	5.13
SnSb	6.13	NaCl	5.64
ThSb	6.32	RbCl	6.58
USb	6.19		
		Cyanides	
Arsenides		KCN	6.53
CeAs	6.06	NaCN	5.89
LaAs	6.13	RbCN	6.82
ScAs	5.49		
SnAs	5.68	*Fluorides*	
ThAs	5.97	AgF	4.92
UAs	5.77	CsF	6.01
		KF	5.35
Borides		LiF	4.03
ZrB	4.65	NaF	4.62
		RbF	5.64
Bromides			
AgBr	5.77	*Hydrides*	
KBr	6.60	CsH	6.38
LiBr	5.50	KH	5.70
NaBr	5.97	LiH	4.08
RbBr	6.85	NaH	4.88
		RbH	6.04
Carbides			
HfC	4.46	*Iodides*	
NbC	4.47	KI	7.07
TaC	4.45	LiI	6.00
TiC	4.32	NH_4I	7.26
UC	4.96	NaI	6.47
VC	4.18	RbI	7.34
ZrC	4.68		

Crystal	a_0 (Å)	Crystal	a_0 (Å)
Nitrides		CeSe	5.98
CeN	5.01	LaSe	6.06
CrN	4.14	MgSe	5.45
LaN	5.30	MnSe	5.45
NbN	4.70	PbSe	6.12
ScN	4.44	SnSe	6.02
TiN	4.23	SrSe	6.23
UN	4.88	ThSe	5.87
VN	4.13	USe	5.75
ZrN	4.61		
		Sulphides	
Oxides		BaS	6.39
BaO	5.52	CaS	5.69
CaO	4.81	CeS	5.78
CdO	4.70	LaS	5.84
CoO	4.27	MgS	5.20
FeO	4.28–4.31	MnS	4.45
MgO	4.21	PbS	5.94
MnO	4.45	SrS	6.02
NbO	4.21	ThS	5.68
NiO	4.17	US	5.48
SrO	5.16	ZrS	5.25
TaO	4.42–4.44		
UO	4.92	*Tellurides*	
ZrO	4.62	BaTe	6.99
		BiTe	6.47
Phosphides		CaTe	6.34
CeP	5.90	CeTe	6.35
LaP	6.01	LaTe	6.41
ThP	5.82	PbTe	6.45
UP	5.59	SnTe	6.31
ZrP	5.27	SrTe	6.47
		UTe	6.16
Selenides			
BaSe	6.60		
CaSe	5.91		

A7.3   Crystals with the Caesium Chloride Structure

Crystal	a_0 (Å)
AgCd	3.33
AgMg	3.28
CsBr	4.29
CsCl	4.12
CsCN	4.25
CsI	4.57
CuBe	2.70
CuZn	2.94
FeAl	2.91
NiTi	3.02
TlBr	3.97
TlCl	3.83
TlCN	3.82
TlI	4.20

A7.4   Crystals with the Sphalerite Structure

Crystal	a_0 (Å)	Crystal	a_0 (Å)
AgI	6.47	GaP	5.45
AlAs	5.66	GaSb	6.12
BeS	4.85	HgS	5.85
BeSe	5.07	HgSe	6.08
BeTe	5.54	HgTe	6.43
BN	3.62	InAs	6.06
CdS	5.82	InP	5.87
CdTe	6.48	InSb	6.48
CuBr	5.69	SiC	4.35
CuCl	5.40	ZnS	5.41
CuF	4.26	ZnSe	5.67
CuI	6.04	ZnTe	6.09
GaAs	5.65		

A7.5   Crystals with the Wurtzite Structure

Crystal	a_0 (Å)	c_0 (Å)
AgI	4.58	7.49
AlN	3.11	4.98
BeO	2.70	4.38
CdS	4.14	6.75
CdSe	4.30	7.02
NH_4F	4.39	7.02
SiC	3.08	5.05
TaN	3.05	4.94
ZnO	3.25	5.21
ZnS	3.81	6.23
ZnSe	3.98	6.53
ZnTe	4.27	6.99

A7.6   Crystals with the Nickel Arsenide Structure

Crystal	a_0 (Å)	c_0 (Å)	Crystal	a_0 (Å)	c_0 (Å)
CoS	3.37	5.16	MnTe	4.14	6.70
CoSb	3.87	5.19	NiAs	3.60	5.01
CoSe	3.63	5.30	NiSb	3.94	5.14
CoTe	3.89	5.36	NiSe	3.66	5.36
CrS	3.45	5.75	NiSn	4.05	5.12
CrSb	4.11	5.44	NiTe	3.96	5.35
CrSe	3.68	6.02	PtB^a	3.36	4.06
FeS	3.44	5.88	PtBi	4.31	5.49
FeSb	4.06	5.13	PtSb	4.13	5.47
FeSe	3.64	5.96	PtSn	4.10	5.43
FeTe	3.80	5.65	VP	3.18	6.22
MnAs	3.71	5.69	VS	3.36	5.81
MnBi	4.30	6.12	VSe	3.58	5.98
MnSb	4.12	5.78	VTe	3.94	6.13

[a] Anti-nickel arsenide structure.

A7.7 Crystals with the Fluorite Structure

Crystal	a_0 (Å)	Crystal	a_0 (Å)
$BaCl_2$	7.34	Li_2S	5.71
BaF_2	6.20	Li_2Se	6.01
Be_2B	4.67	Li_2Te	6.50
Be_2C	4.33	Na_2O	5.55
CaF_2	5.46	Na_2S	6.53
CdF_2	5.39	Na_2Se	6.81
CeH_2	5.59	Na_2Te	7.31
CeO_2	5.41	NbH_2	4.56
$CoSi_2$	5.36	$NiSi_2$	5.39
HfO_2	5.12	Rb_2O	6.74
HgF_2	5.54	Rb_2S	7.65
K_2O	6.44	$SrCl_2$	6.98
K_2S	7.39	SrF_2	5.80
K_2Se	7.68	ThO_2	5.60
K_2Te	8.15	UO_2	5.47
Li_2O	4.62	ZrO_2	5.07

A7.8 Crystals with the Rutile Structure

Crystal	a_0 (Å)	c_0 (Å)	Crystal	a_0 (Å)	c_0 (Å)
CoF_2	4.70	3.18	NiF_2	4.65	3.08
FeF_2	4.70	3.31	PbO_2	4.95	3.38
GeO_2	4.40	2.86	SnO_2	4.74	3.19
MgF_2	4.62	3.05	TaO_2	4.71	3.06
MnF_2	4.87	3.31	TiO_2	4.59	2.96
MnO_2	4.40	2.87	WO_2	4.86	2.77
MoO_2	4.86	2.79	ZnF_2	4.70	3.13
NbO_2	4.77	2.96			

Appendix 8

Further Resources

A8.1   Useful Web Sites

Useful databases pertinent to crystal structures are listed at the end of Chapter 3. For basic, rather than in-depth, introductions to the topics discussed in this book, Wikipedia, www.wikipedia.org, is a good resource; Google or other search engines will also produce results rapidly.

In addition, the following Web sites contain a lot of helpful material, but for many there may be a charge for accessing the material. In addition, web sites by their very nature can be ephemeral – while some Web sites are maintained actively, others exist for finite periods of time, for example as long as research funding permits, or change their Web site name, such as Web of Science.

A8.2   Educational and Information Resources

www.doitpoms.ac.uk – the Web site for dissemination of information technology for the promotion of materials science, based at the University of Cambridge. This has a number of easy-to-use open-access teaching and learning packages which are relevant to crystallography and crystal defects.

www.iucr.org – the Web site of the International Union of Crystallography.

www.materials.ac.uk – maintained at the University of Liverpool, this Web site is a provider of computer-based learning software for materials science, engineering and related disciplines.

Crystallography and Crystal Defects, Third Edition. Anthony Kelly and Kevin M. Knowles.
© 2020 John Wiley & Sons Ltd. Published 2020 by John Wiley & Sons Ltd.
Companion website: www.wiley.com/go/kelly/crystallography3e

For more specialized information, a good database for searching the scientific literature is the Web of Science at www.wokinfo.com. Scopus, www.scopus.com, and Google Scholar, scholar.google.com, are also useful for such purposes.

A8.3   Computer Software Packages

www.jcrystal.com – this has software packages helpful for visualizing crystal shapes and nanogeometries, as well as a software package for plotting stereographic projections.

www.crystalmaker.com – this has software packages helpful for visualizing crystal structures and for plotting stereographic projections.

zig.onera.fr/mm_home_page/index.html – this is the home page of the 'mM' (microMegas) software. This is an open-source program which simulates dislocation dynamics.

In addition to these software package providers, there are many other software packages developed for molecular dynamics simulations and used by academic researchers. One example is CASTEP, www.castep.org, which uses density functional theory to model atomic systems.

Brief Solutions to Selected Problems[1]

Chapter 1

1.2 (b) Two, (c) six: four Ti–O distances of $\sqrt{2a^2\left(\frac{1}{2}-u\right)^2+c^2/4} \approx 1.94$ Å and two Ti–O distances of $\sqrt{2}\,ua \approx 2$ Å.

1.3 (a) I, (b) P, (c) I.

1.4 Orthorhombic F. The sides of the orthorhombic cell are of length $2a\cos(\beta/2)$, b and $2a\sin(\beta/2)$.

1.5 (a) $[1\bar{1}0]$; $(111),(11\bar{1}),(\bar{1}\bar{1}\bar{1}),(\bar{1}\bar{1}1)$, (c) $a/\sqrt{2}$, (d) $70.53°$ or $109.47°$ $(=180°-70.53°)$.

1.6 (a) $0°$, (b) $40.2°$.

1.7 (a) $[2\bar{1}\bar{1}] = [\bar{2}11]$, (b) $35.26°$.

1.8 (a) $c/a = 0.985$, (b) $31.42°$.

1.9 See Table 1.2.

1.11 See diagrams in Figure 2.24.

1.12 See diagrams in Figures 1.28 and A4.2.

1.14 See diagrams in Figure 2.24.

1.15 $60°$, $\frac{2}{a\sqrt{3}}$, $\frac{2}{a\sqrt{3}}$, $1/c$.

Chapter 2

2.3 Trigonal, orthorhombic, monoclinic, tetragonal.

2.5 (a) $\bar{3}m$, 32; $\bar{3}$. (b) $\bar{4}$, $4/m$, 422, $\bar{4}2m$, $4/mmm$, (c) (i) $m\bar{3}$, 432 and $m\bar{3}m$, (ii) 23 and $\bar{4}3m$, (iii) $m\bar{3}$, 432 and $m\bar{3}m$.

2.6 (302), $72.08°$; $[2\bar{2}3]$, $90°$.

2.7 (a) $32.6°$, $(1\bar{2}11)$, (b) $(21\bar{3}4)$.

2.8 (122).

2.9 $a{:}b{:}c = 0.690 : 1 : 0.412$; $\beta = 99.3°$.

2.10 $c/a = 1.141$, $[010]$, $[\bar{1}2\bar{1}0]$.

---

[1] Detailed worked solutions for all the problems in this book are available on the Wiley Web page accompanying it at http://booksupport.wiley.com

---

Crystallography and Crystal Defects, Third Edition. Anthony Kelly and Kevin M. Knowles.
© 2020 John Wiley & Sons Ltd. Published 2020 by John Wiley & Sons Ltd.
Companion website: www.wiley.com/go/kelly/crystallography3e

2.11 (b) 62.25°, 39.05°, (c) all are special apart from {111} and {113}.

2.12 (b) One choice of matrix is $\begin{pmatrix} 0 & -\frac{1}{2} & \frac{1}{2} \\ \frac{1}{2} & 0 & -\frac{1}{2} \\ 1 & 1 & 1 \end{pmatrix}$.

(c) The hexagonal cell has 0.75 of the volume of the cubic cell.

(d) $(1\bar{1}08)$, $(01\bar{1}2)$, $(11\bar{2}0)$.

Chapter 3

3.2 (a) $\bar{4}3m$, (b) $m\bar{3}m$, (c) $\bar{4}3m$, (d) $3m$, (e) $\bar{3}m$, (f) $\bar{6}m2$.

3.3 (a) $a/\sqrt{3}$, (b) $a/(2\sqrt{3})$.

3.7 (b) ... $\beta A\beta\alpha B\alpha\beta A\beta\alpha B\alpha$ ..., (c) $\left(\frac{1}{3}, \frac{1}{6}, \frac{1}{4}\right)$, (d) Mo at $\pm\left(\frac{1}{3}, \frac{1}{6}, \frac{1}{4}\right)$, S at $\pm\left(\frac{1}{3}, \frac{1}{6}, -\frac{1}{4} + z\right)$ and $\pm\left(\frac{1}{3}, \frac{1}{6}, -\frac{1}{4} - z\right)$.

3.8 (a) One, (b) 2.94 Å, 3.26 Å, 2.52 Å, 2.55 Å, (c) 6.47 mg m^{-3}.

3.9 Interstitial.

3.10 4.54×10^{27} iron vacancies per m^3.

3.11 $p6m$ (No. 17). See Figure 2.24.

Chapter 4

4.1 Rhombic dodecahedron.

4.3 Two tetrahedra sharing a common face would produce such a shape: a triangular bipyramid.

4.7 The MoAl$_{12}$ icosahedron has $m\bar{3}\bar{5}$ noncrystallographic point group symmetry. In MoAl$_{12}$ the icosahedra pack together in a cubic I arrangement; the space group of the crystal structure is $Im\bar{3}$.

Chapter 5

5.1 D_{12} is the flux of atoms in direction 1 when a unit (negative) concentration gradient is imposed along axis 2.

5.2 Although there are nine components of the array a_{ij}, the two subscripts each refer to a different coordinate system – the 'i' suffix refers to the new coordinate system and the 'j' suffix refers to the old coordinate system. For a tensor T_{ij} of the second rank, the suffices 'i' and 'j' refer to the *same* coordinate system. Only linear operators that transform one vector to another in the same vector space (coordinate system) are second-order tensors.

5.3 $\begin{pmatrix} 12 & 5 & 0 \\ 5 & 7 & 0 \\ 0 & 0 & 3 \end{pmatrix} + \begin{pmatrix} 0 & 1 & 0 \\ -1 & 0 & 0 \\ 0 & 0 & 0 \end{pmatrix}$.

5.5 $\begin{pmatrix} S_{11} & S_{12} & 0 \\ -S_{12} & S_{11} & 0 \\ 0 & 0 & S_{33} \end{pmatrix}$; S_{12} has to be zero.

5.7 (a) $\begin{pmatrix} 1.78 & 0 & 0 \\ 0 & 5.35 & 0 \\ 0 & 0 & 4 \end{pmatrix}$, (b) $3.1 \times 10^7 \Omega^{-1} m^{-1}$.

5.8 (a) $\begin{array}{c} \\ Ox_1' \\ Ox_2' \\ Ox_3' \end{array} \begin{array}{ccc} Ox_1 & Ox_2 & Ox_3 \end{array} \begin{pmatrix} \frac{1}{2} & -\frac{\sqrt{3}}{2} & 0 \\ \frac{\sqrt{3}}{2} & \frac{1}{2} & 0 \\ 0 & 0 & 1 \end{pmatrix}$, (b) $\begin{pmatrix} 25 & 0 & 0 \\ 0 & 16 & 0 \\ 0 & 0 & 9 \end{pmatrix} \times 10^8 \Omega^{-1} m^{-1}$, (d) At 30° to x_1',

$22.75 \times 10^7 \Omega^{-1} m^{-1}$; at 60° to x_1', $18.25 \times 10^7 \Omega^{-1} m^{-1}$, (f) At 48° to x_1'.

Chapter 6

6.2 $\begin{pmatrix} e & 0 & 0 \\ 0 & -e & 0 \\ 0 & 0 & 0 \end{pmatrix}$; $\begin{pmatrix} 0 & -e & 0 \\ -e & 0 & 0 \\ 0 & 0 & 0 \end{pmatrix}$.

6.3 $\begin{pmatrix} e & 0 & 0 \\ 0 & -ve & 0 \\ 0 & 0 & -ve \end{pmatrix}$; $\begin{pmatrix} \frac{e}{2}(1-v) & -\frac{e}{2}(1+v) & 0 \\ -\frac{e}{2}(1+v) & \frac{e}{2}(1-v) & 0 \\ 0 & 0 & -ve \end{pmatrix}$.

6.4 $\sigma/2$

6.8 The principal stresses are 100 MPa, 200 MPa, 400 MPa along [010], [$\bar{1}$01] and [101], respectively. The maximum shear stress is 300 MPa.

6.12 $\left(\frac{1}{4}, \frac{1}{4}, \frac{1}{4}\right)$. All the zinc atoms move in the [00$\bar{1}$] direction, inducing electrical polarization. Diamond is centrosymmetric – piezoelectricity cannot occur in diamond.

6.18 (a) $\frac{1}{s_{11}}$, (b) $\frac{s_{11}+s_{12}}{s_{11}^2+s_{11}s_{12}-2s_{12}^2}$, (c) $\frac{s_{11}}{s_{11}^2+s_{11}s_{12}-2s_{12}^2}$, (d) $\frac{1}{s_{11}+2s_{12}}$.

Chapter 7

7.1 (a) No, (b) yes.

7.3 12 (four distinct {111} planes, each of which contains three distinct $\langle 1\bar{1}0 \rangle$ directions), (a) $\langle 100 \rangle$, (b) $\langle 111 \rangle$, (c) $\langle 101 \rangle$, $\phi = \lambda = 45°$.

7.4 3; 3.

7.5 (a) $\frac{1}{2}\begin{pmatrix} 0 & 0 & 0 \\ 0 & \gamma & 0 \\ 0 & 0 & -\gamma \end{pmatrix}$,   (b) $\frac{\gamma'}{2\sqrt{2}}\begin{pmatrix} 0 & 1 & -1 \\ 1 & 0 & 0 \\ -1 & 0 & 0 \end{pmatrix}$,   (c) 2,   (d) 1.

7.6 [100], [010] and [001] in a crystal belonging to one of the orthogonal systems (cubic, tetragonal, orthogonal).

7.7 (a) $[101](\bar{1}11)$, (b) $[101](\bar{1}11)$ and $[011](1\bar{1}1)$, (c) $[101](\bar{1}11)$ and $[110](\bar{1}11)$.

7.8 (a) $[\bar{1}01](101)$ and $[101](\bar{1}01)$, (b) $[0\bar{1}1](011)$ and $[011](0\bar{1}1)$, (c) $[\bar{1}01](101)$ and $[101](\bar{1}01)$.

7.9 (a) $\langle 111 \rangle$, (b) no, (c) $\langle 0001 \rangle$ and $\langle uvt0 \rangle$.

7.11 (a) $\begin{pmatrix} \sigma & \tau & 0 \\ \tau & 0 & 0 \\ 0 & 0 & 0 \end{pmatrix}$,   (b) $\sigma/2$,   (c) $\left(\frac{\sigma}{2} + \tau \right)$.

7.13 (a) $[\bar{1}2\bar{1}0]$, (b) 0.83 MPa.

7.14 (a) $[\bar{1}01](111)$, (b) $[\bar{1}01](111)$ and $[011](\bar{1}\bar{1}1)$, (c) $[\bar{1}12]$, (d) 13.1 cm.

Chapter 8

8.1 The material inside the cylindrical surface is extruded out from the rest of the crystal.

8.3 2.82×10^{-16} m^3.

8.4 (a) $2 \times 10^{-10} \sigma$ N m^{-1}, (b) zero.

8.5 (a) $(1\bar{1}0)$, (b) (001).

8.6 Couple $= \frac{\mu b}{2}(R^2 - r_0^2)$.

8.7 $\varepsilon_{11} = \varepsilon_{22} = \frac{-b}{2\pi x_2}\left(\frac{\mu}{\lambda + 2\mu} \right)$; dilatation $= \pm 0.0153$.

8.10 Width $= \frac{a}{2A^{1/2}}(a = $ lattice parameter).

8.11 $\frac{1}{l}\frac{dl}{dt} = \frac{1}{2}\rho b \bar{v}$.

8.13 Force $= \frac{\mu b_1 b_2}{2}$.

8.15 (a) 4, (b) 3, (c) 3, (d) 4.

Chapter 9

9.1 $L/10$.

9.2 σbl, μb^2.

9.4 Jointed extrinsic fault.

9.5 Two.

9.6 (a) 1.6×10^{15} m^{-2}, (b) 1.6×10^{14} m^{-2}.

9.7 26.7 Å, 3.1 Å, 12.3 Å.

9.8 Simple cubic: $\langle 110 \rangle$, $\langle 111 \rangle$; b.c.c.: none; c.c.p.: $\langle 100 \rangle$; hexagonal: $\frac{1}{3}\langle 11\bar{2}3 \rangle$.

9.13 Identical $\frac{1}{2}\langle 111 \rangle$ dislocations will form pairs.

Chapter 10

10.1 290.

10.2 1.25 eV, 2.1k in units of J K^{-1}.

10.3 The concentration in the impure crystal is 300 times that in the pure crystal.

10.4 0.06.

10.5 1.08 eV; 0.02 eV, so that the true $E_f = 1.10$ eV.

10.6 24.

10.7 1475; 4.

10.8 2.

10.10 Fractional increase $= 1.3 \times 10^{-4}$.

10.11 (a) 4mm, (b) $\bar{3}m$; (a) Tensile stress along $\langle 111 \rangle$, (b) along $\langle 100 \rangle$.

10.12 (a) 48, (b) 16, the number of orientations equals the ratio of the multiplicities.

10.13 (a) 8, (b) the group $[111], [1\bar{1}1], [11\bar{1}], [1\bar{1}\,\bar{1}]$ can be distinguished from the group $[\bar{1}11], [\bar{1}1\bar{1}], [\bar{1}\,\bar{1}\,\bar{1}], [\bar{1}\,\bar{1}1]$, (c) the four lying in $(1\bar{1}0)$ can be distinguished from the remaining four.

Chapter 11

11.1 $\begin{pmatrix} \bar{1} & 1 & 1 \\ 1 & \bar{1} & 1 \\ 2 & 2 & 0 \end{pmatrix}$.

11.2 70.53°.

11.5 (110), $(1\bar{1}2)$ and $(\bar{1}12)$, $\begin{pmatrix} \bar{1} & 1 & 1 \\ 1 & \bar{1} & 1 \\ 2 & 2 & 0 \end{pmatrix}$, (011), referring to axes reflected in K_1.

11.7 $p_1 = N(100) * \cdot (h_1 k_1 l_1) * d^2_{h_1 k_1 l_1} - p_2$.

   $q_1 = N(010) * \cdot (h_1 k_1 l_1) * d^2_{h_1 k_1 l_1} - q_2$.

   $r_1 = N(001) * \cdot (h_1 k_1 l_1) * d^2_{h_1 k_1 l_1} - r_2$.

11.9 $[110], [1\bar{1}1]$ and $[\bar{1}11]$, $\begin{pmatrix} \bar{1} & 1 & 2 \\ 1 & \bar{1} & 2 \\ 1 & 1 & 0 \end{pmatrix}$, [010], referring to axes reflected in K_1.

11.11 $\tan^{-1} \frac{c}{\sqrt{3}a}$; elongation for metals with $c/a < \sqrt{3}$, compression for metals with $c/a > \sqrt{3}$; for practical values of c/a seen in h.c.p. metals ($1.5 < c/a < 1.9$), elongation would be possible if the only twinning mode were $(11\bar{2}1)$ and compression would be possible if the only twinning mode were $(11\bar{2}2)$.

11.12 Yes.

11.14 4°.

11.15 10%.

Chapter 12

12.1 4, 2 (twins). Component of **b** normal to (111) must equal $2a/\sqrt{3}$.

12.3 12 $\langle 110 \rangle$ vectors. Yes. $\begin{pmatrix} 1 & 0 & 0 \\ -\frac{1}{2} & \frac{1}{2} & -\frac{1}{2} \\ 0 & 1 & 1 \end{pmatrix}$.

12.5 1.57.

12.6 0.18°, 0.003.

12.7 2, 4, 6.

12.10 $\begin{pmatrix} 1 & 0 & 0 \\ 0 & \frac{1}{2} & -\frac{1}{2} \\ 0 & \frac{1}{2} & \frac{1}{2} \end{pmatrix}$.

Chapter 13

13.2 Step heights are 1, 3, 1 and 2 lattice plane spacings, respectively.

13.3 $\gamma_{(0\,0\,1)} = 0.894\ \gamma_{(2\,1\,0)}$, $\gamma_{(1\,1\,1)} = 0.775\ \gamma_{(2\,1\,0)}$, $\gamma_{(1\,1\,0)} = 0.949\ \gamma_{(2\,1\,0)}$ (as in Figure 13.5).

13.4 $1.16\,\mathrm{J\,m^{-2}}$, 15%.

13.5 (a) Attract, (b) energy increases as $b^2 \ln b$.

13.6 Boundary plane: (a) {110}, (b) {110}, (c) {11$\bar{2}$0}. Axis of tilt: (a) $\langle 001 \rangle$, (b) $\langle 211 \rangle$, (c) $\langle \bar{1}100 \rangle$. Angle of tilt: (a) $2\sin^{-1}\frac{a}{2\sqrt{2}d}$, (b) $2\sin^{-1}\frac{a}{2\sqrt{2}d}$, (c) $2\sin^{-1}\frac{a}{2d}$.

13.7 (a) (1$\bar{1}$0), (b) $3.1 \times 10^7\,\mathrm{m^{-1}}$, (d) about 8°.

13.8 See Freise, E.J. and Kelly, A. (1961). Twinning in graphite. *Proc. Roy. Soc. Lond. A* 264: 269–276.

13.10 70.53° about the vector [$\bar{1}$01]. No.

13.11 $\mathbf{R}_2 = \frac{1}{3}\begin{pmatrix} -1 & 2 & 2 \\ 2 & -1 & 2 \\ 2 & 2 & -1 \end{pmatrix}$. $\mathbf{R}_3 = \begin{pmatrix} 0 & 0 & 1 \\ 0 & -1 & 0 \\ 1 & 0 & 0 \end{pmatrix}$; this represents a rotation of 180° about [101] or a rotation of 180° about [$\bar{1}0\bar{1}$], i.e. a symmetry operation of a holosymmetric cube. The disorientation description of $\mathbf{R}_2$ is a rotation of 60° about [111].

13.12 **p** describes a rotation angle of 180° about the vector [117]; **q** describes a rotation of 22.84° about [$\bar{1}$10]. **p.q** describes a rotation of 180° about [001], i.e. a symmetry operation of a holosymmetric cubic material. **q. p** describes a rotation of 180° about the [14, 14, 47] axis.

13.13 $\mathbf{R} = \frac{1}{9}\begin{pmatrix} 2 & 2 & 1 \\ -1 & 2 & -2 \\ -2 & 1 & 2 \end{pmatrix}\begin{pmatrix} 2 & 2 & -1 \\ -1 & 2 & 2 \\ 2 & -1 & 2 \end{pmatrix} = \frac{1}{9}\begin{pmatrix} 4 & 7 & 4 \\ -8 & 4 & 1 \\ -1 & -4 & 8 \end{pmatrix}$; this describes a rotation of $a\cos^{-1}(7/18) = 67.11°$ about a unit vector parallel to [$\bar{1}1\bar{3}$].

13.15 The difference in angle in radians is of the order of the ratio of the twin boundary free energy to the surface free energy.

13.17 Effective surface free energy $= 0.96\gamma$.

Chapter 14

14.1 4%.

14.2 Orientation 1: $[\bar{1}10]_F \parallel [\bar{1}11]_B$, $[\bar{1}\bar{1}2]_F \parallel [1\bar{1}2]_B$ and $[111]_F \parallel [110]_B$.
Orientation 2: $[01\bar{1}]_F \parallel [\bar{1}1\bar{1}]_B$, $[\bar{2}11]_F \parallel [\bar{1}12]_B$ and $[111]_F \parallel [110]_B$.

14.7 Two.

Chapter 15

15.1 (a) $\mathbf{R}_1 = \begin{pmatrix} \cos\theta & -\sin\theta & 0 \\ \sin\theta & \cos\theta & 0 \\ 0 & 0 & 1 \end{pmatrix}$. (b) $\mathbf{R}_2 = \begin{pmatrix} \cos\theta & \sin\theta & 0 \\ -\sin\theta & \cos\theta & 0 \\ 0 & 0 & 1 \end{pmatrix}$; $\mathbf{R}_2$ is both the inverse and the transpose of $\mathbf{R}_1$ in the special case considered in (a) and (b) and also in the general case.

15.3 $(001)_{Ag} \parallel (001)_{Substrate}$ and $[100]_{Ag} \parallel [110]_{Substrate}$.

Index

Note: Page numbers in *italics* refer to Figures; those in **bold** to Tables

aggregation of point defects, 313, 315
albite twinning, 365
alkali halides, point defects, 313–14, 319, 323
alkali metals, crystal structure, 96, 379
allotropy (elements), 90, 114
alloys, metallic
   amorphous (noncrystalline), 130
   phase transformations, 374, 385–92
   quasicrystalline phases (i-phases), 142–3
   strained layer epitaxy, 456
   topological close-packing (TCP), 141
   twinning frequency, 346–7, 357
aluminium
   dislocation strain energies, 273
   elastic isotropy, 258
   vacancy generation, with heating, 322, *322*
amorphous materials, 130–1
anelastic strain, 330–1, 333
anisotropy
   Hooke's law application, dislocation strains, 258–60
   of mesogenic molecules (liquid crystals), 132
   and polycrystalline textures, 464
annealing, 327, *327*, 329
   twins, 340
antimony, crystal structure, 101–2, **102**, *102*

antiphase domain boundaries, *116*, 117
antisymmetric tensors, 156–7
arsenic, crystal structure, 101–2, **102**, *102*
atactic polymers, 119
atomic radii, 114
atomistic modelling of twin boundaries, 365–7
austenite, 373, 389–91
axes of rotational symmetry, 16, 43, *44*
   permissible combinations, 21–5, **25**
   pure (proper rotation), 43–4, *44*
axial angles, 6, *7*, 9, 67
axial glide planes, 71
axial *(c/a)* ratios, 52
   hexagonal close-packed metals, 94–5, **95**
      and slip planes, 215, **218**, 292
   and martensitic transformation, 385
   wurtzite structure compounds, 104, **105**

Bain correspondence, 388–91, *389*, 392
Bain orientation relationship, 451, 452
barium titanate ($BaTiO_3$)
   crystal structure, 109
   phase transitions and ferroelectric properties, 394
b.c.c. metals, interphase boundaries, 450–6
binding energy, divacancies, 314
bismuth, crystal structure, 101–2, **102**, *102*
body-centred Bravais lattices, 451

Crystallography and Crystal Defects, Third Edition. Anthony Kelly and Kevin M. Knowles.
© 2020 John Wiley & Sons Ltd. Published 2020 by John Wiley & Sons Ltd.
Companion website: www.wiley.com/go/kelly/crystallography3e

body-centred cubic (b.c.c.) structure
  dislocation geometry, 296–7, *297*
  lattice symmetry and unit cells, 36, *36,
    96–7, 96–7*
  point defects
    interstitials, *318,* 318–19
    vacancies, **317,** 319, *320*
  transformation to h.c.p. structure, 379,
    380, 384, 385, 395–6
  twinning (metals), 346–7, *347*
Boltzmann's constant, 255, 309
Born–Meyer potential, 316
boron nitride, hexagonal form, 101, *101*
brass, texture, 467, 469, *469*
Bravais (space) lattices, 26–37, 70, 72
brittleness, 209, 412
bulk metallic glasses (BMGs), 130
Burgers circuit, 247, 248, *248,* 249
Burgers vector
  general description methods, 247–9, *248*
  related to strain energy, 271–5
  as slip plane displacement, 243–8
  stability of dislocations, 274, **275**

caesium chloride
  crystal structure, 103, *104,* **542**
  unit cell, 6, *7*
calcite
  crystal structure, *113,* 113–14
  deformation twinning, 349, *349,* 358
calcium fluoride (fluorite), crystal structure,
    106–7, *107*
c/a ratios *see* axial *(c/a)* ratios
carbon
  iron impurity atomic arrangement, 330,
    330, 333
  in steel, and martensitic transformation,
    373, 390, *390*
  see also diamond; graphite
cassiterite structure *see* rutile crystal
    structure
cellular solid materials, 144–5
centred rectangular cells, *19,* 21
centre of symmetry, 16, 45, 161, 183, *183*
charged dislocations, *290,* 291
chemical force, on dislocations, 254–5

cholesteric phases (liquid crystals), 133,
    134, 135
chrysoberyl crystal structure, 111
cleavage planes, 296, 412, *413*
climb (dislocation motion), *246,* 250, 254–5
closed forms, 49–50
close packing (of spheres), 90–3, *91, 93*
  Bernal's RCPS model, 130, 131
  interstices, 92–3, *94,* 108
  plane stacking, 91–2, *92*
  stacking faults, 279–81, *280,* 282
    Frank notation, 282
  topological (TCP) structures, icosahedral
    packing, 141
cobalt
  martensitic transformation, 376–9, *378,*
    446
  stacking fault energy, 292
coherency strain, 447, 459
coincidence site lattices (CSL), 421, **422,**
    422
compliances (elastic constants), 188–90,
    195, 196
composition plane, K_1 (twinning), 339, 341,
    343, 352–7
compounds, simple inorganic (*MX* and
    MX_2), 102–8, **540–44**
compound (degenerate) twins, 344
compression, and crystal lattice
    deformation, 235–7, *236*
conductivity (electrical)
  direction, referred to tensor principal
    axes, 157–60, *159*
  measurement, and point defect mobility,
    323–5, *324, 325*
  tensor components, 149–50, 152
constancy of angle, law of, 8, *8*
converse effect (piezoelectricity), 187
coordination number, 4, 90, 114
copper
  alloys, martensitic transformation, 392
  point defects, mobility, 312–13
  rolled, texture, 464, 467, *467*
correspondence matrix, 380, 389
corundum *see* sapphire
covalent bonding

dislocation geometry, 297–301, *301*
dislocation width (Peierls model), 263
in polymers, 119
critical resolved shear stress, law of, 228
cross-slip (dislocations), 264, *264*
crowdions (Paneth), 318–19, *319,* 321
crystallographic glide strain
definition, 212, *213*
in displacement analysis, 219
and lattice orientation, in
tension/compression, 231, 233,
235
crystals, elasticity
independent elastic constants, 193–8
isotropic media, 194, 196, 199
crystal structure
descriptive methods, 89–90, 120
growth, 437
crystal systems, 26, **26**
elastic constants, 187–94, **194–7**
enantiomorphous, 69
Internet information sources, 545–6
of named elements, **535–9**
noncentrosymmetric, piezoelectric
moduli, 184–7
second-rank tensor limitations, 162, **163,**
166–7
special forms, 68, **68**
cubic close-packed (c.c.p.) metals
epitaxial deposition of, **449**
interphase boundaries, 450–6
cubic close-packed (c.c.p.) structure, 90–3,
91–3
coincidence lattices, 421, *422*
dislocation geometry, 279–88, *280,* **289**
double slip, *234,* 234–5
and icosahedral packing, 140–2, *141,*
142
point defects
interstitials, 318, *318*
vacancies, 316–17, **317,** 320, *320,*
321
slip system stereograms, 228–9, *229,*
230–1, *231*
transformation to h.c.p. structure, 376–9,
378, 395

twinning (metals), 339–40, *340, 341,*
346
see also face-centred cubic (f.c.c.)
structure
cubic crystals, elasticity
shear stiffnesses and compliances, 200–1
transformation of the stiffness and
compliance tensors, 200
cubic crystal system
elastic constants, 192–3, **194,** 200–1
glide systems, pure strain and rotation,
221–5, *222, 224*
lattice symmetry and unit cells, *35,*
35–6, *36*
planar spacing and interplanar angles,
513–14
point group symmetry elements, *53,*
53–5
property representation quadric, 166
shear stress and piezoelectricity, 186,
186
stereograms, 55, *55,* 495–7, *496*
tetragonal defect orientation, *332,* 333
cyclotrimethylene trinitramine, 302

damping, 331
defect structures, 103
deformation twinning, 341–6, *343,* 358–61
dendrites, 437
density
dislocation, 246, 263
of solid solutions, 115
deviatoric stress, 181
devitrite, 365–6
diad (two-fold) axis of symmetry, 18, *19,*
21, 30
diagonal glide planes, 72
diamond
crystal structure, 99, *99, 100*
dislocations, 297–8, *298*
diamond glide planes, 72
Diehl's rule, 231, 527–8, 529–30
diffraction patterns
in crystal structure determination, 89,
506
Laue method, 69, **69**

diffusional phase transformations, 445
diffusion-based phase transformations, 374
dihedral angles, 445, **446,** *446*
dilatation, 175–6, *176,* 263
dipoles
    dislocation, 265, 285
    electric, single crystal (piezoelectricity),
        183, 186
directions
    closest-packed (lattice), 90–3, *91–3*
    glide (slip), *211,* 211–12, 214, 234, 275
    indices, transformation for unit cell
        changes, 517–18
    physical properties, tensor
        representation, 158–9, *159,*
        162–6
    relative displacement, in dislocations,
        249, *250*
    shear, in twinning (η), 352
    specification of, 9, *9*
dislocation glide motion, 250–1, *251*
dislocations
    Burgers vector, general description,
        247–9, *248*
    core, atomic positions, 260–3
    crystal distortion, 255–60
    forces and crystal stress, 251–5, 414–15
    geometry, for specific crystal structures,
        303
        body-centred cubic (b.c.c.) crystals,
            296–7, *297*
        covalent solids, 297–301
        cubic close-packed (c.c.p.) metals,
            281–8, **289**
        hexagonal metals, 291–6, *292*
        rock salt, 288–91, *290*
    glide mechanisms, 214, 243–7
    interactions and multiplication, 263–7,
        264, 274
    loop motion, 249–51, *251*
    low symmetry crystal structures, 302–3
    partial, and stacking faults, 279–81, *280*
    pile-up, in deformation twinning, 359,
        360
    strain energy and stability, 271–9, **275**
dislocation spacing, 447

disorientation descriptions, 426
displacive phase transitions, 394
domains, in ordered crystals, *116,* 117
double hexagonal close-packed structure, 96
double (duplex) slip, 74, 234, 237
ductility, 209, 463
dummy suffix notation (tensor
        components), 153–4
dyadic operators *see* second-rank tensors

edge dislocations
    atomic positions, 246, *246*
        bonding and Peierls force, 263
    and Burgers vector, 247, *248*
    description, for cubic lattice, 245, *245*
    distortion stresses and strains, 257–8
        strain energy calculation, 272–3
    electric charge (NaCl), *290,* 291
    parallel, 264–5, *265*
    stair rods, 287, *288,* 299
    stress and force components, 253–5, *254*
Einstein summation convention, 154, 479
elastic constants
    in crystal classes, 187–94, **194–7**
    of cubic crystals, 193–4
    of hexagonal crystals, 194, 197
elasticity
    crystal deformation, compared with
        plastic glide, 210
    of crystals
        independent elastic constants, 193–8
        isotropic media, 194, 196, 199
    of cubic crystals
        shear stiffness and compliances,
            200–1
        transformation of the stiffness and
            compliance tensors, 200
    dislocation elastic energy, 275, 279
    elastic constants
        in crystal classes, 187–94, **194–7**
        in isotropic medium, 194, 199–200
    in isotropic materials, 194, 199–200, 258
    strain and stress relationships, 187–91
elasticity theory, 448
elastic strain energy, 458, 459
elastic twinning, 359

electrical conductivity *see* conductivity
electric fields *see* space charge regions
electron back-scattered diffraction (EBSD),
    428, 471
electron irradiation, 328, *329*
ellipsoids
    as effect of homogeneous strain on
        spheres, 381, *381,* 532
    physical property representation, 164–7,
        165, 167
emissary dislocations, 359, *360*
enantiomorphs
    crystal classes, 69
    related to symmetry, 43
engineering shear strain, 171, *172,* 176
entropy, related to point defects, 309–10,
    313
epitaxial growth, 449, **449,** *450*
equilibrium point defect concentration, 309,
    311, 312, **323**
Euler angles, *469,* 469–71
Euler's theorem, related to regular
    polyhedra, 137, 139
extension, 171, *172,* 532
extrinsic (double) stacking faults, *280,* 282,
    285
extrusion (of dislocations), *277,* 277–9, *278*

face-centred Bravais lattices, 451
face-centred cubic (f.c.c.) structure
    Bravais lattice, 36, *36,* 90
    transformation to face-centred tetragonal
        (f.c.t.), 385–6
F-centres (anion vacancies), 314
ferroelectric materials, 339, 394
ferrous oxide, point defects, 314
fibre composite materials, 145
field tensors, 150
fluorite crystal structure, 106–7, *107,* **544**
foams, *144,* 144–5
forces, on dislocations
    climb and glide components, *254,*
        254–5, 265–6
    Peierls model, 262–3
    work and energy relationships, 251–3,
        412–13

Frank partials, 282–4, 287, *288*
Frank–Read mechanism, dislocation source,
    264, 277
Frank's rule, *274,* 274–5, 287
Frenkel defects, 310, *329*
Friedel's classification of twinning, 361–2
    by merohedry, 362–4
    by pseudo-merohedry, 365
    by reticular merohedry, 364
    by reticular pseudomerohedry, 365
FS/RH convention (Burgers vector), 248,
    248, 249, 285
functional fatigue, 399

garnets, crystal structure, 111–13, *112*
glasses, 129, 130, 199
glide (slip)
    conditions and process, 209–10, *210*
    displacement tensor components
        analysis as pure strain and rotation,
            221–2, *222*
        for different slip systems, 221–4, *222*
        in three dimensions, 219–21, *220*
        in two dimensions, 218–19, *219*
    fine structure, 210–12, *212*
    independent slip systems, 218–27, **225,**
        525–7
    large strains in single crystals
        glide system choice, 227–31
        lattice orientation changes, 231–7
    measurement of glide strain, 212, *213*
    plane symmetry elements, in space
        groups, 74
    polycrystalline grain deformation, 227
    system elements, *213,* 213–15, **216–17**
        for named crystal types, 214
    see also dislocations
glide cylinder (dislocation motion), 250,
    251
glide planes, 71–2
grain boundaries
    energy related to tilt angle, 414–16, *415*
    good fit and coincidence lattices, 420–3,
        421, **422**
    high-angle, geometry of, 425–30
    interface junctions, 429–34

grain boundaries (*Continued*)
   source lattice (dislocation array), 424, *424*
   see also phase boundaries; tilt boundaries
grains
   lattice rotation, 227
   polycrystal arrangement, froth analogy, 437–9, *439*
   preferred lattice orientation (texture), 463–4
   role in structure of crystalline solids, 413
graphene, unit cell definition, 3–7, *4*
graphite
   basal plane dislocation, *300,* 300–1, *301*
   basal plane dislocations, 351
   crystal structures, 99–101, *101*
   twinning, 351–2, *352, 355*
   vacancies, 321
grossular (natural garnet), 112
group theory, 80

habit plane, 375, 381–3, *383*
   in steels, 391, *391*
   variants, 397
halite *see* sodium chloride
H-centres (anion interstitial defects), 319, *319*
Heidenreich–Shockley partials *see* Shockley partials
Hermann–Mauguin notation, 79–81
Herring's construction, crystal shape, 434–6, *435*
hexad (six-fold) axis of symmetry, 18, *19–20*
hexagonal close-packed (h.c.p.) structure, 91, 93–6, *94, 95, 95, 96, 96*
   second-order pyramidal glide, 293, *294*
   at twin boundary, 340
hexagonal crystal system, 32, 34, *34, 35*
   dislocation geometry, 291–6, *292*
   elastic constants, 194, **197**
   orthohexagonal cell indices, *520,* 520–1
   planar spacing and interplanar angles, 513

point group symmetry elements, 55–9, *56*
   twinning (metals), *347,* 349–51
hexahydro-1,3,5-trinitro-1,3,5-triazinane, 302
high-angle grain boundaries, geometry of, 425–30
Hirth lock configuration, 288, *288*
holosymmetric class (of crystal systems), 50
homogeneous strain, 172, 210, 218, 332, 486
   mathematical properties, 531–2
   see also extension; shear
Hooke's law, 187, 487
   anisotropic media, calculations, 258–60
   dislocation stresses and strains, 256, 257
   for isotropic solids, 199–200
hydrostatic pressure, 181, 209

icosahedral packing, 140–2, *141, 142*
ilmenite (FeTiO$_3$), crystal structure, 110
improper rotations, *44*
impurities, and point defects, 309, 314, 326, 330
incommensurate structures, 143–4
indium
   crystal structure, 97–8
   thallium (In–Tl) alloys, martensitic transformation, 385–8, *386*
inert gases
   crystal structure, 90
   secondary bonding, 119
infinitesimal strain, 172
   in three dimensions, 174–6, *176*
   in two dimensions, 172–4, *173–5*
inorganic crystals and minerals, 83
in-plane stress-strain relationship, 457, 458
interatomic distances (in crystals), 114, **535–9**
interface junctions, 429–34, *430, 432*
interfacial misfit dislocations, 456
internal friction, 331
International Tables for Crystallography
   arithmetic crystal classes and space groups, 73–5, *76, 77*

entry example, 3-D space group, 80–3, *81–2*
conventional symbols, 48
interphase boundaries
  boundaries between different phases, 445–50
  between c.c.p. and b.c.c. phases, 450–6
  strained layer epitaxy of semiconductors, 456–60
interstices
  in amorphous solids, 130
  body-centred cubic crystals, 96–7, *96–7*
  cubic close-packed crystals, 92–3, *94*
  filled, in compounds, 102, 104, 105, 108
  hexagonal close-packed crystals, 95, *96*
interstitial defects, 309, 313
  configurations, 317–19, *318, 319*
  mobility, 321
interstitial solid solutions, 114–15, *115*
intrinsic (single) stacking faults, 282, 283, 285–7
invariant line concept, 455
invariant planes, 375–6, 381
inverse matrices, 487
inverse spinel structure, 110–11
inversion twinning, 364
ionic radii, **535–9**
i-phases, 142–3
isomorphic groups, 80
isotactic polymers, 119
isotropic crystals, 166
  elastic properties, 194, 199–200, 258
isotropic media, 194, 196, 199

jogs, 267, *267,* 289, 291

Kevlar (PPTA), 131, 133
kinks, 267, 298, *298*
Kronecker delta, 183, 193, 200, 481
Kurdjumow–Sachs orientation, 445, 451

Lamé constants, 199, 200, 258
lamellae, twinned crystals, 341, *358,* 358–9, *359,* 360–1
  twin boundary surface grooving, 432, *433*

lattice parameters, 6, 9
  derivation of interatomic distances, 114
  ratio (axial ratios), 52
  of steel (austenite/martensite) phases, 389, *390*
  and stiffness constants, 458
lattice points, 6, 9, 11
lattices
  Bravais (space), 26–37, *27,* 70, 72
  cellular material structures, 144, 145
  definition, formal, 6
  plane spacing and direction, 8–11, *8–11,* 483–4, 511–14
  vectors and lattice planes, 11–14, *12, 482,* 482–4, 519
  see also symmetry elements
Laue groups, 69, **69,** 165, 191
line tension, dislocation loops, *276,* 276–9
liquid crystals, 131–5, *133, 134*
lithium niobate (LiNbO$_3$), crystal structure, 109–10, *110*
Lomer–Cottrell lock configuration, 287, *287,* 299
long-range order (solid solutions), 117, 397
low-angle grain boundary, 447
low symmetry crystal structures, 302–3

magnesioferrite (MgFe$_2$O$_4$), 446
magnetic groups, 83–85, *84, 85*
magnetic shape memory alloys, 393
martensite (quenched steel), 373, 388–91, *391*
martensitic transformations
  in alloys, 374–83, 385–8, 392–4
  crystallographic features, 374–6, *375, 377*
  in metal crystals, 395–7
  non-linear elasticity model, 401
  in non-metals, 393–4
  nucleation and growth of plates, 395–7
  shape memory effect and superelasticity, 397–400
  topological model, 400–1

massive transformation, 374
matrices
  applications
    for martensitic lattice
      transformations, 380
    notation for piezoelectric moduli,
      184–7
    for stiffness and compliance, 190, **195**
    for twinning index transformations,
      345
  mathematical operations, 484–9
  rotation, 22, 151, 489
matter tensors, 150, **150**
mercury, crystal structure, 98, *98*
merohedry, 362–4
meshes *see* nets, two-dimensional; porous
    solid materials
mesogenic units, in polymers, 133
mesophase materials (liquid crystals),
    131–5, *133, 134*
metallic crystal structures, 90–7
microtexture, 471–2
Miller–Bravais indices, 56–7, *57,* 61
  three-to four-index transformation, 57–8,
    58
  twinning shear transformation, 345
Miller index notation, 10, *11,* 52, 482, 514
mirror plane, 16, *16,* 18, *20,* 43
  related to space groups, 79
  related to twinning, 339–40
misfits (interface strain), 446–9, *447,* 459,
    459
mixed dislocations
  description, for cubic lattice, *245,* 246
  distortion stresses and strains, 258
  strain energy calculation, 273
mobility, point defects, **312,** 312–13, 321–4
molecular beam epitaxy (MBE), 449, 456
monoclinic crystal system
  elastic constants, 191–2
  lattice symmetry and unit cells, *28,* 30
  piezoelectric moduli, 184–5
  planar spacing, 512
  point group symmetry elements, 63–5,
    64, 65
  property representation quadric, 166

twin boundary, 423, *423*
Morse function, 316
multiplicity, 49–50

nematic phases (liquid crystals), 134, *134,*
    135, *136*
nets, two-dimensional
  consistent with crystal symmetry, 18,
    19–20, 21
  lattice point arrays, 6
  rectangular, *19–20,* 21
  stacking, 27, *27,* 31, *31–3,* 33
Neumann bands, 358
Neumann's principle, 161–2, 183, 191
nickel
  arsenide (niccolite), crystal structure,
    105–6, **106,** *106,* **543**
  titanium (Ni–Ti) alloys, martensitic
    transformation, 392–3
Nishiyama–Wassermann orientation, 445,
    451, 452
  see also Pitsch orientation
noble metals, crystal structure, 90
node, dislocation lines, 249, 285, 286, *286*
noncrystallographic symmetry, 143
non-linear elasticity model, 401
nonprimitive unit cell, 21
nucleation, martensite, *395,* 395–7

octahedral interstices, 93, *94,* 95, 96, *96*
OILS rule (Hutchings), 231, 529–30
one-way shape memory effect, 397–8, *398*
open forms, 49–50
ordering, in solid solutions, 115–17, *116,*
    118
orientation distribution function (ODF),
    467, 468, *469*
(001)-oriented strained layer, 456–7
orthorhombic crystal system, 447
  elastic constants, 192
  example of space group description,
    80–3, *81–2*
  lattice symmetry and unit cells, 30–1,
    31, 33
  piezoelectric moduli, 184–5

planar spacing and interplanar angles, 512–13

point group symmetry elements, *49,* 49–52, *51*

property representation quadric, 166

PAA (*para*-azoxyanisole), 131, *133*

packing fraction, 91, 130, 131

partial dislocations, 279–81, *280,* 347

partially coherent boundary, 446

Pearson symbol (crystal structure), 123

Peierls dislocation model, 262–3

Peierls model, 448

pencil glide, 215, *215,* 225

pericline twinning, 365

perovskite crystal structure, 108–9, *109,* 140, 142–3

phase boundaries, 445–50, **446,** *446*

phase transformations, 374, 385–92

see also martensitic transformations

piezoelectricity, 182–7, 456

converse effect, 187

Pitsch orientation, 451

plagioclase feldspars, twin structures, 355

plane groups (2-D space groups), 74–5, *76–8,* 79, **79**

plastic deformation *see* glide

Platonic solids, 137, *138,* 139

point defects

aggregation, 314–15

configurations, 316–21

definition and types, 309

energy characteristics, 309–13

experimental production and studies, 321–5

in ionic crystals (alkali halides), 313–14

produced by radiation, 328–9

in quenched metals, 325–8

symmetry and anelasticity, 329–3

point groups, 43

general and special forms, 49–50

related to space groups, 70

symmetry elements, 43–8, *46–7*

terminology for cubic systems, 54

Poisson's ratio *(v),* 199, 258, 458

polarization, electrical, 182, 183, 186

pole figures, 464, 467, *467*

poles

location on stereographic projections, 50–2, *51,* 65, 493–7, *494*

construction geometry, *500,* 500–1, *501, 502*

using Wulff net, 502–3, *504, 505*

orientation and rolling texture, 467–8, *468*

twin growth dislocation, 360, *360*

polycrystalline materials, 144, 199, 413

glide deformation, 227

grain boundary tensions, 437, *438*

texture development, during working, 463–9, **465–6**

polyethylene, structure, 119, *120, 121*

polygonization, 266

polyhedra, geometry of, 135, 137, *138,* 139–40

Voronoi construction, 131, *132*

polymers, crystal structures, 118–20, *120, 121,* **122,** *123,* **544**

imperfections and amorphous regions, 119, 123

lyotropic and thermotropic liquid crystals, 131, 133

martensitic transformation, 394

unit cell, 119, *121*

polymorphism (compounds), 90

porous solid materials, 144–5

primary glide plane, 231, *231*

primitive unit cell, 6, 21, 482

principal axes and components, symmetric tensors, 157, 175, *176*

prismatic dislocation loops, 250, *251,* 255, 292

projections, geometry of, 493–7, *494, 495*

pseudo-merohedry, 365

pseudomorphic strained layers, 456–9

pure shear

related to simple shear, 533, *533*

stresses, 181, 182

pure strain tensors, 174, 175, *175*

and independent slip systems, 224–6, **225,** 525–7

pyramidal glide, 293, *294,* 295

quasicrystals, 142–3
quaternion algebra, 22, 490–1
quaternion geometry, 427
quenched metals, 325–8, 373–4

racemic twinning, 364
radiation damage, 311, 314, 328–9
radius–normal property, representation
    ellipsoid, 166–7, *167*
random close-packed sphere model (RCPS),
    130, 131
randomly oriented crystals, 445
rare earth metals
    crystal structure, 93, 96
    ion substitution, in garnets, 112–13
rational indices, law of, 11
rational line/plane, 9
reciprocal lattice vectors, 13, *482, 482–4,*
    511
reciprocal twins, 344
reconstructive transformation, 394
recrystallization, 340, 464
rectangular nets
    diad axes and mirror planes, *19–20,* 21
    stacking, and orthorhombic lattices,
        30–1, *32*
reflection symmetry, 16, *16*
reflection (normal, type I) twins, 343, *344*
relative displacement (strain) tensors,
    173–5, 219–21
relaxation time (elastic compliance), *331*
repeat units (distance), polymers, 119, *120*
replacement collisions (point defects), 328
representation quadric, tensors and property
        measurement
    radius–normal property, 166–7, *167*
    representation surfaces, equations,
        163–7, *167*
representative volume elements, 145
resistivity, quenched metals, 325, *327*
resolved shear stress
    evaluation, *228,* 228–9, *229*
    largest, associated glide systems,
        229–31, 527–9
reticular merohedry, 364
reticular pseudomerohedry, 365

rhombus nets and unit cells, 30–1, *33*
rigidity (shear) modulus, 199, 255, 256
rock salt *see* sodium chloride
Rodrigues space, 472
Rodrigues vectors, 472
rolled metals, crystal orientation (texture),
    464, 467, *467,* 468, *469*
rotational symmetry, 15–16, 485, 489
    combinations, possible, 21–5, **25,** 71
    and crystallographic point groups, 44–8,
        46, 47
    occurrence in crystal nets, 17–21, *19, 20*
    in quasicrystals, 142–3
rotation (parallel, type II) twins, 344, *344*
rotoinversion axes, 44, 45, 48
rutile crystal structure, 107, *108,* 140, **544**

sapphire (corundum, α-Al_2O_3)
    conditions for glid, 209
    crystal structure, 109–10, *110*
scalar product
    mathematical operation, 478–80
    scalar triple product, 480–1
    and Weiss zone law, 13
Schläfli symbols, 137, *138*
Schlegel diagrams, 137, *138,* 139
Schmid criterion, 228
Schmid factor, 227, 231, 527–30
Schoenflies notation, 79–80
Schottky defect, 313–14, 323–4
screw axes (symmetry elements), in space
        groups, 70–1, *71*
screw dislocations
    atomic positions, 246, *247,* 260–1
        dislocation width, 261, *261, 262*
    description, for cubic lattice, 243–6, *244*
    distortion stresses and strains, 255–7,
        256, 257
        strain energy calculation, 271–2
    orthogonal, 266, *266,* 267
    parallel, 264, 266
secondary bonding, in polymers, 119
second-rank (dyadic) tensors
    definition and symmetry, 156–7
    limitations imposed by crystal
        symmetry, 160–3, **163**

representation surfaces, 163–7, *167*

physical properties represented, 149–50, **150**

referred to principal axes, 157–60, *159,* 175, 179

vector components, transformation of, 150–6

self-accommodation, martensite variants, 385, *387,* 388, 393

self-diffusion, 312, 326, **326,** 328

semiconductors

crystal growth and structure, 456–60

lattice parameters and stiffness constants, **458**

shape memory alloys, magnetic, 393

shape memory effect

one-way shape memory effect, 397–8, *398*

superelasticity, 399–400

two-way shape memory effect, 398–9, *399*

shape memory materials, 373, 393

shear

engineering shear strain (γ), 171, *172,* 176

forces, plastic yield (glide) responses, 200–13

modulus (rigidity), 199, 255, 256

pure strain tensor components, 174, 175, *175*

resolved shear stress calculation, 227–9, *228*

simple and pure, compared, 532–3, *533*

stress components (σ), 176, 178–9, *180–2*

and twinning mechanism, 340–2, *341, 342*

shear magnitude, *s,* 342, 356

shear, in twinning (η), 342, *343,* 352

sheet textures, 468, *469*

Shockley partials

compared with Frank partials, 282–4

separation and stacking fault energy, 279–81, *280,* 281, **281**

short-range order (solid solutions), 117

shuffle (atomic displacement), 350, 357

silicon

stacking faults and dislocations, 297, 299, *299*

states and properties, 130, 144

silver, rolled (crystal orientation), 467, *469*

slip *see* glide

slip lines (bands), observation of, 210–12, *212,* 264

smectic phases (liquid crystals), *134,* 135

soaps, 133

sodium chloride (rock salt, halite)

crystal structure, 102–3, *103,* **540–1**

and discovery of glide, 210

dislocation geometry, 288–91, *290*

glide planes and directions, 214

Schottky defect, 313–14, 323–4

vacancies, ionic displacement, 317, *317*

solid solutions, 114–17, *115, 116,* **117,** *118*

see also alloys, metallic

space charge regions, rock salt, *290,* 291, 313

space groups, 70–9

combinations of symmetry operations to form, 72–4

glide planes, 71–2

in inorganic crystals and minerals, 83

vs. point group symmetry, 74

screw axes, 70–1, *71,* **72,** *73*

three-dimensional, example, 80–3, *81–2*

two-dimensional (plane) groups, 74–5, *76–8,* 79, **79**

space lattices *see* Bravais lattices

sphalerite (zinc blende, α-ZnS)

crystal structure, 104, *105,* **542**

dislocations, *298,* 298–9, *299–300*

piezoelectricity, 186

plane stacking sequence, 99, *100*

twin structures, 347–8, *348*

spheres *see* close packing

spinel crystal structure, 110–11, *111*

stacking faults, 279–81, *280,* 340

see also dislocations

stair rod dislocations, 287, *288,* 299

steel, martensitic transformation, 373, 388–92, *390, 391*

nucleation and growth of plates, 395–7

stereograms
  construction
    general diagrams and triclinic
      application, 65–7, *66*
    hexagonal system, 58, *58*
    monoclinic system, axis setting
      conventions, 63–5, *64, 65*
    orthorhombic system example, 50–2,
      51
    pole opposites and angles, 500–2
    small circle, 497–9, *498, 499*
    Wulff net graphical aid, 502–5
  conventional symbols, 44–5, *46,* 48
  glide system depiction, *229,* 229–31, *230*
  projection geometry, 493–7, *495*
  properties, proof of, 507–10, *508, 510*
stiffness constants, 188, 191–6, **195, 196,**
  198
strain
  anelastic, 330–1, 333
  definitions, 171, *172*
  elastic, produced by applied stress,
    187–90
  energy, stored in dislocation, 271–3, *272*
  homogeneous, 210, 218, 332
  homogeneous, 172
  infinitesimal, 172–6
  in martensite plates and lattices, 376
    combined with rotation, 382–4, *383,*
      384
    pure and observed lattice strains, *381,*
      382, *382*
  in multiple twinning, 361
strained layer epitaxy, 456–60, *457*
strain-induced piezoelectric effect, 456
stress
  calculation, *178,* 178–9
  chemical (in dislocation), 254–5
  components, definition and notation,
    176–8, *177,* 180, *181, 182*
  contracted notation, 184, 189
  fields, on intersecting dislocations, 266,
    279
  homogeneous, 179–80
  hydrostatic and deviatoric, 181
  pure shear, 181, *181,* 182

related to strain (Hooke's law), 187, 256,
  257
Strukturbericht designation (crystal
  structure), 123
subgroups, 80
substitution
  of cations, in garnet structures, 112–13
  of metals, in solid solutions, 114–15, *115*
superconducting materials, crystal
  structures, 109
superelasticity, 388, 393, 399–400
supergroups, 80
superlattices, 116–17, **117,** *118*
surface free energy, 410, **412**
  at interface junctions, *429,* 429–34
  measurement, 410–13, *411*
  related to equilibrium crystal shape
    (Wulff theorem), 434–6, *435*
  and twin boundaries, 432–3, **434**
surfaces
  atomic bonding and energy plots, 408–9,
    409
    interfacial simulation, 423–4
  definition and alternative structures, *407,*
    407–8
  growth rates, 437
  outer, reconstruction, 407
  see also grain boundaries
symmetric tensors, 156–157, 162
symmetry elements
  crystals, macroscopic level, 3, 43–8
  of defects, 332
  groups, subgroups and supergroups, 80
  limitations on crystal physical
    properties, 160–163, **163**
  notation and nomenclature, 79–80
  operation types, 43–4
  slip system families, 213–215
  space group descriptions, 70–9
symmetry operations
  operation types, 14–16
  restrictions, in crystals, 17–21, **18**
syndiotactic polymers, 119

tactic polymers, 119
tensile strains, 175, 199, 330

tension
  dislocation line, *276,* 276–279
  and glide, in metal crystals, 231–235, *232, 234*
tensors
  principal axes and components, 157
  transformation operations, 150–156
  types and ranks, 149–150, 168, 173, 183
  see also second-rank tensors
tetrad (four-fold) axis of symmetry, *19,* 21
tetragonal crystal system, *28,* 30
  elastic constants, 199
  planar spacing, 513
  point group symmetry elements, 52–3
tetrahedral interstices, 93, 95, *96*
texture, 463–9, **465–6**
thermal faceting, 436, *437*
thermoelastic martensite, 396
Thompson's tetrahedron, *283,* 283–5, 289
tilt boundaries
  asymmetrical, 417–18, *418*
  rotated (twist), *419,* 419–20
  symmetrical, 413–17, *414, 417*
tin (*α*-and *β*-Sn), crystal structures, 98–9
titanium
  dioxide (rutile), crystal structure, 107, *108,* 140
  martensitic transformation, metal and alloys, 379, *384,* 385, 392–3
topologically close-packed (TCP) structures, 141
topological model, 400–1
torque terms, grain interface junctions, 430, *430,* 432
transformation formula (tensor components), 159, 160, 162, 176
transformation toughening, 394
translational symmetry, 6, 15, 43
  combined with rotation/inversion (space groups), 70–9
translation glide, 200–13, *210*
triad (three-fold) axis of symmetry, *19,* 21
triclinic crystal system, 26, *28*
  elastic constants, 191
  planar spacing, 512

point group symmetry elements, 65–7, *66*
property representation quadric, 166
trigonal crystal system
  lattice symmetry and unit cells, 26, **26,** 33–6, *34, 35*
  planar spacing, 513
  point group symmetry elements, *60–2,* 60–3
    special forms, 62, *62*
    use of rhombohedral/hexagonal axes, 60–2, 72
    use of rhombohedral/hexagonal axes, 522–4, *523*
tungsten, elastic isotropy, 258
twin boundaries, 445
twinned crystals
  crystal morphology, 358–61
  geometric elements, *343,* 343–6, 352–7, **353–4**
  martensitic lamellae, *386,* 386–8
  occurrence and causes, 339, 340
  structure, examples of, 346–52
twinning
  atomistic modelling of twin boundaries, 365–7
  Friedel's classification of, 361–5
twin orientation descriptions, 426
twist boundaries, *419,* 419–20
two-dimensional Poisson's ratio, 458
two-surface analysis, 211, 505–7, *507*
two-way shape memory effect, 398–9, *399*

unit cells, 3–7
  of cellular solid materials, 145
  conventional, in crystal systems, 26, **26**
  electron density, 89
  of ordered solid solutions, 116, *116*
  rhombohedral and hexagonal, in trigonal crystals, *60,* 60–2, *61,* 72, 522–4
  of tactic polymers, 119, *121*
  transformation of indices (directions and planes), 517–20
unit parallelogram, 6
unit quaternions, 21, 22, 491
uranium, twinning, 344

vacancies, 309
  binding energy, 314
  configuration models, 316–17, **317,** *317*
  and dislocation forces, 254–5
  dislocation loops, in hexagonal metals, 292, *293*
  formation energy calculation, 311
  positions and generation, 310–11, *311*
  related to electrical charge, *290,* 291
vapour deposition, 340, 449, 464
vectors
  definition and components, 477–8, *478*
  director, in liquid crystal alignment, 133, 135
  lattice translation, 15
  related to crystal lattice planes, 11–14, *482,* 482–4, *483*
  scalar product, 478–80
  transformation of components, 150–3, *151,* 154
    dummy suffix notation, 153–4
  vector product operations, *480,* 480–2
    triple product formula, 235
von Mises condition, 227
Voronoi polyhedra, 131, *132*

wavy glide, 215, *215,* 225
Weaire–Phelan structure, 438, *439*
Weiss zone law, 11–14, 62
white tin (β-Sn), crystal structure, 98–9
Wulff net, 502–3, *503, 504, 505*
Wulff plot, crystal/grain shape, 434–6, *435*
wurtzite (β-ZnS)
  crystal structure, 104–5, **105,** *105,* **543**
    surfaces, 407, *407*
    stacking faults and dislocations, 299–300

Young's modulus *(E),* 199, 200, 201, 458

Zener anisotropy factor, 200
Zener ratio, 200
zeolites, 144
zinc-blende *see* sphalerite
zinc, twinning, 350, *351*
zirconium
  martensitic transformations, 379–85, *384*
  oxide (zirconia), transformation toughening, 394
  twinning, 350, *350*
  zone addition rule, 14, 51, 52
  zones, of crystal lattice planes, 12–14